GAME THEORY

Second Edition

GAME THEORY

Second Edition

Leon A. Petrosyan
Nikolay A. Zenkevich

St. Petersburg State University, Russia

World Scientific

NEW JERSEY • LONDON • SINGAPORE • BEIJING • SHANGHAI • HONG KONG • TAIPEI • CHENNAI • TOKYO

Published by

World Scientific Publishing Co. Pte. Ltd.
5 Toh Tuck Link, Singapore 596224
USA office: 27 Warren Street, Suite 401-402, Hackensack, NJ 07601
UK office: 57 Shelton Street, Covent Garden, London WC2H 9HE

Library of Congress Cataloging-in-Publication Data
Names: Petrosi͡an, L. A. (Leon Aganesovich), author. | Zenkevich, N. A. (Nikolaĭ Anatol'evich), author.
Title: Game theory / Leon A. Petrosyan (St. Petersburg State University, Russia) & Nikolay Zenkevich (St. Petersburg State University, Russia).
Description: 2nd edition. | New Jersey : World Scientific, 2016. | Includes bibliographical references and index.
Identifiers: LCCN 2015042776 | ISBN 9789814725385 (hc : alk. paper)
Subjects: LCSH: Game theory. | Probabilities.
Classification: LCC QA269 .P47 2016 | DDC 519.3--dc23
LC record available at http://lccn.loc.gov/2015042776

British Library Cataloguing-in-Publication Data
A catalogue record for this book is available from the British Library.

In-house Editors: Chandrima Maitra/Philly Lim

Typeset by Stallion Press
Email: enquiries@stallionpress.com

Printed in Singapore by B & Jo Enterprise Pte Ltd

Preface

Game theory is a branch of modern applied mathematics that aims to analyze various problems of conflict between parties that have opposed, similar or simply different interests. A theory of games, introduced in 1921 by Émile Borel, was established in 1928 by John von Neumann and Oskar Morgenstern, to develop it as a means of decision making in complex economic systems. In their book "The Theory of Games and Economic Behaviour", published in 1944, they asserted that the classical mathematics developed for applications in mechanics and physics fail to describe the real processes in economics and social life. They have also seen many common factors such as conflicting interests, various preferences of decision makers, the dependence of the outcome for each individual from the decisions made by other individuals both in actual games and economic situations. Therefore, they named this new kind of mathematics game theory.

Games are grouped into several classes according to some important features. In our book we consider zero-sum two-person games, strategic n-person games in normal form, cooperative games, games in extensive form with complete and incomplete information, differential pursuit games and differential cooperative and non-cooperative n-person games.

There is no single game theory which could address such a wide range of "games". At the same time there are common optimality principles applicable to all classes of games under consideration, but the methods of effective computation of solutions are very different. It is also impossible to cover in one book all known optimality principles and solution concepts. For instance only the set of different "refinements" of Nash equilibria generates more than 20 new optimality principles. In this book we try to explain the principles which from our point of view are basic in game theory, and bring the reader to the ability to solve problems in this field of mathematics. We have included results published before in Petrosyan (1965), (1968), (1970), (1972), (1977), (1992), (1993); Petrosyan and Zenkevich (1986); Zenkevich and Marchenko (1987), (1990); Zenkevich and Voznyuk (1994); Kozlovskaya and Zenkevich (2010); Gladkova, Sorokina and Zenkevich (2013); Gao, Petrosyan and Sedakov (2014); Zenkevich and Zyatchin (2014); Petrosyan and Zenkevich (2015); Yeung and Petrosyan (2006), (2012); Petrosyan and Sedakov (2014); Petrosyan and Zaccour (2003); Zenkevich, Petrosyan and Yeung (2009).

The book is the second revised edition of Petrosyan and Zenkevich (1996).

Acknowledgments

We begin by acknowledging our debts to our teacher Nikolay Vorobjev who started in the former Soviet Union teaching us game theory, the time when this subject was not a necessary part of applied mathematics, economics and management science curriculum.

We thank Ekaterina Gromova, Artem Sedakov, Elena Semina, Elena Parilina, and Sergey Voznyuk for their effective assistance.

Many thanks to Anna Melnik and Andrey Ovsienko for preparation of the manuscript in LaTeX.

We acknowledge Saint Petersburg State University (research project 26 9.38.245.2014) and Russian Foundation for Basic Research (project 16-01-00805).

Contents

Chapter 1

Matrix Games

1.1 Definition of a Two-Person Zero-Sum Game in Normal Form

1.1.1 Definition. The system

$$\Gamma = (X, Y, K), \tag{1.1.1}$$

where X and Y are nonempty sets, and the function $K : X \times Y \to R^1$, is called a two-person zero-sum game in normal form.

The elements $x \in X$ and $y \in Y$ are called the *strategies* of Players 1 and 2, respectively, in the game Γ, the elements of the Cartesian product $X \times Y$ (i.e. the pairs of strategies (x, y), where $x \in X$ and $y \in Y$) are called *situations*, and the function K is the payoff of Player 1. Player 2's payoff in situation (x, y) is equal to $[-K(x, y)]$; therefore the function K is also called the *payoff function* of the game Γ and the game Γ is called a *zero-sum game*. Thus, in order to specify the game Γ, it is necessary to define the sets of strategies X, Y for Players 1 and 2, and the payoff function K given on the set of all situations $X \times Y$.

The game Γ is interpreted as follows. Players simultaneously and independently choose strategies $x \in X, y \in Y$. Thereafter Player 1 receives the payoff equal to $K(x, y)$ and Player 2 receives the payoff equal to $(-K(x, y))$.

Definition. The game $\Gamma' = (X', Y', K')$ is called a subgame of the game $\Gamma = (X, Y, K)$ if $X' \subset X, Y' \subset Y$, and the function $K' : X' \times Y' \to R^1$ is a restriction of function K on $X' \times Y'$.

This chapter focuses on two-person zero-sum games in which the strategy sets of the players' are finite.

1.1.2. Definition. Two-person zero-sum games in which both players have finite sets of strategies are called matrix games.

Suppose that Player 1 in matrix game (1.1.1) has a total of m strategies. Let us order the strategy set X of the first player, i.e. set up a one-to-one correspondence between the sets $M = \{1, 2, \ldots, m\}$ and X. Similarly, if Player 2 has n strategies, it is possible to set up a one-to-one correspondence between the sets $N = \{1, 2, \ldots, n\}$ and Y. The game Γ is then fully defined by specifying the matrix $A = \{a_{ij}\}$, where $a_{ij} = K(x_i, y_i)$, $(i, j) \in M \times N$, $(x_i, y_j) \in X \times Y, i \in M, j \in N$ (whence comes the name of the game — the matrix game). In this case the game Γ is realized as follows. Player 1 chooses row $i \in M$ and Player 2 (simultaneously and independently from Player 1) chooses column $j \in N$. Thereafter Player 1 receives the payoff (a_{ij}) and Player 2 receives the payoff $(-a_{ij})$. If the payoff is equal to a negative number, then we are dealing with the actual loss of Player 1.

Denote the game Γ with the payoff matrix A by Γ_A and call it the $(m \times n)$ game according to the dimension of matrix A. We shall drop index A if the discussion makes it clear what matrix is used in the game.

Strategies in the matrix game can be enumerated in different ways; therefore to each order relation, strictly speaking, corresponds its matrix. Accordingly, a finite two-person zero-sum game can be described by distinct matrices different from one another only by the order of rows and columns.

1.1.3. *Example 1* (Dresher, 1961). This example is known in literature as Colonel Blotto game. Colonel Blotto has m regiments and his enemy has n regiments. The enemy is defending two posts. The post will be taken by Colonel Blotto if when attacking the post he is more powerful in strength on this post. The opposing parties are two separate regiments between the two posts.

Define the payoff to the Colonel Blotto (Player 1) at each post. If Blotto has more regiments than the enemy at the post (Player 2), then his payoff at this post is equal to the number of the enemy's regiments plus one (the occupation of the post is equivalent to capturing of one regiment). If Player 2 has more regiments than Player 1 at the post, Player 1 loses his regiments at the post plus one (for the lost of the post). If each side has the same number of regiments at the post, it is a draw and each side gets zero. The total payoff to Player 1 is the sum of the payoffs at the two posts.

The game is zero-sum. We shall describe strategies of the players. Suppose that $m > n$. Player 1 has the following strategies: $x_0 = (m, 0)$ — to place all of the regiments at the first post; $x_1 = (m - 1, 1)$ — to place $(m - 1)$ regiments at the first post and one at the second; $x_2 = (m - 2, 2), \dots, x_{m-1} = (1, m - 1), x_m = (0, m)$. The enemy (Player 2) has the following strategies: $y_0 = (n, 0)$, $y_1 = (n - 1, 1), \dots, y_n = (0, n)$.

Suppose that the Player 1 chooses strategy x_0 and Player 2 chooses strategy y_0. Compute the payoff a_{00} of Player 1 in this situation. Since $m > n$, Player 1 wins at the first post. His payoff is $n + 1$ (one for holding the post). At the second post it is draw. Therefore, $a_{00} = n + 1$. Compute a_{01}. Since $m > n - 1$, then in the first post Player 1's payoff is $n - 1 + 1 = n$. Player 2 wins at the second post. Therefore, the loss of Player 1 at this post is one. Thus, $a_{01} = n - 1$. Similarly, we obtain $a_{0j} = n - j + 1 - 1 = n - j,\ 1 \le j \le n$. Further, if $m - 1 > n$ then $a_{10} = n + 1 + 1 = n + 2,\ a_{11} = n - 1 + 1 = n,\ a_{1j} = n - j + 1 - 1 - 1 = n - j - 1,\ 2 \le j \le n$. In a general case (for any m and n) the elements $a_{ij},\ i = \overline{0, m},\ j = \overline{0, n}$, of the payoff matrix are computed as follows:

$$a_{ij} = K(x_i, y_j) = \begin{cases} n+2 & \text{if } m-i > n-j, \quad i > j, \\ n-j+1 & \text{if } m-i > n-j, \quad i = j, \\ n-j-i & \text{if } m-i > n-j, \quad i < j, \\ -m+i+j & \text{if } m-i < n-j, \quad i > j, \\ j+1 & \text{if } m-i = n-j, \quad i > j, \\ -m-2 & \text{if } m-i < n-j, \quad i < j, \\ -i-1 & \text{if } m-i = n-j, \quad i < j, \\ -m+i-1 & \text{if } m-i < n-j, \quad i = j, \\ 0 & \text{if } m-i = n-j, \quad i = j. \end{cases}$$

Thus, with $m = 4, n = 3$, considering all possible situations, we obtain the payoff matrix A of this game:

$$A = \begin{array}{c} \\ x_0 \\ x_1 \\ x_2 \\ x_3 \\ x_4 \end{array} \begin{array}{c} \begin{array}{cccc} y_0 & y_1 & y_2 & y_3 \end{array} \\ \begin{bmatrix} 4 & 2 & 1 & 0 \\ 1 & 3 & 0 & -1 \\ -2 & 2 & 2 & -2 \\ -1 & 0 & 3 & 1 \\ 0 & 1 & 2 & 4 \end{bmatrix} \end{array}.$$

Example 2. Game of Evasion [Gale (1960)]. Players 1 and 2 choose integers i and j from the set $\{1, \ldots, n\}$. Player 1 wins the amount $|i - j|$. The game is zero-sum. The payoff matrix is square $(n \times n)$ matrix, where $a_{ij} = |i - j|$. For $n = 4$, the payoff matrix A has the form

$$A = \begin{array}{c} \\ 1 \\ 2 \\ 3 \\ 4 \end{array} \begin{array}{c} \begin{array}{cccc} 1 & 2 & 3 & 4 \end{array} \\ \begin{bmatrix} 0 & 1 & 2 & 3 \\ 1 & 0 & 1 & 2 \\ 2 & 1 & 0 & 1 \\ 3 & 2 & 1 & 0 \end{bmatrix} \end{array}.$$

Example 3. Discrete Duel Type Game [Gale (1960)]. Players approach one another by taking n steps. After each step a player may or may not fire a bullet, but during the game he may fire only once. The probability that the player will hit his opponent (if he shoots) on the kth step is assumed to be k/n $(k \leq n)$.

A strategy for Player 1 (2) consists in taking a decision on shooting at the ith (jth) step. Suppose that $i < j$ and Player 1 makes a decision to shoot at the ith step and Player 2 makes a decision to shoot at the jth step. The payoff a_{ij} to Player 1 is then determined by

$$a_{ij} = \frac{i}{n} - \left(1 - \frac{i}{n}\right)\frac{j}{n} = \frac{n(i-j)+ij}{n^2}.$$

Thus the payoff a_{ij} is the difference in the probabilities of hitting the opponent and failing to survive. In the case $i > j$, Player 2 is the first to fire and $a_{ij} = -a_{ji}$. If however, $i = j$, then we set $a_{ij} = 0$. Accordingly, if we set $n = 5$, the game matrix multiplied by 25 has the form

$$A = \begin{bmatrix} 0 & -3 & -7 & -11 & -15 \\ 3 & 0 & 1 & -2 & -5 \\ 7 & -1 & 0 & 7 & 5 \\ 11 & 2 & -7 & 0 & 15 \\ 15 & 5 & -5 & -15 & 0 \end{bmatrix}.$$

Example 4. Attack-Defense Game. Suppose that Player 1 wants to attack one of the targets $c_1, \ldots, c_n$ having positive values $\tau_1 > 0, \ldots, \tau_n > 0$. Player 2 defends one of these targets. We assume that if the undefended target c_i is attacked, it is necessarily destroyed (Player 1 wins τ_i) and the defended target is hit with probability $1 > \beta_i > 0$ (the target c_i withstands the attack with probability $1 - \beta_i > 0$), i.e. Player 1 wins (on the average) $\beta_i \tau_i$, $i = 1, 2, \ldots, n$.

The problem of choosing the target for attack (for Player 1) and the target for defense (for Player 2) reduces to the game with the payoff matrix

$$A = \begin{bmatrix} \beta_1\tau_1 & \tau_1 & \cdots & \tau_1 \\ \tau_2 & \beta_2\tau_2 & \cdots & \tau_2 \\ \cdots & \cdots & \cdots & \cdots \\ \tau_n & \tau_n & \cdots & \beta_n\tau_n \end{bmatrix}.$$

Example 5. Discrete Search Game. There are n cells. Player 2 hide an object in one of n cells and Player 1 wishes to find it. In examining the ith cell, Player 1 exerts $\tau_i > 0$ efforts, and the probability of finding the object in the ith cell (if it is concealed there) is $0 < \beta_i \leq 1$, $i = 1, 2, \ldots, n$. If the object is found, Player 1 receives the amount α. The players' strategies are the numbers of cells wherein the players respectively hide and search for the object. Player 1's payoff is equal to the difference in the expected receipts and the efforts made in searching for the object. Thus, the problem of hiding and searching for the object reduces to the game with the payoff matrix

$$A = \begin{bmatrix} \alpha\beta_1 - \tau_1 & -\tau_1 & -\tau_1 & \ldots & -\tau_1 \\ -\tau_2 & \alpha\beta_2 - \tau_2 & -\tau_2 & \ldots & -\tau_2 \\ \ldots & \ldots & \ldots & \ldots & \ldots \\ -\tau_n & -\tau_n & -\tau_n & \ldots & \alpha\beta_n - \tau_n \end{bmatrix}.$$

Example 6. Noisy Search. Suppose that Player 1 is searching for a mobile object (Player 2) for the purpose of detecting it. Player 2's objective is the opposite one (i.e. he seeks to avoid being detected). Player 1 can move with velocities $\alpha_1 = 1$, $\alpha_2 = 2$, $\alpha_3 = 3$ and Player 2 with velocities $\beta_1 = 1$, $\beta_2 = 2$, $\beta_3 = 3$, respectively. The range of the detecting device used by Player 1, depending on the velocities of the players is determined by the matrix

$$D = \begin{matrix} & \beta_1 & \beta_2 & \beta_3 \\ \alpha_1 & 4 & 5 & 6 \\ \alpha_2 & 3 & 4 & 5 \\ \alpha_3 & 1 & 2 & 3 \end{matrix}.$$

Strategies of the players are the velocities, and Player 1's payoff in the situation (α_i, β_j) is assumed to be the search efficiency $a_{ij} = \alpha_i \delta_{ij}$, $i = \overline{1,3}, j = \overline{1,3}$, where δ_{ij} is an element of the matrix D.

Then the problem of selecting velocities in a noisy search can be represented by the game with matrix

$$A = \begin{array}{c} \\ \alpha_1 \\ \alpha_2 \\ \alpha_3 \end{array} \begin{array}{c} \begin{array}{ccc} \beta_1 & \beta_2 & \beta_3 \end{array} \\ \begin{bmatrix} 4 & 5 & 6 \\ 6 & 8 & 10 \\ 3 & 6 & 9 \end{bmatrix} \end{array}.$$

Example 7. The Battle of the Bismarck See. The conflict can be modeled as the following 2×2 matrix game

$$\begin{array}{c} \\ N \\ S \end{array} \begin{array}{c} \begin{array}{cc} N & S \end{array} \\ \begin{bmatrix} 2 & 2 \\ 1 & 3 \end{bmatrix} \end{array}.$$

The first player is US Admiral Kenney and the second Japanese Admiral Imamura. The conflict happens in the South Pacific in 1943. Imamura has to transport troops across the Bismarck See to New Guinea, and his opponent Kenney wants to bomb the transport. Imamura has two possible choices: a shorter Northern route (N, 2 days) or a longer Southern route (S, 3 days). Kenney must choose one of this routs (N or S) to send his planes to. If he chooses the wrong route he can call back the planes and send them to another route, but the number of bombing days is reduced by 1. We assume, that the number of bombing days represents the payoff to Kenney in a positive sense to Imamura in negative sense.

1.2 Maximin and Minimax Strategies

1.2.1. Consider a two-person zero-sum game $\Gamma = (X, Y, K)$. In this game each of the players seeks to maximize his payoff by choosing a proper strategy. But for Player 1 the payoff is determined by the function $K(x, y)$, and for Player 2 it is determined by $(-K(x, y))$, i.e. the players' objectives are directly opposite. Note that the payoff of Player 1 (2) (the payoff function) is determined on the set of situations $(x, y) \in X \times Y$. Each situation, and hence the player's payoff do not depend only on his own choice, but also on what strategy will be chosen by his opponent whose objective is directly opposite. Therefore, seeking to obtain the maximum possible payoff, each player must take into account the opponent's behavior.

Colonel Blotto game provides a good example of the foregoing. If Player 1 wants to obtain the maximum payoff, he must adopt the strategy x_0 (or x_4). In this case, if Player 2 uses strategy $y_0(y_3)$ he receives the payoff 4, but if the Player 2 uses strategy y_3 (corresponding y_0), then the first player receives the payoff 0, i.e. he loses 4 units. Similar reasonings are applicable to Player 2.

In the theory of games it is supposed that the behavior of both players is rational, i.e. they wish to obtain the maximum payoff, assuming that the opponent is acting in the best (for himself) possible way. What maximal payoff can Player 1 guarantee himself? Suppose Player 1 chooses strategy x. Then, at worst case he will win $\min_y K(x, y)$. Therefore, Player 1 can always guarantee himself the payoff $\max_x \min_y K(x, y)$. If the max and min are not reached, Player 1 can guarantee himself obtaining the payoff arbitrarily close to the quantity

$$\underline{v} = \sup_{x \in X} \inf_{y \in Y} K(x, y), \tag{1.2.1}$$

which is called the *lower value* of the game. The principle of constructing strategy x based on the maximization of the minimal payoff is called the *maximin principle*, and the strategy x selected by this principle is called the *maximin strategy* of Player 1.

For Player 2 it is possible to provide similar reasonings. Suppose he chooses strategy y. Then, at worst, he will lose $\max_x K(x, y)$. Therefore, the second player can always guarantee himself the payoff $-\min_y \max_x K(x, y)$. The number

$$\overline{v} = \inf_{y \in Y} \sup_{x \in X} K(x, y) \tag{1.2.2}$$

is called the *upper value of the game* Γ. The principle of constructing a strategy y, based on the minimization of maximum losses, is called the *minimax principle*, and the strategy y selected for this principle is called the *minimax strategy* of Player 2. It should be stressed that the existence of the minimax (maximin) strategy is determined by the reachability of the extremum in (1.2.2), (1.2.1).

Consider the $(m \times n)$ matrix game Γ_A. Then the extrema in (1.2.1) and (1.2.2) are reached and the lower and upper values of the game are, respectively equal to

$$\underline{v} = \max_{1 \le i \le m} \min_{1 \le j \le n} a_{ij}, \tag{1.2.3}$$

$$\overline{v} = \min_{1 \le j \le n} \max_{1 \le i \le m} a_{ij}. \tag{1.2.4}$$

The minimax and maximin for the game Γ_A can be found by the following scheme

$$\begin{array}{c} \underbrace{\begin{array}{cccc} \left[\begin{array}{cccc} a_{11} & a_{12} & \dots & a_{1n} \\ a_{21} & a_{22} & \dots & a_{2n} \\ \dots & \dots & \dots & \dots \\ a_{m1} & a_{m2} & \dots & a_{mn} \end{array}\right] \\ \max_i a_{i1} \quad \max_i a_{i2} \quad \dots \quad \max_i a_{in} \end{array}}_{\min\limits_j \max\limits_i a_{ij}} \end{array} \left.\begin{array}{c} \min_j a_{1j} \\ \min_j a_{2j} \\ \dots \\ \min_j a_{mj} \end{array}\right\} \max_i \min_j a_{ij}.$$

Thus, in the game Γ_A with the matrix

$$A = \begin{bmatrix} 1 & 0 & 4 \\ 5 & 3 & 8 \\ 6 & 0 & 1 \end{bmatrix}$$

the lower value (maximin) $\underline{v}$ and the maximin strategy i_0 of the first player are $\underline{v} = 3$, $i_0 = 2$, respectively, and the upper value (minimax) $\overline{v}$ and the minimax strategy j_0 of the second player are $\overline{v} = 3$, $j_0 = 2$, respectively.

1.2.2. The following assertion holds for any game $\Gamma = (X, Y, K)$.

Lemma. *In two-person zero-sum game* Γ

$$\underline{v} \leq \overline{v} \tag{1.2.5}$$

or

$$\sup_{x \in X} \inf_{y \in Y} K(x, y) \leq \inf_{y \in Y} \sup_{x \in X} K(x, y). \tag{1.2.6}$$

Proof. Let $x \in X$ be an arbitrary strategy of Player 1. Then we have

$$K(x, y) \leq \sup_{x \in X} K(x, y).$$

Hence we get

$$\inf_{y \in Y} K(x, y) \leq \inf_{y \in Y} \sup_{x \in X} K(x, y).$$

Note that we have a constant on the right-hand side of the latter inequality, and the value $x \in X$ has been chosen arbitrarily. Therefore, the following inequality holds

$$\sup_{x \in X} \inf_{y \in Y} K(x, y) \leq \inf_{y \in Y} \sup_{x \in X} K(x, y).$$

1.3 Saddle Points

1.3.1. Consider the optimal behavior of players in a two-person zero-sum game. In the game $\Gamma = (X, Y, K)$ it is natural to consider as optimal a situation $(x^*, y^*) \in X \times Y$ the deviation from which gives no advantage for both players. Such a point (x^*, y^*) is called the *equilibrium point* and the optimality principle based on constructing an equilibrium point is called the *equilibrium principle*. For two-person zero-sum games, as will be shown later, the equilibrium principle is equivalent to the principles of minimax and maximin. This, of course, requires the existence of an equilibrium (i.e. that the optimality principle be applicable).

Definition. In the two-person zero-sum game $\Gamma = (X, Y, K)$ the point (x^*, y^*) is called an equilibrium point, or a saddle point, if

$$K(x, y^*) \leq K(x^*, y^*), \tag{1.3.1}$$

$$K(x^*, y) \geq K(x^*, y^*) \tag{1.3.2}$$

for all $x \in X$ and $y \in Y$.

The set of all equilibrium points in the game Γ will be denoted as

$$Z(\Gamma),\ \ Z(\Gamma) \subset X \times Y.$$

In the matrix game Γ_A the equilibrium points are the *saddle points* of the payoff matrix A, i.e. the points (i^*, j^*) for which for all $i \in M$ and $j \in N$ the following inequalities are satisfied

$$a_{ij^*} \leq a_{i^*j^*} \leq a_{i^*j}.$$

The element of the matrix $a_{i^*j^*}$ at the saddle point is simultaneously the minimum of its row and the maximum of its column. For example, in the game with the matrix $\begin{bmatrix} 1 & 0 & 4 \\ 5 & 3 & 8 \\ 6 & 0 & 1 \end{bmatrix}$ the point $\alpha_{22} = 2$ is a saddle point (equilibrium).

1.3.2. The set of saddle points in the two-person zero-sum game Γ has the properties which enable one to deal with the optimality of a saddle point and the strategies involved.

Theorem. *Let* (x_1^*, y_1^*), (x_2^*, y_2^*) *be two arbitrary saddle points in the two-person zero-sum game* Γ. *Then*:

1. $K(x_1^*, y_1^*) = K(x_2^*, y_2^*)$;
2. $(x_1^*, y_2^*) \in Z(\Gamma)$, $(x_2^*, y_1^*) \in Z(\Gamma)$.

Proof. From the definition of a saddle point for all $x \in X$ and $y \in Y$ we have

$$K(x, y_1^*) \leq K(x_1^*, y_1^*) \leq K(x_1^*, y); \tag{1.3.3}$$

$$K(x, y_2^*) \leq K(x_2^*, y_2^*) \leq K(x_2^*, y). \tag{1.3.4}$$

We substitute x_2^* into the left-hand side of the inequality (1.3.3), y_2^* into the right-hand side, x_1^* into the left-hand side of the inequality (1.3.4) and y_1^* into the right-hand side. Then we get

$$K(x_2^*, y_1^*) \le K(x_1^*, y_1^*) \le K(x_1^*, y_2^*) \le K(x_2^*, y_2^*) \le K(x_2^*, y_1^*).$$

From this it follows that:

$$K(x_1^*, y_1^*) = K(x_2^*, y_2^*) = K(x_2^*, y_1^*) = K(x_1^*, y_2^*). \tag{1.3.5}$$

Show the validity of the second statement. Consider the point (x_2^*, y_1^*). From (1.3.3)–(1.3.5), we then have

$$K(x, y_1^*) \le K(x_1^*, y_1^*) = K(x_2^*, y_1^*) = K(x_2^*, y_2^*) \le K(x_2^*, y), \tag{1.3.6}$$

for all $x \in X, y \in Y$. The inclusion $(x_1^*, y_2^*) \in Z(\Gamma)$ can be proved in much the same way.

From the theorem it follows that the payoff function takes the same values at all saddle points. Therefore, it is meaningful to introduce the following definition.

Definition. Let (x^*, y^*) be a saddle point in the game Γ. Then the number

$$v = K(x^*, y^*) \tag{1.3.7}$$

is called the value of the game Γ.

The second statement of the theorem suggests, in particular, the following fact. Denote by X^* and Y^* projections of the set $Z(\Gamma)$ onto X and Y, respectively, i.e.

$$X^* = \{x^* \, | x^* \in X, \exists y^* \in Y, (x^*, y^*) \in Z(\Gamma)\},$$
$$Y^* = \{y^* | y^* \in Y, \exists x^* \in X, (x^*, y^*) \in Z(\Gamma)\}.$$

The set $Z(\Gamma)$ may then be represented as Cartesian product

$$Z(\Gamma) = X^* \times Y^*. \tag{1.3.8}$$

The proof of (1.3.8) is a corollary of the second statement of the theorem, and is left to the reader.

Definition. The set $X^*(Y^*)$ is called the set of optimal strategies of Player 1 (2) in the game Γ, and their elements-optimal strategies of the 1(2) player.

Note that from (1.3.5) it follows that any pair of optimal strategies forms a saddle point, and the corresponding payoff is the value of the game.

1.3.3. Optimality of the players' behavior remains unaffected if the strategy sets in the game remain the same and the payoff function is multiplied by a positive constant, or a constant number is added thereto.

Lemma (on Scale). *Let* $\Gamma = (X, Y, K)$ *and* $\Gamma' = (X, Y, K')$ *be two zero-sum games and*

$$K' = \beta K + \alpha, \;\; \beta > 0, \alpha = const, \beta = const. \tag{1.3.9}$$

Then

$$Z(\Gamma') = Z(\Gamma), \;\; v_{\Gamma'} = \beta v_\Gamma + \alpha. \tag{1.3.10}$$

Proof. Let (x^*, y^*) be a saddle point in the game Γ. Then we have

$$K'(x^*, y^*) = \beta K(x^*, y^*) + \alpha \le \beta K(x^*, y) + \alpha = K'(x^*, y),$$
$$K'(x, y^*) = \beta K(x, y^*) + \alpha \le \beta K(x^*, y^*) + \alpha = K'(x^*, y^*),$$

for all $x \in X$ and $y \in Y$.

Therefore $(x^*, y^*) \in Z(\Gamma'), Z(\Gamma) \subset Z(\Gamma')$.

Conversely, let $(x, y) \in Z(\Gamma')$. Then

$$K(x, y) = (1/\beta)K'(x, y) - \alpha/\beta$$

and, by the similar reasoning, we have that $(x, y) \in Z(\Gamma)$. Therefore, $Z(\Gamma) = Z(\Gamma')$ and we have

$$v_{\Gamma'} = K'(x^*, y^*) = \beta K(x^*, y^*) + \alpha = \beta v_\Gamma + \alpha.$$

Conceptually, this lemma states strategic equivalence of the two games differing only by the payoff origin and the scale of their measurements.

1.3.4. We shall now establish a link between the principle of equilibrium and the principles of minimax and maximin in a two-person zero-sum game.

Theorem. *For the existence of the saddle point in the game* $\Gamma = (X, Y, K)$, *it is necessary and sufficient that the quantities*

$$\min_y \sup_x K(x, y), \quad \max_x \inf_y K(x, y), \tag{1.3.11}$$

exist and the following equality holds:

$$\underline{v} = \max_x \inf_y K(x, y) = \min_y \sup_x K(x, y) = \overline{v}. \tag{1.3.12}$$

Proof. Necessity. Let $(x^*, y^*) \in Z(\Gamma)$. Then for all $x \in X$ and $y \in Y$ the following inequality holds:

$$K(x, y^*) \le K(x^*, y^*) \le K(x^*, y) \tag{1.3.13}$$

and hence

$$\sup_x K(x, y^*) \le K(x^*, y^*). \tag{1.3.14}$$

At the same time, we have

$$\inf_y \sup_x K(x, y) \le \sup_x K(x, y^*). \tag{1.3.15}$$

Comparing (1.3.14) and (1.3.15), we get

$$\inf_y \sup_x K(x, y) \le \sup_x K(x, y^*) \le K(x^*, y^*). \tag{1.3.16}$$

In the similar way we get the inequality

$$K(x^*, y^*) \le \inf_y K(x^*, y) \le \sup_x \inf_y K(x, y). \tag{1.3.17}$$

On the other hand, the inverse inequality (1.2.6) holds. Thus, we get

$$\sup_x \inf_y K(x, y) = \inf_y \sup_x K(x, y), \tag{1.3.18}$$

and finally we get

$$\min_y \sup_x K(x,y) = \sup_x K(x,y^*) = K(x^*,y^*),$$
$$\max_x \inf_y K(x,y) = \inf_y K(x^*,y) = K(x^*,y^*),$$

i.e. the exterior extrema of the min sup and max inf are reached at the points y^* and x^* respectively.

Sufficiency. Suppose there exist the min sup and max inf

$$\max_x \inf_y K(x,y) = \inf_y K(x^*,y); \tag{1.3.19}$$
$$\min_y \sup_x K(x,y) = \sup_x K(x,y^*) \tag{1.3.20}$$

and the equality (1.3.12) holds. We shall show that (x^*,y^*) is a saddle point. Indeed,

$$K(x^*,y^*) \geq \inf_y K(x^*,y) = \max_x \inf_y K(x,y); \tag{1.3.21}$$
$$K(x^*,y^*) \leq \sup_x K(x,y^*) = \min_y \sup_x K(x,y). \tag{1.3.22}$$

By (1.3.12) the min sup is equal to the max inf, and from (1.3.21), (1.3.22) it follows that the min sup is also equal to the $K(x^*,y^*)$, i.e. the inequalities in (1.3.21), (1.3.22) are satisfied as equalities. Now we have

$$K(x^*,y^*) = \inf_y K(x^*,y) \leq K(x^*,y),$$
$$K(x^*,y^*) = \sup_x K(x,y^*) \geq K(x,y^*)$$

for all $x \in X$ and $y \in Y$, i.e. $(x^*,y^*) \in Z(\Gamma)$.

The proof shows that the common value of the min sup and max inf is equal to $K(x^*,y^*) = v$, the value of the game, and any min sup (max inf) strategy $y^*(x^*)$ is optimal in terms of the theorem, i.e. the point (x^*,y^*) is a saddle point.

The proof of the theorem yields the following statement.

Corollary 1. *If the* min sup *and* max inf *in* (1.3.11) *exist and are reached on* $\overline{y}$ *and* $\overline{x}$, *respectively, then*

$$\max_x \inf_y K(x,y) = K(\overline{x},\overline{y}) = \min_y \sup_x K(x,y). \tag{1.3.23}$$

The games, in which saddle points exist, are called *strictly determined.* Therefore, this theorem establishes the criterion for strict determination of the game and can be restated as follows. For the game to be strictly determined it is necessary and sufficient that the min sup and max inf in (1.3.11) exist and the equality (1.3.12) is satisfied.

Note that, in the game Γ_A, the extrema in (1.3.11) are always reached and the theorem may be reformulated in the following form.

Corollary 2. *For the* $(m \times n)$ *matrix game to be strictly determined it is necessary and sufficient that the following equalities hold*

$$\min_{j=1,2,\dots,n} \max_{i=1,2,\dots,m} \alpha_{ij} = \max_{i=1,2,\dots,m} \min_{j=1,2,\dots,n} \alpha_{ij}. \tag{1.3.24}$$

For example, in the game with the matrix $\begin{bmatrix} 1 & 4 & 1 \\ 2 & 3 & 4 \\ 0 & -2 & 7 \end{bmatrix}$ the situation (2,1) is a saddle point. In this case

$$\max_i \min_j a_{ij} = \min_j \max_i a_{ij} = 2.$$

On the other hand, the game with the matrix $\begin{bmatrix} 1 & 0 \\ 0 & 1 \end{bmatrix}$ does not have a saddle point, since

$$\min_j \max_i a_{ij} = 1 > \max_i \min_j a_{ij} = 0.$$

Note that the games formulated in Examples 1–3 are not strictly determined, whereas the game in Example 6 is strictly determined and its value is $v = 6$. In Example 7 (The Battle of the Bismarck See) the lower value of the game is equal to $\underline{v} = 2$ and the upper value is also $\overline{v} = 2$. The saddle point is $(N,\, N)$ with payoff of Kenney equal to 2, but in reality Kenney chooses N and Imamura chooses S, with the same payoff equal to 2.

1.4 Mixed Extension of a Game

1.4.1. Consider the matrix game Γ_A. If the game has a saddle point, then the minimax is equal to the maximin; and each of the players can, by the definition of the saddle point, inform the opponent of his optimal (maximin, minimax) strategy and hence no player can receive extra benefits. Now assume that the game Γ_A has no saddle point. Then, by Theorem 1.3.4, and Lemma 1.2.2, we have

$$\min_j \max_i \alpha_{ij} - \max_i \min_j a_{ij} > 0. \tag{1.4.1}$$

In this case the maximin and minimax strategies are not optimal. Moreover, it is not advantageous for the players to play such strategies, as he can obtain a larger payoff. The information about a choice of a strategy supplied to the opponent, however, may cause greater losses than in the case of the maximin or minimax strategy.

Indeed, let the matrix A be of the form

$$A = \begin{bmatrix} 7 & 3 \\ 2 & 5 \end{bmatrix}.$$

For this a matrix $\min_j \max_i \alpha_{ij} = 5$, $\max_i \min_j \alpha_{ij} = 3$, i.e. the saddle point does not exist. Denote by i^* the maximin strategy of Player 1 ($i^* = 1$), and by j^* the minimax strategy of Player 2 ($j^* = 2$). Suppose Player 2 adopts strategy $j^* = 2$ and Player 1 chooses strategy $i = 2$. Then the latter receives the payoff 5, i.e. 2 units greater than the maximin. If, however, Player 2 guesses the choice by Player 1, he alters his strategy to $j = 1$ and then Player 1 receives a payoff of 2 units only, i.e. 1 unit less than in the case of the maximin. Similar reasonings apply to the second player.

How to keep the information about the choice of the strategy in secret from the opponent? To answer this question, it may be wise to choose the strategy using some random device. In this case the opponent cannot learn the player's particular strategy in advance, since the player does not know it himself until strategy will actually be chosen at random.

1.4.2. Definition. The random variable whose values are strategies of a player is called a mixed strategy of the player.

Thus, for the matrix game Γ_A, a mixed strategy of Player 1 is a random variable whose values are the row numbers $i \in M$, $M = \{1, 2, \ldots, m\}$. A similar definition applies to Player 2's mixed strategy whose values are the column numbers $j \in N$ of the matrix A.

Considering the above definition of *mixed strategies*, the former strategies will be referred to as *pure strategies*. Since the random variable is characterized by its distribution, the mixed strategy will be identified in what follows with the probability distribution over the set of pure strategies (Feller, 1971). Thus, Player 1's mixed strategy x in the game is the m-dimensional vector

$$x = (\xi_1, \ldots, \xi_m), \sum_{i=1}^{m} \xi_i = 1, \xi_i \geq 0, i = 1, \ldots, m. \tag{1.4.2}$$

Similarly, Player 2's mixed strategy y is the n-dimensional vector

$$y = (\eta_1, \ldots, \eta_n), \sum_{j=1}^{n} \eta_j = 1, \eta_j \geq 0, j = 1, \ldots, n. \tag{1.4.3}$$

In this case, $\xi_i \geq 0$ and $\eta_j \geq 0$ are the probabilities of choosing the pure strategies $i \in M$ and $j \in N$, respectively, when the players use mixed strategies x and y.

Denote by X and Y the *sets of mixed strategies for the first and second players*, respectively. It can easily be seen that the set of mixed strategies of each player is a compact set in the corresponding finite Euclidean space (closed, bounded set).

Definition. Let $x = (\xi_1, \ldots, \xi_m) \in X$ be a mixed strategy of Player 1. The set of indices

$$M_x \triangleq \{i | i \in M, \xi_i > 0\}, \tag{1.4.4}$$

where $M = \{1, 2, \ldots, m\}$, is called the spectrum of strategy x.

Similarly, for the mixed strategy $y = (\eta_1, \ldots, \eta_n) \in Y$ of Player 2 the spectrum N_y is determined as follows:

$$N_y \triangleq \{j | j \in N, \eta_j > 0\}, \tag{1.4.5}$$

where $N = \{1, 2, \ldots, n\}$. Thus, the spectrum of mixed strategy is composed of such pure strategies that are chosen with positive probabilities.

For any mixed strategy x the spectrum $M_x \neq \oslash$, since the vector x has non-negative components with the sum equal to 1 (see (1.4.2)).

Consider a mixed strategy $u_i \triangleq (\xi_1, \ldots, \xi_m) \in X$, where $\xi_i = 1$, $\xi_j = 0, j \neq i, i = 1, 2, \ldots, m$. Such a strategy prescribes a selection of the ith row of the matrix A with probability 1. It would appear natural to identify a mixed strategy $u_i \in X$ with the choice of ith row, i.e. the pure strategy $i \in M$ of Player 1. In a similar manner, we shall identify the mixed strategy $w_j \triangleq (\eta_1, \ldots, \eta_j, \ldots, \eta_n) \in Y$, where $\eta_j = 1, \eta_i = 0, i \neq j, j = 1, \ldots, n$ with the pure strategy $j \in N$ of Player 2. Thus we have that the player's set of mixed strategies is the extension of his set of pure strategies.

Definition. The pair (x, y) of mixed strategies in the matrix game Γ_A is called the situation in mixed strategies.

We shall define the payoff of Player 1 at the point (x, y) in mixed strategies for the $(m \times n)$ matrix game Γ_A as the mathematical expectation of his payoff provided that the players use mixed strategies x and y, respectively. The players choose their strategies independently; therefore the mathematical expectation of payoff $K(x, y)$ in mixed strategies $x = (\xi_1, \ldots, \xi_m)$, $y = (\eta_1, \ldots, \eta_n)$ is equal to

$$K(x, y) = \sum_{i=1}^{m} \sum_{j=1}^{n} a_{ij} \xi_i \eta_j = (xA)y = x(Ay). \tag{1.4.6}$$

The function $K(x, y)$ is continuous in $x \in X$ and $y \in Y$. Notice that when one player uses a pure strategy (i or j, respectively) and the other uses a mixed strategy (y or x), the payoffs $K(i, y), K(x, j)$ are computed by formulas

$$K(i,y) \triangleq K(u_i,y) = \sum_{j=1}^{n} a_{ij}\eta_j = a_i y, i = 1,\dots,m,$$

$$K(x,j) \triangleq K(x,w_j) = \sum_{i=1}^{m} a_{ij}\xi_i = xa^j, j = 1,\dots,n,$$

where a_i, a^j are respectively the ith row and the jth column of the $(m \times n)$ matrix A.

Thus, from the matrix game $\Gamma_A = (M, N, A)$ we have arrived at a new game $\overline{\Gamma}_A = (X, Y, K)$, where X and Y are the sets of mixed strategies in the game Γ_A and K is the payoff function in mixed strategies (mathematical expectation of the payoff). The game $\overline{\Gamma}_A$ will be called a *mixed extension* of the game Γ_A. The game Γ_A is a subgame for $\overline{\Gamma}_A$, i.e. $\Gamma_A \subset \overline{\Gamma}_A$.

1.4.3. Definition. The point (x^*, y^*) in the game $\overline{\Gamma}_A$ forms a saddle point and the number $v = K(x^*, y^*)$ is the value of the game $\overline{\Gamma}_A$ if for all $x \in X$ and $y \in Y$

$$K(x, y^*) \le K(x^*, y^*) \le K(x^*, y). \tag{1.4.7}$$

The strategies (x^*, y^*) appearing in the saddle point are called optimal. Moreover, by Theorem 1.3.4, the strategies x^* and y^* are respectively the maximin and minimax strategies, since the exterior extrema in (1.3.11) are reachable (the function $K(x, y)$ is continuous on the compact sets X and Y).

Lemma 1.3.3 shows that the two games differing by the payoff reference point and the scale of payoff measurements (Lemma on Scale) are strategically equivalent. It turns out that if two matrices games Γ_A and $\Gamma_{A'}$ are subject to this lemma, their mixed extensions are strategically equivalent. This fact is formally established by the following lemma.

Lemma. *Let* Γ_A *and* $\Gamma_{A'}$ *be two* $(m \times n)$ *matrix games, where*

$$A' = \alpha A + B,\ \ \alpha > 0, \alpha = const,$$

and B is the matrix with the same elements β, i.e. $\beta_{ij} = \beta$ for all i and j. Then $Z(\overline{\Gamma}_{A'}) = Z(\overline{\Gamma}_A)$, $\overline{v}_{A'} = \alpha\overline{v}_A + \beta$, where $\overline{\Gamma}_{A'}$ and $\overline{\Gamma}_A$ are the mixed extensions of the games $\Gamma_{A'}$ and Γ_A, respectively, and $\overline{v}_{A'}, \overline{v}_A$ are the values of the games $\overline{\Gamma}_{A'}$ and $\overline{\Gamma}_A$.

Proof. Both matrices A and A' are of dimension $m \times n$; therefore the sets of mixed strategies in the games $\Gamma_{A'}$ and Γ_A coincide. We shall show that for any situation in mixed strategies (x, y) the following equality holds

$$K'(x, y) = \alpha K(x, y) + \beta, \tag{1.4.8}$$

where K' and K are Player 1's payoffs in the games $\overline{\Gamma}_{A'}$ and $\overline{\Gamma}_A$, respectively.

Indeed, for all $x \in X$ and $y \in Y$ we have

$$K'(x, y) = xA'y = \alpha(xAy) + xBy = \alpha K(x, y) + \beta.$$

From Scale Lemma it then follows that $Z(\overline{\Gamma}_{A'}) = Z(\overline{\Gamma}_A)$, $\overline{v}_{A'} = \alpha\overline{v}_A + \beta$.

Example 8. Verify that the strategies $y^* = (\frac{1}{2}, \frac{1}{4}, \frac{1}{4})$, $x^* = (\frac{1}{2}, \frac{1}{4}, \frac{1}{4})$ are optimal and $v = 0$ is the value of the game $\overline{\Gamma}_A$ with matrix

$$A = \begin{bmatrix} 1 & -1 & -1 \\ -1 & -1 & 3 \\ -1 & 3 & -1 \end{bmatrix}.$$

We shall simplify the matrix A (to obtain the maximum number of zeros). Adding a unity to all elements of the matrix A, we get the matrix

$$A' = \begin{bmatrix} 2 & 0 & 0 \\ 0 & 0 & 4 \\ 0 & 4 & 0 \end{bmatrix}.$$

Each element of the matrix A' can be divided by 2. The new matrix is of the form

$$A'' = \begin{bmatrix} 1 & 0 & 0 \\ 0 & 0 & 2 \\ 0 & 2 & 0 \end{bmatrix}.$$

By the lemma we have $v_{A''} = \frac{1}{2}v_{A'} = \frac{1}{2}(v_A + 1)$. Verify that the value of the game Γ_A is equal to 1/2. Indeed, $K(x^*, y^*) = Ay^* = 1/2$. On the other hand, for each strategy $y \in Y, y = (\eta_1, \eta_2, \eta_3)$ we have $K(x^*, y) = \frac{1}{2}\eta_1 + \frac{1}{2}\eta_2 + \frac{1}{2}\eta_3 = \frac{1}{2} \cdot 1 = \frac{1}{2}$, and for all $x = (\xi_1, \xi_2, \xi_3), x \in X$, $K(x, y^*) = \frac{1}{2}\xi_1 + \frac{1}{2}\xi_2 + \frac{1}{2}\xi_3 = \frac{1}{2}$. Consequently, the above-mentioned strategies x^*, y^* are optimal and $v_A = 0$.

In what follows, whenever the matrix game Γ_A is mentioned, we shall mean its mixed extension $\overline{\Gamma}_A$.

1.5 Convex Sets and Systems of Linear Inequalities

This section is auxiliary in nature and can be omitted by reader with no loss in continuity. In order to obtain an understanding of the proofs of the following assertions, however, it may be reasonable to recall certain widely accepted definitions and theorems. Most of the theorems are given without proofs and special references are provided when needed.

1.5.1. The set $M \subset R$ is called *convex* if for any two points of this set $x_1, x_2 \in M$ all the points of the interval $\lambda x_1 + (1-\lambda)x_2, 0 \leq \lambda \leq 1$ are contained in M. The notion of a *convex set* can be formulated in a more general, but equivalent form.

The set $M \subset R^m$ is convex if, together with points $x_1, \dots, x_k$ from M, it also contains all points of the form

$$x = \sum_{i=1}^{k} \lambda_i x_i, \ \lambda_i \geq 0, \ \sum_{i=1}^{k} \lambda_i = 1,$$

referred to as convex linear combinations of the points $x_1, \dots, x_k$.

Intersection of convex sets is a convex set.

Consider a system of linear inequalities

$$xA \leq b$$

or

$$xa^j \leq \beta_j,\ j \in N, N = \{1, \ldots, n\}, \tag{1.5.1}$$

where $A = [a^j, j \in N]$ is the $(m \times n)$ matrix, $b = (\beta_1, \ldots, \beta_n) \in R$.

Denote a set of solutions of (1.5.1) as $\tilde{X} \triangleq \{x | xA \leq b\}$. From the definition it immediately follows that $\tilde{X}$ is a convex set. The set $\tilde{X}$ is called a *convex polyhedral set* given by the system of constraints (1.5.1).

1.5.2. The point $X \in M$, where M is the convex set is called an *extreme point* if from the condition $x = \lambda x_1 + (1-\lambda)x_2$, $x_1 \in M, x_2 \in M$ and $0 < \lambda < 1$ it follows that $x_1 = x_2 = x$. Conceptually, the definition implies that $x \in M$ is an extreme point if there is no line segment with two end points in M to which x is an interior point.

Notice that the extreme point of the convex set is always a boundary point and the converse is not true.

Let X be a convex polyhedral set that is given by the system of constraints (1.5.1). Then the following assertions are true.

Theorem [Ashmanov (1981)]. *The set $\tilde{X}$ has extreme points if and only if $rankA = rank[a^j, j \in N] = m$.*

Theorem [Ashmanov (1981)]. *For the point $x_0 \in X$ to be extreme, it is necessary and sufficient that this point be a solution of the system*

$$x_0 a^j = \beta_j, j \in N_1; \tag{1.5.2}$$

$$x_0 a^j \leq \beta_j, i \in N \setminus N_1, \tag{1.5.3}$$

where $N_1 \subset N$, $rank[a^i, j \in N_1] = m$.

The latter theorem yields an algorithm for finding extreme points of the set $\tilde{X}$. To this end, we need to consider column bases of the matrix A, to solve the system of linear equations (1.5.2), and to verify that the inequalities (1.5.3) hold. However, this method of searching for extreme points of the polyhedral set is of little practical significance, since its application involves complete exhaustion of all possible column bases of the matrix A.

1.5.3. An intersection of all convex sets containing P will be called the *convex hull of the set* P and denoted as $conv(P)$. This definition is equivalent to the following statement. The convex hull of the set P is composed of all convex linear combinations of all possible finite systems of points from P, i.e.

$$conv(P) = \left\{ x | x = \sum_{i=1}^{n} \lambda_i x_i, \sum_{i=1}^{n} \lambda_i = 1, \lambda_i \geq 0, x_i \in P \right\}.$$

The convex hull of a finite number of points is called a *convex polyhedron* generated by these points. The convex polyhedron is generated by its extreme points. Thus, if we consider the set X of Player 1's mixed strategies in the $(m \times n)$ game, then $X = conv\{u_1, \ldots, u_m\}$, where $u_i = (0, \ldots, 0, 1, 0, \ldots, 0)$ are unit vectors of the space R^m of pure strategies of Player 1. The set X is a convex polyhedron of $(m-1)$ dimension and is also called the $(m-1)$-*dimensional simplex* (or the fundamental simplex). In this case, all vectors u_i (pure strategies) are extreme points of the polyhedron X. Similar assumptions apply to the Player 2's set Y of mixed strategies.

The *cone* C is called a set of such points that if $x \in C, \lambda \geq 0$, then $\lambda x \in C$.

Conceptually, the cone C, which is the subset R^m, contains, together with the point x, the entire half-line (x), such that

$$(x) \triangleq \{y \,|\, y = \lambda x, \lambda \geq 0\}.$$

The cone C is also called a *convex cone* if the following condition is satisfied: $x+y \in C$ holds for all $x, y \in C$. In other words, the cone C is convex if it is closed with respect to addition. Another equivalent definition may also be given. The cone is called convex if it is a convex set.

The sum of convex cones $C_1 + C_2 \triangleq \{c \,|c = c_1 + c_2, c_1 \in C_1, c_2 \in C_2\}$ and their intersection $C_1 \cap C_2$ are also convex cones.

Immediate verification of the definition may show that the set $C \triangleq \{x|xA \leq 0\}$ of solutions to a homogeneous system of linear inequalities corresponding to (1.5.1) is a convex cone.

Let $\tilde{X}$ be a convex polyhedral set given in the equivalent form

$$\sum_{i=1}^{m} \xi_i a_i \leq b, \tag{1.5.4}$$

where $x = (\xi_1, \ldots, \xi_m) \in R^m$, a_i is the ith row of the matrix A, $i = 1, \ldots, m$. Now suppose that $rankA = r, r \leq m$, and vectors $a_1, \ldots, a_r$ form the row basis of the matrix A. Decompose the remaining rows with respect to the basis

$$a_j = \sum_{i=1}^{r} \delta_{ij} a_j, \ j = r+1, \ldots, m. \tag{1.5.5}$$

Substituting (1.5.5) into (1.5.4), we obtain the following system of inequalities (equivalent to (1.5.4))

$$\sum_{i=1}^{r} \left(\xi_i + \sum_{j=r+1}^{m} \xi_j \delta_{ij} \right) a_i \leq b. \tag{1.5.6}$$

Denote by X_0 the set of vectors $x = (\xi_1, \ldots, \xi_m) \in R$ satisfying the inequalities (1.5.6) and condition $\xi_j = 0, j = r+1, \ldots, m$. By Theorem in 1.5.2, the set X_0 has extreme points. The following theorem holds [Ashmanov (1981)].

Theorem on representation of a polyhedral set. *Let $\tilde{X}$ be the polyhedral set given by the system of constraints* (1.5.4). *Then*

$$\tilde{X} = M + C,$$

where $M + C \triangleq \{x|x = y + z, y \in M, z \in C\}$, M is a convex polyhedron generated by extreme points of the polyhedral set X_0 given by (1.5.6), *and $C = \{x|xA \leq 0\}$ is convex cone.*

This theorem, in particular, implies that if the solution set $\tilde{X}$ of (1.5.4) is bounded, then $\tilde{X}$ is a convex polyhedron.

1.5.4. Recall that the problem of finding $\min cx$ under constraints

$$xA \geq b,\ x \geq 0, \tag{1.5.7}$$

where A is an $(m \times n)$ matrix, $c \in R^m$, $x \in R^m$, $b \in R^n$ is called a direct problem of linear programming in standard form, and the problem of determining $\max by$ the constraints

$$Ay \leq c,\ y \geq 0, \tag{1.5.8}$$

where $y \in R^n$, is called a dual linear programming problem for (1.5.7).

The vector $x \in R^n$, satisfying the system (1.5.7) is a feasible solution of problem (1.5.7). The notion of a feasible solution $y \in R^n$ of problem (1.5.8) is introduced in a similar manner. The feasible solution $\overline{x}(\overline{y})$ is an optimal solution of problem (1.5.7) [(1.5.8)] if $c\overline{x} = \min cx$ $(b\overline{y} = \max by)$ and the minimum (maximum) of the function $cx(by)$ is achieved on the set of all feasible solutions.

The following statement is true [Ashmanov (1981)].

Duality Theorem. *If both problems* (1.5.7), (1.5.8) *have feasible solutions then both of them have optimal solutions* $\overline{x}, \overline{y}$, *respectively, and*

$$c\overline{x} = b\overline{y}.$$

1.5.5. In closing this section, we give one property of convex functions. First recall that the function $\varphi : M \to R^1$, where $M \subset R^m$ is a convex set, is convex if

$$\varphi(\lambda x_1 + (1-\lambda)x_2) \leq \lambda\varphi(x_1) + (1-\lambda)\varphi(x_2) \tag{1.5.9}$$

for any $x_1, x_2 \in M$ and $\lambda \in [0,1]$. If the inverse inequality holds in (1.5.9), then the function φ is called concave.

Let $\varphi_i(x)$ be the family of functions convex on $M, i = 1, \dots, n$. Then the upper envelope $\psi(x)$ of this family of functions

$$\psi(x) \triangleq \max_{i=1,\dots,n} \varphi_i(x) \tag{1.5.10}$$

is convex on M.

Indeed, by the definition of the convex function for $x_1, x_2 \in M$ and $\alpha \in [0, 1]$ we have

$$\begin{aligned}\varphi_i(\alpha x_1 + (1-\alpha)x_2) &\le \alpha\varphi_i(x_1) + (1-\alpha)\varphi_i(x_2) \\ &\le \alpha \max_i \varphi_i(x_1) + (1-\alpha)\max_i \varphi_i(x_2).\end{aligned}$$

Hence we get

$$\begin{aligned}\psi(\alpha x_1 + (1-\alpha)x_2) &= \max_i \varphi_i(\alpha x_1 + (1-\alpha)x_2) \\ &\le \alpha\psi(x_1) + (1-\alpha)\psi(x_2),\end{aligned}$$

which is what we set out to prove.

In a similar manner, we may show that the lower envelope (in (1.5.10) the minimum is taken in i) of the family of concave functions is concave.

1.6 Existence of a Solution of the Matrix Game in Mixed Strategies

We shall prove that an arbitrary matrix game is strictly determined in mixed strategies.

1.6.1. Theorem [von Neumann (1928)]. *Any matrix game has a saddle point in mixed strategies.*

Proof. Let Γ_A be an arbitrary $(m \times n)$ game with a strictly positive matrix $A = \{a_{ij}\}$, i.e. $a_{ij} > 0$ for all $i = \overline{1, m}$ and $j = \overline{1, n}$. Show that in this case the theorem is true. To do this, we shall consider an auxiliary linear programming problem

$$\min xu, \;\; xA \geq w, \;\; x \geq 0 \tag{1.6.1}$$

and its dual problem

$$\max yw, \;\; Ay \leq u, \;\; y \geq 0, \tag{1.6.2}$$

where $u = (1, \ldots, 1) \in R^m$, $w = (1, \ldots, 1) \in R^n$. From the strict positivity of the matrix A it follows that there exists a vector $x > 0$ for which $xA > w$, i.e. problem (1.6.1) has a feasible solution. On the other hand, the vector $y = 0$ is a feasible solution of problem (1.6.2). And it can by easily seen that there exist a feasible solution of (1.6.2) y' with $|y'| > 0$. Therefore, by the duality theorem of linear programming (see 1.5.4), both problems (1.6.1) and (1.6.2) have optimal solutions $\overline{x}, \overline{y}$, respectively, and

$$\overline{x}u = \overline{y}w \stackrel{\triangle}{=} \Theta > 0. \tag{1.6.3}$$

Consider vectors $x^* \stackrel{\triangle}{=} \overline{x}/\Theta$ and $y^* \stackrel{\triangle}{=} \overline{y}/\Theta$ and show that they are optimal strategies for the Players 1 and 2 in the game $\overline{\Gamma}_A$, respectively and the value of the game is equal to $1/\Theta$.

Indeed, from (1.6.3) we have

$$x^*u = (\overline{x}u)/\Theta = (\overline{y}w)/\Theta = y^*w = 1,$$

and from feasibility of $\overline{x}$ and $\overline{y}$ for problems (1.6.1), (1.6.2), it follows that $x^* = \overline{x}/\Theta \geq 0$ and $y^* = \overline{y}/\Theta \geq 0$, i.e. x^* and y^* are the mixed strategies of Players 1 and 2 in the game Γ_A.

Let us compute a payoff to Player 1 at (x^*, y^*):

$$K(x^*, y^*) = x^*Ay^* = (\overline{x}A\overline{y})/\Theta^2. \tag{1.6.4}$$

On the other hand, from the feasibility of vectors $\overline{x}$ and $\overline{y}$ for problems (1.6.1),(1.6.2) and equality (1.6.3), we have

$$\Theta = w\overline{y} \leq (\overline{x}A)\overline{y} = \overline{x}(A\overline{y}) \leq \overline{x}u = \Theta. \tag{1.6.5}$$

Thus, $\overline{x}A\overline{y} = \Theta$ and (1.6.4) implies that

$$K(x^*, y^*) = 1/\Theta. \qquad (1.6.6)$$

Let $x \in X$ and $y \in Y$ be arbitrary mixed strategies for Players 1 and 2. The following inequalities hold:

$$K(x^*, y) = (x^*A)y = (\overline{x}A)y/\Theta \geq (wy)/\Theta = 1/\Theta, \qquad (1.6.7)$$

$$K(x, y^*) = x(Ay^*) = x(A\overline{y})/\Theta \leq (xu)/\Theta = 1/\Theta. \qquad (1.6.8)$$

Comparing (1.6.6)–(1.6.8), we have that (x^*, y^*) is a saddle point and $1/\Theta$ is the value of the game Γ_A with a strictly positive matrix A.

Now consider the $(m \times n)$ game $\Gamma_{A'}$ with an arbitrary matrix $A' = \{a'_{ij}\}$. Then there exists such constant $\beta > 0$ that the matrix $A = A' + B$ is strictly positive, where $B = \{\beta_{ij}\}$ is an $(m \times n)$ matrix, $\beta_{ij} = \beta$, $i = \overline{1, m}$, $j = \overline{1, n}$. In the game Γ_A there exists a saddle point (x^*, y^*) in mixed strategies, and the value of the game equals $v_A = 1/\Theta$, where Θ is determined as in (1.6.3).

From Lemma 1.4.3, it follows that $(x^*, y^*) \in Z(\overline{\Gamma}_{A'})$ is a saddle point in the game $\Gamma_{A'}$ in mixed strategies and the value of the game is equal to $v_{A'} = v_A - \beta = 1/\Theta - \beta$. This completes the proof of Theorem.

Informally, the existence of a solution in the class of mixed strategies implies that, by randomizing the set of pure strategies, the players can always eliminate uncertainty in choosing their strategies they have encountered before the game starts. Note that the mixed strategy solution does not necessarily exist in zero-sum games. Examples of such games with an infinite number of strategies are given in Secs. 2.3, 2.4.

Notice that the proof of theorem is constructive, since the solution of the matrix game is reduced to a linear programming problem (Danskin, 1967), and the solution algorithm for the game $\Gamma_{A'}$ is as follows.

1. By employing the matrix A', construct a strictly positive matrix $A = A' + B$, where $B = \{\beta_{ij}\}, \beta_{ij} = \beta > 0$.

2. Solve the linear programming problems (1.6.1),(1.6.2). Find vectors $\overline{x}, \overline{y}$ and a number Θ [see (1.6.3)].
3. Construct optimal strategies for Players 1 and 2, respectively,

$$x^* = \overline{x}/\Theta,\ y^* = \overline{y}/\Theta.$$

4. Compute the value of the game $\Gamma_{A'}$

$$v_{A'} = 1/\Theta - \beta.$$

Example 9. Consider the matrix game Γ_A determined by the matrix

$$A = \begin{bmatrix} 4 & 0 \\ 2 & 3 \end{bmatrix}.$$

Associated problems of linear programming are of the form

$$\begin{array}{ll} \min \xi_1 + \xi_2, & \max \eta_1 + \eta_2, \\ 4\xi_1 + 2\xi_2 \geq 1, & 4\eta_1 \leq 1, \\ 3\xi_2 \geq 1, & 2\eta_1 + 3\eta_2 \leq 1, \\ \xi_1 \geq 0, \xi_2 \geq 0, & \eta_1 \geq 0, \eta_2 \geq 0. \end{array}$$

Note that, these problems may be written in the equivalent form with constraints in the form of equalities

$$\begin{array}{ll} \min \xi_1 + \xi_2, & \max \eta_1 + \eta_2, \\ 4\xi_1 + 2\xi_2 - \xi_3 = 1, & 4\eta_1 + \eta_3 = 1, \\ 3\xi_2 - \xi_4 = 1, & 2\eta_1 + 3\eta_2 + \eta_4 = 1, \\ \xi_1 \geq 0, \xi_2 \geq 0, \xi_3 \geq 0, \xi_4 \geq 0, & \eta_1 \geq 0, \eta_2 \geq 0, \eta_3 \geq 0, \eta_4 \geq 0. \end{array}$$

Thus, any method of solving the linear programming problems can be used to solve the matrix games. The simplex method is most commonly used to solve such problems. Its systematic discussion may be found in Ashmanov (1981), Gale (1960), and Hu (1970).

1.6.2. In a sense, the linear programming problem is equivalent to the matrix game Γ_A. Indeed, consider the following direct and dual problems of linear programming

$$\begin{aligned} &\min xu \\ &xA \geq w, \\ &x \geq 0, \end{aligned} \tag{1.6.9}$$

$$\begin{aligned} &\max yw \\ &Ay \leq u, \\ &y \geq 0. \end{aligned} \tag{1.6.10}$$

Let $\overline{X}$ and $\overline{Y}$ be the sets of optimal solutions of the problems (1.6.9) and (1.6.10), respectively. Denote $(1/\Theta)\overline{X} \triangleq \{\overline{x}/\Theta \mid \overline{x} \in X\}$, $(1/\Theta)\overline{Y} \triangleq \{\overline{y}/\Theta \mid \overline{y} \in Y\}$, $\Theta > 0$.

Theorem. *Let Γ_A be the $(m \times n)$ game with the positive matrix A (all elements are positive) and let there be given two dual problems of linear programming* (1.6.9) *and* (1.6.10). *Then the following statements hold.*

1. *Both linear programming problems have a solution ($\overline{X} \neq \oslash$ and $\overline{Y} \neq \oslash$), in which case*

$$\Theta = \min_x xu = \max_y yw.$$

2. *The value v_A of the game Γ_A is*

$$v_A = 1/\Theta,$$

and the strategies

$$x^* = \overline{x}/\Theta, \ y^* = \overline{y}/\Theta$$

are optimal, where $\overline{x} \in \overline{X}$ is an optimal solution of the direct problem (1.6.9) *and $\overline{y} \in \overline{Y}$ is the solution of the dual problem* (1.6.10).

3. *Any optimal strategies $x^* \in X^*$ and $y^* \in Y^*$ of the players can be constructed as shown above, i.e.*

$$X^* = (1/\Theta)\overline{X}, \ Y^* = (1/\Theta)\overline{Y}.$$

Proof. Statements 1, 2 and inclusions $(1/\Theta)\overline{X} \subset X^*$, $(1/\Theta)\overline{Y} \subset Y^*$, immediately follow from the proof of Theorem 1.6.1.

Show the inverse inclusion. To do this, consider the vectors $x^* = (\xi_1^*, \ldots, \xi_m^*) \in X^*$ and $\overline{x} = (\overline{\xi}_1, \ldots, \overline{\xi}_m)$, where $\overline{x} = \Theta x^*$. Then for all $j \in N$ we have

$$\overline{x}a^j = \Theta x^* a^j \geq \Theta(1/\Theta) = 1,$$

in which case $\overline{x} \geq 0$, since $\Theta > 0$ and $x^* \geq 0$. Therefore, $\overline{x}$ is a feasible solution to problem (1.6.9).

Let us compute the value of the objective function

$$\overline{x}u = \Theta x^* u = \Theta = \min_x xu,$$

i.e. $\overline{x} \in \overline{X}$ is an optimal solution to problem (1.6.9).

The inclusion $Y^* \subset (1/\Theta)\overline{Y}$ can be proved in a similar manner. This completes the proof of the theorem.

1.7 Properties of Optimal Strategies and Value of the Game

Consider the properties of optimal strategies which, in some cases, assist in finding the value of the game and a saddle point.

1.7.1. Let $(x^*, y^*) \in X \times Y$ be a saddle point in mixed strategies for the game Γ_A. It turns out that, to test the point (x^*, y^*) for a saddle, it will suffice to test the inequalities (1.4.7) only for $i \in M$ and $j \in N$, not for all $x \in X$ and $y \in Y$, since the following statement is true.

Theorem. *For the situation (x^*, y^*) to be an equilibrium (saddle point) in the game Γ_A, and the number $v = K(x^*, y^*)$ be the value, it is necessary and sufficient that the following inequalities hold for all $i \in M$ and $j \in N$:*

$$K(i, y^*) \leq K(x^*, y^*) \leq K(x^*, j). \tag{1.7.1}$$

Proof. Necessity. Let (x^*, y^*) be a saddle point in the game Γ_A. Then

$$K(x, y^*) \leq K(x^*, y^*) \leq K(x^*, y)$$

for all $x \in X$, $y \in Y$. Hence, in particular, for $u_i \in X$ and $w_j \in Y$ we have

$$K(i, y^*) \stackrel{\triangle}{=} K(u_i, y^*) \leq K(x^*, y^*) \leq K(x^*, w_j) \stackrel{\triangle}{=} K(x^*, j),$$

for all $i \in M$ and $j \in N$.

Sufficiency. Let (x^*, y^*) be a pair of mixed strategies for which the inequalities (1.7.1) hold. Also, let $x = (\xi_1, \ldots, \xi_m) \in X$ and $y = (\eta_1, \ldots, \eta_n) \in Y$ be arbitrary mixed strategies for Players 1 and 2, respectively. Multiplying the first and second inequalities (1.7.1) by ξ_i and η_j, respectively, and summing, we get

$$\sum_{i=1}^{m} \xi_i K(i, y^*) \leq K(x^*, y^*) \sum_{i=1}^{m} \xi_i = K(x^*, y^*), \qquad (1.7.2)$$

$$\sum_{j=1}^{n} \eta_j K(x^*, j) \geq K(x^*, y^*) \sum_{j=1}^{n} \eta_j = K(x^*, y^*). \qquad (1.7.3)$$

In this case, we have

$$\sum_{i=1}^{m} \xi_i K(i, y^*) = K(x, y^*), \qquad (1.7.4)$$

$$\sum_{j=1}^{n} \eta_j K(x^*, j) = K(x^*, y). \qquad (1.7.5)$$

Substituting (1.7.4), (1.7.5) into (1.7.2) and (1.7.3), respectively, and taking into account the arbitrariness of strategies $x \in X$ and $y \in Y$, we obtain saddle point conditions for the pair of mixed strategies (x^*, y^*).

Corollary. *Let (i^*, j^*) be a saddle point in the game Γ_A. Then the situation (i^*, j^*) is also a saddle point in the game $\overline{\Gamma}_A$.*

Example 10. Solution of the Evasion-type Game. Suppose the players select integers i and j between 1 and n, and Player 1 wins the amount $a_{ij} = |i-j|$, i.e. the distance between the numbers i and j.

Suppose the first player uses strategy $x^* = (1/2, 0, \dots, 0, 1/2)$. Then

$$\begin{aligned} K(x^*, j) &= 1/2|1-j| + 1/2|n-j| \\ &= 1/2(j-1) + 1/2(n-j) = (n-1)/2, \end{aligned}$$

for all $1 \le j \le n$.

a) Let $n = 2k+1$ be odd. Then Player 2 has a pure strategy $j^* = (n+1)/2 = k+1$ such that

$$a_{ij^*} = |i - (n+1)/2| = |i-k-1| \le k = (n-1)/2,$$

for all $i = 1, 2, \dots, n$.

b) Let $n = 2k$ be even. Then Player 2 has a strategy $y^* = (0, 0, \dots, 1/2, 1/2, 0, \dots, 0)$, where $\eta_k^* = 1/2$, $\eta_{k+1}^* = 1/2$, $\eta_j^* = 0, j \ne k+1$, $j \ne k$, and

$$\begin{aligned} K(j, y^*) &= 1/2|i-k| + 1/2|i-k-1| \\ &\le 1/2k + 1/2(k-1) = (n-1)/2, \end{aligned}$$

for all $1 \le i \le n$.

Now, using Theorem, it can be easily seen that the value of the game is $v = (n-1)/2$, Player 1 has optimal strategy x^*, and Player 2's optimal strategy is j^* if $n = 2k+1$, and y^* if $n = 2k$.

1.7.2. Provide the results that immediately follow from the Theorem in 1.7.1.

Theorem. *Let Γ_A be an $(m \times n)$ game. For the situation in mixed strategies, let (x^*, y^*) be an equilibrium (saddle point) in the game $\overline{\Gamma}_A$, it is necessary and sufficient that the following equality holds*

$$\max_{1 \le i \le m} K(i, y^*) = \min_{1 \le j \le n} K(x^*, j). \tag{1.7.6}$$

Proof. Necessity. If (x^*, y^*) is a saddle point, then, by Theorem in 1.7.1, we have

$$K(i, y^*) \leq K(x^*, y^*) \leq K(x^*, j),$$

for all $i \in \{1, \dots, m\}$, $j \in \{1, \dots, n\}$. Therefore,

$$K(i, y^*) \leq K(x^*, j),$$

for each i and j. Suppose the opposite is true, i.e. (1.7.6) is not satisfied. Then

$$\max_{1 \leq i \leq m} K(i, y^*) < \min_{1 \leq j \leq n} K(x^*, j).$$

Consequently, the following inequalities hold

$$K(x^*, y^*) = \sum_{i=1}^{m} \xi_i^* K(i, y^*) \leq \max_{1 \leq i \leq m} K(i, y^*) < \min_{1 \leq j \leq n} K(x^*, j)$$
$$\leq \sum_{j=1}^{n} \eta_j^* K(x^*, j) = K(x^*, y^*).$$

The obtained contradiction proves the necessity of the Theorem assertion.

Sufficiency. Let a pair of mixed strategies $(\tilde{x}, \tilde{y})$ be such that $\max_i K(i, \tilde{y}) = \min_j K(\tilde{x}, j)$. Show that in this case $(\tilde{x}, \tilde{y})$ is a saddle point in the game $\overline{\Gamma}_A$.

The following relations hold

$$\min_{1 \leq j \leq n} K(\tilde{x}, j) \leq \sum_{j=1}^{n} \tilde{\eta}_j K(\tilde{x}, j) = K(\tilde{x}, \tilde{y})$$
$$= \sum_{i=1}^{m} \tilde{\xi}_i K(i, \tilde{y}) \leq \max_{1 \leq i \leq m} K(i, \tilde{y}).$$

Hence we have

$$K(i, \tilde{y}) \leq \max_{1 \leq i \leq m} K(i, \tilde{y}) = K(\tilde{x}, \tilde{y}) = \min_{1 \leq j \leq n} K(\tilde{x}, j) \leq K(\tilde{x}, j),$$

for all $1 \le i \le m$ and $1 \le j \le n$, then, by the Theorem in 1.7.1, $(\tilde{x}, \tilde{y})$ is the saddle point in the game $\overline{\Gamma}_A$.

From the proof it follows that any one of the numbers in (1.7.6) is the value of the game.

1.7.3. Theorem. *The following relation holds for the matrix game* Γ_A

$$\max_x \min_j K(x, j) = v_A = \min_y \max_i K(i, y), \tag{1.7.7}$$

in which case the extrema are achieved on the players' optimal strategies.

This theorem follows from the Theorems 1.3.4, 1.7.2, and its proof is left to the reader.

1.7.4. Theorem. *In the matrix game* Γ_A *the players' sets of optimal mixed strategies* X^* *and* Y^* *are convex polyhedra.*

Proof. By Theorem 1.7.1, the set X^* is the set of all solutions of the system of inequalities

$$xa^j \ge v_A, \ j \in N,$$

$$xu = 1,$$

$$x \ge 0,$$

where $u = (1, \dots, 1) \in R^m$, v_A is the value of the game. Thus, X^* is a convex polyhedral set (1.5.1). On the other hand, $X^* \subset X$, where X is a convex polyhedron (1.5.3). Therefore, X^* is bounded. Consequently, by Theorem 1.5.3, the set X^* is a convex polyhedron.

In a similar manner, it may be proved that Y^* is a convex polyhedron.

1.7.5. As an application of Theorem 1.7.3, we shall provide a geometric solution to the games with two strategies for one of the players $(2 \times n)$ and $(m \times 2)$ games. This method is based on the property that the optimal strategies x^* and y^* deliver exterior extrema in the equality

$$v_A = \max_x \min_j K(x, j) = \min_y \max_i K(i, y).$$

Example 11. $(2 \times n)$ *game.* We shall examine the game in which Player 1 has two strategies and Player 2 has n strategies. The matrix is of the form

$$A = \begin{bmatrix} \alpha_{11} & \alpha_{12} & \dots & \alpha_{1n} \\ \alpha_{21} & \alpha_{22} & \dots & \alpha_{2n} \end{bmatrix}.$$

Suppose Player 1 chooses mixed strategy $x = (\xi, 1-\xi)$ and Player 2 chooses pure strategy $j \in N$. Then a payoff to Player 1 at (x, j) is

$$K(x,j) = \xi\alpha_{1j} + (1-\xi)\alpha_{2j}. \tag{1.7.8}$$

Geometrically, the payoff is a straight line segment with coordinates (ξ, K). Accordingly, to each pure strategy j corresponds a straight line. The graph of the function

$$H(\xi) = \min_j K(x,j)$$

is the lower envelope of the family of straight lines (1.7.8). This function is concave as the lower envelope of the family of concave (linear in the case) function (1.5.5). The point ξ^*, at which the maximum of the function $H(\xi)$ is achieved with respect to $\xi \in [0,1]$, yields the required optimal solution $x^* = (\xi^*, 1-\xi^*)$ and the value of the game $v_A = H(\xi^*)$.

For definiteness, we shall consider the game with the matrix

$$A = \begin{bmatrix} 1 & 3 & 1 & 4 \\ 2 & 1 & 4 & 0 \end{bmatrix}.$$

For each $j = 1,2,3,4$ we have: $K(x,1) = -\xi + 2$, $K(x,2) = 2\xi + 1$, $K(x,3) = -3\xi + 4$, $K(x,4) = 4\xi$. The lower envelope $N(\xi)$ of the family of straight lines $\{K(x,j)\}$ and the lines themselves, $K(x,j)$, $j = 1,2,3,4$ are shown in Fig. 1.1. The maximum $H(\xi^*)$ of the function $H(\xi)$ is found as the intersection of the first and the fourth lines. Thus, ξ^* is a solution of the equation.

$$4\xi^* = -\xi^* + 2 = v_A.$$

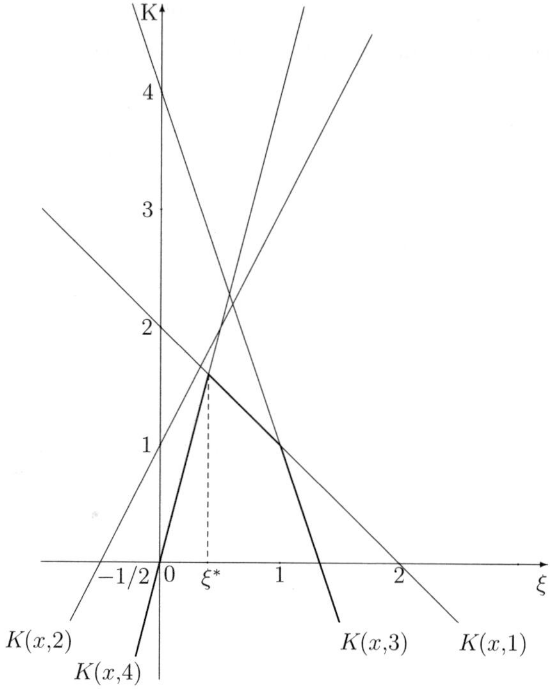

Figure 1.1

Hence we get the optimal strategy $x^* = (2/5, 3/5)$ of Player 1 and the value of the game is $v_A = 8/5$. Player 2's optimal strategy is found from the following reasonings. Note that in the case studied $K(x^*, 1) = K(x^*, 4) = v_A = 8/5$.

For the optimal strategy $y^* = (\eta_1^*, \eta_2^*, \eta_3^*, \eta_4^*)$ the following equality must hold

$$v_A = K(x^*, y^*) = \eta_1^* K(x^*, 1) + \eta_2^* K(x^*, 2) + \eta_3^* K(x^*, 3) + \eta_4^* K(x^*, 4).$$

In this case $K(x^*, 2) > 8/5$, $K(x^*, 3) > 8/5$; therefore, $\eta_2^* = \eta_3^* = 0$, and η_1^*, η_4^* can be found from the conditions

$$\eta_1^* + 4\eta_4^* = 8/5,$$
$$2\eta_1^* = 8/5.$$

Thus, $\eta_1^* = 4/5$, $\eta_4^* = 1/5$ and the optimal strategy of Player 2 is $y^* = (4/5, 0, 0, 1/5)$.

Example 12. $(m \times 2)$ *game.* In this example, Player 2 has two strategies and Player 1 has m strategies. The matrix A is of the form

$$A = \begin{bmatrix} \alpha_{11} & \alpha_{12} \\ \alpha_{21} & \alpha_{22} \\ \dots & \dots \\ \alpha_{m1} & \alpha_{m2} \end{bmatrix}.$$

This game can be analyzed in a similar manner. Indeed, let $y = (\eta, 1-\eta)$ be an arbitrary mixed strategy of Player 2. Then Player 1's payoff in situation (i, y) is

$$K(i, y) = \alpha_{i1}\eta + \alpha_{i2}(1 - \eta) = (\alpha_{i1} - \alpha_{i2})\eta + \alpha_{i2}.$$

The graph of the function $K(i, y)$ is a straight line. Consider the upper envelope of these straight lines, i.e. the function

$$H(\eta) = \max_i [(\alpha_{i1} - \alpha_{i2})\eta + \alpha_{i2}].$$

The function $H(\eta)$ is convex (as the upper envelope of the family of convex functions).

The point of minimum η^* of the function $H(\eta)$ yields the optimal strategy $y^* = (\eta^*, 1-\eta^*)$ and the value of the game is $v_A = H(\eta^*) = \min_{\eta \in [0,1]} H(\eta)$.

1.7.6. We shall provide a theorem that is useful in finding a solution of the game.

Theorem. *Let $x^* = (\xi_1^*, \dots, \xi_m^*)$ and $y^* = (\eta_1^*, \dots, \eta_n^*)$ be optimal strategies in the game $\overline{\Gamma}_A$ and v_A be the value of the game. Then for any i, for which $K(i, y^*) < v_A$, there must be $\xi_i^* = 0$, and for any j such that $v_A < K(x^*, j)$ there must be $\eta_j^* = 0$.*

Conversely, if $\xi_i^* > 0$, then $K(i, y^*) = v_A$, and if $\eta_j^* > 0$, then $K(x^*, j) = v_A$.

Proof. Suppose that for some $i_0 \in M$, $K(i_0, y^*) < v_A$ and $\xi^*_{i0} \neq 0$. Then we have

$$K(i_0, y^*)\xi^*_{i0} < v_A \xi^*_{i0}.$$

For all $i \in M$, $K(i, y^*) \leq v_A$, therefore

$$K(i, y^*)\xi^*_i \leq v_A \xi^*_i.$$

Consequently, $K(x^*, y^*) < v_A$, which contradicts to the fact that v_A is the value of the game. The second part of the Theorem can be proved in a similar manner.

This result is a counterpart of the complementary stackness theorem [Hu (1970)] or, as it is sometimes called the canonical equilibrium theorem for the linear programming problem [Gale (1960)].

Definition. Player 1's (2's) pure strategy $i \in M (j \in N)$ is called an essential or active strategy if there exists the player's optimal strategy $x^* = (\xi^*_1, \ldots, \xi^*_m)$ $(y^* = (\eta^*_1, \ldots, \eta^*_n))$ for which $\xi^*_i > 0$ $(\eta^*_j > 0)$.

From the definition, and from the latter theorem, it follows that for each essential strategy i of Player 1 and any optimal strategy $y^* \in Y^*$ of Player 2 in the game Γ_A the following equality holds:

$$K(i, y^*) = a_i y^* = v_A.$$

A similar equality holds for any essential strategy $j \in N$ of Player 2 and for the optimal strategy $x^* \in X^*$ of Player 1

$$K(x^*, j) = a^j x^* = v_A.$$

If the equality $a_i y = v_A$ holds for the pure strategy $i \in M$ and mixed strategy $y \in Y$, then the strategy i is the best reply to the mixed strategy y in the game Γ_A.

Thus, using this terminology, the theorem can be restated as follows. If the pure strategy of the player is essential, then it is the best reply to any optimal strategy of the opponent.

A knowledge of the optimal strategy spectrum simplifies to finding a solution of the game. Indeed, let M_{X^*} be the spectrum of

Player 1's optimal strategy x^*. Then each optimal strategy $y^* = (\eta_1^*, \ldots, \eta_n^*)$ of Player 2 and the value of the game v satisfy the system of inequalities

$$a_i y^* = v, \ i \in M_{x^*},$$
$$a_i y^* \leq v, \ i \in M \setminus M_{x^*},$$
$$\sum_{j=1}^{n} \eta_j^* = 1, \ \eta_j^* \geq 0, \ j \in N.$$

Thus, only essential strategies may appear in the spectrum M_{x^*} of any optimal strategy x^*.

1.7.7. To conclude this section, we shall provide the analytical solution of Attack and Defence game (see Example 4, 1.1.3).

Example 13. [Sakaguchi (1973)]. Let us consider the game with the $(n \times n)$ matrix A.

$$A = \begin{bmatrix} \beta_1\tau_1 & \tau_1 & \ldots & \tau_1 \\ \tau_2 & \beta_2\tau_2 & \ldots & \tau_2 \\ \ldots & \ldots & \ldots & \ldots \\ \tau_n & \tau_n & \ldots & \beta_n\tau_n \end{bmatrix}.$$

Here $\tau_i > 0$ is the value and $0 < \beta_i < 1$ is the probability of hitting the target $C_i, i = 1, 2, \ldots, n$ provided that it is defended.

Let $\tau_1 \leq \tau_2 \leq \ldots \leq \tau_n$. We shall define the function φ, of integers $1, 2, \ldots, n$ as follows:

$$\varphi(k) = \left\{ \sum_{i=k}^{n} (1 - \beta_i)^{-1} - 1 \right\} \Big/ \sum_{i=k}^{n} (\tau_i(1 - \beta_i))^{-1} \tag{1.7.9}$$

and let $l \in \{1, 2, \ldots, n\}$ be an integer which maximize the function $\varphi(k)$, i.e.

$$\varphi(l) = \max_{k=1,2,\ldots,n} \varphi(k). \tag{1.7.10}$$

We shall establish properties of the function $\varphi(k)$. Denote by R one of the signs of the order relation $\{>, =, <\}$. In this case

$$\varphi(k) R \varphi(k+1) \tag{1.7.11}$$

if and only if

$$\tau_k R \varphi(k),\ k = 1, 2, \ldots, n-1,\ \tau_0 \equiv 0. \tag{1.7.12}$$

Indeed, from (1.7.9) we obtain

$$\begin{aligned}\frac{\varphi(k)}{\tau_k} & \frac{(1-\beta_k)^{-1}}{\sum_{i=k+1}^{n}(\tau_i(1-\beta_i))^{-1}} + \varphi(k) \\ &= \varphi(k+1) + \frac{(1-\beta_k)^{-1}}{\sum_{i=k+1}^{n}(\tau_i(1-\beta_i))^{-1}}.\end{aligned}$$

Then we have

$$\left(\frac{\varphi(k)}{\tau_k} - 1\right)\frac{(1-\beta_k)^{-1}}{\sum_{i=k+1}^{n}(\tau_i(1-\beta_i))^{-1}} + \varphi(k) = \varphi(k+1). \tag{1.7.13}$$

Note that the coefficient in (1.7.13) placed after brackets, is positive. Therefore, from (1.7.13) we obtained equivalence of relations (1.7.11) and (1.7.12).

Now, since $\varphi(l) \geq \varphi(l-1)$ or $\varphi(l) \geq \varphi(l+1)$, (in this case $\tau_{l-1} \leq \varphi(l-1)$ or $\tau_l \geq \varphi(l)$), then from relations (1.7.10), (1.7.11) we have

$$\tau_{l-1} \leq \varphi(l) \leq \tau_l. \tag{1.7.14}$$

Find optimal strategies in the game Γ_A. Recall that we have inequalities $\tau_1 \leq \tau_2 \leq \ldots \leq \tau_n$. Then the optimal strategies $x^* = (\xi_1^*, \ldots, \xi_n^*)$ and $y^* = (\eta_1^*, \ldots, \eta_n^*)$ for Players 1 and 2 respectively, are as follows:

$$\xi_i^* = \begin{cases} 0,\ i = 1, \ldots, l-1, \\ (\tau_i(1-\beta_i))^{-1} \Big/ \sum_{j=l}^{n}(\tau_j(1-\beta_j))^{-1},\ i = l, \ldots, n, \end{cases} \tag{1.7.15}$$

$$\eta_j^* = \begin{cases} 0,\ j = 1, \ldots, l-1, \\ (\tau_j - \varphi(l))/(\tau_j(1-\beta_j)),\ j = l, \ldots, n, \end{cases} \tag{1.7.16}$$

and the value of the game is

$$v_A = \varphi(l).$$

We have that $\xi_i^* \geq 0, i = 1, 2, \ldots, n$ and $\sum_{i=1}^n \xi_i^* = 1$. From the definition of $\varphi(l)$ and (1.7.14) we have that $\eta_j^* \geq 0$, $j = 1, 2, \ldots, n$ and $\sum_{j=1}^n \eta_j^* = 1$.

Let $K(x^*, j)$ be a payoff of Player 1 at (x^*, j). Similarly, let $K(i, y^*)$ be a payoff at (i, y^*). Substituting (1.7.15), (1.7.16) into the payoff function and using the assumption that the values of targets do not decrease, and using (1.7.14), we obtain

$$K(x^*, j) = \begin{cases} \sum\limits_{i=l}^{n} \tau_i \xi_i^* = \varphi(l) + \sum\limits_{j=l}^{n} (\tau_j(1-\beta_j))^{-1} > \varphi(l), \ j = \overline{1, l-1}, \\ \sum\limits_{i=l}^{n} \tau_i \xi_i^* - (1-\beta_j)\tau_j \xi_j^* = \varphi(l), \ j = \overline{l, n}, \end{cases}$$

$$K(i, y^*) = \begin{cases} \tau_i \leq \varphi(l), \ i = \overline{1, l-1}, \\ \tau_i - \tau_i(1-\beta_i)\eta_i^* = \varphi(l), \ i = \overline{l, n}. \end{cases}$$

Thus, for all $i, j = 1, \ldots, n$ the following inequalities hold

$$K(i, y^*) \leq \varphi(l) \leq K(x^*, j).$$

Then, by Theorem, 1.7.1, x^* and y^* are optimal and $v_A = \varphi(l)$. This completes the solution of the game.

Example 14. Search game. This game is a special case of previous example, where matrix A has the form,

$$A = \begin{bmatrix} \beta_1 & 0 & \ldots & 0 \\ 0 & \beta_2 & \ldots & 0 \\ \ldots & \ldots & \ldots & \ldots \\ 0 & 0 & \ldots & \beta_n \end{bmatrix},$$

$$1 > \beta_1 > \beta_2 > \ldots > \beta_n > 0.$$

First player is called Searcher (S), second player Hider (H). The game proceeds as follows. Both players simultaneously choose one of boxes from the set $N = \{1, \dots, n\}$. If H chooses box $i \in N$, he hides in the box i, if S chooses the box $j \in N$, he searches in the box j. If Searcher searches in the box $i \in N$, if Hider hides in $j \in N$ and $i \neq j$, the payoff of both players is equal to 0. If $i = j$, the Searcher wins β_i (the probability to find hider in the box under condition that Hider is there). In this game $\underline{v} = 0$, and $\overline{v} = \beta_n$ ($\underline{v} \neq \overline{v}$), and there is only mixed strategy saddle point. Denote optimal mixed strategies of players by $\overline{x} = (\overline{\xi_1}, \dots, \overline{\xi_i}, \dots, \overline{\xi_n})$ and $\overline{y} = (\overline{\eta_1}, \dots, \overline{\eta_j}, \dots, \overline{\eta_n})$. Then if v is the value of the game we must have

$$\overline{\xi_i}\beta_i \geq v \geq \overline{\eta_i}\beta_i, \;\; i \in N.$$

Suppose that $\overline{\xi_i} > 0$, $\overline{\eta_i} > 0$, $i \in N$. Then we have

$$\overline{\xi_i}\beta_i = v = \overline{\eta_i}\beta_i,$$
$$\overline{\xi_i} = \frac{v}{\beta_i}, \;\; \overline{\eta_i} = \frac{v}{\beta_i},$$

since $\sum\limits_{i=1}^{n} \overline{\xi_i} = \sum\limits_{i=1}^{n} \overline{\eta_i} = 1$, we get

$$\overline{\xi_i} = \frac{\frac{1}{\beta_i}}{\sum\limits_{k=1}^{n} \frac{1}{\beta_k}} = \overline{\eta_i}, \;\; \textit{and} \;\; v = \frac{1}{\sum\limits_{k=1}^{n} \frac{1}{\beta_k}}.$$

1.8 Dominance of Strategies

The complexity of solving a matrix game increases as the dimensions of the matrix A increase. In some cases, however, the analysis of payoff matrices permits a conclusion that some pure strategies do not appear in the spectrum of optimal strategy. This can result in replacement of the original matrix by the payoff matrix of a smaller dimension.

1.8.1. Definition. Strategy x' of Player 1 is said to dominate strategy x'' in the $(m \times n)$ game Γ_A if the following inequalities hold for all pure strategies $j \in \{1, \dots, n\}$ of Player 2

$$x'a^j \geq x''a^j. \tag{1.8.1}$$

Similarly, strategy y' of Player 2 dominates his strategy y'' if for all pure strategies $i \in \{1, \dots, m\}$ of Player 1

$$a_i y' \leq a_i y''. \tag{1.8.2}$$

If inequalities (1.8.1), (1.8.2) are satisfied as strict inequalities, then we are dealing with a *strict dominance.* A special case of the dominance of strategies is their equivalence.

Definition. Strategies x' and x'' of Player 1 are equivalent in the game Γ_A if for all $j \in \{1, \dots, n\}$

$$x'a^j = x''a^j.$$

We shall denote this fact by $x' \sim x''$.

For two equivalent strategies x' and x'' the following equality holds (for every $y \in Y$)

$$K(x', y) = K(x'', y).$$

Similarly, strategies y' and y'' of Player 2 are equivalent ($y' \sim y''$) in the game Γ_A if for all $i \in \{1, \dots, m\}$

$$y'a_i = y''a_i.$$

Hence we have that for any mixed strategy $x \in X$ of Player 1 the following equality holds

$$K(x, y') = K(x, y'').$$

For pure strategies the above definitions are transformed as follows. If Player 1's pure strategy i' dominates strategy i'' and Player 2's pure strategy j' dominates strategy j'' of the same player, then for all $i = 1, \dots, m$; $j = 1, \dots, n$ the following inequalities hold

$$a_{i'j} \geq a_{i''j}, \; a_{ij'} \leq a_{ij''}.$$

This can be written in vector form as follows:

$$a_{i'} \geq a_{i''},\ a^{j'} \leq a^{j''}.$$

Equivalence of the pairs of strategies $i', i''(i' \sim i'')$ and $j', j''(j' \sim j'')$ implies that the conditions $a_{i'} = a_{i''}(a^{j'} = a^{j''})$ are satisfied.

Definition. The strategy $x''(y'')$ of Player 1(2) is dominated if there exists a strategy $x' \neq x''$ $(y' \neq y'')$ of this player which dominates $x''(y'')$; otherwise strategy $x''(y'')$ is an undominated strategy.

Similarly, strategy $x''(y'')$ of Player 1(2) is *strictly dominated* if there exists a strategy $x'(y')$ of this player which strictly dominates $x''(y'')$, i.e. for all $j = \overline{1,n}(i = \overline{1,m})$ the following inequalities hold

$$x'a^j > x''a^j,\ a_i y' < a_i y'';$$

otherwise strategy $x''(y'')$ of Player 1(2) is not strictly dominated.

1.8.2. Show that players playing optimally do not use dominated strategies. This establishes the following assertion.

Theorem. *If, in the game Γ_A, strategy x' of one of the players dominates an optimal strategy x^*, then strategy x' is also optimal.*

Proof. Let x' and x^* be strategies of Player 1. Then, by dominance,

$$x'a^j \geq x^*a^j,$$

for all $j = \overline{1,n}$. Hence, using the optimality of strategy x^* (see 1.7.3), we get

$$v_A = \min_j x^*a^j \geq \min_j x'a^j \geq \min_j x^*a^j = v_A,$$

for all $j = \overline{1,n}$. Therefore, by Theorem 1.7.3, strategy x' is also optimal.

Thus, an optimal strategy can be dominated only by another optimal strategy. On the other hand, no optimal strategy is strictly dominated; hence the players when playing optimally must not use strictly dominated strategies.

Theorem. *If, in the game $\overline{\Gamma}_A$, strategy x^* of one of the players is optimal, then strategy x^* is not strictly dominated.*

Proof. For definiteness, let x^* be an optimal strategy of Player 1. Assume that x^* is strictly dominated, i.e. there exist such strategy $x' \in X$ that

$$x'a^j > x^*a^j, \; j = 1, 2, \ldots, n.$$

Hence

$$\min_j x'a^j > \min_j x^*a^j.$$

However, by the optimality of $x^* \in X$, the equality $\min_j x^*a^j = v_A$ is satisfied. Therefore, the strict inequality

$$\max_x \min_j xa^j > v_A$$

holds and this contradicts to the fact that v_A is the value of the game (1.7.3). The contradiction proves the theorem.

It is clear that the reverse assertion is generally not true. Thus, in the game with the matrix $\begin{bmatrix} 1 & 0 \\ 0 & 2 \end{bmatrix}$ the first and second strategies of Player 1 are not strictly dominated, although they are not optimal.

On the other hand, it is intuitively clear that if the ith row of the matrix A (the jth column) is dominated, then there is no need to assign positive probability to it. Thus, in order to find optimal strategies instead of the game Γ_A, it suffices to solve a subgame $\Gamma_{A'}$, where A' is the matrix obtained from the matrix A by deleting the dominated rows and columns.

Before proceeding to a precise formulation and proof of this result, we will introduce the notion of an *extension of mixed strategy x at the ith place.* If $x = (\xi_1, \ldots, \xi_m) \in X$ and $1 \leq i \leq m + 1$, then the extension of strategy x at the ith place is called the vector $\overline{x}_i = (\xi_1, \ldots, \xi_{i-1}, 0, \xi_i, \ldots, \xi_m) \in R^{m+1}$. Thus the extension of vector $(1/3, 2/3, 1/3)$ at the 2nd place is the vector $(1/3, 0, 2/3, 1/3)$; the extension at the 4th place is the vector $(1/3, 2/3, 1/3, 0)$; the extension at the 1st place is the vector $(0, 1/3, 2/3, 1/3)$.

1.8.3. Theorem. *Let Γ_A be an $(m \times n)$ game. We assume that the ith row of matrix A is dominated (i.e. Player 1's pure strategy i is dominated) and let $\Gamma_{A'}$ be the game with the matrix A' obtained from A by deleting the ith row. Then the following assertions hold.*

1. $v_A = v_{A'}$.
2. *Any optimal strategy y^* of Player 2 in the game $\Gamma_{A'}$ is also optimal in the game Γ_A.*
3. *If x^* is an arbitrary optimal strategy of Player 1 in the game $\Gamma_{A'}$ and $\overline{x}_i^*$ is the extension of strategy x^* at the ith place, then $\overline{x}_i^*$ is an optimal strategy of that player in the game Γ_A.*
4. *If the ith row of the matrix A is strictly dominated, then an arbitrary optimal strategy $\overline{x}^*$ of Player 1 in the game Γ_A can be obtained from an optimal strategy x^* in the game $\Gamma_{A'}$ by the extension at the ith place.*

Proof. Without loss of generality, we may assume, that the last mth row is dominated. Let $x = (\xi_1, \dots, \xi_m)$ be a mixed strategy which dominates the row m. If $\xi_m = 0$, then from the dominance condition for all $j = 1, 2, \dots, n$ we get

$$\sum_{i=1}^{m} \xi_i \alpha_{ij} = \sum_{i=1}^{m-1} \xi_i \alpha_{ij} \geq \alpha_{mj},$$

$$\sum_{i=1}^{m-1} \xi_i = 1, \;\; \xi_i \geq 0, \;\; i = 1, \dots, m-1. \tag{1.8.3}$$

Otherwise ($\xi_m > 0$), consider the vector $x' = (\xi'_1, \dots, \xi'_m)$, where

$$\xi'_i = \begin{cases} \xi_i/(1-\xi_m), & i \neq m, \\ 0, & i = m. \end{cases} \tag{1.8.4}$$

Components of the vector x are non-negative, ($\xi'_i \geq 0, i = 1, \dots, m$) and $\sum_{i=1}^{m} \xi'_i = 1$. On the other hand, for all $i = 1, \dots, n$ we have

$$\frac{1}{1-\xi_m} \sum_{i=1}^{m} \xi_i \alpha_{ij} \geq \alpha_{mj} \frac{1}{1-\xi_m} \sum_{i=1}^{m} \xi_i$$

or

$$\frac{1}{1-\xi_m}\sum_{i=1}^{m-1}\xi_i\alpha_{ij} \geq \alpha_{mj}\frac{1}{1-\xi_m}\sum_{i=1}^{m-1}\xi_i.$$

Considering (1.8.4) we get

$$\sum_{i=1}^{m-1}\xi'_i\alpha_{ij} \geq \alpha_{mj}\sum_{i=1}^{m-1}\xi'_i = \alpha_{mj},\ j=1,\dots,n,$$

$$\sum_{i=1}^{m-1}\xi'_i = 1,\ \xi'_i \geq 0,\ i=1,\dots,m-1. \tag{1.8.5}$$

Thus, from the dominance of the mth row it always follows that it does not exceed a convex linear combination of the remaining $m-1$ rows [(1.8.5)].

Let $(x^*, y^*) \in Z(\Gamma_{A'})$ be a saddle point in the game $\Gamma_{A'}$, $x^* = (\xi_1^*, \dots, \xi_{m-1}^*)$, $y^* = (\eta_1^*, \dots, \eta_n^*)$. To prove assertions 1,2,3 of the theorem, it suffices to show that $K(x_m^*, y^*) = v_{A'}$ and

$$\sum_{j=1}^{n}\alpha_{ij}\eta_j^* \leq v_{A'} \leq \sum_{i=1}^{m-1}\alpha_{ij}\xi_i^* + 0\cdot\alpha_{mj}, \tag{1.8.6}$$

for all $i = 1, \dots, m,\ j = 1, \dots, n$.

The first equality is straightforward, and the optimality of strategies (x^*, y^*) in the game $\Gamma_{A'}$ implies that the following inequalities are satisfied

$$\sum_{j=1}^{n}\alpha_{ij}\eta_j^* \leq v_{A'} \leq \sum_{i=1}^{m-1}\alpha_{ij}\xi_i^*,\ i=\overline{1,m-1},\ j=\overline{1,n}. \tag{1.8.7}$$

The second of the inequalities (1.8.6) is evident from (1.8.7). We shall prove the first inequality. To do this, it suffices to show that

$$\sum_{j=1}^{n}\alpha_{mj}\eta_j^* \leq v_{A'}.$$

From inequalities (1.8.3), (1.8.5) we obtain

$$\sum_{j=1}^{n} \alpha_{mj}\eta_j^* \leq \sum_{j=1}^{n}\sum_{i=1}^{m-1} \alpha_{ij}\xi_i'\eta_j^* \leq \sum_{i=1}^{m-1} v_{A'}\xi_i' = v_{A'},$$

which proves the first part of the theorem.

To prove the second part of the theorem (assertion 4), it suffices to note that in the case of strict dominance of the mth row the inequalities (1.8.3), (1.8.5) are satisfied as strict inequalities for all $j = \overline{1,n}$; hence

$$\sum_{j=1}^{n} \alpha_{mj}\eta_j^* < \sum_{j=1}^{n}\sum_{i=1}^{m-1} \alpha_{ij}\xi_i'\eta_j^* \leq v_{A'}.$$

From Theorem 1.7.6, we then have that the mth component of any optimal strategy of Player 1 in the game Γ_A is zero. This completes the proof.

Formulate now the dominance theorem for the second player without providing any proof.

Theorem. *Let Γ_A be an $(m \times n)$ game. Assume that the jth column of the matrix A is dominated and $\Gamma_{A'}$ is the game having the matrix A' obtained from A by deleting the jth column. Then the following assertions are true.*

1. $v_A = v_{A'}$.
2. *Any optimal strategy x^* of Player 1 in the game $\Gamma_{A'}$ is also optimal in the game Γ_A.*
3. *If y^* is an arbitrary optimal strategy of Player 2 in the game $\Gamma_{A''}$ and $\overline{y}_j^*$ is the extension of strategy y at the jth place, then $\overline{y}_j^*$ is an optimal strategy of Player 2 in the game Γ_A.*
4. *Further, if the jth column of the matrix A is strictly dominated, then an arbitrary optimal strategy $\overline{y}^*$ of Player 2 in the game Γ_A can be obtained from an optimal strategy y^* in the game $\Gamma_{A'}$ by extension at the jth place.*

1.8.4. To summarize: The theorems in 1.8.3 yield an algorithm for reducing the dimension of a matrix game. Thus, if the matrix row (column) is not greater (not smaller) than a convex linear combination of the remaining rows (columns) of the matrix, then to find a solution of the game, this row (column) can be deleted. In this case, an extension of optimal strategy in the truncated matrix game yields an optimal solution of the original game. If the inequalities are satisfied as strict inequalities, the set of optimal strategies in the original game can be obtained by extending the set of optimal strategies in the truncated game; otherwise this procedure may cause a loss of optimal strategies. An application of these theorems is illustrated by the following example.

Example 15. Consider the game with the matrix

$$A = \begin{bmatrix} 2 & 1 & 1 & 0 \\ 5 & 3 & 1 & 3 \\ 0 & 1 & 2 & 0 \\ 7 & 3 & 0 & 6 \end{bmatrix}.$$

Since the 3rd row a_3 dominates the 1st row ($a_3 \geq a_1$), then, by deleting the 1st row, we obtain

$$A_1 = \begin{bmatrix} 5 & 3 & 1 & 3 \\ 0 & 1 & 2 & 0 \\ 7 & 3 & 0 & 6 \end{bmatrix}.$$

In this matrix the 1st column a^3 dominates the 3rd column a^1. Hence we get

$$A_2 = \begin{bmatrix} 3 & 1 & 3 \\ 1 & 2 & 0 \\ 3 & 0 & 6 \end{bmatrix}.$$

In the latter matrix no row (column) is dominated by the other row (column). At the same time, the 1st column a^1 is dominated by the convex linear combination of columns a^2 and a^3, i.e.

$a^1 \geq 1/2a^2 + 1/2a^3$, since $3 > 1/2 + 1/2 \cdot 3$, $1 = 1/2 \cdot 2 + 1/2 \cdot 0$, $3 = 0 \cdot 1/2 + 1/2 \cdot 6$. By eliminating the 1st column, we obtain

$$A_3 = \begin{bmatrix} 1 & 3 \\ 2 & 0 \\ 0 & 6 \end{bmatrix}.$$

In this matrix the 1st row is equal to the linear convex combination of the second and third rows with a mixed strategy $x = (0, 1/2, 1/2)$, since $1 = 1/2 \cdot 2 + 0 \cdot 1/2$, $3 = 0 \cdot 1/2 + 6 \cdot 1/2$. Thus, by eliminating the 1st row, we obtain the matrix

$$A_4 = \begin{bmatrix} 2 & 0 \\ 0 & 6 \end{bmatrix}.$$

The players' optimal strategies x^* and y^* in the game with this matrix are $x^* = y^* = (3/4, 1/4)$, in which case the game value v is $3/2$.

The latter matrix is obtained by deleting the first two rows and columns; hence the players' optimal strategies in the original game are extensions of these strategies at the 1st and 2nd places, i.e. $\overline{x}^*_{12} = \overline{y}^*_{12} = (0, 0, 3/4, 1/4)$.

1.9 Completely Mixed and Symmetric Games

A knowledge of the optimal strategy spectrum simplifies solving games. The optimal strategy spectrum includes only essential pure strategies of a player. Thus no essential strategy is strictly dominated which follows immediately from the theorems in 1.8.

1.9.1. We consider the class of games in which a knowledge of the spectrum suffices to find a solution of the game.

Definition. Strategy $x(y)$ of Player 1(2) is completely mixed if its spectrum consists of the set of all strategies of the player, i.e. $M_x = M$ $(N_y = N)$.

A saddle point (x^*, y^*) is completely mixed if the strategies x^* and y^* are completely mixed. The game Γ_A is completely mixed if each saddle point therein is completely mixed.

The following theorem states that a completely mixed game has a unique solution.

Theorem. *A completely mixed* $(m \times n)$ *game* Γ_A *has a unique solution* (x^*, y^*) *and a square matrix* $(m = n)$. *If* $v_A \neq 0$, *then the matrix* A *is nonsingular and*

$$x^* = \frac{uA^{-1}}{uA^{-1}u}, \tag{1.9.1}$$

$$y^* = \frac{A^{-1}u}{uA^{-1}u}, \tag{1.9.2}$$

$$v_A = \frac{1}{uA^{-1}u}. \tag{1.9.3}$$

Proof. Let $x^* = (\xi_1^*, \ldots, \xi_m^*) \in X^*$ and $y^* = (\eta_1^*, \ldots, \eta_n^*) \in Y^*$ be the players arbitrary optimal strategies and let v_A be the value of the game Γ_A. Since Γ_A is completely mixed game, x^* and y^* are completely mixed strategies that (and only them) are solutions of the systems of linear inequalities in 1.7.6:

$$xa^j = v_A, \ xu = 1, \ x > 0, \ j = 1, \ldots, n, \tag{1.9.4}$$

$$ya_i = v_A, \ yw = 1, \ y > 0, \ i = 1, \ldots, m, \tag{1.9.5}$$

where $u = (1, \ldots, 1) \in R^m, \ w = (1, \ldots, 1) \in R^n$.

We shall show that the solution of the completely mixed game (x^*, y^*) is unique. The sets X^*, Y^* given by (1.9.4) and (1.9.5) are nonempty convex polyhedra and, consequently, have extreme points. By the second of the theorems in 1.5.2, we have

$$m \leq rank[a^1, \ldots, a^n, u] = rank[A, u] \leq m, \tag{1.9.6}$$

$$n \leq rank[a_1, \ldots, a_m, w] = rank[A, w] \leq n. \tag{1.9.7}$$

This theorem now implies that the sets X^*, Y^* have one extreme point each and hence consist only of such points (as convex polyhedra containing a unique extreme point). This completes the proof of uniqueness of the solution (x^*, y^*).

Let $v_A = 0$. Then a homogeneous system

$$xa^j = v_A,\ j = \overline{1, n}$$

has a nonzero solution; hence $rank(A) < m$. Since $rank[A, u] = m$, we have: $rank(A) = m - 1$. Similarly, from (1.9.5) and (1.9.7) it follows that $rank(A) = n - 1$. Hence $n = m$.

Let $v_A \neq 0$. Then

$$rank(A) = rank[A, v_A u] = rank[A, u] = m,$$

$$rank(A) = rank[A, v_A w] = rank[A, w] = n.$$

Hence we have $n = m = rank(A)$, i.e. A is a nonsingular matrix. The system of equations $x^* A = v_A u$ has a solution

$$x^* = v_A u A^{-1}.$$

Write a solution of the system $Ay^* = v_A u$:

$$y^* = v_A A^{-1} u,$$

then

$$v_A = \frac{1}{uA^{-1}u}.$$

This completes the proof of Theorem.

The reverse is also true, although the proof will be left to the reader.

Theorem. *Suppose the matrix A is nonsingular in the $(m \times m)$ game Γ_A. If Player 2 has in Γ_A a completely mixed optimal strategy, then Player 1 has a unique optimal strategy x^* (1.9.1). If Player 1 has in the game Γ_A a completely mixed optimal strategy, then Player 2 has a unique optimal strategy y^* (1.9.2), the value of the game v_A is defined by (1.9.3).*

Example 16. (2 × 2) *game.* Consider the (2 × 2) game with the matrix

$$A = \begin{bmatrix} \alpha_{11} & \alpha_{12} \\ \alpha_{21} & \alpha_{22} \end{bmatrix}.$$

Arbitrary mixed strategy x of Player 1 can be written as $x = (\xi, 1-\xi)$, where $0 \le \xi \le 1$. Similarly, a mixed strategy of Player 2 is of the form $y = (\eta, 1-\eta)$, where $0 \le \eta \le 1$. The payoff at (x, y) is

$$K(x, y) = \xi[\alpha_{11}\eta + \alpha_{12}(1-\eta)] + (1-\xi)[\alpha_{21}\eta + \alpha_{22}(1-\eta)].$$

We now assume that the game Γ_A has no saddle point in pure strategies (a solution is then found from the maximin minimax condition) and $x^* = (\xi^*, 1-\xi^*)$, $y^* = (\eta^*, 1-\eta^*)$ are arbitrary optimal strategies of the first and second players respectively. In this case, the saddle point (x^*, y^*) and game Γ_A are completely mixed ($\xi^* > 0$ and $\eta^* > 0$). Therefore, by Theorem 1.9.1, the game has a unique pair of optimal mixed strategies which are a solution of the system of equations

$$\alpha_{11}\eta^* + (1-\eta^*)\alpha_{12} = v_A,$$

$$\alpha_{21}\eta^* + (1-\eta^*)\alpha_{22} = v_A,$$

$$\alpha_{11}\xi^* + (1-\xi^*)\alpha_{21} = v_A,$$

$$\alpha_{12}\xi^* + (1-\xi^*)\alpha_{22} = v_A.$$

If we ensure that $v_A \neq 0$ (e.g. if all the elements of the matrix A are positive, this inequality is satisfied), then the solution of the game is

$$v_A = \frac{1}{uA^{-1}u}, \; x^* = v_A u A^{-1}, \; y^* = v_A A^{-1} u,$$

where $u = (1,1)$. Thus, it can be readily verified that the matrix $A = \begin{bmatrix} 1 & 0 \\ -1 & 1 \end{bmatrix}$ has no saddle point. The inverse matrix A^{-1} is

$$A^{-1} = \begin{bmatrix} 1 & 0 \\ 1 & 1 \end{bmatrix}.$$

Then $v_A = 1/3,\ x^* = (2/3, 1/3),\ y^* = (1/3, 2/3)$.

1.9.2. We now examine a class of the games having matrices of a special form.

Definition. The game Γ_A with the square matrix A is symmetric if the matrix A is skew-symmetric, i.e. if $a_{ij} = -a_{ji}$ for all i and j.

In this case, all diagonal elements of the matrix A are zero, i.e. $a_{ii} = 0$ for all i. For the skew-symmetric matrix A we have $A^T = -A$. Since the matrix A is square, the players' sets of mixed strategies coincide, i.e. $X = Y$.

We shall prove the theorem on properties of a solution to the skew-symmetric game Γ_A which may be useful in finding a saddle point.

Theorem. *Let Γ_A be a symmetric game. Then*

$$v_A = 0$$

and the players' sets of optimal strategies coincide, i.e.

$$X^* = Y^*.$$

Proof. Let A be the game matrix and let $x \in X$ be an arbitrary strategy. Then $xAx = xA^T x = -xAx$. Hence $xAx = 0$.

Let $(x^*, y^*) \in Z(A)$ be a saddle point, and let v_A be the value of the game. Then

$$v_A = x^* A y^* \leq x^* A y,$$
$$v_A = x^* A y^* \geq x A y^*$$

for all $x \in X, y \in Y$. Consequently,

$$v_A \leq x^* A x^* = 0,\ v_A \geq y^* A y^* = 0.$$

Hence we get $v_A = 0$.

Let the strategy x^* be optimal in the game Γ_A. Then (see Theorem 1.7.1)

$$x^* A \geq 0.$$

It follows, however, that $x^*(-A^T) \geq 0$, hence $x^* A^T \leq 0$. Thus we get

$$Ax^* \leq 0.$$

By the same Theorem 1.7.1, this means that x^* is the optimal strategy of Player 2. We have thus proved that $X^* \subset Y^*$. The inverse inclusion is proved in a similar manner.

In what follows, dealing with the player's optimal strategy in the symmetric game, because of the equality $X^* = Y^*$ we shall not indicate which of the players is concerned.

Example 17. Let us solve the game with the matrix

$$A = \begin{bmatrix} 0 & -1 & 1 \\ 1 & 0 & -1 \\ -1 & 1 & 0 \end{bmatrix}.$$

Let $x^* = (\xi_1^*, \xi_2^*, \xi_3^*)$ be an optimal strategy in the game Γ_A. Then the following inequalities are satisfied

$$\begin{aligned} &\xi_2^* - \xi_3^* \geq 0, \ -\xi_1^* + \xi_3^* \geq 0, \ \xi_1^* - \xi_2^* \geq 0, \\ &\xi_1^* + \xi_2^* + \xi_3^* = 1, \ \xi_1^* \geq 0, \ \xi_2^* \geq 0, \ \xi_3^* \geq 0. \end{aligned} \tag{1.9.8}$$

We shall show that this game is completely mixed. Indeed, let $\xi_1^* = 0$. From the system of inequalities (1.9.8) we then obtain the system

$$\begin{gathered} \xi_2^* - \xi_3^* \geq 0, \ \xi_3^* \geq 0, \ -\xi_2^* \geq 0, \\ \xi_1^* + \xi_2^* + \xi_3^* = 1, \end{gathered}$$

which has no non-negative solution. Similar reasoning show that the cases $\xi_2^* = 0$ and $\xi_3^* = 0$ are impossible. Therefore, the game Γ_A is completely mixed. Consequently, the components $\xi_1^*, \xi_2^*, \xi_3^*$ are solutions of the system

$$\xi_2^* - \xi_3^* = 0,\ -\xi_1^* + \xi_3^* = 0,\ \xi_1^* - \xi_2^* = 0,$$
$$\xi_1^* + \xi_2^* + \xi_3^* = 1,\ \xi_i > 0,\ i = 1, 2, 3.$$

This system has a unique solution. The vector $x^* = (1/3, 1/3, 1/3)$ is an optimal strategy.

Example 18. Solve a discrete five-step duel game in which each duelist has one bullet. This game was formulated in 1.1.4 (see Example 3). The payoff matrix A of Player 1 is symmetric and is of the form

$$A = \begin{bmatrix} 0 & -3 & -7 & -11 & -15 \\ 3 & 0 & 1 & -2 & -5 \\ 7 & -1 & 0 & 7 & 5 \\ 11 & 2 & -7 & 0 & 15 \\ 15 & 5 & -5 & -15 & 0 \end{bmatrix}.$$

Note that the first strategy of each player (first row and first column of the matrix) is strictly dominated; hence it cannot be essential and can be deleted. In the resulting truncated matrix

$$A' = \begin{bmatrix} 0 & 1 & -2 & -5 \\ -1 & 0 & 7 & 5 \\ 2 & -7 & 0 & 15 \\ 5 & -5 & -15 & 0 \end{bmatrix}$$

not all strategies are essential.

Indeed, symmetry of the game $\Gamma_{A'}$ implies that $v_{A'} = 0$. If all strategies were essential, the optimal strategy x^* would be a solution of the system of equations

$$x^* a'^j = 0,\ j = 2, 3, 4, 5,$$
$$\sum_{i=2}^{5} \xi_i^* = 1,$$

which define the solution. Exhausting different possibilities, we dwell on the essential submatrix A'' composed of the rows and columns of the matrix A that are labeled as 2,3, and 5:

$$A'' = \begin{bmatrix} 0 & 1 & -5 \\ -1 & 0 & 5 \\ 5 & -5 & 0 \end{bmatrix}.$$

The game with the matrix A'' is completely mixed and has a unique solution $\overline{y} = \overline{x} = (5/11, 5/11, 1/11)$.

In the original game, we now consider the strategies $x^* = y^* = (0, 5/11, 5/11, 0, 1/11)$ which are optimal.

Thus, we finally have that $v_A = 0$ and the saddle point (x^*, y^*) is unique. As far as the rules of the game are concerned, the duelist should not fire at the 1st step, he must fire with equal probability after the 2nd or 3rd step, never after the 4th step, and only with small probability may he fire when the duelists are breast to breast.

1.10 Iterative Methods of Solving Matrix Games

The popular method of solving a matrix game by reducing it to a linear programming problem has a disadvantage that the process of solving the linear programming problem is complicated for matrices of large dimensions.

In such cases it is standard practice to employ decomposition methods for linear programming problems when, instead of solving the problem with the original matrix, we construct the corresponding problem with the matrix which has few rows but many columns. A set of auxiliary linear programming problems with matrices of smaller dimensions is solved at each iteration of the corresponding problem.

Unfortunately, decomposition methods are only effective for matrices of a special type (e.g. block-diagonal matrices).

1.10.1. [Robinson (1950)]. *Brown–Robinson iterative method* (*fictitious play method*). This method employs a repeated fictitious play of game having a given payoff matrix. One repetition of the game is called a play. Suppose the game is played with an $(m \times n)$ matrix $A = \{a_{ij}\}$. In the 1st play both players choose arbitrary pure strategies. In the kth play each player chooses the pure strategy which maximizes his expected payoff against the observed empirical probability distribution of the opponents pure strategies for $(k-1)$ plays.

Thus, we assume that in the first k plays Player 1 uses the ith strategy ξ_i^k times $(i = 1, \ldots, m)$ and Player 2 uses the jth strategy η_j^k times $(j = 1, \ldots, n)$. In the $(k+1)$ play, Player 1 will then use i_{k+1} strategy and Player 2 will use his j_{k+1} strategy, where

$$\overline{v}^k = \max_i \sum_j a_{ij}\eta_j^k = \sum_j a_{i_{k+1}j}\eta_j^k$$

and

$$\underline{v}^k = \min_j \sum_i a_{ij}\xi_i^k = \sum_i a_{ij_{k+1}}\xi_i^k.$$

Let v be the value of the matrix game Γ_A. Consider the expressions

$$\overline{v}^k/k = \max_i \sum_j \alpha_{ij}\eta_j^k/k = \sum_j \alpha_{i_{k+1}j}\eta_j^k/k,$$

$$\underline{v}^k/k = \min_j \sum_i \alpha_{ij}\xi_i^k/k = \sum_i \alpha_{ij_{k+1}}\xi_i^k/k.$$

The vectors $x^k = (\xi_1^k/k, \ldots, \xi_m^k/k)$ and $y^k = (\eta_1^k/k, \ldots, \eta_n^k/k)$ are mixed strategies for Players 1 and 2, respectively; hence, by the definition of the value of the game we have

$$\max_k \underline{v}^k/k \leq v \leq \min_k \overline{v}^k/k.$$

We have thus obtained an iterative process which enables us to find an approximate solution of the matrix game, the degree of approximation to the true value of the game being determined by the length of the interval $[\max_k \underline{v}^k/k, \min_k \overline{v}^k/k]$. Convergence of the algorithm is guaranteed by the Theorem [Robinson (1950)].

Theorem.

$$\lim_{k\to\infty}(\min_k \overline{v}^k/k) = \lim_{k\to\infty}(\max_k \underline{v}^k/k) = v.$$

Example 19. Find an approximate solution to the game having the matrix

$$A = \begin{array}{c} \\ \alpha \\ \beta \\ \gamma \end{array}\begin{array}{c} \begin{array}{ccc} a & b & c \end{array} \\ \begin{bmatrix} 2 & 1 & 3 \\ 3 & 0 & 1 \\ 1 & 2 & 1 \end{bmatrix} \end{array}.$$

Denote Player 1's strategies by α, β, γ, and Player 2's strategies by a, b, c. Suppose the players first choose strategies α and a, respectively. If Player 1 chooses strategy α, then Player 1 can receive one of the payoffs (2,1,3). If Player 2 chooses strategy a, then Player 1 can receive one of the payoffs (2,3,1). In the 2nd and 3rd plays, Player 1 chooses strategy β and Player 2 chooses strategy b, since these strategies ensure the best result, etc.

Table 1.1 shows the results of plays, the players' strategies, the accumulated payoff, and the average payoff.

Thus, for 12 plays, we obtain an approximate solution

$$x^{12} = (1/4, 1/6, 7/12), \quad y^{12} = (1/12, 7/12, 1/3)$$

and the accuracy can be estimated by the number $5/12$. The principal disadvantage of this method is its low speed of convergence which decreases as the matrix dimension increases. This also results from the nonmonotonicity of sequences $\overline{v}^k/k$ and $\underline{v}^k/k$.

Consider another iteration algorithm which is free of the above-mentioned disadvantages.

1.10.2. [Sadovsky (1978)]. *Monotonic iterative method of solving matrix games.* We consider a mixed extension $\Gamma_A = (X, Y, K)$ of the matrix game with the $(m \times n)$ matrix A.

Denote by $x^N = (\xi_1^N, \ldots, \xi_m^N) \in X$ the approximation of Player 1's optimal strategy at the Nth iteration, and by $c^N \in R^N$,

Table 1.1

Play No	Player 1's choice	Player 2's choice	Player 1's payoff			Player 2's payoff			$\frac{\overline{v}^k}{k}$	$\frac{\underline{v}^k}{k}$
			α	β	γ	a	b	c		
1	α	a	2	3	1	2	1	3	3	1
2	β	b	3	3	3	5	1	4	3/2	1/2
3	β	b	4	3	5	8	1	5	5/3	1/3
4	γ	b	5	3	7	9	3	6	7/4	3/4
5	γ	b	6	3	9	10	5	7	9/5	5/5
6	γ	b	7	3	11	11	7	8	11/6	7/6
7	γ	b	8	3	13	12	9	9	13/7	9/7
8	γ	c	14	4	14	13	12	10	14/8	10/8
9	γ	c	14	5	15	14	12	11	15/9	11/9
10	γ	c	17	6	16	15	14	12	17/10	12/10
11	α	c	20	7	17	17	15	15	20/11	15/11
12	α	b	21	7	19	19	16	18	21/12	16/12

$c^N = (\gamma_1^N, \dots, \gamma_n^N)$ an auxiliary vector. Algorithm makes it possible to find (exactly and approximately) an optimal strategy for Player 1 and a value of the game v.

At the start of the process, Player 1 chooses an arbitrary vector of the form $c^0 = a_{i_0}$, where a_{i_0} is the row of the matrix A having the number i_0.

Iterative process is constructed as follows. Suppose the $N-1$ iteration is performed and vectors x^{N-1}, c^{N-1} are obtained. Then x^N and c^N are computed from the following iterative formulas

$$x^N = (1-\alpha_N)x^{N-1} + \alpha_N \tilde{x}^N, \tag{1.10.1}$$

$$c^N = (1-\alpha_N)c^{N-1} + \alpha_N \tilde{c}^N, \tag{1.10.2}$$

where $0 \le \alpha_N \le 1$. Vectors $\tilde{x}^N$ and $\tilde{c}^N$ will be obtained below.

We consider the vector $c^{N-1} = (\gamma_1^{N-1}, \ldots, \gamma_n^{N-1})$ and select indices j_k on which the minimum is achieved

$$\min_{j=1,\ldots,n} \gamma_j^{N-1} = \gamma_{j_1}^{N-1} = \gamma_{j_2}^{N-1} = \ldots = \gamma_{j_k}^{N-1}.$$

Denote

$$\underline{v}^{N-1} \triangleq \min_{j=1,\ldots,n} \gamma_j^{N-1} \tag{1.10.3}$$

and $J^{N-1} = \{j_1, \ldots, j_k\}$ be the set of indices on which (1.10.3) is achieved.

Let $\Gamma^N \subset \Gamma_A$ be a subgame of the game Γ_A with the matrix $A^N = \{a_{ij}^{N-1}\}$, $i = 1, \ldots, m$, and the index $j^{N-1} \in J^{N-1}$. Solve the subgame and find an optimal strategy $\tilde{x}^N \in X$ for Player 1. Let $\tilde{x}^N = (\tilde{\xi}_1^N, \ldots, \tilde{\xi}_m^N)$.

Compute the vector $\tilde{c}^N = \sum_{i=1}^m \tilde{\xi}_i^N a_i$. Suppose the vector $\tilde{c}^N$ has components $\tilde{c}^N = (\tilde{\gamma}_1^N, \ldots, \tilde{\gamma}_n^N)$. Consider the $(2 \times n)$ game with matrix

$$\begin{bmatrix} \gamma_1^{N-1} & \ldots & \gamma_n^{N-1} \\ \tilde{\gamma}_1^N & \ldots & \tilde{\gamma}_n^N \end{bmatrix}.$$

Find Player 1's optimal strategy $(\alpha_N, 1 - \alpha_N)$, $0 \leq \alpha_N \leq 1$ in this subgame.

Substituting the obtained values $\tilde{x}^N, \tilde{c}^N, \alpha_N$ into (1.10.1), (1.10.2), we find x^N and c^N. We continue the process until the equality $\alpha_N = 0$ is satisfied or the required accuracy of computations is achieved. Convergence of the algorithm is guaranteed by the following theorem [Sadovsky (1978)].

Theorem. *Let* $\{\underline{v}^N\}, \{x^N\}$ *be the iterative sequences determined by* (1.10.1), (1.10.3). *Then the following assertions are true.*

1. $\underline{v}^N > \underline{v}^{N-1}$, *i.e. the sequence* $\{\underline{v}^{N-1}\}$ *strictly and monotonically increases,*

2. $\lim_{N\to\infty} \underline{v}^N = \underline{v} = v$, (1.10.4)
3. $\lim_{N\to\infty} x^N = x^*$, *where* $x^* \in X^*$ *is an optimal strategy of Player 1.*

Example 20. By employing a monotonic algorithm, solve the game with the matrix

$$A = \begin{bmatrix} 2 & 1 & 3 \\ 3 & 0 & 1 \\ 1 & 2 & 1 \end{bmatrix}.$$

Iteration 0. Suppose Player 1 chooses the 1st row of the matrix A, i.e. $x^* = (1,0,0)$ and $c^0 = a_1 = (2,1,3)$. Compute $\underline{v}^0 = \min\limits_j \gamma_j^0 = \gamma_2^0 = 1$, $J^0 = 2$.

Iteration 1. Consider the subgame $\Gamma^1 \subset \Gamma$ having the matrix

$$A^1 = \begin{bmatrix} 1 \\ 0 \\ 2 \end{bmatrix}.$$

An optimal strategy $\tilde{x}^1$ of Player 1 is the vector $\tilde{x}^1 = (0,0,1)$. Then $\tilde{c}^1 = a_3 = (1,2,1)$. Solve the (2×3) game with the matrix $\begin{bmatrix} 2 & 1 & 3 \\ 1 & 2 & 1 \end{bmatrix}$. Note that the 3rd column of the matrix is dominated and so we consider the matrix $\begin{bmatrix} 2 & 1 \\ 1 & 2 \end{bmatrix}$. Because of the symmetry, Player 1's optimal strategy in this game is the vector $(\alpha_N, 1-\alpha_N) = (1/2, 1/2)$.

We compute x^1 and c^1 by formulas (1.10.1), (1.10.2). We have

$$x^1 = 1/2x^0 + 1/2\tilde{x}^1 = (1/2, 0, 1/2),$$
$$c^1 = 1/2c^0 + 1/2\tilde{c}^1 = (3/2, 3/2, 2),$$
$$\underline{v}^1 = \min_j \gamma_j^1 = \gamma_1^1 = \gamma_2^1 = 3/2 > \underline{v}^0 = 1.$$

The set of indices is of the form $J^1 = \{1,2\}$.

Iteration 2. Consider the subgame $\Gamma^2 \subset \Gamma$ having the matrix

$$A = \begin{bmatrix} 2 & 1 \\ 3 & 0 \\ 1 & 2 \end{bmatrix}.$$

The first row in this matrix is dominated; hence it suffices to examine the submatrix

$$\begin{bmatrix} 3 & 0 \\ 1 & 2 \end{bmatrix}.$$

Player 1's optimal strategy in this game is the vector $(1/4, 3/4)$; hence $\tilde{x}^2 = (0, 1/4, 3/4)$.

Compute $\tilde{c}^2 = 1/4a_2 + 3/4a_3 = (3/2, 3/2, 1)$ and consider the (2×3) game with the matrix

$$\begin{bmatrix} 3/2 & 3/2 & 1 \\ 3/2 & 3/2 & 2 \end{bmatrix}.$$

The second strategy of Player 1 dominates the first strategy and hence $\alpha_2 = 0$. This completes the computations $x^* = x^1 = (1/2, 0, 1/2)$; the value of the game is $v = \underline{v}^1 = 3/2$, and Player 2's optimal strategy is of the form $y^* = (1/2, 1/2, 0)$ (see Example 19).

1.11 Exercises and Problems

1. Each of the two players shows m fingers, $(1 \leq m \leq n,\ n \leq 5)$, and simultaneously calls his guess of number of fingers his opponent will show. If just one player guesses correctly (the opponent's guess being incorrect), he wins an amount equal to the sum of fingers shown by both players. In all other cases the players' payoffs are zero.

(a) How many strategies will each player have for $n = 3$?

(b) Construct the game matrix for $n = 2$.

2. *Allocation of search efforts.* Player 2 hides in one of the n cells. Player 1 has a team of r searches to be allocated to cells for the purpose of searching Player 2. For example, $(r-1)$ searchers can be allocated to the first cell, one searcher to the second cell, and no searcher to the remaining cells, etc.

In the search conducted by one searcher, the probability of finding the Player 2 in the ith cell (if he is there) is assumed to be known. Finding Player 2 by each of the searchers are independent events.

Player 1's payoff is the probability of finding the Player 2 when the allocation of searchers is given.

(a) Compute the number m of pure strategies for Player 1.
(b) Construct the game matrix.

3. *Searching for many objects.* Player 2 hides m black balls in n containers. The total number of balls (black and white) in the jth container is $l_j, j = 1, \ldots, n$. Player 2 has to allocate m black balls to n containers, with the total number of balls in each container being constant and equal to l_j, $l_j > m$.

The opponent (Player 1) tries to find as many black balls as possible and has an opportunity to examine one of the containers. In examination of the ith container, Player 1 chooses at random (equiprobably) m balls from l_i and his payoff is the mathematical expectation of the number of black balls in the sample from m balls.

(a) Let p_i black balls be hidden in the ith container. Compute the probability β_{ij} that the sample of r balls chosen from the ith container contains exactly j black balls.
(b) Construct the game matrix.

4. *Air defense.* An air defense system can use three types of weapons to hit an air target (1,2,3) which are to be allocated to two launcher units. The enemy (Player 2) has two types of aircraft (type 1 and type 2). The probabilities of hitting the planes by one defense system are summarized in the matrix

$$\begin{array}{c} \\ 1 \\ 2 \\ 3 \end{array}\begin{array}{c} \begin{array}{cc} 1 & \;\;\;2 \end{array} \\ \begin{bmatrix} 0.3 & 0.5 \\ 0.5 & 0.3 \\ 0.1 & 0.6 \end{bmatrix} \end{array}.$$

It is assumed that only one plane is attacking.

Player 1's payoff is the probability of hitting the plane by the air defense system.

(a) Construct the game matrix.

(b) Find out if there exists a solution in pure strategies.

5. Find a saddle point and value of the following games:

(a) $\begin{bmatrix} 3 & 5 \\ 3 & 2 \end{bmatrix}$; (b) $\begin{bmatrix} 1/2 & 0 & 1/2 \\ 1 & 3/2 & 2 \\ 0 & -1 & 7/4 \end{bmatrix}$.

6. Verify that $v = 2$ and the pair (x^*, y^*), where $x^* = (0, 0, 1)$, $y^* = (2/5, 3/5, 0)$ are respectively the value and saddle point in the game with the matrix $\begin{bmatrix} 3 & -2 & 4 \\ -1 & 4 & 2 \\ 2 & 2 & 6 \end{bmatrix}$.

7. Let $A'(A'')$ be the submatrix of the matrix A obtained by deleting a series of rows (columns) of A. Show that the inequalities $v_{A'} \le v_A \le v_{A''}$, where $v_{A'}, v_{A''}$ are the values of the games $\Gamma_{A'}$, $\Gamma_{A''}$, respectively, are satisfied.

8. Consider the game $\Gamma_{A'}$, with the matrix $\begin{bmatrix} -1 & 3 & -3 \\ 2 & 0 & 3 \\ 2 & 1 & 0 \end{bmatrix}$. The value of the game is $v_A = 1$ and Player 1's optimal strategy is $x^* = (1/3, 2/3, 0)$. Find an optimal strategy y^* for Player 2.

9. Solve the matrix game using the graphical method $\begin{bmatrix} -4 & 0 \\ 3 & -2 \\ 5 & -3 \\ -1 & -1 \end{bmatrix}$.

10. Show that a strictly dominated strategy cannot be essential.

11. Show that the 3rd row of the matrix A is dominated, if

$$A = \begin{bmatrix} 20 & 0 \\ 0 & 8 \\ 4 & 5 \end{bmatrix}.$$

12. Show that the choice of the 1st column is equivalent to a mixed strategy $y = (0, 1/3, 2/3)$ in the game with matrix $\begin{bmatrix} 1 & 3 & 0 \\ 2 & 0 & 3 \end{bmatrix}$.

13. Employing the notion of dominance, find a solution of the game with the matrix $\begin{bmatrix} 1 & 7 & 2 \\ 6 & 2 & 7 \\ 5 & 1 & 6 \end{bmatrix}$.

14. Prove Theorem 1.7.3.

15. Solve a discrete search game (Example 5, 1.1.3) with the assumption that $\alpha\beta_i - \tau_i \neq 0$, $i = 1, \ldots, n$.

Hint. Make use of the result from 1.7.7.

16. *Two-object search game.* Player 2 hides two objects in n cells (both objects can be placed in the same cell). Player 1's objective is to find at least one object; he has an opportunity to examine one cell ($\beta_i > 0$ is the probability of finding one object in the ith cell if the object is hidden in this cell). If the ith cell contains simultaneously two objects, the probability of finding them both is β_i^2. Thus, the matrix $A = \{\alpha_{ka}\}, a = (i, j), j = 1, \ldots, n$, is of the form

$$\begin{aligned} \alpha_{ka} &= 0, && i = j,\ i \neq k, \\ \alpha_{ka} &= \beta_i, && i = k,\ i \neq j, \\ \alpha_{ka} &= \beta_j, && j = k,\ i \neq j, \\ \alpha_{ka} &= \beta_i(2 - \beta_i), && i = j = k. \end{aligned}$$

Solve the game.

17. Solve the search game with many hidden objects (see Exercise 3).

18. *Search game for several sets on a plane.* The family of n fixed convex compact sets $K_1, K_2, \ldots, K_n \subset R^2$ and the system m of convex compact congruent sets $T_1, \ldots, T_m \subset R^2$ are given. The simultaneous discrete search game is as follows. Player 2 "hides" m sets T_j ($j = 1, \ldots, m$) in n sets K_i ($i = 1, \ldots, n$) in such a manner that each set intersects one and only one set K_i.

Player 2's pure strategy is of the form

$$a = (p_1, p_2, \ldots, p_n) \in R^n, \ \sum_{i=1}^{n} p_i = m,$$

where p_i is the number of sets T_j hidden in the set K_i.

Player 1 can examine one of the sets K_i by selecting the point x from K_i. Player 1's payoff is the number of sets $\{T_j\}$ whereto x belongs.

Find a solution of the game.

19. *Search game with two trials for a searcher.* Player 2 hides an object in one of the n cells and Player 1 (searcher) searches for it in one of these cells. Player 1 has an opportunity to examine two cells (repeated examination of cells is not allowed).
Player 1's set of pure strategies consists of pairs (i, j), $i = 1, \ldots, n$, $j = 1, \ldots, n$, $i \neq j$ and contains C_n^2 elements. Player 2's set of pure strategies contains n elements $k = 1, \ldots, n$. The payoff matrix is of the form, where

$$\beta_{(i,j)k} = \begin{cases} \delta_k, & \text{if } i = k \text{ or } j = k, \ (\delta_k > 0), \\ 0, & \text{otherwise.} \end{cases}$$

Find a solution of the game.

20. In the search game with two trials for a searcher, consider the case where Player 1's set of pure strategies consists of all possible pairs (i, j) and contains n^2 elements. Solve the game under the assumption that

$$\sum_{k=1}^{n-1} \delta_n/\delta_k < 1.$$

21. In the evasion-type game, (see 1.7.1), show that Player 1 always has a unique optimal strategy.

Chapter 2

Infinite Zero-Sum Two-Person Games

2.1 Infinite Games

2.1.1. This chapter deals with zero-sum two-person games differing from matrix games in that one or two players in such games have an infinite (countable or continuum) sets of strategies. From the game-theoretic point of view this difference is unimportant, since the game continues to be a zero-sum two-person game and the problem is only one of employing more sophisticated analytical techniques of research.

Thus, we shall examine general zero-sum two-person games, i.e. systems of the form

$$\Gamma = (X, Y, H), \tag{2.1.1}$$

where X and Y are arbitrary infinite sets whose elements are the strategies of Players 1 and 2, respectively, and $H : X \times Y \to R^1$ is the payoff function of Player 1. Recall that the rules for zero-sum two-person games are described in 1.1.1. Player 2's payoff in the situation (x, y) is $[-H(x, y)]$, $x \in X, y \in Y$ (the game being zero-sum).

In this chapter, the games with bounded payoff function H are considered.

2.1.2. *Example 1. Simultaneous planar pursuit-evasion game.* Let S_1 and S_2 be the sets on a plane. Player 1 chooses a point $x \in S_1$ and Player 2 chooses a point $y \in S_2$. In making his choice, no player has information on the opponent's actions, and hence such a choice can be conveniently interpreted as simultaneous. In this case, the points $x \in S_1, y \in S_2$ are strategies for Players 1 and 2, respectively. Thus the players' sets of strategies coincide with the sets S_1 and S_2 on the plane.

Player 2's objective is to minimize the distance between himself and Player 1 (Player 1 pursues the opposite objective). Therefore, by Player 1's payoff $H(x, y)$ in this game is equal to the Euclidean distance $\rho(x, y)$ between the points $x \in S_1$ and $y \in S_2$, i.e. $H(x, y) = \rho(x, y), x \in S_1, y \in S_2$. Player 2's payoff is taken to be equal to $[-\rho(x, y)]$ (the game being zero-sum).

Example 2. Search on a closed interval [Diubin and Suzdal (1981)]. The simplest search game with an infinite number of strategies is the following.

Player 2 (Hider) chooses a point $y \in [0, 1]$ and Player 1 (Searcher) chooses simultaneously and independently a point $x \in [0, 1]$. The point y is considered to be " detected" if $|x - y| \leq l$, where $0 < l < 1$. In this case, Player 1 wins an amount $+1$; otherwise his payoff is 0. The game is zero-sum.

Thus the payoff function is

$$H(x, y) = \begin{cases} 1, & \text{if } |x - y| \leq l. \\ 0, & \text{otherwise.} \end{cases}$$

The payoff to Player 2 is $[-H(x, y)]$.

Example 3. Search on a sphere. Let a sphere C of radius R be given in R^3. Player 1 (Searcher) chooses a system of the points $x_1, x_2, \ldots, x_s \in C$ and Player 2 chooses one point $y \in C$. The players make their choices simultaneously and independently of one another. Player 2 is said to be detected if the point $y \in C$ is found in the

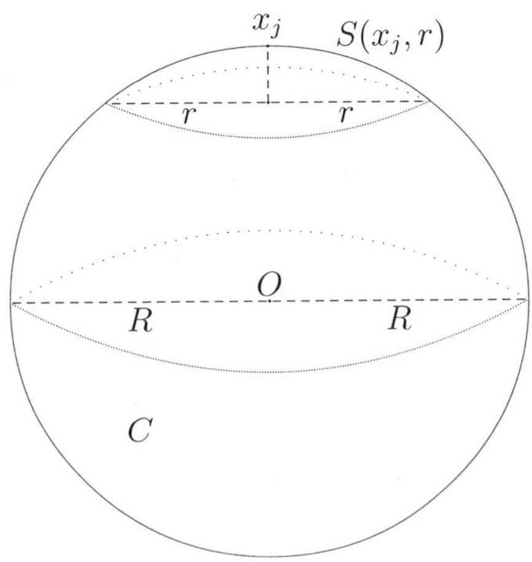

Figure 2.1 Search on a sphere

r-neighborhood of one of the points x_j, $j = 1, \dots, s$. Here by the r-neighborhood of the point x_j is meant the segment of a sphere (cup) having its apex at the point x_j with r as the base radius (Fig. 2.1). In what follows the r-neighborhood of the point x_j is denoted by $S(x_j, r)$.

The objective of Player 1 is to find Player 2, whereas Player 2 pursues the opposite objective. Accordingly the payoff to Player 1 is

$$H(x, y) = \begin{cases} 1, & \text{if } y \in M_x, \\ 0, & \text{otherwise,} \end{cases}$$

where $x = (x_1, \dots, x_s)$ and $M_x = \cup_{j=1}^{s} S(x_j, r)$. The payoff to Player 2 is $[-H(x, y)]$.

Example 4. Noisy duel [Karlin (1959)]. Each duelist has only one bullet to fire. The duel is assumed to be a noisy duel because each duelist is informed of his opponent's action, firing his bullet, as soon as it takes place. Further, it is assumed that the accuracy function $p_1(x)$ (the probability of hitting the opponent at the instant x) for Player 1 is defined on $[0, 1]$, is continuous and increases monotonically in x and $p_1(0) = 0$, $p_1(1) = 1$. Similarly, the accuracy of

Player 2 is described by the function $p_2(y)$ on $[0,1]$ where $p_2(0)=0$, $p_2(1)=1$. If Player 1 hits Player 2, his payoff is $+1$. If Player 2 hits Player 1, his payoff is -1. If, however, both players fire simultaneously and achieve the same result (positive or negative), the payoff to Player 1 is 0.

The information structure in this game (the fact that the weapons are noisy) is taken into account in constructing the payoff function $H(x,y)$. If $x<y$, the probability of Player 1's hitting the opponent is $p_1(x)$ and the probability of Player 1's missing is $1-p_1(x)$. If Player 2 has not fired and knows that Player 1 cannot fire any more, Player 2 can obtain a sure hit by waiting until y is equal to 1. Thus, if Player 1 misses at the instant x, he is sure to be hit by Player 2 provided $x<y$; hence

$$H(x,y) = p_1(x) + (-1)[1-p_1(x)], \quad x<y.$$

Similarly, we have

$$H(x,y) = p_2(y)(-1) + [1-p_2(y)]\cdot 1, \quad x>y$$

and

$$H(x,y) = p_1(x)[1-p_2(y)] + p_2(y)[1-p_1(x)](-1), \quad x=y.$$

Thus, the payoff function $H(x,y)$ in the game is

$$H(x,y) = \begin{cases} 2p_1(x)-1, & x<y, \\ p_1(x)-p_2(y), & x=y, \\ 1-2p_2(y), & x>y, \end{cases}$$

where $x\in[0,1]$, $y\in[0,1]$.

Example 5. Silent duel [Karlin (1959)]. In a silent duel each duelist has one bullet, but neither duelist knows whether his opponent has fired. For simplicity, let the accuracy functions be given by $p_1(x) = p_2(x) = x$. Then the payoff function describing the game is

$$H(x, y) = \begin{cases} x - (1 - x)y, & x < y, \\ 0, & x = y, \\ -y + (1 - y)x, & x > y, \end{cases}$$

where $x \in [0, 1], \ y \in [0, 1]$. In this game the payoff function $H(x, y)$ is constructed in the same manner as in Example 4, except that neither duelist can determine the time of his opponent's firing provided the opponent misses.

2.1.3. In conclusion we will point out a special class of zero-sum two-person game in which $X = Y = [0, 1]$. In these games, situations are the pairs of numbers (x, y), where $x, y \in [0, 1]$. Such games are called the *games on the unit square.* The class of the games on the unit square is basic in examination of infinite games. In particular, the Examples 2, 4, 5 are illustrative of the unit square games. Also, Example 6 is the unit square game if we set $x_0 = y_0 = 0$, $x_1 = y_1 = 1$.

2.2 ϵ-Saddle Points, ϵ-Optimal Strategies

2.2.1. In the infinite game, as in any zero-sum two-person game $\Gamma(X, Y, H)$ the principle of players' optimal behavior is the saddle point (equilibrium) principle. The point (x^*, y^*) for which the inequality

$$H(x, y^*) \leq H(x^*, y^*) \leq H(x^*, y) \qquad (2.2.1)$$

holds for all $x \in X, \ y \in Y$ is called saddle point. This principle may be realized in the game Γ if and only if

$$v = \overline{v} = \underline{v} = H(x^*, y^*),$$

where

$$\underline{v} = \max_x \inf_y H(x, y), \ \overline{v} = \min_y \sup_x H(x, y), \qquad (2.2.2)$$

i.e. the external extrema of maximin and minimax are achieved and the lower value of the game $\underline{v}$, is equal to the upper value of the game $\overline{v}$. The game Γ for which the (2.2.2) holds is called *strictly determined* and the number v is the value of the game (see 1.3.4).

For matrix games, the existence of the saddle point and the equality of maximin to minimax were proved in the class of mixed strategies (see Sec. 1.6) and hence a solution consists in finding of their common value v and those strategies x^*, y^* in terms of which the external extrema are achieved in (2.2.2).

In infinite games, the existence of external extrema in (2.2.2) cannot be proved in general case.

2.2.2. *Example 6.* Suppose each of Players 1 and 2 chooses a number from the open interval $(0, 1)$. Then Player 1 receives a payoff equal to the sum of the chosen numbers. In this manner we obtain the game on the open unit square with the payoff function $H(x, y)$ for Player 1.

$$H(x, y) = x + y, \ x \in (0, 1), \ y \in (0, 1). \tag{2.2.3}$$

Here the situation $(1, 0)$ would be equilibrium if 1 and 0 were among the players' strategies, with the game value v being $v = 1$. Actually the external extrema in (2.2.2) are not achieved but in the same time the upper value is equal to the lower value of the game. Therefore, $v = 1$ and Player 1 can always receive the payoff sufficiently close to the game value by choosing a number $1 - \epsilon$, $\epsilon > 0$ sufficiently close to 1. On the other hand, by choosing $\epsilon > 0$ as a sufficiently small number (close to 0), Player 2 can guarantee that his loss will be arbitrarily close to the value of the game.

2.2.3. Definition. The point (x_ϵ, y_ϵ) in the zero-sum two-person game $\Gamma = (X, Y, H)$ is called the ϵ-equilibrium point if the following inequality holds for any strategies $x \in X$ and $y \in Y$ of Players 1 and 2, respectively:

$$H(x, y_\epsilon) - \epsilon \le H(x_\epsilon, y_\epsilon) \le H(x_\epsilon, y) + \epsilon. \tag{2.2.4}$$

The point (x_ϵ, y_ϵ), for which (2.2.4) holds, is called the ϵ-saddle point and the strategies x_ϵ and y_ϵ are called ϵ-optimal strategies for Players 1 and 2, respectively.

Compare the definitions of the saddle point (2.2.1) and the ϵ-saddle point (2.2.4). A deviation from the optimal strategy reduce the player's payoff, whereas a deviation from the ϵ-optimal strategy may increase the payoff by no more than ϵ.

Thus, the point $(1-\epsilon, \epsilon)$, $0 < \epsilon < 1$, is ϵ-equilibrium in Example 6, and the strategies $x_\epsilon = 1-\epsilon$, $y_\epsilon = \epsilon$ are ϵ-optimal strategies for the players 1 and 2 respectively.

2.2.4. Note that the following results hold for two strategically equivalent games $\Gamma = (X, Y, H)$ and $\Gamma'(X, Y, H')$, where $H' = \beta H + \alpha$, $\beta > 0$. If (x_ϵ, y_ϵ) is the ϵ-equilibrium point in the game Γ, then it is the $(\beta\epsilon)$-equilibrium point in the game Γ' (compare it with Scale Lemma in Sec. 1.3).

2.2.5. The following theorem yields the main property of ϵ-optimal strategies.

Theorem. *For the finite value v of the zero-sum two-person game $\Gamma = (X, Y, H)$ to exist, it is necessary and sufficient that, for any $\epsilon > 0$, there be ϵ-optimal strategies x_ϵ, y_ϵ for Players 1 and 2, respectively, in which case*

$$\lim_{\epsilon \to 0} H(x_\epsilon, y_\epsilon) = v. \tag{2.2.5}$$

Proof. Necessity. Suppose the game Γ has the finite value v. For any $\epsilon > 0$ we choose strategy x_ϵ from the condition

$$\sup_{x \in X} H(x, y_\epsilon) - \frac{\epsilon}{2} \leq v \tag{2.2.6}$$

and strategy x_ϵ from the condition

$$\inf_{y \in Y} H(x_\epsilon, y) + \frac{\epsilon}{2} \geq v. \tag{2.2.7}$$

From (2.2.2), (2.2.6), and (2.2.7), we obtain the inequality

$$H(x, y_\epsilon) - \frac{\epsilon}{2} \leq v \leq H(x_\epsilon, y) + \frac{\epsilon}{2}, \tag{2.2.8}$$

for all strategies x, y. Consequently,

$$|H(x_\epsilon, y_\epsilon) - v| \le \frac{\epsilon}{2}. \tag{2.2.9}$$

The relations (2.2.4), (2.2.5) follow from (2.2.8), (2.2.9).

Sufficiency. If the inequalities (2.2.4) hold for any number $\epsilon > 0$, then

$$\begin{aligned}\overline{v} = \inf_y \sup_x H(x,y) &\le \sup_x H(x, y_\epsilon) \le H(x_\epsilon, y_\epsilon) + \epsilon \\ &\le \inf_y H(x,y) + 2\epsilon \le \sup_x \inf_y H(x,y) + 2\epsilon = \underline{v} + 2\epsilon. \quad (2.2.10)\end{aligned}$$

Hence it follows that $\overline{v} \le \underline{v}$, but according to the Lemma given in 1.2.2, the inverse inequality holds true. Thus, it remains to prove that the value of the game Γ is finite. Let us take such sequence $\{\epsilon_n\}$ that $\lim_{n\to\infty} \epsilon_n = 0$. Let $\epsilon_k \in \{\epsilon_n\}$, $\epsilon_{k+m} \in \{\epsilon_n\}$, where m is any fixed natural number. We have

$$\begin{aligned} H(x_{\epsilon_{k+m}}, y_{\epsilon_k}) + \epsilon_{k+m} \ge H(x_{\epsilon_{k+m}}, y_{\epsilon_{k+m}}) \ge H(x_{\epsilon_k}, y_{\epsilon_{k+m}}) - \epsilon_{k+m}, \\ H(x_{\epsilon_k}, y_{\epsilon_{k+m}}) + \epsilon_k \ge H(x_{\epsilon_k}, y_{\epsilon_k}) \ge H(x_{\epsilon_{k+m}}, y_{\epsilon_k}) - \epsilon_k. \end{aligned}$$

Thus,

$$|H(x_{\epsilon_k}, y_{\epsilon_k}) - H(x_{\epsilon_{k+m}}, y_{\epsilon_{k+m}})| \le \epsilon_k + \epsilon_{k+m} = \delta_{km}.$$

Since $\lim_{k\to\infty} \delta_{km} = 0$ for any fixed value of m, then there exists a finite limit $\lim_{\epsilon\to 0} H(x_\epsilon, y_\epsilon)$. From the relationship (2.2.10) we obtain the inequality $|H(x_\epsilon, y_\epsilon) - v| \le \epsilon$; hence $v = \lim_{\epsilon\to 0} H(x_\epsilon, y_\epsilon)$. This completes the proof of the theorem.

2.2.6. To illustrate the definitions given in this section, we shall consider in greater details Example 1, Ref. 1.1.2.

Example 7. Suppose the sets S_1 and S_2 are the closed circles of radii R_1 and R_2 ($R_1 < R_2$), respectively. Find the lower value of the game

$$\underline{v} = \max_{x\in S_1} \min_{y \in S_2} \rho(x,y).$$

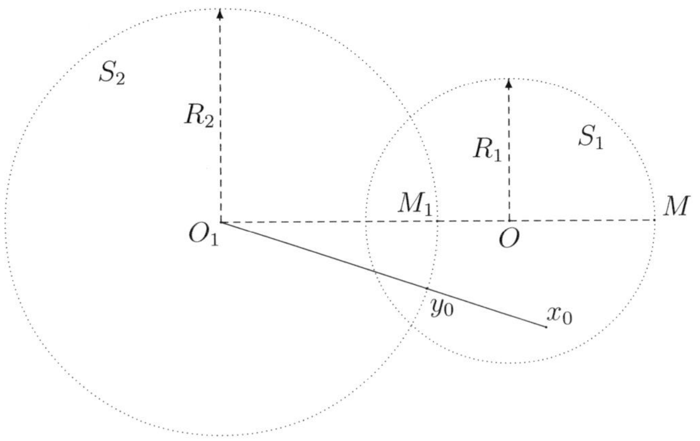

Figure 2.2 Lower value of the game

Let $x_0 \in S_1$. Then $\min_y \rho(x_0, y)$ is achieved at the intersection point y_0 of the straight line, passing through the center O_1 of the circle S_2 and the point x_0, and the boundary of the circle S_2. Evidently, the quantity $\min_{y \in S} \rho(x_0, y)$ is a maximum at the point $M \in S_1$ where the lines of centers OO_1 (Fig. 2.2) intersect the boundary of the circle S_1 that is farthest from the point O_1.

Thus, $\underline{v} = |O_1 M| - R_2$.

In order to compute the upper value of the game

$$\overline{v} = \min_{y \in S_2} \max_{x \in S_1} \rho(x, y)$$

we shall consider two cases.

Case 1. The center O of the circle S_1 belongs to the set S_2 (Fig. 2.3). For each $y_0 \in S_2$ the point x_0 providing $\max_{x \in S_1} \rho(x, y_0)$ is constructed as follows.

Let x_0^1 and x_0^2 be the intersection points of the line $O_1 y_0$ and the boundary of the circle S_1, x_0^3 is the intersection point of the line $O y_0$ with the boundary of the circle S_1, that is farthest from the point y_0. Then x_0 is determined from the condition

$$\rho(x_0, y_0) = \max_{i=1,2,3} \rho(x_0^i, y_0).$$

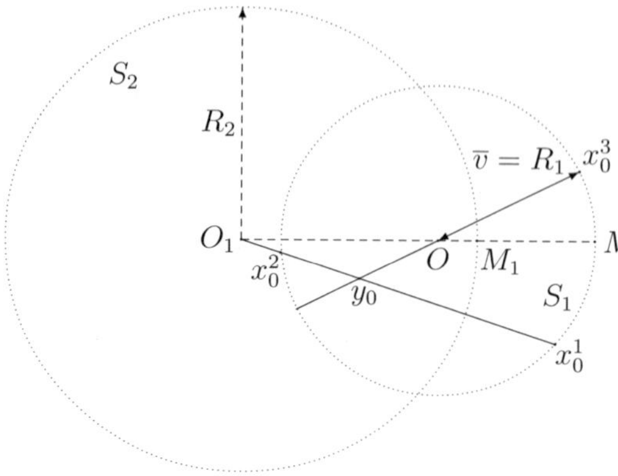

Figure 2.3 Case 1, when $O \in S_2$

By construction, for all $y_0 \in S_2$

$$\max_{x \in S_1} \rho(x, y_0) = \rho(x_0, y_0) \geq R_1.$$

With $y_0 = O$, however, we get

$$\max_{x \in S_1} \rho(x, O) = R_1,$$

hence

$$\min_{y \in S_2} \max_{x \in S_1} \rho(x, y) = \overline{v} = R_1.$$

It can be readily seen that since $O \in S_2$, in Case 1 $\overline{v} = R_1 \geq |\overline{O_1 M}| - R_2 = \underline{v}$. Furthermore, we get the equality provided O belongs to the boundary of the set S_2.

Thus, if in Case 1 the point O does not belong to the boundary of the set S_2, then the game has no saddle point. If, however, the point O belongs to the boundary of the set S_2, then there exists a saddle point and an optimal strategy for Player 1 is to choose the point M lying at the intersection of the line of centers OO_1 with the boundary of the set S_1 that is farthest from the point O_1. An optimal strategy for Player 2 is to choose the point $y \in S_2$ coinciding with the center O of the circle S_1. In this case the value of the game is $v = \underline{v} = \overline{v} = R_1 + R_2 - R_2 = R_1$.

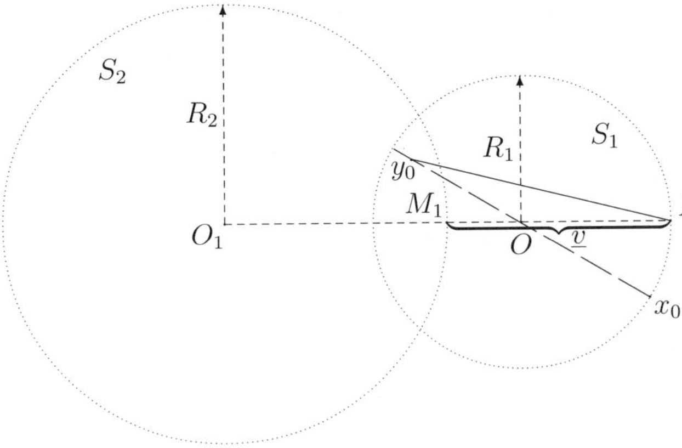

Figure 2.4 Case 2, when $O \notin S_2$

Case 2. The center of the circle $O \notin S_2$. This case is coinsidered in the same way as Case 1 when the center of the circle S_1 belongs to the boundary of the set S_2. Compute the quantity $\overline{v}$ (Fig. 2.4). Let $y_0 \in S_2$. Then the point x_0 providing $\max_{x \in S_1} \rho(x, y_0)$ coincides with the intersection point x_0 of the straight line, passing through y_0 and the center O of the circle S_1, and the boundary of the circle S_1 that is farthest from the point y_0. Indeed, the circle of radius $\overline{x_0 y_0}$ with its center at the point y_0 contains S_1 and its boundary is tangent to the boundary of the circle S_1 at the unique point x_0. Evidently, the quantity $\max_{x \in S_1} \rho(x, y) = \rho(x_0, y)$ takes its minimum value at the intersection point M_1 of the line segment OO_1 and the boundary of the circle S_2. Thus, in the case under study

$$\overline{v} = \min_{y \in S_2} \max_{x \in S_1} \rho(x, y) = |\overline{O_1 M}| - R_2 = \underline{v}.$$

Optimal strategies for Players 1 and 2 are to choose the points $M \in S_1$ and $M_1 \in S_2$, respectively.

If the open circles S_1 and S_2 are considered to be the strategy sets in Example 1, ref. 1.1.2, then in Case 2 the value of the game exists and is equal to

$$\underline{v} = \sup_{x \in S_1} \inf_{y \in S_2} \rho(x, y) = \inf_{y \in S_2} \sup_{x \in S_1} \rho(x, y) = \overline{v} = |\overline{O_1 M}| - R_2 = v.$$

Optimal strategies, however, do not exist, since $M \notin S_1$, $M_1 \notin S_2$. Nevertheless for any $\epsilon > 0$ there are ϵ-optimal strategies that are the points from the ϵ-neighborhood of the points M and M_1 belonging respectively to the sets S_1 and S_2.

2.2.7. In conclusion, it should be noted that the games in the Examples 1–5, generally, do not have an equilibrium point and a game value. Thus in Example 2 Player 1 has an optimal strategy $x^* = 1/2$ when $l \geq 1/2$, and the game value is 1 (any strategy of Player 2 being optimal).

2.3 Mixed Strategies

2.3.1. Consider zero-sum two-person game $\Gamma = (X, Y, H)$. If it has no value, then $\overline{v} \neq \underline{v}$. As noted in Sec. 1.4, in this case in order to increase his payoff for the player it is important to know the opponent's intention. Although the rules of the games do not provide such possibility one can estimate statistically the chances of choosing one or another strategy and take a proper course of action after playing the game with the same opponent sufficiently many times. But what course of action should the player take if he wishes to keep his intentions secret? The only rational course of action here is to choose a strategy by using some chance device, i.e. it is necessary to use mixed strategies.

We shall give a formal definition of the mixed strategy for the infinite game.

2.3.2. Let $\mathcal{X}$ be some σ-algebra of subsets of the set X (including singletons $x \in X$) and let $\mathcal{Y}$ be σ-algebra of subsets of Y ($y \in \mathcal{Y}$ if $y \in Y$). Denote by $\overline{X}$ and $\overline{Y}$ the sets of all probability measures on the σ-algebras $\mathcal{X}$ and $\mathcal{Y}$, respectively, and let the function H be measurable with respect to σ-algebra of $\mathcal{X} \times \mathcal{Y}$. Consider the integral

$$K(\mu, \nu) \triangleq \int_X \int_Y H(x, y) d\mu(x) d\nu(y), \quad \mu \in \overline{X}, \ \nu \in \overline{Y}, \qquad (2.3.1)$$

representing the mathematical expectation of the payoff $H(x, y)$ for measures μ, ν [Prokhorov and Riazanov (1967)].

Definition. A mixed extension of the game $\Gamma = (X, Y, H)$ is a zero-sum two-person game in normal form with the strategy sets $\overline{X}, \overline{Y}$ and the payoff function $K(\mu, \nu)$, i.e. the game $\Gamma = (\overline{X}, \overline{Y}, K)$.

The players behavior in the mixed extension of the game Γ can be interpreted as follows. The players choose independently of each other the measures $\mu \in \overline{X}$ and $\nu \in \overline{Y}$. In accordance with these measures they implement (e.g. by the table of random numbers) a random choice of the strategies $x \in X$ and $y \in Y$. Thereafter Player 1 receives the payoff $H(x, y)$. The strategies $\mu \in \overline{X}$, $\nu \in \overline{Y}$ are called mixed, and $x \in X$, $y \in Y$ are called pure strategies in the game Γ.

Introduction of a mixed extension of the infinite game deserves further comment. The sets $\overline{X}$ and $\overline{Y}$ depend on what σ-algebras $\mathcal{X}$ and $\mathcal{Y}$ are used to consider probability measures. In the case of matrix games (the sets X and Y are finite) in a mixed extension, the players choose their strategies in accordance with probability distributions over the sets X and Y. If X is an infinite set and we are acting just as in the finite case, then we need to consider measures for which all the subsets of the infinite set X are measurable. Such measures, however, are very special, being concentrated on the sets that are, at most, countable. Where only such measures are used the players impoverish their possibilities and cannot always guarantee the existence of an equilibrium point in mixed strategies. Therefore they use less extensive σ-algebras on which the probability measures are determined. This substantially increases the range of probability measures (and, as a rule, guarantees the existence of an equilibrium point in mixed strategies). In this case, however, not every function H on $X \times Y$ proves to be measurable, and hence there is no way to define a mathematical expectation of the payoff, thereby defining the concept of an equilibrium, and the values of the game and optimal strategies. Thus, trade-off is required here. From the point of view of finding a solution it is desirable that the mixed strategies be of the simplest form and there be at least the value of the game in this extension.

Strictly speaking, the integral in (2.3.1) should be taken over the measure $\mu \times \nu$ on the Cartesian product $X \times Y$. By the rules of the zero-sum two-person game, however, the mixed strategies (measures) μ and ν are selected by the players simultaneously and independently, i.e. the probability measures μ and ν are stochastically independent.

Definition. A pair of probability measures $\mu \in \overline{X}$, $\nu \in \overline{Y}$ that are stochastically independent is called the situation (μ, ν) in mixed strategies.

Thus, at the situation (μ, ν) in mixed strategies the payoff $K(\mu, \nu)$ is the iterated integral (2.3.1). Singletons belong to the σ-algebra of subsets of the strategy set on which the probability measures are determined; hence every pure strategy $x(y)$ can be placed in correspondence with the probability measure $\mu_x \in \overline{X}$ $(\nu_y \in \overline{Y})$ concentrated at the point $x \in X$ $(y \in Y)$. Identifying the strategies x and μ_x, y and ν_y we see that pure strategies are a special case of mixed strategies, i.e. the inclusions $X \subset \overline{X}$, $Y \subset \overline{Y}$ hold true. Then the payoffs of Player 1 at the points (x, ν) and (μ, y) are respectively mathematical expectations

$$K(x, \nu) \triangleq K(\mu_x, \nu) = \int_Y H(x, y) d\nu(y), \tag{2.3.2}$$

$$K(\mu, y) \triangleq K(\mu, \nu_y) = \int_X H(x, y) d\mu(x), \tag{2.3.3}$$

where the integrals in (2.3.1), (2.3.2), and (2.3.3) are taken in the sense of Lebesgue–Stieltjes. If, however, the distributions $\mu(x), \nu(y)$ have the densities $f(x)$ and $g(y)$, i.e. $d\mu(x) = f(x)dx$ and $d\nu(y) = g(y)dy$, then the integrals in (2.3.1), (2.3.2), and (2.3.3) are taken in the sense of Riemann–Stieltjes. The game $\Gamma \subset \overline{\Gamma}$ is a subgame of its mixed extension $\overline{\Gamma}$. Whatever the probability measures μ and ν, all integrals in (2.3.1), (2.3.2), (2.3.3) are supposed to exist.

Definition. Let $\Gamma = (X, Y, H)$ be a zero-sum two-person game, and let $\overline{\Gamma} = (\overline{X}, \overline{Y}, K)$ be its mixed extension. Then the point

$(\mu^*, \nu^*) \in \overline{X} \times \overline{Y}$ is called an equilibrium point in the game Γ in mixed strategies if for all $\mu \in \overline{X}$ and $\nu \in \overline{Y}$ the following inequality holds:

$$K(\mu, \nu^*) \leq K(\mu^*, \nu^*) \leq K(\mu^*, \nu), \tag{2.3.4}$$

i.e. (μ^*, ν^*) is an equilibrium point in the mixed extension of the game Γ, and $\mu^*(\nu^*)$ is Player 1's (2's) optimal strategy in $\overline{\Gamma}$.

Similarly, the point $(\mu^*_\epsilon, \nu^*_\epsilon) \in \overline{X} \times \overline{Y}$ is called the ϵ-equilibrium point in the mixed extension of the game Γ if for all $\mu \in \overline{X}$ and $\nu \in \overline{Y}$ the following inequalities hold

$$K(\mu, \nu^*_\epsilon) - \epsilon \leq K(\mu^*_\epsilon, \nu^*_\epsilon) \leq K(\mu^*_\epsilon, \nu) + \epsilon, \tag{2.3.5}$$

i.e. μ^*_ϵ (ν^*_ϵ) is the ϵ-optimal strategy of Player 1(2) in $\overline{\Gamma}$.

2.3.3. As in the case of matrix games, we may show that if the payoff functions of the games $\Gamma = (X, Y, H)$ and $\Gamma' = (X, Y, H')$ are related by the equality $H'(x, y) = \alpha H(x, y) + \beta$, $\alpha > 0$, then the sets of equilibrium points in the mixed strategies in the games Γ and Γ' coincide, and for the game values we have $v(\Gamma') = \alpha v(\Gamma) + \beta$ (see Sec. 1.4).

2.3.4. The equilibrium points in mixed strategies have the same properties as in the case of matrix games, which follows from the theorems given below.

Theorem. *For the pair $(\mu^* \nu^*)$, $\mu^* \in \overline{X}$, $\nu^* \in \overline{Y}$ to be an $(\epsilon-)$ equilibrium point in mixed strategies in the game Γ, it is necessary and sufficient that for all $x \in X, y \in Y$ the following inequalities hold:*

$$K(x, \nu^*) \leq K(\mu^*, \nu^*) \leq K(\mu^*, y), \tag{2.3.6}$$

$$(K(x, \nu^*) - \epsilon \leq K(\mu^*, \nu^*) \leq K(\mu^*, y) + \epsilon. \tag{2.3.7}$$

Proof. Necessity of the theorem is obvious, since the pure strategies are a special case of mixed strategies. Prove the sufficiency. First prove (2.3.6) ((2.3.7) can be proved in the same way). Let μ and ν

be arbitrarily mixed strategies for players 1 and 2, respectively. From (2.3.1), (2.3.2), and (2.3.6), we then get

$$K(\mu, \nu^*) = \int_X K(x, \nu^*) d\mu(x) \leq K(\mu^*, \nu^*),$$

$$K(\mu^*, \nu) = \int_Y K(\mu^*, y) d\nu(y) \geq K(\mu^*, \nu^*).$$

This implies the inequalities (2.3.4), which is the required result.

From the theorem, in particular, it follows that if (x^*, y^*) is the pure strategy (ϵ-) equilibrium point in the game Γ, then it is also the (ϵ-) equilibrium point in the mixed extension $\overline{\Gamma}$ in which case the game value is preserved.

Note that the mixed extension $\overline{\Gamma}$ is a zero-sum two-person game and hence the notion of a strictly determined game (in 2.2.1) holds true for $\overline{\Gamma}$. Although this applies to the theorem in (2.2.5), but we are dealing here with the equilibrium point and value of the game in mixed strategies.

2.3.5. Theorem. *For the game $\Gamma = (X, Y, H)$ to have the value v in mixed strategies, it is necessary and sufficient that the following equation holds:*

$$\sup_\mu \inf_y K(\mu, y) = \inf_\nu \sup_x K(x, \nu) = v. \tag{2.3.8}$$

If the players have optimal strategies, then the exterior extrema in (2.3.8) are achieved and the equations

$$\inf_y K(\mu^*, y) = v, \tag{2.3.9}$$

$$\sup_x K(x, \nu^*) = v \tag{2.3.10}$$

are the necessary and sufficient optimality condition for the mixed strategies $\mu^* \in \overline{X}$ and $\nu^* \in \overline{Y}$.

Proof. Let v be the value of the game. Then, by definition,

$$v = \sup_\mu \inf_\nu K(\mu, \nu). \tag{2.3.11}$$

For a fixed strategy μ, the set $\{K(\mu,\nu)|\ \nu \in \overline{Y}\}$ is a convex hull of numbers $K(\mu,y)$, $y \in \overline{Y}$. Since the exact lower bound of any set of real numbers coincides with that of a convex hull of these numbers, then

$$\inf_{\nu\in\overline{Y}} K(\mu,\nu) = \inf_{y\in Y} K(\mu,y). \tag{2.3.12}$$

Equality (2.3.12) can also be obtained from the following reasonings. Since $Y \subset \overline{Y}$, we have

$$\inf_{\nu\in\overline{Y}} K(\mu,\nu) \le \inf_{y\in Y} K(\mu,y).$$

Suppose the inequality is strict, i.e.

$$\inf_{\nu} K(\mu,\nu) < \inf_{y} K(\mu,y).$$

This means that for a sufficiently small $\epsilon > 0$ the following inequality holds:

$$\inf_{\nu} K(\mu,\nu) + \epsilon < \inf_{y} K(\mu,y).$$

Thus, for all $y \in Y$

$$K(\mu,y) > \inf_{\nu} K(\mu,\nu) + \epsilon. \tag{2.3.13}$$

Passing to the mixed strategies in (2.3.13) we now get

$$\inf_{\nu} K(\mu,\nu) \ge \inf_{\nu} K(\mu,\nu) + \epsilon.$$

The obtained contradiction proves (2.3.12). Let us take a supremum for μ in (2.3.12). Then

$$v = \sup_{\mu}\inf_{y} K(\mu,y).$$

The second equality in (2.3.8) can be proved in the same way. Conversely, if (2.3.8) is satisfied, it follows from (2.3.12) that v is the value of the game.

Now let μ^*, ν^* be optimal strategies for the players 1 and 2, respectively. By the Theorem given in 1.3.4, the exterior extrema in (2.3.8) are achieved, and (2.3.9), (2.3.10) are the necessary and sufficient optimality conditions for the mixed strategies μ^* and ν^*.

As noted in 2.3.2, introduction of mixed strategies in a zero-sum infinite game depends on the way of randomizing the set of pure strategies. From (2.3.8), however, it follows that the game value v is independent of the randomization method. Thus, to prove its existence, it is sufficient to find at least one mixed extension of the game for which (2.3.8) holds.

Corollary. *For any zero-sum two-person game* $\Gamma = (X, Y, H)$ *having the value* v *in mixed strategies, the following inequality holds*:

$$\sup_x \inf_y H(x, y) \leq v \leq \inf_y \sup_x H(x, y). \tag{2.3.14}$$

Proof. Theorem 2.3.5 implies

$$\sup_x \inf_y H(x, y) \leq \sup_\mu \inf_y K(\mu, y) = v = \inf_\nu \sup_x K(x, \nu)$$
$$\leq \inf_y \sup_x H(x, y).$$

2.3.6. From (2.3.14), one of the methods for an approximate solution of the zero-sum two-person game follows. Indeed, suppose the exterior extrema in (2.3.14) are achieved, i.e.

$$v^- = \max_x \inf_y H(x, y) = \inf_y H(x^0, y), \tag{2.3.15}$$

$$v^+ = \min_y \sup_x H(x, y) = \sup_x H(x, y^0), \tag{2.3.16}$$

and let $\alpha = v^+ - v^-$. Then Player 1's maximin strategy x^0 and Player 2's minimax strategy y^0 describe the players' optimal behavior with α accuracy, and can be taken as an approximate solution to the game Γ. In this case the problem thus reduces to that of finding the maximin and minimax strategies for Players 1 and 2, respectively, with the accuracy of the approximate solution determined by $\alpha = v^+ - v^-$. Here, by (2.3.14), the game value v lies in the interval $v \in [v^-, v^+]$. Minimax theory [von Neumann (1928); Demyanov and Malozemov (1990)] is devoted to the methods of finding solutions to problems (2.3.15), (2.3.16).

2.3.7. As in the case of matrix games, the notion of a mixed strategy spectrum is important for infinite games.

Definition. Let $\Gamma = (X, Y, H)$ be a zero-sum two-person game. The pure strategy $x_0 \in X$ ($y_0 \in Y$) of Player 1(2) is then called the concentration point of his mixed strategy $\mu(\nu)$ if $\mu(x_0) > 0$ ($\nu(y_0) > 0$).

Definition. The pure strategy $x_0 \in X$ ($y_0 \in Y$), where X (Y, respectively) is a topological space, is called the point of the mixed strategy spectrum $\mu(\nu)$ given on Borel σ-algebra of subsets of the set $X(Y)$ if, for any measurable neighborhood ω of the point $x_0(y_0)$ the following inequality holds:

$$\mu(\omega) = \int_\omega d\mu(x) > 0 \quad \left(\nu(\omega) = \int_\omega d\nu(y) > 0\right).$$

The least closed set whose μ-measure (ν-measure) is equal to 1 will be called the spectrum of the mixed strategy $\mu(\nu)$.

The mixed strategy concentration points are also spectrum points. The opposite is not true.

The mixed strategy spectrum μ(ν, respectively) will be denoted by $X_\mu(Y_\nu)$. We shall prove the analog of the complementary slackness theorem 1.7.6 for infinite games.

Theorem. *Suppose $\Gamma = (X, Y, H)$ is a zero-sum two-person game having the value v. If $x_0 \in X$ and ν^* is an optimal mixed strategy of Player 2 and*

$$K(x_0, \nu^*) < v, \tag{2.3.17}$$

then x_0 cannot be the concentration point for an optimal strategy of Player 1.

A similar result is true for the concentration points of Player 2's optimal strategies.

Proof. It follows from the optimality of mixed strategy $\nu^* \in \overline{Y}$ that for all $x \in X$ the following inequality holds:

$$K(x, \nu^*) \leq v.$$

Integrating it with respect to Player 1's optimal mixed strategy (measure) μ^* over the set $X \setminus \{x_0\}$, we get

$$\int_{X\setminus\{x_0\}} K(x,\nu^*)d\mu^*(x) \leq v \int_{X\setminus\{x_0\}} d\mu^*(x).$$

Let $\mu^*(x_0) > 0$, i.e. x_0 is concentration point of Player 1's optimal mixed strategy μ^*. Then it follows from (2.3.17) that

$$K(x_0,\nu^*)\mu^*(x_0) < v\mu^*(x_0).$$

Combining the last two inequalities yields a contradiction

$$v = \int_X K(x,\nu^*)d\mu^*(x) = K(\mu^*,\nu^*) < v.$$

Hence $\mu^*(x_0) = 0$ for all optimal strategies $\mu^* \in \overline{X}$.

2.3.8. For zero-sum infinite games, the notion of strategy dominance can be introduced in the same way as in Sec. 1.8.

Definition. Player 1's strategy $\mu_1 \in \overline{X}$ strictly dominates strategy $\mu_2 \in \overline{X}$ $(\mu_1 \succ \mu_2)$ if

$$H(\mu_1, y) > H(\mu_2, y),$$

for all $y \in Y$. Similarly, Player 2's strategy $\nu_1 \in \overline{Y}$ strictly dominates strategy $\nu_2 \in \overline{Y}$ $(\nu_1 \succ \nu_2)$ if

$$H(x,\nu_1) < H(x,\nu_2),$$

for all $x \in X$. Strategies μ_2 and ν_2 are called strictly dominated if there exist $\mu_1 \succ \mu_2$ and $\nu_1 \succ \nu_2$.

If the last inequalities are satisfied as nonstrict, then we say that μ_1 dominates μ_2 $(\mu_1 \succeq \mu_2)$ and ν_1 dominates ν_2 $(\nu_1 \succeq \nu_2)$.

Dominance theorems that are similar to those in 1.8.3 will be given without proof.

Theorem. *For a zero-sum infinite game having a solution, none of the player's strictly dominated pure strategies is contained in the spectrum of his optimal mixed strategies.*

Theorem. *Let $\Gamma = (X, Y, H)$ be a zero-sum infinite game having a solution (X and Y are topological spaces), and each element of the* *open set $X^0 \subset X$ is dominated by a strategy μ^0 whose spectrum does not intersect X^0. Then any solution of the game $\Gamma' = (X \setminus X^0, Y, H)$ is a solution of the game Γ.*

A similar theorem holds for strategies of Player 2.

Dominance theorems are similar to those in 1.8.1 and are given without proof.

2.3.9. This section deals with the properties of optimal (ϵ-optimal) mixed strategies with the assumption that there exists a solution of the game. The matrix game is strictly determined in mixed strategies, i.e. there always exist a value and an equilibrium point, which follows from the Theorem 1.6.1.

The saddle point in zero-sum two-person infinite games does not exist always as it is seen from the next example.

Example 8. The game without a value in mixed strategies [Vorobjev (1984)]. Consider the game $\Gamma = (X, Y, H)$, where $X = Y = \{1, 2, \ldots\}$ is the set of natural numbers, and the payoff function is

$$H(x, y) = \begin{cases} 1, & \text{if } x > y, \\ 0, & \text{if } x = y, \\ -1, & \text{if } x < y. \end{cases}$$

This game has no value in pure strategies. Show that it has no value in mixed strategies as well.

Let μ be an arbitrary mixed strategy of Player 1, and $d\mu(x) = \delta_x$, where $\delta_x \geq 0$ and $\sum_{x=1}^{\infty} \delta_x = 1$. Take $\epsilon > 0$ and find y_ϵ such that

$$\sum_{x \leq y_\epsilon} \delta_x > 1 - \epsilon.$$

Then

$$K(\mu, y_\epsilon) = \sum_{x=1}^{\infty} \delta_x H(x, y_\epsilon) = \sum_{x \leq y_\epsilon} \delta_x H(x, y_\epsilon) + \sum_{x > y_\epsilon} \delta_x H(x, y_\epsilon)$$

$$= -\sum_{x \leq y_\epsilon} \delta_x + \sum_{x > y_\epsilon} \delta_x < -1 + 2\epsilon,$$

is true for arbitrary $\epsilon > 0$ and because $H(x, y)$ does not take values less than -1, we have

$$\inf_y K(\mu, y) = -1.$$

Consequently, since the strategy μ is arbitrary,

$$\underline{v} = \sup_\mu \inf_y K(\mu, y) = -1.$$

By a similar reasoning we may receive

$$\overline{v} = \inf_\nu \sup_x K(x, \nu) = 1.$$

Since $\overline{v} > \underline{v}$, then the game Γ has no value in mixed strategies. As is shown in the next section, the continuity of the payoff function and the compactness of the strategy spaces is sufficient for the existence of a solution (value and optimal strategies) in the mixed extension.

2.4 Games with Continuous Payoff Functions

2.4.1. In this section, the zero-sum two-person games $\Gamma = (X, Y, H)$ are considered with the assumption that the strategy spaces X and Y are metric compact sets (more often they will be the subsets of Euclidean spaces), and the function H is continuous in both variables. The sets $\overline{X}$, $\overline{Y}$ of mixed strategies of Players 1 and 2 mean the sets of probability measures given on the σ-algebras $\mathcal{X}$ and $\mathcal{Y}$ of the Borel subsets of the sets X and Y, respectively. Then Player 1's payoff $K(\mu, \nu)$ in the situation in mixed strategies $(\mu, \nu) \in \overline{X} \times \overline{Y}$ is determined by the integral (2.3.1) and is the mathematical expectation of the payoff over the probability measure $\mu \times \nu$.

The defined game $\Gamma = (X, Y, H)$ will be called a *continuous game*.

2.4.2. Theorem. *If $\Gamma = (X, Y, H)$ is the zero-sum infinite game having the value v and the equilibrium point (μ^*, ν^*), and the functions $K(\mu^*, y), K(x, \nu^*)$ are continuous respectively in y and x,* *then the following equalities hold:*

$$K(\mu^*, y) = v, \quad y \in Y_{\nu^*}, \tag{2.4.1}$$

$$K(x, \nu^*) = v, \quad x \in X_{\mu^*}, \tag{2.4.2}$$

where Y_{ν^*}, X_{μ^*} *are spectrums of mixed strategies* ν^* *and* μ^* *respectively.*

Proof. From the Theorem 2.3.4 it follows that the inequality

$$K(\mu^*, y) \geq v \tag{2.4.3}$$

holds for all points $y \in Y$. If (2.4.1) does not hold, then there exists a point $y_0 \in Y_{\nu^*}$ such that $K(\mu^*, y_0) > v$. By the continuity of the function $K(\mu^*, y)$ there exists such a neighborhood that the inequality (2.4.3) in a neighborhood ω of the point y_0 is strict. From the fact that $y_0 \in Y_{\nu^*}$ is a point of the mixed strategy spectrum ν^*, it follows that $\nu^*(\omega) > 0$. From this, and from inequality (2.4.3) we get

$$v = K(\mu^*, \nu^*) = \int_Y K(\mu^*, y) d\nu^*(y) > v.$$

The contradiction proves the validity of (2.4.1). Equality (2.4.2) can be proved in a similar way.

This result is analog of the complementary stackness theorem, 1.7.6. Recall that the pure strategy x appearing in the optimal strategy spectrum is called *essential.* Thus, the theorem states that (2.4.1) or (2.4.2) must hold for essential strategies.

Theorem 2.4.2 holds for any continuous game since the following assertion is true.

2.4.3. Lemma. *If the function* $H : X \times Y \to R^1$ *is continuous on* $X \times Y$ *then the integrals of* $K(\mu, y)$ *and* $K(x, \nu)$ *are respectively continuous functions of* y *and* x *for any fixed mixed strategies* $\mu \in \overline{X}$ *and* $\nu \in \overline{Y}$.

Proof. The function $H(x, y)$ is continuous on the compact set $X \times Y$, and hence is uniformly continuous.

Take an arbitrary $\epsilon > 0$ and find such $\delta > 0$ that as soon as $\rho(y_1, y_2) < \delta$, then for any x the following inequaility holds:

$$|H(x, y_1) - H(x, y_2)| < \epsilon, \tag{2.4.4}$$

where $\rho(\cdot)$ is a metric in the space Y.

Then

$$\begin{aligned} |K(\mu, y_1) - K(\mu, y_2)| &= \left| \int_X H(x, y_1) d\mu(x) - \int_X H(x, y_2) d\mu(x) \right| \\ &= \left| \int_X [H(x, y_1) - H(x, y_2)] d\mu(x) \right| \\ &\leq \int_X |H(x, y_1) - H(x, y_2)| d\mu(x) \\ &< \epsilon \int_X d\mu(x) = \epsilon. \end{aligned} \tag{2.4.5}$$

Hence the function $K(\mu, y)$ is continuous in y.

The continuity of the function $K(x, \nu)$ in x can be proved in a similar way.

2.4.4. Let us formulate the main theorem of this section.

Theorem. *The zero-sum infinite game $\Gamma = (X, Y, H)$, where X, Y are metric compact sets and H is a continuous function, has a solution in mixed strategies (the value and optimal strategies).*

This theorem can be proved in terms of analytical properties of the mixed extension of the game $\overline{\Gamma} = (\overline{X}, \overline{Y}, K)$ and some auxiliary results.

2.4.5. Recall that the sequence of Borel measures μ_n, $n = 1, 2, \ldots$, given on the Borel σ-algebra X of the compact metric space X is called weakly convergent if

$$\lim_{n \to \infty} \int_X \varphi(x) d\mu_n(x) = \int_X \varphi(x) d\mu(x) \tag{2.4.6}$$

for any continuous function $\varphi(x), x \in X$.

Lemma. *In terms of the Theorem 2.4.4, the mixed strategy sets $\overline{X}$ and $\overline{Y}$ (the sets of Borel probability measures) are compact metric sets in the topology of weak convergence.*

Let us outline the proof for the set of mixed strategies $\overline{X}$ (similar arguments apply to $\overline{Y}$).

The space of Borel measures $\overline{X}$ given on the Borel σ-algebra $\mathcal{X}$ of the compact metric space X will be a metric space if we introduce the metric

$$\rho(\mu', \mu'') = \max(\rho', \rho''),$$

where ρ' and ρ'' are respectively lower bounds of the numbers r' and r'' such that for any closed set $F \subseteq X$

$$\mu'(F) < \mu''(V_{r'}(F)) + r', \quad \mu''(F) < \mu'(V_{r''}(F)) + r'',$$

where $V_r(F) \triangleq \{x \in X : \min_{z \in F} \rho_1(x, z) < r\}$, $r > 0$, and $\rho_1(\cdot)$ is the metric in the space X.

It is known [Prokhorov and Riazanov (1967)] that convergence in this metric space is equivalent to weak convergence, and the set of measures μ defined on a Borel σ-algebra of the subsets of the space X is weakly compact (i.e. compact in terms of the above defined metric space of all Borel measures) if and only if this set is uniformly bounded

$$\mu(X) \le c \tag{2.4.7}$$

and uniformly dense, i.e. for any $\epsilon > 0$ there is such compact set $A \subseteq X$ that

$$\mu(X \setminus A) \le \epsilon. \tag{2.4.8}$$

Condition (2.4.8) follows from compactness of X, and (2.4.7) follows from the fact that measures $\mu \in \overline{X}$ are normed ($\mu(X) = 1$).

2.4.6. Note that under conditions of Theorem 2.4.4, the set of mixed strategies $\overline{X}(\overline{Y})$ of Player 1(2) is also a compact set in the ordinary sense, since in this case the weak convergence of the measure sequence $\{\mu_n\}$, $n = 1, 2, \ldots$ is equivalent to the convergence in the ordinary sense:

$$\lim_{n\to\infty} \mu_n(A) = \mu(A),$$

for any Borel set $A \subseteq X$ with the bound A' having the zero measure $\mu(A') = 0$.

Proof of this result involves certain complexities and can be found, in Prokhorov and Riazanov (1967).

2.4.7. Denote by $\underline{v}$ and $\overline{v}$ respectively the lower and upper values of the game $\Gamma = (X, Y, H)$.

$$\underline{v} = \sup_{\mu} \inf_{y} K(\mu, y), \quad \overline{v} = \inf_{\nu} \sup_{x}(x, \nu). \tag{2.4.9}$$

Lemma. *If the conditions of the* 2.4.4 *are satisfied, the extrema in* (2.4.9) *are achieved, and*

$$\underline{v} = \max_{\mu\in\overline{X}} \min_{y\in Y} K(\mu, y), \quad \overline{v} = \min_{\nu\in\overline{Y}} \max_{x\in X} K(x, \nu). \tag{2.4.10}$$

Proof. Since $H(x, y)$ is continuous, then, by the Lemma 2.4.3, for any measure $\mu \in \overline{X}$ the function

$$K(\mu, y) = \int_X H(x, y) d\mu(x)$$

is continuous in y. Since Y is a compact set, then $K(\mu, y)$ achieves a minimum at a particular point of this set.

By the definition of $\underline{v}$, for any n there exists such a measure $\mu_n \in \overline{X}$ that

$$\min_{y} K(\mu_n, y) \geq \underline{v} - 1/n.$$

Since $\overline{X}$ is a compact set in the topology of weak convergence (Lemma, 2.4.5), then a weakly convergent subsequence can be chosen from the sequence $\{\mu_n\}_{n=1}^{\infty}$, $\mu_n \in \overline{X}$. Suppose the sequence $\{\mu_n\}_{n=1}^{\infty}$

weakly converges to a certain measure $\mu_0 \in \overline{X}$. Then

$$\lim_{n\to\infty} K(\mu_n, y) = \lim_{n\to\infty} \int_X H(x,y)d\mu_n(x)$$
$$= \int_X H(x,y)d\mu_0(x) = K(\mu_0, y),\ \ y \in Y.$$

But $K(\mu_0, y)$ is not less than $\underline{v}$ for every $y \in Y$. Hence $\min_y K(\mu_0, y) \geq \underline{v}$ and the required maximum is achieved on $\mu_0 \in \overline{X}$.

Similarly, inf sup in (2.4.9) can be shown to be replaced by min max.

2.4.8. Turn now to the proof of the Theorem 2.4.4.

Proof. Since X and Y are metric compact sets, then for any integer n there exist a finite $(1/n)$-networks

$$X_n = \{x_1^n, \ldots, x_{r_n}^n\},\ X_n \subset X, \quad Y_n = \{y_1^n, \ldots, y_{s_n}^n\},\ Y_n \subset Y$$

of the sets X and Y, respectively. This means that for any points $x \in X$ and $y \in Y$ there are such points $x_i^n \in X_n$ and $y_j^n \in Y_n$ that

$$\rho_1(x, x_i^n) < 1/n,\ \ \rho_2(y, y_j^n) < 1/n, \tag{2.4.11}$$

where $\rho_1(\cdot), \rho_2(\cdot)$ are metrics of the spaces X and Y, respectively.

For an arbitrary integer n we construct a matrix game with the matrix $A_n = \{\alpha_{ij}^n\}$, where

$$\alpha_{ij}^n = H(x_i^n, y_j^n),\ \ x_i^n \in X_n,\ \ y_j^n \in Y_n. \tag{2.4.12}$$

The game with the matrix A_n has the value θ_n and optimal mixed strategies $p_n = (\pi_1^n, \ldots, \pi_{r_n}^n)$, $t_n = (\tau_1^n, \ldots, \tau_{s_n}^n)$ for Players 1 and 2, respectively (see Theorem in 1.6.1).

The function $H(x, y)$ is continuous on the Cartesian product $X \times Y$ of metric compact sets, and hence it is uniformly continuous, i.e. for a given $\epsilon > 0$ we may find such $\delta > 0$ that as soon as

$$\rho_1(x, x') < \delta, \quad \rho_2(y, y') < \delta,$$

then

$$|H(x,y) - H(x', y')| < \epsilon. \tag{2.4.13}$$

We choose n such that $1/n < \delta$, and then determine the strategy $\mu_n \in \overline{X}$ by the following rule:

$$\mu_n(F) = \sum_{\{i|x_i^n \in F, x_i^n \in X_n\}} \pi_i^n, \tag{2.4.14}$$

for each Borel set F of the space X. Thus, we have

$$K(\mu_n, y_j^n) = \sum_{i=1}^{r_n} \alpha_{ij}^n \pi_i^n \geq \theta_n. \tag{2.4.15}$$

If $\rho_2(y, y_j^n) < \delta$, then by (2.4.4), (2.4.5), and (2.4.13) we get

$$|H(x,y) - H(x, y_j^n)| < \epsilon,$$
$$|K(\mu_n, y) - K(\mu_n, y_j^n)| < \epsilon.$$

Consequently, for any $y \in Y$ (Y_n is $(1/n)$-network of the set Y)

$$K(\mu_n, y) > \theta_n - \epsilon. \tag{2.4.16}$$

Since $\min_y K(\mu_n, y)$ is achieved (Lemma, 2.4.7), then

$$\underline{v} > \theta_n - \epsilon. \tag{2.4.17}$$

Similarly, we may show that

$$\overline{v} < \theta_n + \epsilon. \tag{2.4.18}$$

From (2.4.17) and (2.4.18), we obtain

$$\underline{v} > \overline{v} - 2\epsilon.$$

But, by Lemma, 1.2.2, the inequality $\underline{v} \leq \overline{v}$ always holds. Because $\epsilon > 0$ was arbitrary, we obtain

$$\underline{v} = \overline{v}, \tag{2.4.19}$$

then from Lemma, 2.4.7, and (2.4.19) follows the assertion of the theorem (see 2.2.1).

2.4.9. Corollary. *The following relation holds:*

$$v = \lim_{n\to\infty} \theta_n, \tag{2.4.20}$$

where $\theta_n = v(A_n)$ *is the value of the matrix game with matrix* A_n (2.4.12).

2.4.10. It follows from the proof of the Theorem 2.4.4 that a continuous game can be approximated by finite games to any degree of accuracy. Moreover, the following result holds true.

Theorem. *An infinite two-person zero-sum game* $\Gamma = (X, Y, H)$, *where* X, Y *are metric compact sets and* H *is a continuous function, has* ϵ*-optimal mixed strategies with a finite spectrum for any* $\epsilon > 0$.

Proof of this theorem follows from proof (2.4.8) of the Theorem 2.4.4. Indeed, by the game Γ, we may construct matrix games with matrices A_n and mixed strategies $\mu_n \in \overline{X}$ that are respectively determined by (2.4.12), (2.4.14) for an arbitrary integer n. By analogy, Player 2's strategies $\nu_n \in \overline{Y}$ are determined as follows:

$$\nu_n(G) = \sum_{\{j|y_j^n \in G, y_j^n \in Y_n\}} \nu_j^n, \tag{2.4.21}$$

where $t^n = (\tau_1^n, \ldots, \tau_{s_n}^n)$ is an optimal mixed strategy for Player 2 in the game with matrix A_n and value θ_n.

By construction, we have

$$\theta_n = \sum_{i=1}^{r_n} \sum_{j=1}^{s_n} \alpha_{ij}^n \pi_i^n \tau_j^n = K(\mu_n, \nu_n), \tag{2.4.22}$$

where $K(\mu, \nu)$ is the payoff in mixed strategies (μ, ν) in the game Γ. From (2.4.16) and a similar inequality for the strategy ν_n we have that for an arbitrary $\epsilon > 0$ there exist an index n such that

$$K(x, \nu_n) - \epsilon < \theta_n < K(\mu_n, y) + \epsilon, \tag{2.4.23}$$

for all $x \in X$ and $y \in Y$. Considering that the strategies μ_n and ν_n have respective finite spectra X_n and Y_n, and X_n, Y_n are finite $1/n$-networks of the sets X and Y, respectively, we obtain the assertion of theorem (see 2.3.4).

2.4.11. Summarizing the results of Theorems 2.4.4 and 2.4.10, we may conclude that the infinite two-person zero-sum game with the continuous payoff function and compact strategy sets for any $\epsilon > 0$ has ϵ-optimal strategies of the players that are mixtures of a finite number of pure strategies, and the optimal mixed strategies in the class of Borel probability measures. Specifically, these results hold for the games on the unit square (see 2.1.3) with a continuous payoff function.

2.4.12. There are many papers proving the existence of the value of infinite two-person zero-sum games. The most general result in this line is attributed to Sion (1957). The results are well known for the games with compact strategy spaces and semicontinuous payoff functions [Peck and Dulmage (1957), Yanovskaya (1973)]. We shall show that in some respects they do not lend themselves to generalization.

Example 9. Game on the square with no value in mixed strategies [Sion and Wolfe (1957)]. Consider a two-person zero-sum game $\Gamma = (X, Y, H)$, where $X = Y = [0, 1]$ and the payoff function H is of the form

$$H(x,y) = \begin{cases} -1, & \text{if } x < y < x + 1/2, \\ 0, & \text{if } x = y \text{ or } x = y + 1/2, \\ 1, & \text{if } y < x \text{ or } x + 1/2 < y. \end{cases}$$

This function has the points of discontinuity on the straight lines $y = x$ and $y = x + 1/2$. Show that

$$\sup_{\mu} \inf_{\nu} K(\mu, \nu) = 1/3, \quad \inf_{\nu} \sup_{\mu} K(\mu, \nu) = 3/7. \tag{2.4.24}$$

Let μ be a probability measure on $[0, 1]$. If $\mu([0, 1/2)) \leq 1/3$, then we set $y_\mu = 1$. If, however, $\mu([0, 1/2)) > 1/3$, then we choose $\delta > 0$ so that $\mu([0, 1/2 - \delta]) > 1/3$, and set $y_\mu = 1/2 - \delta$. In each of these cases we have inequalities

$$\inf_{\nu} K(\mu, \nu) \leq K(\mu, y_\mu) \leq 1/3,$$

which can be proved directly.

On the other hand, if μ is chosen such that $\mu(\{0\}) = \mu(\{1/2\}) = \mu(\{1\}) = 1/3$, then for all $y \in [0,1]$ we have

$$\int_0^1 H(x,y)d\mu(x) = 1/3[H(0,y) + H(1/2,y) + H(1,y)] \geq 1/3.$$

We have thus proved the first equality of (2.4.24).

Now let ν be a probability measure on $[0,1]$. If $\nu([0,1)) \geq 3/7$, then we set $x_\nu = 1$. If, however, $\nu([0,1)) < 3/7$, then $\nu(\{1\}) > 4/7$, in which case we set $x_\nu = 0$ provided $\nu([0,1/2)) \leq 1/7$; if $\nu([0,1/2)) > 1/7$, then we choose $\delta > 0$ such that $v([0,1/2-\delta]) > 1/7$, and set $x_\nu = 1/2 - \delta$. In each of these cases we see that

$$\sup_\mu K(\mu,\nu) > K(x_\nu,\nu) > 3/7.$$

On the other hand, if ν is chosen such that

$$\nu(\{1/4\}) = 1/7,\ \ \nu(\{1/2\}) = 2/7,\ \ \nu(\{1\}) = 4/7,$$

then for any $x \in [0,1]$ we have

$$\int_0^1 H(x,y)d\nu(y) = \frac{1}{7}\left[H\left(x,\frac{1}{4}\right) + 2H\left(x,\frac{1}{2}\right) + 4H(x,1)\right] \leq \frac{3}{7}.$$

We have thus proved the second equality of (2.4.24).

2.5 Games with a Convex Payoff Function

In Sec. 2.4, the existence of a solution to the infinite two-person zero-sum games with a continuous payoff function and compact strategy sets was proved under sufficiently general assumptions. At the same time, it may be interesting, both theoretically and practically, to distinguish the classes of games, where one or both players have optimal pure strategies. This section deals with such games.

2.5.1. Definition. Suppose $X \subset R^m$, $Y \subset R^n$ are compact sets, the set Y is convex, the payoff function $H : X \times Y \to R^1$ is continuous in all its arguments and is convex with respect to $y \in Y$ for any fixed value of $x \in X$. Then the game $\Gamma(X, Y, H)$ is called the game with a convex payoff function (a convex game).

A symmetric definition for Player 1 is given as follows:

Definition. If $X \subset R^m$, $Y \subset R^n$ are compact sets, X is convex, the function H is continuous in all its arguments and is concave with respect to $x \in X$ for any fixed $y \in Y$, then the game $\Gamma = (X, Y, H)$ is called the game with a concave payoff function (a concave game).

If, however, $X \subset R^m, Y \subset R^n$ are compact sets and the payoff function $H(x, y)$ which is continuous in all its arguments, is concave with respect to x for any fixed y and is convex with respect to y for each x, then the game $\Gamma(X, Y, H)$ is called the game with a *concave-convex payoff function* (*a concave-convex game*).

We will now consider convex games. Note that similar results are also true for concave games.

Theorem. *Suppose* $\Gamma = (X, Y, H)$ *is a convex game. Then Player* 2 *has an optimal pure strategy, with the game value being equal to*

$$v = \min_{y \in Y} \max_{x \in X} H(x, y). \tag{2.5.1}$$

Proof. Since X and Y are metric compact sets (in the metric of Euclidean spaces R^m and R^n) and the function H is continuous on the product $X \times Y$, then, by the Theorem 2.4.4, in the game Γ there exist the value v and optimal mixed strategies μ^*, ν^*. It is well known that the set of all probability measures with a finite support is everywhere dense in the set of all probability measures on Y [Prokhorov and Riazanov (1967)]. Therefore, there exists a sequence of mixed strategies ν^n with a finite spectrum that is weakly convergent to ν^*. Suppose the strategy spectrum ν^n consists of the points $y_n^1, \ldots, y_n^{k_n}$ that are chosen with probabilities $\eta_1^n, \ldots, \eta_{k_n}^n$. By the convexity of the function H, we then have

$$K(x, \nu^n) = \sum_{j=1}^{k_n} \eta_j^n H(x, y_n^j) \geq H(x, \overline{y}^n), \tag{2.5.2}$$

where $\overline{y}^n = \sum_{j=1}^{k_n} \eta_j^n y_n^j$. Passing to the limit as $n \to \infty$ in (2.5.2) (with the sequence $\{\overline{y}^n\}$ to be considered as required) we obtain

$$K(x, \nu^*) \geq H(x, \overline{y}), \quad x \in X, \tag{2.5.3}$$

where $\overline{y}$ is a limit point of the sequence $\{\overline{y}^n\}$. From (2.5.3) and Lemma 2.4.3, we have

$$\max_x K(x, \nu^*) \geq \max_x H(\overline{y}). \tag{2.5.4}$$

Suppose inequality (2.5.4) is strict. Then

$$v = \max_x K(x, \nu^*) > \max_x H(x, \overline{y}) \geq \min_\nu \max_x H(x, \nu) = v,$$

which is impossible. Thus, $\max_x H(x, \overline{y}) = \max_x K(x, \nu^*) = v$ and it follows from Theorem 2.3.5 that $\overline{y}$ is an optimal strategy for Player 2.

We will now demonstrate the validity of (2.5.1). Since $\overline{y} \in Y$ is an optimal strategy of Player 2, then

$$v = \max_x H(x, \overline{y}) \geq \min_y \max_x H(x, y).$$

On the other hand, the following inequality holds:

$$v = \min_\nu \max_x K(x, \nu) \leq \min_y \max_x H(x, y).$$

Comparing the latter inequalities we obtain (2.5.1).

2.5.2. Recall that the function $\varphi : Y \to R^1$, $Y \subset R_n$, Y being a convex set, is *strictly convex* if the following strict inequality holds for all $\lambda \in (0, 1)$:

$$\varphi(\lambda y_1 + (1 - \lambda) y_2) < \lambda \varphi(y_1) + (1 - \lambda)\varphi(y_2), \quad y_1, y_2 \in Y,\ y_1 \neq y_2.$$

Theorem. *Let $\Gamma = (X, Y, H)$ be a convex game with a strictly convex payoff function. Then Player 2 then has a unique optimal pure strategy.*

Proof. Let μ^* be an optimal strategy for Player 1, $\varphi(y) = K(\mu^*, y)$ and v — the value of the game. If $\overline{y}$ is a point of Player 2's optimal strategy spectrum, then the following relation holds (2.4.2):

$$K(\mu^*, \overline{y}) = v.$$

For all $y \in Y$, however, we have $K(\mu^*, y) \geq v$, and hence

$$\varphi(\overline{y}) = \min_{y \in Y} \varphi(y) = v.$$

The function $\varphi(y)$ is strictly convex since for $\lambda \in (0, 1)$ the following inequality holds:

$$\begin{aligned}
\varphi(\lambda y_1 + (1-\lambda)y_2) &= \int_X H(x, \lambda y_1 + (1-\lambda)y_2) d\mu^*(x) \\
&< \lambda \int_X H(x, y_1) d\mu^*(x) + (1-\lambda) \\
&\times \int_X H(x, y_2) d\mu^*(x) \\
&= \lambda\varphi(y_1) + (1-\lambda)\varphi(y_2). \qquad (2.5.5)
\end{aligned}$$

From (2.5.5), it follows that the function $\varphi(y)$ cannot achieve a minimum at two different points. On the other hand, the existence of a minimum point $\overline{y}$ of the function $\varphi(y)$ is guaranteed by the Theorem 2.5.1, which completes the proof.

2.5.3. The results that are counterparts of the theorems in 2.5.1 and 2.5.2 for concave and concave-convex games will be presented without proof.

Theorem. *Let* $\Gamma = (X, Y, H)$, $X \subset R^m$, $Y \subset R^n$ *be a concave game. The game value* v *equals*

$$v = \max_x \min_y H(x, y) \qquad (2.5.6)$$

and every pure strategy x^**, on which* max min (2.5.6) *is achieved, is optimal for Player 1. Moreover, if the function* $H(x, y)$ *is strictly concave with respect to* x *for every fixed* $y \in Y$*, Player 1's optimal strategy is unique.*

Theorem. *Let $\Gamma = (X, Y, H)$, $X \subset R^m$, $Y \subset R^n$ be a concave–convex game. Then the game value v is equal*

$$v = \min_y \max_x H(x, y) = \max_x \min_y H(x, y). \tag{2.5.7}$$

In the game Γ there always exists a pure strategy saddle point (x^*, y^*), where $x^* \in X$, $y^* \in Y$ are pure strategies for Players 1 and 2, on which exterior extrema in (2.5.7) *are achieved. In this case, if the function $H(x, y)$ is strictly concave (convex) with respect to a variable $x(y)$ for any fixed $y \in Y$ $(x \in X)$, then Player 1(2) has an optimal unique strategy that is pure.*

2.5.4. We will now clarify the structure of Player 1's optimal strategy in the *convex* game $\Gamma = (X, Y, H)$.

Theorem. *In the convex game $\Gamma = (X, Y, H)$, $Y \subset R^n$, Player 1 has an optimal mixed strategy μ^* with the finite spectrum composed of no more than $(n + 1)$ points of the set X.*

The proof of this result is based on the well known Helly theorem of convex sets that is presented below without proof [Rockafellar (1970), Davidov (1978)].

Theorem (Helly). *Let K be a family composed of at least $n+1$ convex set in R^n, each set from K being compact. Then, if each $n+1$ of the sets of the family K have a common point, there exists a point common to all the sets of the family K.*

Before proving the theorem, we shall prove some auxiliary statements.

Suppose the function $H(x, y)$ is continuous on the product $X \times Y$ of the compact sets $X \subset R^m, Y \subset R^n$. Denote by $X^r = X \times \ldots \times X$ the Cartesian product of r sets of X.

Consider the function $\varphi : X^r \times Y \to R^1$:

$$\varphi(x_1, \ldots, x_r, y) = \max_{1 \le i \le r} H(x_i, y).$$

Lemma. *The function $\varphi(x_1, \ldots, x_r, y)$ is continuous on $X^r \times Y$.*

Proof. The function $H(x, y)$ is continuous on the compact set $X \times Y$ and hence is uniformly continuous on it. Then for any $\epsilon > 0$ there is $\delta > 0$ such that from the inequalities $\rho_1(\overline{x}, \overline{\overline{x}}) < \delta$, $\rho_2(y_1, y_2) < \delta$ follows the inequality $|H(\overline{x}, y_1) - H(\overline{\overline{x}}, y_2)| < \epsilon$, where $\rho_1(\cdot)$, $\rho_2(\cdot)$ are distances in R^m and R^n, respectively. We have

$$|\varphi(\overline{x}_1, \ldots, \overline{x}_r, y_1) - \varphi(\overline{\overline{x}}_1, \ldots, \overline{\overline{x}}_r, y_2)|$$

$$= |\max_{1 \le i \le r} H(\overline{x}_i, y_1) - \max_{1 \le i \le r} H(\overline{\overline{x}}_i, y_2)| = |H(\overline{x}_{i_1}, y_1) - H(\overline{\overline{x}}_{i_2}, y_2)|,$$

where

$$H(\overline{x}_{i_1}, y_1) = \max_{1 \le i \le r} H(\overline{x}_i, y_1), \quad H(\overline{\overline{x}}_{i_2}, y_2) = \max_{1 \le i \le r} H(\overline{\overline{x}}_i, y_2).$$

If $\rho_1(\overline{x}_i, x_i) < \delta$ for $i = 1, \ldots, r,\ \rho_2(y_1, y_2) < \delta$ and if $H(\overline{x}_{i_1}, y_1) \ge H(\overline{x}_{i_2}, y_2)$, then

$$0 \le H(\overline{x}_{i_1}, y_1) - H(x_{i_2}, y_{i_2}) \le H(\overline{x}_{i_1}, y_1) - H(x_{i_1}, y_2) < \epsilon.$$

Similar inequalities also hold in the event that $H(\overline{x}_{i_1}, y_1) \le H(x_{i_2}, y_2)$.

Lemma. *In the convex game $\Gamma = (X, Y, H)$, $Y \subset R^n$ the game value v is*

$$v = \min_y \max_x H(x, y) = \max_{x_1, \ldots, x_{n+1}} \min_y \max_{1 \le i \le n+1} H(x_i, y), \tag{2.5.8}$$

where $y \in Y$, $x_i \in X$, $i = 1, \ldots, n+1$.

Proof. Denote

$$\Theta \triangleq \max_{x_1, \ldots, x_{n+1}} \min_y \max_{1 \le i \le n+1} H(x_i, y).$$

Since $\min_y \max_{1 \le i \le n+1} H(x_i, y) \le \min_y \max_x H(x, y) = v$ for each system of points $(x_1, \ldots, x_{n+1}) \in X^{n+1}$, then

$$\Theta \le v. \tag{2.5.9}$$

For an arbitrary fixed set of strategies $x_i \in X$, $i = 1, \ldots, n+1$, we shall consider a system of inequalities with respect to y:

$$H(x_i, y) \le \Theta, \quad y \in Y,\ i = 1, \ldots, n+1. \tag{2.5.10}$$

We shall show that system (2.5.10) has a solution.

Indeed,

$$\Theta \geq \min_y \max_{1\leq i\leq n+1} H(x_i, y) = \max_{1\leq i\leq n+1} H(x_i, \overline{y})$$
$$\geq H(x_i, \overline{y}),\ i = 1, \ldots, n+1.$$

Thus, $\overline{y}$ satisfies system (2.5.10).

Consequently, system (2.5.10) has a solution for any $x_i \in X$, $i = 1, 2, \ldots, n+1$.

Let us fix x and consider the set

$$D_x \triangleq \{y \mid H(x, y) \leq \Theta\}.$$

The function $H(x, y)$ is convex and continuous in y, and hence the set D_x is closed and convex for each x. The sets $\{D_x\}$ forms a system of convex compact sets in R^n. And since the inequalities (2.5.10) always have a solution, any collection in the $(n+1)$ sets of system $\{D_x\}$ has a nonempty intersection. Therefore, by the Helly theorem, there exists a point $y_0 \in Y$ common to all sets D_x. That is there exists a point such that

$$H(x, y_0) \leq \Theta, \tag{2.5.11}$$

for any $x \in X$. Suppose that $\Theta \neq v$. From (2.5.9) and (2.5.11), we then have

$$\Theta < v = \min_y \max_x H(x, y) \leq \max_x H(x, y_0) \leq \Theta,$$

i.e. $\Theta < \Theta$. It is this contradiction that establishes (2.5.8).

We shall now prove the theorem.

Proof. From the preceding lemma we have

$$v = \max_{1\leq i\leq n+1} \min_y \max_{1\leq i\leq n+1} H(x_i, y) = \min_y \max_{1\leq i\leq n+1} H(\overline{x}_i, y)$$
$$= \min_y \max_p \sum_{i=1}^{n+1} H(\overline{x}_i, y)\pi_i, \tag{2.5.12}$$

where $\overline{x}_1, \ldots, \overline{x}_{n+1}$ are vectors on which an exterior maximum in (2.5.8) is achieved.

$$p = (\pi_1, \ldots, \pi_{n+1}) \in R^{n+1}, \quad \pi_i \geq 0, \ \sum_{i=1}^{n+1} \pi_i = 1. \tag{2.5.13}$$

Consider the function

$$K(p, y) = \sum_{i=1}^{n+1} H(x_i, y)\pi_i, \quad y \in Y, \ p \in P,$$

where P is composed of the vectors satisfying (2.5.13). The function $K(p, y)$ is continuous in p and y, convex in y and concave in p, with the sets $Y \subset R^n$, $P \subset R^{n+1}$ compact in the corresponding Euclidean spaces. Therefore, by the Theorem 2.5.3 and from (2.5.12) we have

$$v = \min_y \max_p \sum_{i=1}^{n+1} H(\overline{x}_i, y)\pi_i = \max_p \min_y \sum_{i=1}^{n+1} H(\overline{x}_i, y)\pi_i. \tag{2.5.14}$$

From (2.5.8) and (2.5.14) follows the existence of $p^* \in P$ and $y^* \in Y$ such that for all $x \in X$ and $y \in Y$ the following inequality holds:

$$H(x, y^*) \leq v \leq \sum_{i=1}^{n+1} H(\overline{x}_i, y)\pi_i^*.$$

This completes the proof of the theorem.

Let us state the structure theorem for the optimal strategy used by Player 2 in the concave game $\Gamma = (X, Y, H)$.

Theorem. *In the concave game $\Gamma = (X, Y, H)$, $X \subset R^m$, Player 2 has an optimal mixed strategy ν^* with a finite spectrum composed of no more than $(m + 1)$ points of the set Y.*

Proof of this theorem is similar to that of the preceding theorem.

2.5.5. We shall now summarize the results established in this section.

Theorem. *Let $\Gamma = (X, Y, H)$, $X \subset R^m$, $Y \subset R^n$ be a convex game. Then the value v of the game Γ is equal*

$$v = \min_y \max_x H(x, y).$$

Player 1 has an optimal mixed strategy μ_0 with a finite spectrum composed of no more than $(n+1)$ points of the set X. However, all pure strategies y_0, on which $\min_y \max_x H(x,y)$ *is achieved, are optimal for Player 2. Furthermore, if the function $H(x,y)$ for every fixed $x \in X$ is strictly convex in y, then Player 2's optimal strategy is unique.*

We shall illustrate these results by referring to the example given below.

Example 10. Consider a special case of Example 1 (see 2.1.2). Let $S_1 = S_2 = S$ and the set S be a closed circle on a plane of radius R and centered at the point O.

The payoff function $H(x,y) = \rho(x,y)$, $x \in S$, $y \in S$, with $\rho(\cdot)$ as a distance function in R^2, is strictly convex in y, and S is a convex set. Hence, by theorem 2.5.5, the game value v is

$$v = \min_{y \in S} \max_{x \in S} \rho(x,y). \tag{2.5.15}$$

Computing min max in (2.5.15) we have that $v = R$ (see Example 8 in 2.2.6). In this case, the point $y_0 \in S$, on which a minimum of the expression $\max_{x \in S} \rho(x,y)$ is achieved, is unique and coincides with the center of the circle S (i.e. the point O). Also, this point is an optimal strategy for Player 2 (minimizer). The theorem states that Player 1 (maximizer) has an optimal mixed strategy prescribing a positive probability to no more three points of the set S. Because of the symmetry of the set S, however, Player 1's optimal mixed strategy μ_0 actually prescribes with probability 1/2 the choice of any two diametrically opposite points at the boundary of the set S. To prove the optimality of strategies μ_0, y_0, it is sufficient to establish that $K(x,y_0) \leq K(\mu_0,y_0) \leq K(\mu_0,y)$ for all $x,y \in S$, where K is the expectation of the payoff. $K(\mu_0,y_0) = R/2 + R/2 = R$. Indeed, $K(x,y_0) = \rho(0,x) \leq R$ and $K(\mu_0,y) = \rho(x_1,y)/2 + \rho(x_2,y)/2 \geq R$, where x_1 and x_2 are arbitrary diametrically opposite points at the boundary of the circle S. We have thus proved the optimality of strategies μ_0 and y_0.

2.5.6. Consider a special case of the convex game $\Gamma = (X, Y, H)$ when $X = Y = [0, 1]$, i.e. the convex game on a unit square. From the Theorem 2.5.5, it follows that Player 2 always has an optimal pure strategy $y_0 \in [0, 1]$, and Player 1 has a mixed strategy concentrated at no more than two points. In this case, the value of the game is

$$v = \min_{y \in [0,1]} \max_{x \in [0,1]} H(x, y). \tag{2.5.16}$$

The set of all essential strategies $\{x\} \subset [0, 1]$ of Player 1 is a subset of the solutions of (see 2.4.2)

$$H(x, y_0) = v, \quad x \in [0, 1], \tag{2.5.17}$$

where y_0 is an optimal strategy for Player 2. Player 1's pure strategies satisfying (2.5.17) are sometimes called *balancing strategies*. The set of all balancing strategies of Player 1 is closed and bounded, i.e. it is compact. An optimal pure strategy for Player 2 is any point $y_0 \in [0, 1]$ on which (2.5.16) is achieved.

Denote by $H'_y(x, y)$ a partial derivative of the function H with respect to y (the right-hand and left-hand derivatives mean $y = 0$ and $y = 1$ respectively).

Lemma. *If y_0 is Player 2's optimal strategy in a convex game on a unit square with the payoff function H differentiable with respect to y and $y_0 > 0$, then for Player 1 there is a balancing strategy x' for which*

$$H'_y(x', y_0) \leq 0. \tag{2.5.18}$$

If, however, $y_0 < 1$, then for Player 1 there is a balancing strategy x'' such that

$$H'_y(x'', y_0) \geq 0. \tag{2.5.19}$$

Proof. Let us prove (2.5.18). (The second part of lemma can be proved in a similar way.) Suppose the opposite is true, viz. the inequality $H'_y(\overline{x}, y_0) > 0$ holds for every balancing strategy $\overline{x}$ of Player 1, i.e. the function $H(\overline{x}, \cdot)$ is strictly increasing at point y_0.

This means that there are $\epsilon(\overline{x}) > 0$ and $\Theta(\overline{x}) > 0$, such that for $y \in [0,1]$ satisfying the inequality $\Theta(\overline{x}) > y_0 - y > 0$, the following inequality holds:

$$H(\overline{x}, y) < H(\overline{x}, y_0) - \epsilon(\overline{x}).$$

By the continuity of the function H we have that for every balancing strategy $\overline{x}$ and $\epsilon(\overline{x})/2$ there is $\delta(\overline{x}) > 0$ such that for $\Theta(\overline{x}) > y_0 - y > 0$

$$H(x, y) < H(\overline{x}, y) - \frac{\epsilon(\overline{x})}{2} < H(\overline{x}, y_0) - \frac{\epsilon(\overline{x})}{2} = H(x, y_0) - \frac{\epsilon(\overline{x})}{2},$$

for all balancing strategies x for which $|x - \overline{x}| < \delta(\overline{x})$. The set of balancing strategies is compact, and hence it can be covered by a finite number of such $\delta(\overline{x})$-neighborhoods. Let ϵ be the smallest of all corresponding numbers $\epsilon(\overline{x})$. Then we have an inequality holding for all balancing strategies x (and for all essential strategies)

$$H(x, y) < H(x, y_0) - \frac{\epsilon}{2},$$

where

$$y_0 - \min \Theta(\overline{x}) < y < y_0.$$

Let μ_0 be the optimal mixed strategy of Player 1. The last inequality is valid for all spectrum points of μ_0, thus by integrating we set

$$K(\mu_0, y) \le K(\mu_0, y_0) - \frac{\epsilon}{2} = v - \frac{\epsilon}{2},$$

which is contradiction to the optimality of the strategy μ_0.

Theorem. *Suppose that* Γ *is a convex game on a unit square with the payoff function* H *differentiable with respect to* y *for any* x, y_0 *is an optimal pure strategy of Player* 2, *and* v *is the value of the game. Then:*

1. *if* $y_0 = 1$, *then among optimal strategies of Player* 1 *there is pure strategy* x' *for which* (2.5.18) *holds;*

2. *if* $y_0 = 0$, *then among optimal strategies of Player* 1 *there is a strategy* x'' *for which* (2.5.19) *holds;*
3. *if* $0 < y_0 < 1$, *then among optimal strategies of Player* 1 *there is a strategy that is a mixture of two essential strategies* x' *and* x'' *satisfying* (2.5.18), (2.5.19) *with probabilities* α *and* $1 - \alpha$, $\alpha \in [0, 1]$. *And* α, *is a solution of the equation*

$$\alpha H'_y(x', y_0) + (1 - \alpha) H'_y(x'', y_0) = 0. \qquad (2.5.20)$$

Proof. Let $y_0 = 1$. Then, for Player 1, there is an equilibrium strategy x' for which (2.5.18) holds. Hence it follows from the convexity of the function $H(x', y)$ that it does not increase in y over the entire interval $[0, 1]$, achieving its minimum in $y = 1$. This means that

$$H(x', y_0) \leq H(x', y), \qquad (2.5.21)$$

for all $y \in [0, 1]$. On the other hand, it follows from (2.5.17) that

$$H(x, y_0) \leq H(x', y_0), \qquad (2.5.22)$$

for all $x \in [0, 1]$. The inequalities (2.5.21), (2.5.22) show that (x', y_0) is an equilibrium point.

The case $y_0 = 0$ can be examined in a similar way. We shall now discuss case 3. If $0 < y_0 < 1$, then there are two equilibrium strategies x' and x'' satisfying (2.5.18), (2.5.19), respectively.

Consider the function

$$\varphi(\beta) = \beta H'_y(x', y_0) + (1 - \beta) H'_y(x'', y_0).$$

From (2.5.18), (2.5.19) it follows that $\varphi(0) \geq 0$, $\varphi(1) \leq 0$. The function $\varphi(\beta)$ is continuous, and hence there is $\alpha \in [0, 1]$ for which $\varphi(\alpha) = 0$.

Consider a mixed strategy μ_0 of Player 1 that is to choose strategy x' with probability α and strategy x'' with probability $1 - \alpha$.

The function

$$K(\mu_0, y) = \alpha H(x', y) + (1 - \alpha) H(x'', y)$$

is convex in y. Its derivative in y at the point $y = y_0$ is

$$K'_y(\mu_0, y_0) = \alpha H'_y(x', y_0) + (1 - \alpha) H'_y(x'', y_0) = 0.$$

Consequently, the function $K(\mu_0, y)$ achieves a minimum at the point y_0. Hence considering (2.5.17), we have

$$\begin{aligned} K(\mu_0, y_0) \leq K(\mu_0, y), \ K(\mu_0, y) &= H(x, y_0) = v \\ &= \max_x H(x, y_0) \geq H(x, y_0) \end{aligned}$$

for all $x \in [0, 1]$ and $y \in [0, 1]$, which proves the optimality of strategies μ_0 and y_0.

2.5.7. Theorem 2.5.6 provides a way of finding optimal strategies which can be illustrated by referring to the following example.

Example 11. Consider a game over the unit square with the payoff function $H(x, y) = (x - y)^2$. This is a one-dimensional analogy for Example 10 except that the payoff function is taken to be the square of distance. Therefore, it would appear natural that the game value v would be $v = 1/4$, Player 1's optimal strategy be to choose with probability $1/2$ the extreme points 0 and 1 of the interval $[0, 1]$. We shall show this by employing the Theorem 2.5.6.

Note that $\partial^2 H(x, y)/\partial y^2 = 2 > 0$ so the game Γ is strictly convex, and hence Player 2 has a unique optimal strategy that is pure (Theorem 2.5.5). Let y be a fixed strategy for Player 2. Then

$$\max_x (x - y)^2 = \begin{cases} (1 - y)^2, & y \leq 1/2, \\ y^2, & y \geq 1/2. \end{cases}$$

Thus, from (2.5.16) it follows that

$$v = \min \left\{ \min_{0 \leq y \leq \frac{1}{2}} (1 - y)^2, \quad \min_{\frac{1}{2} \leq y \leq 1} y^2 \right\}.$$

Both interior minima are achieved on $y_0 = 1/2$ and is equal to $1/4$ Therefore, $v = 1/4$, and $y_0 = 1/2$ is the unique optimal strategy of Player 2.

We shall now find an optimal strategy for Player 1. To be noted here is that $0 < y_0 < 1$ $(y_0 = 1/2)$. Equation (2.5.17) now becomes $(x - 1/2)^2 = 1/4$. Hence $x_1 = 0$ and $x_2 = 1$, i.e. the extreme points of the interval $[0, 1]$ are essential for Player 1.

Compute derivatives

$$H'_y(x_1, y_0) = 1 > 0, \quad H'_y(x_2, y_0) = -1 < 0.$$

We now set up equation (2.5.20) for α. We have $2\alpha - 1 = 0$, or $\alpha = 1/2$. Thus, the optimal strategy for Player 1 is to choose pure strategies 0 and 1 with probability 1/2.

2.5.8. To conclude this section, we shall present a result that is similar to 2.5.6. for a concave game.

Theorem. *Suppose that Γ is a concave game on a unit square with the payoff function H differentiable with respect to x for any fixed y, x_0 is an optimal pure strategy of Player 1, and v is the value of the game. Then:*

1. *if $x_0 = 1$, then among optimal strategies of Player 2 there is a pure strategy y' for which the following inequality holds:*

$$H'_x(x_0, y') \geq 0; \tag{2.5.23}$$

2. *if $x_0 = 0$, then among optimal strategies of Player 2 there is a pure strategy y'', for which*

$$H'_x(x_0, y'') \leq 0; \tag{2.5.24}$$

3. *if $0 < x_0 < 1$, then among optimal strategies of Player 2 there is a strategy that is a mixture of two essential strategies y' and y'' satisfying (2.5.23), (2.5.24) with probability β and $1 - \beta$. Here the number $\beta \in [0, 1]$ is a solution of the equation*

$$\beta H'_x(x_0, y') + (1 - \beta) H'_x(x_0, y'') = 0.$$

2.6 Simultaneous Games of Pursuit

This section provides a solution to certain games of pursuit whose payoff function or the players' strategy sets are nonconvex. The results obtained in Sec. 2.5 are not applicable to such games, and hence a solution for both players is in the class of mixed strategies. The existence of solution is guaranteed by the Theorem 2.4.4.

2.6.1. *Example 12* [Petrosjan, Azamov and Satimov (1974)]. *Simultaneous game of pursuit inside the ring.* This game is a special case of Example 1 in 2.1.2 where the sets $S_1 = S_2 = S$ and S is a ring. Radii of external and internal circles of the ring S are respectively denoted by R and r, $R > r$.

We shall show that optimal strategies of Players 1 and 2 are to choose the points uniformly distributed over the internal (for Player 2) and external (for Player 1) circles of the ring S. Denote these strategies by μ^* (for Player 1) and ν^* (for Player 2). When using this strategies, the mean payoff (distance) is equal to

$$K(\mu^*, \nu^*) = \frac{1}{4\pi^2} \int_0^{2\pi} \int_0^{2\pi} \sqrt{R^2 + r^2 - 2Rr\cos(\varphi - \psi)}d\varphi d\psi$$

$$= \frac{1}{2\pi} \int_0^{2\pi} \sqrt{R^2 + r^2 - 2Rr\cos\xi}d\xi \triangleq \Phi(r, R), \quad (2.6.1)$$

where ψ and φ are polar angles of pure strategies for Players 1 and 2, respectively. If Player 1 chooses the point x with polar coordinates ρ, ψ then the distance expectation (Player 2 uses strategy ν^*) is

$$K(x, \nu^*) = \Phi(r, \rho) = \frac{1}{2\pi} \int_0^{2\pi} \sqrt{r^2 + \rho^2 - 2r\rho\cos\xi}d\xi.$$

For $r \le \rho \le R$, the function $\varphi(\rho) = \rho^2 + r^2 - 2\rho r\cos\xi$ is monotonically increasing. In particular, $\varphi(\rho) \le \varphi(R)$ for $r \le \rho \le R$. Hence we have $\Phi(r, \rho) \le \Phi(r, R)$. Therefore, for any strategy of Player 1 the distance expectation is at most $\Phi(r, R)$.

We shall now consider the situation (μ^*, y) at which $y \in S$, and ρ and φ are polar coordinates of the point y. We have

$$K(\mu^*, y) = \Phi(\rho, R)$$
$$= \frac{1}{2\pi}\int_0^{2\pi} \sqrt{R^2 + \rho^2 - 2R\rho\cos\xi}\, d\xi(\rho), \quad r \leq \rho \leq R.$$

Let us fix R and consider the function $\Phi(\rho, R)$ on the interval $0 \leq \rho \leq R$. Differentiation with respect to ρ shows that

$$\frac{\partial\Phi(0, R)}{\partial\rho} = 0, \quad \frac{\partial^2\Phi(\rho, R)}{\partial\rho^2} > 0, \quad 0 \leq \rho \leq R.$$

Therefore, the function $\Phi(\rho, R)$ is monotonically increasing in ρ, and hence $\Phi(r, R) \leq \Phi(\rho, R)$

$$K(x, \nu^*) \leq K(\mu^*, \nu^*) \leq K(\mu^*, y),$$

for all $x, y \in S$. We have thus proved the optimality of strategies μ^* and ν^*. Here the game value v is $v = K(\mu^*, \nu^*)$, where $K(\mu^*, \nu^*)$ is determined by (2.6.1). Specifically, if S is a circle of radius R (the case $r = R$), then the value of the game is $4R/\pi$.

2.6.2. *Example 13.* Consider a simultaneous game in which Player 2 chooses a pair of points $y = \{y_1, y_2\}$, where $y_1 \in S$, $y_2 \in S$, and Player 1 having no information about Player 2's choice chooses a point $x \in S$. The payoff to Player 1 is assumed to be $\min_{i=1,2} \rho^2(x, y_i)$. We shall provide a solution for the case where the set S is a circle of radius R centered at the origin of coordinates (the point O): $S = S(O, R)$.

Consider the function $\Phi(r, \rho) = r^2 + \rho^2 - 4r\rho/\pi$, where r and ρ take values from the interval r, $\rho \in [0, R]$. We shall establish properties of the function $\Phi(r, \rho)$.

Lemma 1. *The function $\Phi(r, R)$ (as a function of the variable r) is strictly convex and achieves the absolute minimum at the unique point $r_0 = 2R/\pi$.*

Proof. We have $\partial^2\Phi/\partial r^2 = 2 > 0$. Hence the function $\Phi(r,\rho)$, $r \in [0,R]$ is strictly convex and the derivative

$$\frac{\partial\Phi(r,R)}{\partial r} = 2r - \frac{4R}{\pi} \tag{2.6.2}$$

is strictly monotone, it is evident that the function (2.6.2) is equal to zero at the unique point $r_0 = 2R/\pi$. By the strict convexity of $\Phi(r,R)$, the point r_0 is the unique point of an absolute minimum. This completes the proof of the lemma.

Lemma 2. *The function $\Phi(r_0,\rho)$ is strictly convex in ρ and achieves an absolute maximum at the point $\rho_0 = R$.*

Proof. By the symmetry the function $\Phi(r,\rho)$ is strictly convex in ρ. Therefore, the maximum of this function is achieved at one of the points 0 or R. We have

$$\Phi(r_0,R) - \Phi(r_0,0) = r_0^2 + R^2 - 4r_0R/\pi - r_0^2 = R^2 - \frac{4}{\pi}\left(\frac{2R}{\pi}\right)R = \frac{R^2(\pi^2-8)}{\pi^2} > 0.$$

This completes the proof of the lemma.

From Lemmas 1 and 2 it follows that the pair (r_0,R) is a saddle point of the function Φ:

$$\Phi(r_0,\rho) \le \Phi(r_0,R) \le \Phi(r,R).$$

Theorem. *An optimal mixed strategy for Player 2 is to choose a point y_1 uniformly distributed over the circle $S(O,r_0)$ with center at the point O and radius r_0 ($y_1 = -y_2$). An optimal mixed strategy for Player 1 is to choose a point x uniformly distributed over the circle $S(O,R)$. The value of the game is $\Phi(r_0,R)$.*

Proof. The strategies specified by theorem are denoted by μ^* and ν^* for the players 1 and 2, respectively. Suppose Player 1 uses strategy μ^*, and Player 2 uses an arbitrary pure strategy $y = \{y_1,y_2\}$, $y_i = (r_i\cos\varphi_i, r_i\sin\varphi_i)$, $i = 1,2$. First we consider the case where $y_1 = y_2$.

Denote by r a number $r_1 + r_2$, and by φ an angle $\varphi_1 = \varphi_2$. The payoff to Player 1 is

$$K(\mu^*, y) = \frac{1}{2\pi} \int_0^{2\pi} [R^2 + r^2 - 2Rr \cos(\psi - \varphi)] d\psi$$
$$= R^2 + r^2 \geq R^2 + r^2 - \frac{4}{\pi}(Rr) = \Phi(r, R). \quad (2.6.3)$$

Then, by Lemma 1, we have $K(\mu^*, y) \geq \Phi(r_0, R)$.

In what follows we assume $y_1 \neq y_2$. The polar coordinate system is introduced on a plane as follows. We take the origin of coordinates to be the point O, and the polar axis to be the ray emanating from the point O perpendicular to the chord AB (the set of the points of the circle $S(O, R)$ that are equidistant from y_1 and y_2). For simplicity assume that, for the new coordinate system, the points y_i have the same coordinates $(r_i \cos \varphi_i,\ r_i \sin \varphi_i)$. Then (Fig. 2.5) the payoff to Player 1 is

$$K(\mu^*, y) = \frac{1}{2\pi} \int_0^{2\pi} \min_{i=1,2} [R^2 + r_i^2 - 2Rr_i \cos(\psi - \varphi_i)] d\psi$$
$$= \frac{1}{2\pi} \int_{-\beta}^{\beta} [R^2 + r_2^2 - 2Rr_2 \cos(\psi - \varphi_2)] d\psi$$
$$+ \frac{1}{2\pi} \int_{\beta}^{2\pi-\beta} [R^2 + r_1^2 - 2Rr_1 \cos(\psi - \varphi_1)] d\psi.$$

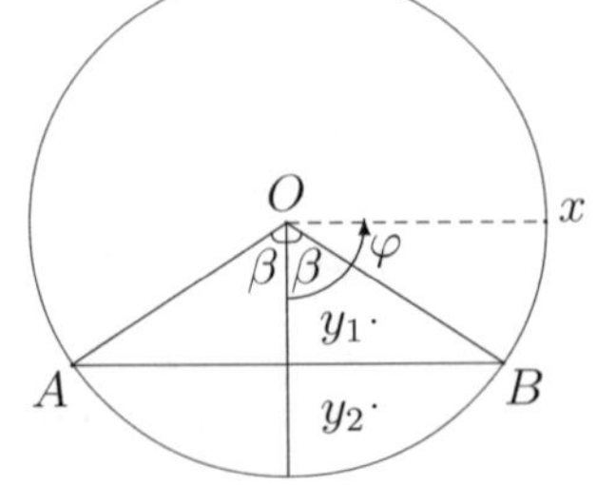

Figure 2.5 Simultaneous pursuit game

Let

$$F_1(\varphi) \triangleq [(R^2 + r_2^2)\beta - 2Rr_2 \sin\beta \cos\varphi]/\pi, \ -\beta \leq \varphi \leq \beta,$$

$$F_2(\varphi) \triangleq [(R^2 + r_1^2)(\pi - \beta) + 2Rr_1 \sin\beta \cos\varphi]/\pi, \ \beta \leq \varphi \leq 2\pi - \beta.$$

Stationary points of the functions F_1 and F_2 are respectively 0 and π since $0 < \beta < \pi/2$ and $F_1'(\varphi) = \frac{2}{\pi} Rr_2 \sin\beta \sin\varphi$, $F_2'(\varphi) = -\frac{2}{\pi} Rr_1 \sin\beta \sin\varphi$, with 0 and π as the points of the absolute minimum of the functions F_1 and F_2, ($F_1'(\varphi) < 0$ for $\varphi \in (-\beta, 0)$, $F_1'(\varphi) > 0$ for $\varphi \in (0, \beta)$; $F_2'(\varphi) < 0$ for $\varphi \in (\beta, \pi), F_2'(\varphi) > 0$ for $\varphi \in (\pi, 2\pi - \beta)$). Consequently,

$$\begin{aligned} K(\mu^*, y) &= F_1(\varphi_2) + F_2(\varphi_1) \geq F_1(0) + F_2(\pi) \\ &= \frac{1}{2\pi} \int_{-\beta}^{\beta} (R^2 + r_2^2 - 2Rr_2 \cos\psi) d\psi \\ &\quad + \frac{1}{2\pi} \int_{\beta}^{2\pi - \beta} (R^2 + r_1^2 - 2Rr_1 \cos(\psi - \pi)) d\psi, \qquad (2.6.4) \end{aligned}$$

i.e. payoff of Player 1, with Player 2 using a strategy $y_1 = \{-r_1, 0\}$, $y_2 = \{r_2, 0\}$, is less than that to be obtained by using a strategy

$$\overline{y}_i = |r_i \cos\varphi_i, r_i \sin\varphi_i|, \quad i = 1, 2.$$

Suppose the points y_1 and y_2 are lying on the diameter of the circle $S(O, R)$ and the distance between them is $2\overline{r}$. Denote by 2α the central angle basing on the arc spanned by the chord AB (Fig. 2.6). Suppose that $y_1 = \{R\cos\alpha - \overline{r}, 0\}$, $y_2 = \{R\cos\alpha + \overline{r}, 0\}$. Then the payoff to Player 1 is

$$\begin{aligned} \psi(\alpha, \overline{r}) &\triangleq \frac{1}{2\pi} \int_{-\alpha}^{\alpha} [(R\cos\psi - R\cos\alpha - \overline{r})^2 + R^2 \sin^2\psi] d\psi \\ &\quad + \frac{1}{2\pi} \int_{\alpha}^{2\pi - \alpha} [(R\cos\psi - R\cos\alpha + \overline{r})^2 + R^2 \sin^2\psi] d\psi \end{aligned}$$

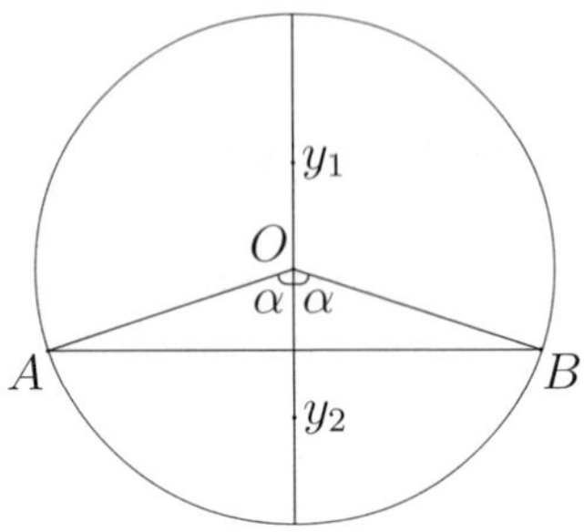

Figure 2.6 Points y_1 and y_2 are lying on the diameter of the circle

$$= \frac{1}{2\pi}\int_{-\alpha}^{\alpha}[R^2 - 2R\cos\psi(R\cos\alpha + \overline{r}) + (R\cos\alpha + \overline{r})^2]d\psi$$

$$+\frac{1}{2\pi}\int_{\alpha}^{2\pi-\alpha}[R^2 - 2R\cos\psi(R\cos\alpha - \overline{r}) + (R\cos\alpha - \overline{r})^2]d\psi$$

$$= \frac{1}{\pi}\{[R^2 + (R\cos\alpha + \overline{r})^2]\alpha - 2R\sin\alpha(R\cos\alpha + \overline{r})$$

$$+[R^2 + (R\cos\alpha - \overline{r})^2](\pi - \alpha) + 2R\sin\alpha(R\cos\alpha - \overline{r})\}.$$

We shall show that, for a fixed $\overline{r}$, the function $\psi(\alpha, \overline{r})$ achieves a minimum in α when $\alpha = \pi/2$. Elementary computations shows $\partial\psi/\partial\alpha = \{2R\sin\alpha[(\pi - 2\alpha)\overline{r} - \pi R\cos\alpha]\}/\pi$, and hence for sufficiently small values of α we have $\partial\psi(\alpha, \overline{r})/\partial\alpha < 0$ since $\sin\alpha > 0$, $\overline{r}(\pi - 2\alpha) - \pi R\cos\alpha < 0$ (in the limiting case $\overline{r}\pi - \pi R < 0$). At the same time $\partial\psi(\pi/2, r)/\partial\alpha = 0$.

For every fixed r the function $\partial\psi(\alpha, r)/\partial\alpha$ has no zeros except for $\alpha = \pi/2$. Suppose the opposite is true. Let α_1 be a zero of this function in the interval $(0, \pi/2)$. Then the function $G(\alpha) = (\pi - 2\alpha)r - \pi R\cos\alpha$ vanishes at $\alpha = \alpha_1$. Thus, $G(\alpha_1) = G(\pi/2) = 0$.

It is evident that $G(\alpha) > 0$ for all $\alpha \in (\alpha_1, \pi/2)$. This is a contradiction to the convexity of the function $G(\alpha)$ ($G''(\alpha) = \pi R\cos \alpha > 0$). Thus $\partial\psi(\alpha, r)/\partial\alpha < 0$ for $\alpha \in (0, \pi/2)$ and $\partial\psi(\pi/2, r)/\partial\alpha = 0$. Consequently, the function $\psi(\alpha, r)$ achieves an absolute minimum in α when $\alpha = \pi/2 : \psi(\alpha, r) \geq \psi(\pi/2, r)$. The implication here is that

$$K(\mu^*, y) = \psi(\alpha, r) \geq \psi(\pi/2, r) = \Phi(r, R) \geq \Phi(r_0, R). \tag{2.6.5}$$

From (2.6.3)–(2.6.5) it follows that for any pure strategy $y = \{y_1, y_2\}$ the following inequality holds:

$$K(\mu^*, y) \geq \Phi(r_0, R). \tag{2.6.6}$$

Suppose Player 2 uses strategy ν^* and Player 1 uses an arbitrary pure strategy $x = \{\rho \cos \psi, \rho \sin \psi\}$. Then Player 1 receives the payoff

$$\begin{aligned} K(x, \nu^*) &= \frac{1}{2\pi} \int_0^{2\pi} \min[\rho^2 + r_0^2 - 2\rho r_0 \cos(\psi - \varphi),\ \rho^2 \\ &\quad + r_0^2 + 2\rho r_0 \cos(\psi - \varphi)] d\varphi \\ &= \frac{1}{2\pi} \int_0^{2\pi} \min(\rho^2 + r_0^2 - 2\rho r_0 \cos \xi,\ \rho^2 \\ &\quad + r_0^2 + 2\rho r_0 \cos \xi) d\xi = \Phi(r_0, \rho) \end{aligned}$$

and by Lemma 2 we have

$$K(x, \nu^*) = \Phi(r_0, \rho) \leq \Phi(r_0, R). \tag{2.6.7}$$

From (2.6.6) and (2.6.7), we have that μ^* and ν^* are optimal strategies of the players, and $\Phi(r_0, R)$ is the value of the game. This completes the proof of the theorem.

2.6.3. *Example 14.* Suppose Player 2 chooses a set of m points $y = \{y_1, \ldots, y_m\}$, where $y_i \in S$, $i = 1, \ldots, m$, and Player 1 simultaneously chooses a point $x \in S$. The payoff to Player 1 is assumed to be $\min_{i=1,\ldots,m} \rho(x, y_i)$. We shall solve the game for the case where the set S coincides with the interval $[-1, 1]$.

Theorem. *Player 2's optimal mixed strategy ν^* is the equiprobable choice of two sets of m points*

$$\left\{-1 + \frac{4i}{2m-1},\ i = 0, 1, \ldots, m-1\right\},$$

$$\left\{1 - \frac{4i}{2m-1},\ i = 0, 1, \ldots, m-1\right\}.$$

The optimal strategy μ^* for Player 1 is to choose the points

$$\left\{\frac{2m-2i-1}{2m-1},\ i = 0, 1, \ldots, 2m-1\right\}$$

with probabilities $1/(2m)$. The value of the game is $1/(2m-1)$.

Proof. Suppose μ^* and ν^* are the respective mixed strategies for Players 1 and 2 whose optimality is to be proved. Introduce the following notation:

$$l_i = \left[\frac{2m-2i-1}{2m-1}, \frac{2m-2i+1}{2m-1}\right], \quad i = 1, 2, \ldots, 2m-1.$$

First show that $K(x, \nu^*) \le \frac{1}{(2m-1)}$ for all $x \in [-1, 1]$. In fact, for all $x \in l_i$ we have

$$\begin{aligned} K(x, \nu^*) &= \frac{1}{2} \min_i \left| \frac{2m-4i-1}{2m-1} - x \right| \\ &\quad + \frac{1}{2} \min_i \left| \frac{-2m+4i+1}{2m-1} - x \right| \\ &= \frac{1}{2}\left(x - \frac{2m-2j-1}{2m-1}\right) + \frac{1}{2}\left(\frac{2m-2j+1}{2m-1} - x\right) \\ &= \frac{1}{2m-1}. \end{aligned} \tag{2.6.8}$$

Now suppose Player 1 chooses a mixed strategy μ^* and Player 2 chooses an arbitrary pure strategy $y = \{y_1, \ldots, y_m\}$.

Denote

$$x_j = \frac{2m-2j-1}{2m-1}, \quad j = 0, 1, \ldots, 2m-1.$$

Then

$$K(\mu^*, y) = \sum_{j=0}^{2m-1} \min_{1\le i\le m} \rho(x_j, y_i)\frac{1}{2m}$$

$$= \frac{1}{2m}\sum_{j=1}^{m}[\min_{1\le i\le m} \rho(x_{2j-1}, y_i) + \min_{1\le i\le m} \rho(x_{2j-2}, y_i)]$$

$$\ge \frac{1}{2m} m \frac{2}{2m-1} = \frac{1}{2m-1}.$$

The statement of the theorem follows from inequalities (2.6.8), (2.6.9).

2.7 One Class of Games with a Discontinuous Payoff Function

The existence of the game value in mixed strategies cannot be guaranteed for the games whose payoff functions are discontinuous (see, e.g. Example, 2.4.12). But sometimes in the case of discontinuity it is possible to obtain optimal strategies and the value of the game. Empirical assumptions of the form of players' optimal strategies can also assist in finding a solution.

2.7.1. This section deals with games of timing or duel type games (see Examples 4, 5 in 2.1.2). The main feature of this class of games on a square is the discontinuity of the payoff function $H(x, y)$ along the diagonal $x = y$.

We shall consider the game on a unit square with the payoff function [Karlin (1959)]

$$H(x, y) = \begin{cases} \psi(x, y), & \text{if } x < y, \\ \varphi(x), & \text{if } x = y, \\ \theta(x, y), & \text{if } x > y, \end{cases} \tag{2.7.1}$$

where $\psi(x, y)$ is defined and continuous on the set $0 \leq x \leq y \leq 1$, the function φ is continuous on $[0, 1]$, and $\theta(x, y)$ is defined and continuous on the set $0 \leq y \leq x \leq 1$. Suppose the game $\Gamma = (X, Y, H)$, where $X = Y = [0, 1]$, with H given (2.7.1), has optimal mixed strategies μ^* and ν^* for Players 1 and 2, respectively. Moreover, the optimal mixed strategies μ^*, ν^* are assumed to be the probability distributions which have continuous densities $f^*(x)$ and $g^*(x)$, respectively.

In what follows in 2.7 we shall consider as strategies of players the corresponding f and g densities.

Let f be a strategy for Player 1. For $y \in [0, 1]$ we have

$$K(f, y) = \int_0^y \psi(x, y) f(x) dx + \int_y^1 \theta(x, y) f(x) dx. \tag{2.7.2}$$

Suppose that f and g are optimal strategies for Players 1 and 2, respectively. Then for any point y_0, at which

$$g(y_0) > 0 \tag{2.7.3}$$

(that is the point of the strategy spectrum g), the following equation holds:

$$K(f, y_0) = v, \tag{2.7.4}$$

where v is the value of the game. But (2.7.3) is strict and hence there is $\delta > 0$ such that inequality (2.7.3) holds for all $y : |y - y_0| < \delta$. Thus, inequality (2.7.4) also holds for these y, i.e. the equality $K(f, y) = v$ is satisfied. This means that

$$\frac{\partial K(f, y)}{\partial y} = 0. \tag{2.7.5}$$

Equation (2.7.5) can be rewritten as

$$[\theta(y, y) - \psi(y, y)] f(y) = \int_0^y \psi_y(x, y) f(x) dx + \int_y^1 \theta_y(x, y) f(x) dx, \quad y \in S(y_0, \delta). \tag{2.7.6}$$

We have thus obtained the integral equation (2.7.6) for the required strategy f.

2.7.2. *Example 15.* Consider the noisy duel formulated in Example 5, 2.1.2. The payoff function $H(x, y)$ in the game is of the form (2.7.1), where

$$\psi(x, y) = x - y + xy, \tag{2.7.7}$$

$$\theta(x, y) = x - y - xy, \tag{2.7.8}$$

$$\varphi(x) = 0. \tag{2.7.9}$$

Note that this game is symmetric, because $H(x, y) = -H(y, x)$ (a skew-symmetric payoff function). Therefore, analysis (similar to the analysis given in 1.9.2) shows that the game value v, if any, is zero, and players' optimal strategies, if any, must coincide.

We have: $\psi_y(x, y) = -1 + x$, $\theta_y(x, y) = -1 - x$, $\theta(y, y) - \psi(y, y) = -2y^2$ and the integral equation (2.7.6) becomes

$$-2y^2 f(y) = \int_0^y (x-1)f(x)dx - \int_y^1 (x+1)f(x)dx. \tag{2.7.10}$$

We seek a strategy f in the class of the differentiable distribution densities taking positive values in the interval $(\alpha, \beta) \subset [0, 1]$, with the interval (α, β) taken as a strategy spectrum of f. Then (2.7.10) can be written as follows:

$$-2y^2 f(y) = \int_\alpha^y (x-1)f(x)dx - \int_y^\beta (x+1)f(x)dx. \tag{2.7.11}$$

Differentiating both sides in (2.7.11) with respect to y we obtain a differential equation of the form

$$-4yf - 2y^2 f' = (y-1)f + (y+1)f$$

or

$$yf' = -3f \quad (y \neq 0). \tag{2.7.12}$$

Integrating equation (2.7.12) yields

$$f(y) = \gamma y^{-3}, \tag{2.7.13}$$

where γ is a constant.

It remains to find α, β, and γ. Recall that the players' optimal strategies in the game are the same. From the assumption on a strategy spectrum of f it follows that

$$K(f, y) = 0, \tag{2.7.14}$$

for all $y \in (\alpha, \beta)$.

Let $\beta < 1$. Since the function $K(f, y)$ is continuous in y, from (2.7.14) we have $K(f, \beta) = 0$. Consequently,

$$\int_\alpha^\beta (x - \beta + \beta x) f(x) dx = 0. \tag{2.7.15}$$

But in the case $\beta < 1$ it follows from (2.7.15) that

$$K(f, 1) = \int_\alpha^\beta (x - 1 + x) f(x) dx < 0$$

which contradicts the optimality of the strategy f. Thus $\beta = 1$ and $K(f, 1) = 0$. By substituting (2.7.13) into (2.7.15), with $\beta = 1$, we obtain

$$\gamma \int_\alpha^1 \frac{2x - 1}{x^3} dx = 0, \quad \gamma \neq 0.$$

Hence it follows that

$$3\alpha^2 - 4\alpha + 1 = 0. \tag{2.7.16}$$

Solving equation (2.7.16) we find two roots, $\alpha = 1$ and $\alpha = 1/3$, the first root lies out of open interval $(0, 1)$. Consequently, $\alpha = 1/3$. The coefficient γ is found from the property of density function $f(y)$

$$\int_{1/3}^1 f(y) dy = \gamma \int_{1/3}^1 y^{-3} dy = 1,$$

whence follows $\gamma = 1/4$.

We have thus obtained the solution of the game given in Example 5, 2.1.2: the value of the game is $v = 0$, and the players' optimal strategies f and g (as distribution densities) are equal to one another and are of the form

$$f(x) = \begin{cases} 0, & x < 1/3, \\ 1/(4x^3), & x > 1/3. \end{cases}$$

2.7.3. *Example 16.* Find a solution of a "noisy duel" game (see Example 4, 2.1.2) for the accuracy functions $p_1(x) = x$ and $p_2(y) = y$. The payoff function $H(x, y)$ in the game is of the form (2.7.1), where

$$\psi(x, y) = 2x - 1, \tag{2.7.17}$$

$$\theta(x, y) = 1 - 2y, \tag{2.7.18}$$

$$\varphi(x) = 0. \tag{2.7.19}$$

The game is symmetric, hence $v = 0$, and the players' optimal strategies coincide. Here both players have an optimal pure strategy $x^* = y^* = 1/2$. In fact, $H(1/2, y) = \theta(1/2, y) = 1 - 2y > 0$ if $y < 1/2$, $H(1/2, y) = \varphi(1/2) = 0$ if $y = 1/2$, $H(1/2, y) = \psi(1/2, y) = 0$ if $y > 1/2$.

As we see, the solution for the duelists is to fire their bullets simultaneously after having advanced half the distance to the barrier.

The class of games of timing has been studied in Davidov (1978), Karlin (1959), Vorobjev (1984).

2.8 Infinite Simultaneous Search Games

This section provides a solution of the games of search with the infinite number of strategies formulated in 2.1.2. It is of interest that, in the first of the games considered, both players have optimal mixed strategies with a finite spectrum.

2.8.1. *Example 17. Search on a closed interval* [Diubin and Suzdal (1981)]. Consider the game of search on closed interval (see Example 2 in 2.1.1) which is modelled by the game on a unit square with the payoff function $H(x,y)$ of the form

$$H(x,y) = \begin{cases} 1, & \text{if } |x-y| \le l, \ l \in (0,1), \\ 0, & \text{otherwise.} \end{cases} \tag{2.8.1}$$

Note that for $l \ge 1/2$ Player 1 has an optimal pure strategy $x^* = 1/2$ and the value of the game is 1; in this case $H(x^*, y) = H(1/2, y) = 1$, since $|y - 1/2| \le 1/2 \le l$ for all $y \in [0,1]$. Let $l < 1/2$. Note that the strategy $\overline{x} = l$ dominates all pure strategies $x < l$, and the strategy $\overline{x} = 1 - l$ dominates all strategies $x > 1 - l$. Indeed,

$$H(\overline{x}, y) = H(l, y) = \begin{cases} 1, & y \in [0, 2l], \\ 0, & \text{otherwise,} \end{cases}$$

and if $x < l$, then

$$H(x,y) = \begin{cases} 1, & y \in [0, l + x], \\ 0, & \text{otherwise.} \end{cases}$$

Thus, with $x < l : H(x,y) \le H(l,y)$ for all $y \in [0,1]$. Similarly, we have

$$H(x,y) = H(1-l, y) = \begin{cases} 1, & y \in [1-2l, 1], \\ 0, & \text{otherwise,} \end{cases}$$

and if $x \in [1-l, 1]$, then

$$H(x,y) = \begin{cases} 1, & y \in [x-l, 1], \\ 0, & \text{otherwise.} \end{cases}$$

Thus, with $x \in [1-l, 1], \ H(x,y) \le H(1-l, y)$ for all $y \in [0,1]$.

Consider the following mixed strategy μ^* of Player 1. Let $l = x_1 < x_2 < \ldots < x_m = 1 - l$ be the points for which the distance between any pair of adjacent points does not exceed $2l$. Strategy μ^* selects each of these points with equal probabilities $1/m$. Evidently, in this case any point $y \in [0,1]$ falls within l-neighborhood of at least one point x_k. Hence

$$K(\mu^*, y) \geq 1/m. \tag{2.8.2}$$

Now let ν^* be a strategy of Player 2 that is the equiprobable choice of points $0 = y_1 < y_2 < \ldots < y_n = 1$, the distance between a pair of adjacent points exceeding $2l$. Then there apparently exists at most one point y_k whose l-neighborhood contains the point x. Consequently,

$$K(x, \nu^*) \leq 1/n. \tag{2.8.3}$$

If strategies μ^*, ν^* were constructed so that $m = n$, the quantity $1/n$ would be the value of the game with strategies μ^*, ν^* as the players' optimal strategies.

It turns out that such strategies can actually be constructed. To do this, it suffices to take

$$m = n = \begin{cases} 1/(2l), & \text{if } 1/(2l) \text{ is an integer,} \\ [1/(2l)] + 1, & \text{otherwise.} \end{cases} \tag{2.8.4}$$

Here $[a]$ is the integer part of the number a. The points

$$x_i = l + \frac{1 - 2l}{n - 1}(i - 1), \quad i = 1, 2, \ldots, n, \tag{2.8.5}$$

are spaced at most $2l$, and the distance between adjacent points

$$y_j = \frac{j - 1}{n - 1}, \quad j = 1, 2, \ldots, n, \tag{2.8.6}$$

exceeds $2l$. Thus, $1/n$ is the value of the game, and the optimal strategies μ^*, ν^* are the equiprobable mixtures of pure strategies determined by (2.8.5), (2.8.6).

2.8.2. *Example 18.* Consider an extension of the preceding problem to the case where Player 1 (Searcher) chooses a system of s points $x_1, \ldots, x_s$, $x_i \in [0, 1]$, $i = 1, \ldots, s$, and Player 2 (Hider) chooses independently and simultaneously with Player 1 a point $y \in [0, 1]$. Player 2 is considered to be discovered if there is $j \in \{1, \ldots, s\}$ such that

$|y - x_j| \le l$, $l > 0$. Accordingly, the payoff function (the payoff to Player 1) is defined as follows:

$$H(x_1, \dots, x_s, y) = \begin{cases} 1, & \min_j |y - x_j| \le l, \\ 0, & \text{otherwise}. \end{cases} \tag{2.8.7}$$

Suppose Player 1 places the points $x_1, \dots, x_s$ at the points $\overline{x}_i = l + (1-2l)(i-1)/(n-1)$, $1 \le i \le n$ that are the points of the strategy spectrum μ^* from the preceding example. Evidently, arrangement of two points x_{j_1}, x_{j_2} at one point of the interval $[0,1]$ (i.e. selection of coincident points) provides no advantage. Let μ_s^* be Player 1's strategy selecting equiprobably any s-collections of different points $\{\overline{x}_i\}$. If $s \ge n$, then, by placing a point x_j at each of the points $\overline{x}_i$, Player 1 covers the entire interval $[0,1]$ with the segments of length $2l$ centered at points $\overline{x}_i$ and thus ensures that for any point $y \in [0,1]$ there is $\min_j |x_j - y| \le l$, i.e. in this case the value of the game is 1. Therefore, we assume that $s < n$. The number of all possible distinct selections of s-collections of points from the set $\{\overline{x}_i\}$ is C_n^s. We have

$$K(\mu_s^*, y) = \sum H(\overline{x}_{i_1}, \dots, \overline{x}_{i_s}, y)(\frac{1}{C_n^s}) \ge \frac{C_{n-1}^{s-1}}{C_n^s} = \frac{s}{n}.$$

In fact, the point y is discovered if it falls within l-neighborhood of at least one of the points $\{\overline{x}_i\}$ selected by strategy μ_s^*. In order for this to occur, Player 1 needs to select the point $\overline{x}_i$ from l-neighborhood of the point y. The number of collections satisfying this requirement is at least C_{n-1}^{s-1}.

We now suppose that Player 2 uses strategy ν^* from the preceding example and Player 1 uses an arbitrary pure strategy $x = (x_1, \dots, x_s)$. Then

$$K(x_1, \dots, x_s, \nu^*) = \sum_{j=1}^{n} H(x_1, \dots, x_s, y_j)\frac{1}{n} \le \frac{s}{n}.$$

Thus, the value of the game is s/n, and μ_s^*, ν^* are the players' optimal strategies. The value of the game is linearly dependent on the number of points to be chosen by the Searcher.

2.8.3. *Example 19. Search on a sphere.* Consider the game of search on a sphere (see Example 3 in 2.1.2). The payoff function $H(x, y)$ is

$$H(x,y) = \begin{cases} 1, & y \in M_x, \\ 0, & \text{otherwise}, \end{cases} \tag{2.8.8}$$

where $x = (x_1, \ldots, x_s)$ is a collection of s points on a sphere C and $M_x = \cup_{j=1}^s S(x_j, r)$; $S(x_j, r)$ is the r-spherical neighborhood (cup) of the point x_j. The set of mixed strategies for Player 1 is the family of probability measures determined on the Cartesian product of s spheres $C \times C \times \ldots \times C \triangleq \Omega$, i.e. on $\Omega = C^s$.

Define the set of mixed strategies for Player 2 as the family of probability measures $\{\nu\}$ determined on the sphere C.

Consider a specific pair of strategies (μ^*, ν^*). We choose a uniform measure on the sphere C to be the strategy ν^*, i.e. we require that

$$\int_A d\nu^* = \frac{L(A)}{4\pi R^2}, \tag{2.8.9}$$

where $L(A)$ is Lebesgue measure (area) of the set A.

Parameters of the game, s, r, and R, are taken such as to permit selection of the system of points $x = (x_1, x_2, \ldots, x_s)$ satisfying condition

$$L(M_x) = \sum_{j=1}^{s} L(S(x_j, r)) \tag{2.8.10}$$

(spherical segments $S(x_j, r)$ do not intersect).

Let us fix a figure M_x on some sphere C'. The mixed strategy μ^* is then generated by throwing at random this figure M_x onto sphere C. To do this, we set in the figure M_x an interior point z whereto rigidly connected are two noncollinear vectors a, b (with an angle $\varphi > 0$ between them) lying in a tangent plane to M_x at point z.

Point z is "thrown" onto sphere C in accordance with uniform distribution (i.e. density $1/(4\pi R^2)$). Suppose this results in realization of point $z' \in C$. Figure M_x with the vectors set thereon, is transferred to sphere C in a parallel way so that the points z and z' coincide. Thus, vectors a, b are lying in a tangent plane to sphere C at point z'.

An angle φ' is then chosen in accordance with uniform distribution in the interval $[0, 2\pi]$, and vector b lying in a tangent plane is turned clockwise together with its associated figure M_x through an angle φ'. This results in the transition of figure M_x and vector b to a new position on sphere C. Random positioning of the set M_x on a sphere in accordance with this two-stage procedure described, generates a random choice of the points $x'_1, x'_2, \ldots, x'_s \in C$ whereas the centers $x_1, \ldots, x_s$ of the spherical neighborhoods $S(x_j, r)$ making up the set M_s are located.

Measure μ^* is so constructed that it is invariant, i.e. the probability of covering the set M_x of any point $y \in C$ is independent of y. Indeed, find the probability of this event. Let $\overline{\Omega} = \{\omega\}$ be the space of all possible positions of M_x on sphere C. Then the average area covered on sphere C by throwing the set M_x thereon (area expectation) is equal to $L(M_x)$. At the same time

$$L(M_x) = \int_{\overline{\Omega}} \int_C J(y, \omega) dy d\mu^*, \tag{2.8.11}$$

where $J(y, \omega)$ is the characteristic function of the set on sphere C covered by the domain M_x. By Fubini theorem, we have

$$\int_{\overline{\Omega}} \int_C J(y, \omega) dy d\mu^* = \int_C \int_{\overline{\Omega}} J(y, \omega) d\mu^* dy. \tag{2.8.12}$$

By the invariance of measure μ^*, however, the integral $\int_{\overline{\Omega}} J(y, \omega) d\mu^*$, which coincides with the probability of covering the point y by the set M_x is independent of y and equals p. Then, from (2.8.11), (2.8.12), we have

$$p = \frac{L(M_x)}{4\pi R^2} = \frac{\sum_{j=1}^{s} L(S(x_j, r))}{4\pi R^2}. \tag{2.8.13}$$

Denote by $K(\mu, \nu)$ the payoff expectation when using mixed strategies $\mu \in \{\mu\}$ and $\nu \in \{\nu\}$. If one of the players uses a pure strategy, then

$$K(x, \nu) = \int_C H(x, y) d\nu = \int_{M_x} d\nu = \Pr(y \in M_x),$$

$$K(\mu, y) = \int_{\overline{\Omega}} H(x, y) d\mu = \int_{\overline{\Omega}} J(x, y) d\mu = \Pr(y \in M_x),$$

in which case mathematical expectation signify respective probabilities that a random point falls within a fixed region and a random region to cover a fixed point. For all y and $x = (x_1, \ldots, x_s)$, under conditions (2.8.9) and (2.8.13), we have

$$K(x, \nu^*) = \frac{L(M_x)}{4\pi R^2} \leq \frac{\sum_{j=1}^{s} L(S(x_j, r))}{4\pi R^2} = \frac{s}{2}\left(1 - \sqrt{1 - \left(\frac{r}{R}\right)^2}\right),$$

$$K(\mu^*, y) = \frac{\sum_{j=1}^{s} L(S(x_j, r))}{4\pi R^2} = \frac{s}{2}\left(1 - \sqrt{1 - \left(\frac{r}{R}\right)^2}\right),$$

since $L(S(x_j, r)) = 2\pi R(R - \sqrt{(R^2 - r^2)})$.

From the definition of a saddle point and the resulting inequality $K(\mu^*, y) \geq K(x, \nu^*)$ it follows that the mixed strategies μ^* and ν^* are optimal and

$$K(\mu^*, \nu^*) = \frac{s}{2}\left(1 - \sqrt{1 - \left(\frac{r}{R}\right)^2}\right)$$

is the value of the game of search discussed above.

2.8.4. Consider an alternative to the preceding game assuming that Player 2, chooses a simply connected set $Y \subset C$ and Player 1 aims to maximize an intersection area

$$L(Y \cap M_x) = L(Y \cap \cup_{j=1}^{s} S(x_j, r)).$$

The objective of Player 2 is the opposite one. Otherwise the game coincides with that considered at the previous section. Player 1's strategy μ^* coincides with that given in the preceding game. Player 2's mixed strategy ν^* is constructed analogously to strategy μ^* and is the random throwing of the set Y onto a sphere (in the preceding case Player 2 choose points $y \in C$ at random). Thus, ν^* is constructed as an invariant measure which is to choose at random (in accordance with a uniform distribution over C) one of the fixed points of the set Y on C and to turn Y around this point through a random angle (in accordance with a uniform distribution over $[0, 2\pi]$). Let $K(x, \nu), K(\mu, y)$ be the mathematical expectation of the intersection area $L(Y \cap M_x)$. Then

$$K(\mu^*, y) = K(x, \nu^*) = K(\mu^*, \nu^*) = \frac{L(Y)L(M_x)}{4\pi R^2}.$$

If Y is the r-neighborhood of the point y, then the value of the game is

$$K(\mu^*, \nu^*) = \pi s(R - \sqrt{R^2 - r^2})^2.$$

2.9 A Poker Model

2.9.1. *A poker model with one round of betting and one size of bet.* [Bellman (1952), Karlin (1959)]. The model examined is a special case of the model treated in 2.9.2, which permits n possible sizes of bet. In this section we follow Karlin (1959).

The model. Two players, A and B, ante one unit each at the beginning of the game. After each draws a card, A acts first: he may either bet a more units or fold and forfeit his initial bet. If A bets, B has two choices: he may either fold (losing his initial bet) or bet a units and "see" A's hand. If B sees, the two players compare hands, and the one with the better hand wins the pot.

We shall denote A's hand by ξ, whose distribution is assumed to be the uniform distribution on the unit interval, and B's hand by η, also distributed uniformly on the unit interval. We shall write $L(\xi, \eta) = sign(\xi - \eta)$ as before.

Strategies and payoff. The strategies are composed as follows: let $\phi(\xi)$ = probability that if A draws ξ he will bet a, $1 - \phi(\xi)$ = probability that if A draws ξ he will fold, $\psi(\eta)$ = probability that if B draws η he will see, $1 - \psi(\eta)$ = probability that if B draws η he will fold.

If the two players follow these strategies, the expected net return $K(\phi, \psi)$ is the sum of the returns corresponding to three mutually exclusive possibilities: A folds; A bets a units and B sees; A bets and B folds. Thus

$$K(\phi, \psi) = (-1)\int [1 - \phi(\xi)]d\xi + (a+1)\iint \phi(\xi)\psi(\eta)L(\xi, \eta)d\xi d\eta$$
$$+ \iint \phi(\xi)[1 - \psi(\eta)]d\xi d\eta.$$

The yield to A may also be more transparently written as

$$K(\phi, \psi) = -1 + \int_0^1 \phi(\xi)\left(2 + a\int_0^\xi \psi(\eta)d\eta - (a+2)\int_\xi^1 \psi(\eta)d\eta\right)d\xi \tag{2.9.1}$$

or

$$K(\phi, \psi) = -1 + 2\int_0^1 \phi(\xi)d\xi + \int_0^1 \psi(\eta)\left(-(a+2)\int_0^\eta \phi(\xi)d\xi\right.$$
$$\left. + a\int_\eta^1 \phi(\xi)d\xi\right)d\eta. \tag{2.9.2}$$

Method of analysis. We begin by observing that the existence of a pair of optimal strategies is equivalent to the existence of two functions ϕ^* and ψ^* satisfying the inequalities

$$K(\phi, \psi^*) \leq K(\phi^*, \psi^*) \leq K(\phi^*, \psi) \tag{2.9.3}$$

for all strategies ϕ and ψ, respectively. Thus ϕ^* maximizes $K(\phi, \psi^*)$ while ψ^* minimizes $K(\phi^*, \psi)$. We shall therefore search for the strategy ϕ^* that maximizes (2.9.1) with ψ replaced by ψ^*; and we shall also search for the strategy ψ^* that minimizes (2.9.2) with ϕ replaced by ϕ^*. Since the constant terms are not important, the problem is to find

$$\max_{\phi} \int_0^1 \phi(\xi)\left(2 + a\int_0^{\xi} \psi^*(\eta)d\eta - (a+2)\int_{\xi}^1 \psi^*(\eta)d\eta\right)d\xi \quad (2.9.4)$$

and

$$\min_{\psi} \int_0^1 \psi(\eta)\left(-(a+2)\int_0^{\eta} \phi^*(\xi)d\xi + a\int_{\eta}^1 \phi^*(\xi)d\xi\right)d\eta. \quad (2.9.5)$$

The crux of the argument is to verify that our results are consistent, i.e. that the function ϕ^* that maximizes (2.9.4) is the same function ϕ^* that appears in (2.9.5), and similarly for ψ^*; if these assertions are valid, then (2.9.3) is satisfied and we have found a solution.

At this point intuitive considerations suggest what type of solution we search for. Since B has no chance to bluff, $\psi^*(\eta) = 1$ for η greater than some critical number c, and $\psi^*(\eta) = 0$ otherwise; also, since B is minimizing, $\psi^*(\eta)$ should be equal to 1 when the coefficient of $\psi(\eta)$ in (2.9.5) is negative. But this coefficient expresses a decreasing function of η, and thus c is the value at which it first becomes zero. Hence

$$-(a+2)\int_0^c \phi^*(\xi)d\xi + a\int_c^1 \phi^*(\xi)d\xi = 0. \quad (2.9.6)$$

With this choice for $\psi^*(\eta)$, we find that the coefficient of $\phi(\xi)$ in (2.9.4) is constant for $\xi \leq c$. If we assume that this constant is 0, we obtain at $\xi = c$

$$2 + 0 - (a+2)(1-c) = 0,$$

or

$$c = \frac{a}{a+2}. \quad (2.9.7)$$

The reason we determine the constant c so as to make the coefficient zero in the interval $[0, c]$ is as follows. In maximizing (2.9.4) we are obviously compelled to make $\phi^*(\xi) = 1$ whenever its coefficient is positive, and $\phi^*(\xi) = 0$ whenever its coefficient is negative. The only arbitrariness allowed in the values of $\phi^*(\xi)$ occurs when its coefficient is zero. But we expect A to attempt some partial bluffing on low hands, which means that probably $0 < \phi^*(\xi) < 1$ for these hands. As pointed out, this is feasible if the coefficient of $\phi(\xi)$ is zero. With the determination of c according to (2.9.7), the coefficient of $\phi(\xi)$ in (2.9.4) is zero for $\xi \leq c$ and positive for $\xi > c$. Under these circumstances it follows from (2.9.4) that the maximizing player is obligated to have $\phi^*(\xi) = 1$ for $\xi > c$ while the values of $\phi^*(\xi)$ for $\xi \leq c$ are irrelevant, in the sense that they do not contribute to the payoff. However, in order to satisfy (2.9.6) with this choice of ϕ^* we must have

$$-(a+2)\int_0^c \phi^*(\xi)d\xi + a(1-c) = 0,$$

or

$$\int_0^c \phi^*(\xi)d\xi = \frac{a(1-c)}{a+2} = \frac{2a}{(a+2)^2},$$

and this can be accomplished with $\phi^*(\xi) < 1$. It is easy to verify now that if

$$\psi^*(\eta) = \begin{cases} 0, & 0 \leq \eta \leq \dfrac{a}{a+2}, \\ 1, & \dfrac{a}{a+2} < \eta \leq 1 \end{cases}$$

and

$$\phi^*(\xi) = \begin{cases} \text{arbitrary between 0 and 1 but satisfying} & \\ \int_0^c \phi^*(\xi)d\xi = \dfrac{2a}{(a+2)^2}, & 0 \leq \xi \leq \dfrac{a}{a+2}, \\ 1, & \dfrac{a}{a+2} < \xi < 1, \end{cases}$$

then ϕ^* maximizes (2.9.4) for ψ^* and ψ^* minimizes (2.10.5) for ϕ^*.

The interpretation of the solution is of some interest:

(1) Both players bet or see on high hands. What is of special significance is that both players use the identical critical number $a/(a+2)$ to distinguish high and low hands.

(2) The element of bluffing shows up for Player 1 only to the extent that the proportion of hands on which he should bluff is determined; he may choose the actual hands in an arbitrary manner from $[0, a/(a+2)]$ subject only to the restriction

$$\int_0^{a/(a+2)} \phi^*(\xi)d\xi = \frac{2a}{(a+2)^2}.$$

2.9.2. *A poker model with several sizes of bet.* [Karlin and Restrepo (1957)]. The model examined here is an extension of the one just analyzed.

As before, the unit interval is taken in the representation of all possible hands that can be dealt to a player. Each hand is considered equally likely, and therefore the operation of dealing a hand to a player may be considered as equivalent to selecting a random number from the unit interval according to the uniform distribution. Of course, a hand ξ_1 is inferior to a hand ξ_2 if and only if $\xi_1 < \xi_2$. The game proceeds as follows. Two players A and B select points ξ and η, respectively, from the unit interval according to the uniform distribution. Both players ante one unit. A, knowing his value ξ, acts first and has the option of either folding immediately, thus forfeiting his ante to B, or betting any one of the amounts $a_1, a_2, \ldots, a_n$, where $1 < a_1 < a_2 < \ldots < a_n$. B must then respond by either passing immediately or seeing. In the first circumstance A wins B's ante. If B sees, the hands ξ and η are compared and the player with the better hand wins the pot. If $\xi = \eta$, no payment is made.

A strategy for A can be described as an n-tuple of functions

$$\phi(\xi) = \langle \phi_1(\xi), \phi_2(\xi), \ldots, \phi_n(\xi) \rangle,$$

where $\phi_i(\xi)$ expresses the probability that A will bet the amount a_i when his hand is ξ. The $\phi_i(\xi)$ must satisfy $\phi_i(\xi) \geq 0$ and

$$\sum_{i=1}^{n} \phi_i(\xi) \leq 1.$$

The probability that A will fold immediately is

$$1 - \sum_{i=1}^{n} \phi_i(\xi).$$

A strategy for B can be represented by the n-tuple

$$\psi(\eta) = \langle \psi_1(\eta), \psi_2(\eta), \ldots, \psi_n(\eta) \rangle,$$

where $\psi_i(\eta)$ expresses the probability that B will see a bet of a_i units when he holds the value η. The probability that B will pass after A has bet a_i is $1 - \psi_i(\eta)$. Each $\psi_i(\eta)$ is subject only to the condition that

$$0 \leq \psi_i(\eta) \leq 1.$$

If A uses the strategy ϕ and B uses the strategy ψ, the expected gain to A is denoted by $K(\phi, \psi)$. Enumerating all the possibilities, we find that

$$\begin{aligned} K(\phi, \psi) = (-1) \int_0^1 \left[1 - \sum_{i=1}^{n} \phi_i(\xi) \right] d\xi \\ + \int_0^1 \int_0^1 \sum_{i=1}^{n} \phi_i(\xi)[1 - \psi_i(\eta)] d\xi d\eta \\ + \sum_{i=1}^{n} (a_i + 1) \int_0^1 \int_0^1 \phi_i(\xi) \psi_i(\eta) L(\xi, \eta) d\xi d\eta, \end{aligned}$$

where $L(\xi, \eta) = sign(\xi - \eta)$.

Any pair of optimal strategies ϕ^* and ψ^* satisfy the inequalities

$$K(\phi^*, \psi) \geq K(\phi^*, \psi^*) \text{ for all } \psi \tag{2.9.8}$$

and

$$K(\phi, \psi^*) \leq K(\phi^*, \psi^*) \text{ for all } \phi. \tag{2.9.9}$$

Conversely, if the inequalities are satisfied, the * strategies are optimal. Thus, ϕ^* maximizes $K(\phi, \psi^*)$ and ψ^* minimizes $K(\phi^*, \psi)$. In this case, rearranging the expression for $K(\phi, \psi)$, we can write

$$K(\phi, \psi^*) = -1 + \sum_{i=1}^{n} \int \phi_i(\xi) \left[2 + a_i \int_0^{\xi} \psi_i^*(\eta) d\eta \right.$$
$$\left. -(a_i + 2) \int_{\xi}^{1} \psi_i^*(\eta) d\eta \right] d\xi \tag{2.9.10}$$

and

$$K(\phi^*, \psi) = -1 + 2 \sum_{i=1}^{n} \int \phi_i^*(\xi) d\xi + \sum_{i=1}^{n} \int_0^1 \psi_i(\eta)$$
$$\times \left[-(a_i + 2) \int_0^{\eta} \phi_i^*(\xi) d\xi + a_i \int_{\eta}^{1} \phi_i^*(\xi) d\xi \right] d\eta. \tag{2.9.11}$$

Thus (2.9.10) and (2.9.11) have the form

$$K(\phi, \psi^*) = C_1 + \sum_{i=1}^{n} \int_0^1 \phi_i(\xi) L_i(\xi) d\xi \tag{2.9.12}$$

and

$$K(\phi^*, \psi) = C_2 + \sum_{i=1}^{n} \int_0^1 \psi_i(\eta) K_i(\eta) d\eta, \tag{2.9.13}$$

where C_1 and C_2 are independent of ϕ and ψ, respectively, and L_i and K_i stand for the bracketed expressions in (2.9.10) and (2.9.11), respectively. In view of the constraints on $\phi_1, \ldots, \phi_n$, it is clear that in order to maximize (2.9.10) or (2.9.12), A must choose $\phi_i(\xi) = 1$

wherever $L_i(\xi)$ is positive and greater than all $L_j(\xi) < 0$; and finally, if $L_i(\xi) = 0$ and if the remaining coefficients $L_j(\xi)(j \neq i)$ are nonpositive, he can maximize $K(\phi, \psi^*)$ by choosing $\phi_i(\xi)$ arbitrary consistent with $0 \leq \phi_i(\xi) \leq 1$ (or, if more than one $L_i(\xi)$ is zero, $\sum \phi_i(\xi) \leq 1$, where the sum is extended over those indices corresponding to $L_i(\xi) = 0$). Similarly, in order to minimize (2.9.11) or (2.9.13) B must choose $\psi_i(\eta) = 1$ wherever $K_i(\eta) < 0$, and $\psi_i(\eta) = 0$ wherever $K_i(\eta) > 0$. Where $K_i(\eta) = 0$, the values of $\psi_i(\eta)$ will not affect the payoff.

Guided to some extent by intuitive considerations, we shall construct two strategies ϕ^* and ψ^*. It will be easy to verify that these strategies satisfy (2.9.8) and (2.9.9) and are therefore optimal. The main problem in the construction is to make sure that the strategies are consistent — i.e. that the function ϕ^* that maximizes (2.9.10) is the same function that appears in (2.9.11), and similarly for ψ^*. This is another illustration of the fixed-point method. We now proceed with the details.

Since B has no opportunity to bluff, we may expect that

$$\psi_i^*(\eta) = \begin{cases} 0, & \eta < b_i, \\ 1, & \eta > b_i \end{cases} \tag{2.9.14}$$

for some b_i. This is in fact the case, since each $K_j(\eta)$ is nonincreasing.

On the other hand, we may expect that A will sometimes bluff when his hand is low. In order to allow for this possibility, we determine the critical numbers b_i which define $\psi_i^*(\eta)$ so that the coefficient $L_i(\xi)$ of ϕ_i is zero for $\xi < b_i$. This can be accomplished, since $L_i(\xi)$ is constant on this interval. Hence we choose

$$b_i = \frac{a_i}{2 + a_i}, \tag{2.9.15}$$

and thus $b_1 < b_2 \ldots < b_n < 1$. The coefficient $L_i(\xi)$ of ϕ_i is zero in the interval $(0, b_i)$, and thereafter it is a linear function of ξ such that

$$L_i(1) = 4\left(1 - \frac{1}{a_i + 2}\right).$$

From this we deduce that the functions $L_i(\xi)$ and $L_j(\xi)$ intersect at the point

$$c_{ij} = 1 - \frac{2}{(2+a_i)(2+a_j)}. \tag{2.9.16}$$

Clearly, c_{ij} is a strictly increasing function of i and j. Define $c_1 = b_1$ and $c_i = c_{i-1,i}$ for $i = 2, \ldots, n$ and $c_{n+1} = 1$. For ξ in the interval (c_i, c_{i+1}), it is clear that $L_i(\xi) > L_j(\xi) \geq 0$ for $j \neq i$. Consequently, according to our previous discussion, if ϕ^* maximizes $K(\phi, \psi^*)$, then $\phi_i^*(\xi) = 1$ for $c_i < \xi < c_{i+1}$. For definiteness we also set $\phi_i^*(c_i) = 1$; this is of no consequence, since if a strategy is altered only at a finite number of points (or on a set of Lebesgue measure zero), the yield $K(\phi, \psi)$ remains unchanged.

Summarizing, we have shown that if ψ^* is defined as in (2.9.14), with

$$b_i = \frac{a_i}{2+a_i},$$

then $K(\phi, \psi^*)$ is maximized by any strategy ϕ^* of the form

$$\phi_i^*(\xi) = \begin{cases} \text{arbitrary}, & \xi < c_1 = b_1, \\ 0, & c_1 \leq \xi < c_i, \\ 1, & c_i \leq \xi < c_{i+1}, \\ 0, & c_{i+1} \leq \xi \leq 1, \end{cases} \tag{2.9.17}$$

where

$$\sum_{i=1}^{n} \phi_i^*(\xi) \leq 1, \;\; \phi_i^*(\xi) \geq 0.$$

The values of $\phi_i^*(\xi)$ in the interval $0 \leq \xi < c_1$ are still undetermined because of the relations $L_i(\xi) \equiv 0$ which are valid for the same interval.

It remains to show that ψ^* as constructed actually minimizes $K(\phi^*, \psi)$. In order to guarantee this property for ψ^*, it might be necessary to impose some further conditions on the ϕ_i^*; for this purpose, we shall utilize the flexibility present in the definition of ϕ^* as ξ ranges over the interval $[0, c_1)$. In order to show that ψ^* minimizes $K(\phi^*, \psi)$, we must show that the coefficient $K_i(\eta)$ of $\psi_i(\eta)$ is non-negative for $\eta < b_i$ and nonpositive for $\eta > b_i$. Since $K_i(\eta)$ is a continuous monotone-decreasing function, the last condition is equivalent to the relation

$$-(a_i + 2) \int_0^{b_i} \phi_i^*(\xi) d\xi + a_i \int_{b_i}^1 \phi_i^*(\xi) d\xi = 0. \tag{2.9.18}$$

Inserting the special form (2.9.17) of ϕ^* into (2.9.18) leads to the equations

$$2 \int_0^{b_1} \phi_1^*(\xi) d\xi = b_1(1 - b_1)(b_2 + 1),$$

$$2 \int_0^{b_1} \phi_n^*(\xi) d\xi = b_n(1 - b_n)(1 - b_{n-1}), \tag{2.9.19}$$

and

$$2 \int_0^{b_1} \phi_i^*(\xi) d\xi = b_i(1 - b_i)(b_{i+1} - b_{i-1}), \ i = 2, \dots, n-1. \tag{2.9.20}$$

Since

$$\sum_{i=1}^{n} \phi_i^*(\xi) \leq 1,$$

these equations can be satisfied if and only if

$$2 \int_0^{b_1} \sum_{i=1}^{n} \phi_i^*(\xi) d\xi \leq 2b_1. \tag{2.9.21}$$

But the sum of the right-hand sides of (2.9.19) and (2.9.21) is at most $(2 + b_n - b_1)/4$, since always $b_i(1 - b_i) \leq 1/4$. As $b_1 \geq 1/3$, we get

$$\frac{1}{4}(2 + b_n - b_1) \leq \frac{1}{4}(3 - b_1) \leq \frac{2}{3} \leq 2b_1.$$

Thus the requirements in (2.9.19)–(2.9.21) can be fulfilled. We have consequently established the inequalities (2.9.8) and (2.9.9) for the strategies ϕ^* and ψ^*. In summary, we display the optimal strategies as follows:

$$\psi_i^*(\eta) = \begin{cases} 0, & \eta < b_i, \\ 1, & \eta \geq b_i, \end{cases} \tag{2.9.22}$$

$$\phi_i^*(\xi) = \begin{cases} \text{arbitrary but satisfying} & \\ \displaystyle\int_0^{b_1} \phi_i^*(\xi)d\xi = b_i(c_{i+1} - c_i), & 0 \leq \xi < b_1 \\ 0, & b_1 \leq \xi < c_i \text{ or } c_{i+1} \leq \xi \leq 1, \\ 1, & c_i \leq \xi < c_{i+1}, \end{cases} \tag{2.9.23}$$

where

$$b_i = \frac{a_i}{2 + a_i}, \quad c_1 = b_1,$$

$$c_i = 1 - \frac{2}{(2 + a_i)(2 + a_{i-1})}, \quad i = 2, \ldots, n, \quad c_{n+1} = 1$$

and

$$\sum_{i=1}^{n} \phi_i^*(\xi) \leq 1.$$

2.9.3. *Poker model with two rounds of betting.* [Bellman (1952), Karlin and Restrepo (1957)]. In this section, we generalize the poker model of the preceding section to include two rounds of betting, but at the same time we restrict it by permitting only one size of bet. We assume again, for convenience, that hands are dealt at random to each player from the unit interval according to the uniform distribution.

Payoff and strategies. After making the initial bet of one unit, A acts first and has two choices: he may fold or bet a units. B acts next and has three choices: he may fold, he may see, or he may raise by betting $a + b$ units. If B has raised, A must either fold or see.

If A and B draw cards ξ and η, respectively, their strategies may be described as follows: $\phi_1(\xi)$ = probability that A bets a and folds later if B raises; $\phi_2(\xi)$ = probability that A bets a and sees if B raises; $1 - \phi_1(\xi) - \phi_2(\xi)$ = probability that A folds initially; $\psi_1(\eta)$ = probability that B sees the initial bet; $\psi_2(\eta)$ = probability that B raises; $1 - \psi_1(\eta) - \psi_2(\eta)$ = probability that B folds.

The expected return is

$$\begin{aligned}
K(\phi, \psi) = &- \int [1 - \phi_1(\xi) - \phi_2(\xi)]d\xi + \int\int [\phi_1(\xi) + \phi_2(\xi)] \\
&\times [1 - \psi_1(\eta) - \psi_2(\eta)]d\xi d\eta + (a+1) \\
&\times \int\int \phi_1(\xi)\psi_1(\eta)L(\xi,\eta)d\xi d\eta \\
&- (a+1)\int\int \phi_1(\xi)\psi_2(\eta)d\xi d\eta \\
&+ (a+1)\int\int \phi_2(\xi)\psi_1(\eta)L(\xi,\eta)d\xi d\eta \\
&+ (1+a+b)\int\int \phi_2(\xi)\psi_2(\eta)L(\xi,\eta)d\xi d\eta,
\end{aligned}$$

where $L(\xi, \eta) = sign(\xi - \eta)$. (This expected yield is derived by considering mutually exclusive possibilities: A folds; A bets and B folds; A acts according to ϕ_1 and B sees or raises; A acts according to ϕ_2 and B sees or raises.)

If we denote the optimal strategies by

$$\phi^*(\xi) = \langle \phi_1^*(\xi), \phi_2^*(\xi) \rangle, \psi^*(\eta) = \langle \psi_1^*(\eta), \psi_2^*(\eta) \rangle$$

and rearrange the terms as we have done in the previous examples, we obtain

$$\begin{aligned} K(\phi,\psi^*) = -1 &+ \int_0^1 \phi_1(\xi)\Big[2 + a\int_0^\xi \psi_1^*(\eta)d\eta \\ &- (a+2)\int_\xi^1 \psi_1^*(\eta)d\eta - (a+2)\int_0^1 \psi_2^*(\eta)d\eta\Big]d\xi \\ &+ \int_0^1 \phi_2(\xi)\Big[2 + a\int_0^\xi \psi_1^*(\eta)d\eta - (a+2)\int_\xi^1 \psi_1^*(\eta)d\eta \\ &+ (a+b)\int_0^\xi \psi_2^*(\eta)d\eta - (a+b+2)\int_\xi^1 \psi_2^*(\eta)d\eta\Big]d\xi \end{aligned} \tag{2.9.24}$$

and

$$\begin{aligned} K(\phi^*,\psi) = \int_0^1 &[-1 + 2\phi_1^*(\xi) + 2\phi_2^*(\xi)]d\xi + \int_0^1 \psi_1(\eta) \\ &\times\Big[-(a+2)\int_0^\eta [\phi_1^*(\xi) + \phi_2^*(\xi)]d\xi \\ &+ a\int_\eta^1 [\phi_1^*(\xi) + \phi_2^*(\xi)]d\xi\Big]d\eta \\ &+ \int_0^1 \psi_2(\eta)\Big[-(a+2)\int_0^1 \phi_1^*(\xi)d\xi \\ &- (a+b+2)\int_0^\eta \phi_2^*(\xi)d\xi + (a+b)\int_\eta^1 \phi_2^*(\xi)d\xi\Big]d\eta. \end{aligned} \tag{2.9.25}$$

Search for the optimal strategies. We shall search again for the functions $\phi^*(\xi)$ that maximize (2.9.24) and the functions $\psi^*(\eta)$ that minimize (2.9.25). Intuitive considerations suggest that A may do a certain amount of bluffing with low cards with the intention of folding if raised; he will also choose $\phi_1^*(\xi) = 1$ in an intermediate range, say $c < \xi < e$, and he will select the strategy $\phi_2^*(\xi) = 1$ for $e \le \xi \le 1$.

B should sometimes bluff by raising hands in the interval $0 \leq \eta < c$; he should choose $\psi_1^* = 1$ in an interval $c \leq \eta < d$, and $\psi_2^* = 1$ in $d \leq \eta \leq 1$.

This is not to imply that this is the only possible form of the optimal strategies. In fact, we shall see later that optimal strategies of different character exist. However, once a pair of optimal strategies is determined, it is then relatively easy to calculate all solutions; hence we first concentrate on determining optimal strategies of the form indicated. To this end we shall attempt to find values of c, d, and e that produce a solution of the given type. In view of the construction of ψ_1^* we see immediately that the coefficient of $\phi_1(\xi)$ in (2.9.24) is constant for $\xi \leq c$; evaluating this constant and setting it equal to zero, we obtain

$$2 - (a+2)\int_0^1 \psi_2^*(\eta)d\eta - (a+2)(d-c) = 0. \tag{2.9.26}$$

The coefficients of ψ_1^* and ψ_2^* should be equal at the point d, where B changes the character of his action; this requires

$$(2a+2)\int_d^1 \phi_1^*(\xi)d\xi = -b\int_0^d \phi_2^*(\xi)d\xi + b\int_d^1 \phi_2^*(\xi)d\xi. \tag{2.9.27}$$

A similar condition at $\xi = e$ requires

$$(2a+b+2)\int_0^e \psi_2^*(\eta)d\eta = b\int_e^1 \psi_2^*(\eta)d\eta. \tag{2.9.28}$$

At the point $\eta = c$, where B begins to play ψ_1^* and ψ_2^* without bluffing, the corresponding coefficients (which are decreasing functions) should change from nonnegative to nonpositive, i.e. they should be zero at $\eta = c$. Hence

$$\begin{aligned} -(a+2)\textstyle\int_0^c [\phi_1^*(\xi) + \phi_2^*(\xi)]d\xi + a\int_c^1 [\phi_1^*(\xi) + \phi_2^*(\xi)]d\xi &= 0, \\ -(a+2)\textstyle\int_0^1 \phi_1^*(\xi)d\xi + (a+b)\int_c^1 \phi_2^*(\xi)d\xi &= 0. \end{aligned} \tag{2.9.29}$$

(In writing (2.9.29) we postulate that $\phi_2^*(\xi) = 0$ for $0 \le \xi \le c$; this is intuitively clear.) At this point we introduce the notation

$$m_1 = \int_0^c \phi_1^*(\xi)d\xi, \quad m_2 = \int_0^c \psi_2^*(\eta)d\eta.$$

Recalling the assumption made on the form of the solution and assuming that $c < e < d$, equations (2.9.26)–(2.9.29) may be written as follows:

$$2 = (a+2)(m_2 + 1 - c), \tag{2.9.30}$$

$$1 - d = d - e \text{ or } 2(1-d) = (1-e), \tag{2.9.31}$$

$$(2a+b+2)m_2 = b(1-d), \tag{2.9.32}$$

$$(a+2)m_1 = a(1-c), \tag{2.9.33}$$

$$(a+2)(m_1 + e - c) = (a+b)(1-c). \tag{2.9.34}$$

We have obtained a system of five equations in the five unknowns m_1, m_2, c, d, e; we now prove that this system of equations has a solution which is consistent with the assumptions made previously, namely that $0 < c < e < d < 1$, $0 < m_1 < 0$, $0 < m_2 < c$.

Solution of equations (2.9.30)–(2.9.34). The system of equations may be solved explicitly as follows. We write the last equation as:

$$(a+2)(m_1 + 1 - c) = (2a+b+2)(1-e).$$

Eliminating m_1 and $1-e$ by means of (2.9.33) and (2.9.31), we obtain

$$(a+1)(1-c) = (2a+b+2)(1-d). \tag{2.9.35}$$

From the remaining equations we eliminate m_2; then

$$\frac{2}{a+2} - (1-c) = \frac{b}{2a+b+2}(1-d). \tag{2.9.36}$$

Therefore

$$(1-d)\left(\frac{b}{2a+b+2} + \frac{2a+b+2}{a+1}\right) = \frac{2}{a+2}. \tag{2.9.37}$$

Having obtained $1-d$, we can solve (2.9.36) for $1-c$, and the remaining unknowns are then calculated from the original equations.

In order to show that the solution is consistent, we first note that $1-d>0$. Equation (2.9.35) shows that $1-c>1-d$, and therefore $c<d$. Also, from (2.9.36),

$$c = \left(\frac{b}{2a+b+2}\right)(1-d) + \left(1 - \frac{2}{a+2}\right) > 0. \tag{2.9.38}$$

Since $2(a+1)(1-c) = (2a+b+2)(1-e)$, we infer that $1-e<1-c$, or $c<e$; and since $2d = 1+e$, we must have $e<d$. Summing up, we have shown that $0<c<e<d<1$. For the two remaining conditions we note that (2.9.30) implies that

$$m_2 = c - \left(1 - \frac{2}{a+2}\right),$$

so that $m_2 < c$, and by (2.9.38), $m_2 > 0$. Finally, using (2.9.33) and (2.9.30), we conclude that

$$\begin{aligned} m_1 &= \frac{a}{a+2}(1-c) = (1-c) - \frac{2}{a+2}(1-c) \\ &= (1-c)[1-(m_2+1-c)] = (1-c)(c-m_2), \end{aligned}$$

so that $0 < m_1 < c$.

Optimality of the strategies ϕ^* and ψ^*. We summarize the representation of ϕ^* and ψ^* in terms of the values of c, e, d, m_1, and m_2 as computed above:

$$\phi_1^*(\xi) = \begin{cases} 1, & c \le \xi < e, \\ 0, & e \le \xi \le 1, \end{cases} \qquad \phi_2^*(\xi) = \begin{cases} 0, & 0 \le \xi < e, \\ 1, & e \le \xi \le 1, \end{cases}$$

$$\psi_1^*(\eta) = \begin{cases} 0, & 0 \le \eta < c, \\ 1, & c \le \eta < d, \\ 0, & d \le \eta \le 1, \end{cases} \qquad \psi_2^*(\eta) = \begin{cases} 0, & c \le \eta < d, \\ 1, & d \le \eta \le 1. \end{cases} \tag{2.9.39}$$

In the remaining interval $0 \leq \eta < c$, the functions $\phi_1^*(\xi)$ and $\psi_2^*(\eta)$ are chosen arbitrarily but bounded between 0 and 1, satisfying

$$\int_0^c \phi_1^*(\xi)d\xi = m_1 \text{ and } \int_0^c \psi_2^*(\eta)d\eta = m_2,$$

respectively.

It remains to verify that the strategies ϕ^* and ψ^* prescribed above maximize $K(\phi, \psi^*)$ of (2.9.24), and minimize $K(\phi^*, \psi)$ of (2.9.25), respectively. To do this, we first examine the coefficients $M_1(\xi)$ and $M_2(\xi)$ of ϕ_1 and ϕ_2 in $K(\phi, \psi^*)$. By construction, the coefficient $M_1(\xi)$ of ϕ_1 is identically zero on $[0, c)$, increases linearly on $[c, d)$, and afterwards remains constant. Also, $M_1(\xi)$ is continuous throughout $[0, 1]$. Next we notice that $M_2(\xi)$ is linear on $[c, d]$ with the same slope as $M_1(\xi)$. Furthermore, they agree at $\xi = e$ in $[c, d]$ by (2.9.28), and hence $M_1 = M_2$ for ξ in $[c, d]$.

We may also deduce immediately from the definition of ψ^* that M_2 increases strictly throughout $[0, 1]$ (see Fig. 2.7). With these facts the maximization of $K(\phi, \psi^*)$ is now easy to perform. Clearly, the maximum is achieved for any ϕ with the properties (a) $\phi_2 = 0$ and ϕ_1 arbitrary ($0 \leq \phi_1 \leq 1$) for ξ in $[0, c)$; (b) $\phi_1 + \phi_2 = 1$ and otherwise

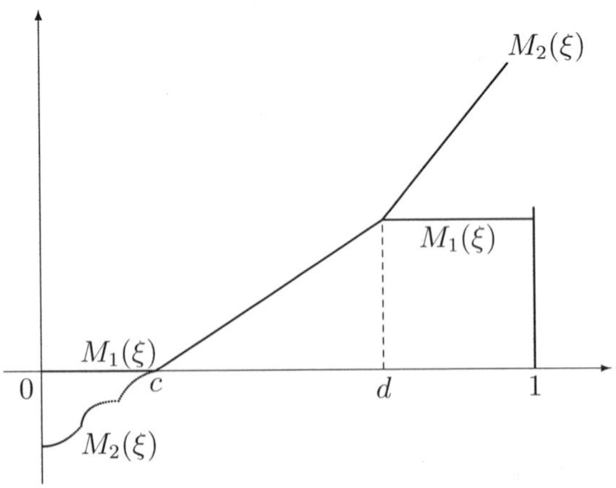

Figure 2.7

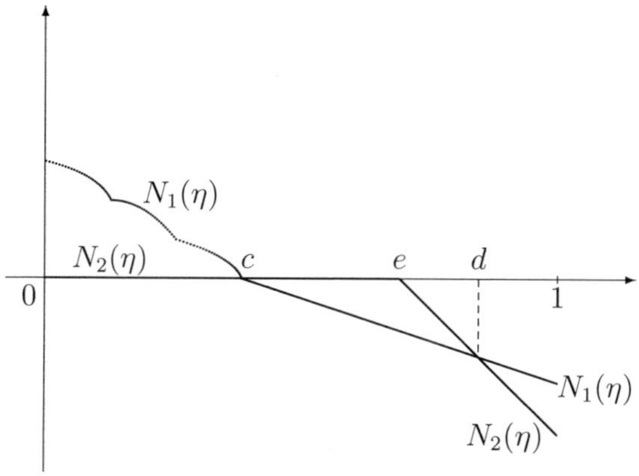

Figure 2.8

arbitrary for ξ in $[c, d)$; (c) $\phi_2 = 1$ for ξ in $[d, 1]$. It is clear that ϕ^* as specified above fulfills these conditions.

A study of the coefficients $N_1(\eta)$ and $N_2(\eta)$ of $\psi_1(\eta)$ and $\psi_2(\eta)$ in $K(\phi^*, \psi)$ shows that they are as indicated in Fig. 2.8. We observe that $K(\phi^*, \psi)$ is minimized by any ψ with the properties (a') $\psi_1 = 0$ and ψ_2 arbitrary ($0 \leq \psi_2 \leq 1$) for η in $[0, c)$; (b') $\psi_1 = 1$ for η in $[c, d)$; (c') $\psi_2 = 1$ for η in $[d, 1]$. Clearly, ψ^* as specified above obeys these requirements. The proof of the optimality of ϕ^* and ψ^* is now complete.

For illustrative purposes we append the following example. Let $a = b = 2$; then

$$\phi_1^*(\xi) = \begin{cases} \int_0^{19/35} \phi_1^* d\xi = 8/35, & 0 \leq \xi < 19/35, \\ 1, & 19/35 \leq \xi < 23/35, \\ 0, & 23/35 \leq \xi \leq 1, \end{cases}$$

$$\phi_2^*(\xi) = \begin{cases} 0, & 0 \leq \xi < 23/35, \\ 1, & 23/35 \leq \xi \leq 1, \end{cases}$$

$$\psi_1^*(\eta) = \begin{cases} 0, & 0 \le \eta < 19/35, \\ 1, & 19/35 \le \eta < 29/35, \\ 0, & 29/35 < \eta \le 1, \end{cases}$$

$$\psi_2^*(\eta) = \begin{cases} \displaystyle\int_0^{19/35} \psi_2^* d\xi = 3/70, & 0 \le \eta < 19/35, \\ 0, & 19/35 \le \eta < 29/35, \\ 1, & 29/35 \le \eta \le 1. \end{cases}$$

The value is $11/35$.

The general solution. It follows readily from Fig. 2.8 that the minimizing solution cannot have any other form than that given in (2.9.39). However, in maximizing $K(\phi, \psi^*)$ we found that on the interval $[c, d]$ the only necessary condition for ϕ to qualify as a maximum is that $\phi_1 + \phi_2 = 1$. Altering ϕ_1 and ϕ_2 in this interval subject to this condition, we can determine all possible optimal strategies for A. To this end, we specify $\tilde{\phi}_1(\xi)$ on $[0, c]$ such that

$$\int_0^c \tilde{\phi}_1 d\xi = m_1$$

(calculated above) and $\tilde{\phi}_2(\xi) = 1$ on $[d, 1]$. For ξ in $[c, d]$ we require only that $\tilde{\phi}_1 + \tilde{\phi}_2 = 1$. Writing out the conditions under which $K(\tilde{\phi}, \psi)$ is minimized for ψ^*, we obtain

$$1 - d = \int_c^d \tilde{\phi}_2(\xi) d\xi \text{ and } \int_\eta^d \tilde{\phi}_1(\xi) d\xi \le \frac{b(d - \eta)}{a + b + 1} \text{ for } \eta \in [c, d], \tag{2.9.40}$$

where c and d are the same as before. We obtain these constraints by equating the coefficients of $\psi_1(\eta)$ and $\psi_2(\eta)$ in (2.9.25) at $\eta = d$ and by requiring that $N_1(\eta) \le N_2(\eta)$ for η in $[c, d]$.

This relation is easily seen to be necessary and sufficient for $\tilde{\phi}$ to be optimal.

2.9.4. *Poker model with k raises.* [Karlin and Restrepo (1957)]. In this section, we indicate the form of the optimal strategies of a poker model with several rounds of betting. The methods of analysis are in principle extensions of those employed in the preceding section, but far more complicated in detail. We omit the proofs, referring the reader to the references.

Rules, strategies, and payoff. The two players ante one unit each and receive independently hands ξ and η (which are identified with points of the unit interval) according to the uniform distribution. There are $k+1$ rounds of betting ("round" in this section means one action by one player). In the first round A may either fold (and lose his ante) or bet a units. A and B act alternately. In each subsequent round a player may either fold or see (whereupon the game ends), or raise the bet by a units. In the last round the player can only fold or see. If k is even, the last possible round ends with A; if k is odd, the last possible round ends with B.

A strategy for A can be described by a k-tuple of functions $\phi = \langle \phi_1(\xi), \phi_2(\xi), \ldots, \phi_k(\xi) \rangle$. These functions indicate A's course of action when he receives the hand ξ. Explicitly,

$$1 - \sum_{i=1}^{k} \phi_i(\xi)$$

is the probability that A will fold immediately, and

$$\sum_{i=1}^{k} \phi_i(\xi)$$

is the probability that A will bet at his first opportunity. Further, $\phi_1(\xi) =$ probability that A will fold in his second round, $\phi_2(\xi) =$ probability that A will see in his second round, $\sum_{i=3}^{k} \phi_i(\xi) =$ probability that A will raise in his second round, if the occasion arises, i.e. if B has raised in his first round and kept the game going. Similarly, if the game continues until A's rth round, then $\phi_{2r-3}(\xi) =$ probability that A will fold in his rth round, $\phi_{2r-2}(\xi) =$ probability that A will see in his rth round, $\sum_{i=2r-1}^{k} \phi_i(\xi)$ = probability that A will raise in his rth round.

Analogously, a strategy for B can be expressed as a k-tuple

$$\psi = \langle \psi_1(\eta), \ldots, \psi_k(\eta) \rangle$$

which indicates B's course of action when he receives the hand η. The probability that B will fold at his first opportunity is

$$\psi_0(\eta) = 1 - \sum_{j=1}^{k} \psi_j(\eta).$$

If the game continues until B's rth round, then $\psi_{2r-2}(\eta)$ = probability that B will fold in his rth round, $\psi_{2r-1}(\eta)$ = probability that B will see in his rth round, $\sum_{j=2r}^{k} \psi_j(\eta)$ = probability that B will raise in his rth round.

If the two players receive hands ξ and η and choose the strategies ϕ and ψ, respectively, then the payoff to A can be computed as in the previous examples by considering the mutually exclusive ways in which the betting may terminate. The payoff to A is as follows:

$$\begin{aligned}
P[\phi(\xi), \psi(\eta)] = (-1)\left(1 - \sum_{i=1}^{k} \phi_i(\xi)\right) \\
+ \sum_{i=1}^{k} \phi_i(\xi)\left[1 - \sum_{j=1}^{k} \psi_j(\eta) + (a+1)\psi_1(\eta)L(\xi,\eta)\right] \\
+ \sum_{j=2}^{k} \psi_j(\eta)\{-(a+1)\phi_1(\xi) + (2a+1)\phi_2(\xi)L(\xi,\eta)\} \\
+ \sum_{j=2r-2}^{k} \psi_j(\eta)\,\{-[(2r-3)a+1]\phi_{2r-3}(\xi) \\
+[(2r-2)a+1]\phi_{2r-2}(\xi)L(\xi,\eta)\}
\end{aligned}$$

$$+\sum_{i=2r-1}^{k}\phi_i(\xi)\{[(2r-2)a+1]\psi_{2r-2}(\eta)$$
$$+[(2r-1)a+1]\psi_{2r-1}(\eta)L(\xi,\eta)\}$$
$$+\sum_{j=2r}^{k}\psi_j(\eta)\{-[(2r-1)a+1]\phi_{2r-1}(\xi)$$
$$+(2ra+1)\phi_{2r}(\xi)L(\xi,\eta)\}$$
$$+\sum_{i=2r+1}^{k}\phi_i(\xi)\{(2ra+1)\psi_{2r}(\eta)$$
$$+[(2r+1)a+1]\psi_{2r+1}(\eta)L(\xi,\eta)\},$$

where $L(\xi,\eta)$ is the function already defined. The expected payoff is

$$K(\phi,\psi)=\int_0^1\int_0^1 P[\phi(\xi),\psi(\eta)]d\xi d\eta. \tag{2.9.41}$$

Description of the optimal strategies. There exist optimal strategies ϕ^* and ψ^* characterized by $2k+1$ numbers $b, c_1, \ldots, c_k, d_1, \ldots, d_k$. When a player gets a hand ξ in $(0,b)$ he will bluff part of the time and fold part of the time. We shall write

$$m_i=\int_0^b \phi_i^*(\xi)d\xi, \quad i=1,3,5,\ldots \tag{2.9.42}$$

and

$$n_j=\int_0^b \psi_j^*(\eta)d\eta, \quad j=2,4,6,\ldots \tag{2.9.43}$$

to represent the probabilities of bluffing in the various rounds of betting. If A receives a hand ξ in (c_{i-1},c_i), where $c_0=b$, he will choose $\phi_i^*(\xi)=1$ and $\phi_l^*(\xi)=0$ for $l\neq i$. Similarly, if B gets a hand η in (d_{j-1},d_j), where $d_0=b$, he will choose $\psi_j^*(\eta)=1$ and $\psi_l^*(\eta)=0$ for $l\neq j$. The solution is represented by Fig. 2.9. The fact that

$$c_{2r-1}<d_{2r-1}<d_{2r}<c_{2r},\ r=1,2,\ldots$$

is important.

The constants c_i, d_j, m_i, and n_j are determined by solving an elaborate system of equations analogous to (2.9.30)–(2.9.34). Explicitly, if k is even, b, c_i, d_j, m_i, and n_j are evaluated as solutions of the following equations:

$$[(4r-1)a+2]\sum_{j=2r}^{k} n_j = a(1-d_{2r-1}), \quad 2r = 2,4,\ldots,k,$$

$$a(c_{2r-2}-d_{2r-2}) = a(1-c_{2r-2}) + [(4r-3)a+2] \times \sum_{j=2r}^{k} n_j,\ 2r = 4,6,\ldots,k,$$

$$[(4r-3)a+2]\sum_{i=2r-1}^{k} m_i = a(1-c_{2r-2}), \quad 2r = 2,4,\ldots,k,$$

$$a(d_{2r-1}-c_{2r-1}) = a(1-d_{2r-1}) + [(4r-1)a+2] \times \sum_{i=2r+1}^{k} m_i,\ 2r = 2,4,\ldots,k,$$

$$(4ra+2)(c_{2r}-d_{2r-1}) = [(4r+2)a+2](c_{2r}-d_{2r}), \quad 2r = 2,4,\ldots,k,$$

$$[(4r-2)a+2](d_{2r-1}-c_{2r-2}) = (4ra+2)(d_{2r-1}-c_{2r-1}), \quad 2r = 2,4,\ldots,k,$$

$$2 = (a+2)\int_0^1 \sum_{j=1}^{k} \psi_j^*(\eta)d\eta.$$

An analogous system applies for k odd. The solutions obtained are consistent with the requirements of Fig. 2.9.

2.10.5. *Poker with simultaneous moves.* [von Neumann and Morgenstern (1944), Karlin (1959)]. Two players, A and B, make simultaneous bets after drawing hands according to the uniform distribution. The initial bet can be either b (the low bet) or a (the high bet). If both bets are equal, the player with the higher hand wins.

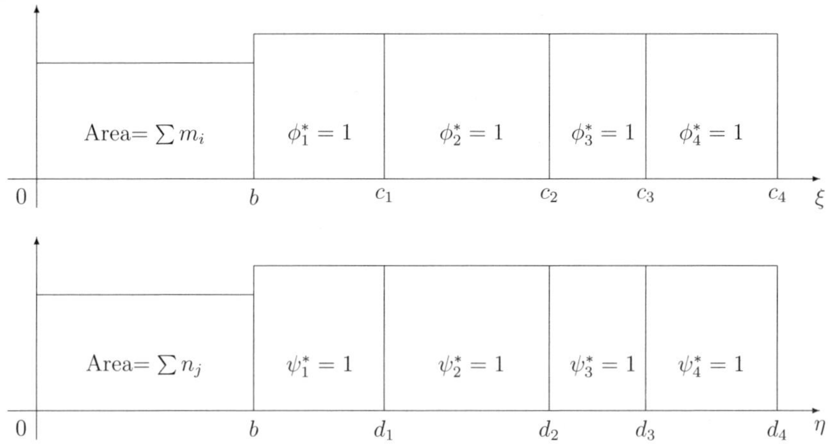

Figure 2.9

If one player bets high and the other bets low, the low bettor has a choice: he may either fold (losing the low bet) or see by making an additional bet of $a - b$. If the low bettor sees, the player with the higher hand wins the pot.

Since the game is symmetric, we need only describe the strategies of one player. If A draws ξ, we shall write $\phi_1(\xi)$ = probability that A will bet low and fold if B bets high, $\phi_2(\xi)$ = probability that A will bet low and subsequently see, $\phi_3(\xi)$ = probability that A will bet high.

These functions are of course subject to the constraints

$$\phi_i(\xi) \geq 0, \ \sum_{i=1}^{3} \phi_i(\xi) = 1,$$

The expected yield to A if he uses strategy ϕ while B employs strategy ψ reduces to

$$K(\phi, \psi) = b \int_0^1 \int_0^1 [\phi_1(\xi) + \phi_2(\xi)][\psi_1(\eta) + \psi_2(\eta)] L(\xi, \eta) d\xi d\eta$$

$$-b \int_0^1 \int_0^1 [\phi_1(\xi)\psi_3(\eta) d\xi d\eta + b \int_0^1 \int_0^1 \phi_3(\xi)\psi_1(\eta) d\xi d\eta$$

$$+a \int_0^1 \int_0^1 \phi_2(\xi)\psi_3(\eta)L(\xi,\eta)d\xi d\eta$$

$$+a \int_0^1 \int_0^1 \phi_3(\xi)\psi_2(\eta)L(\xi,\eta)d\xi d\eta$$

$$+a \int_0^1 \int_0^1 \phi_3(\xi)\psi_3(\eta)L(\xi,\eta)d\xi d\eta.$$

Because of the symmetry of the game, we may replace B's strategy $\psi(\eta)$ in this expression by an optimal strategy $\phi^*(\eta)$. We make the plausible assumption that in this strategy $\phi_2^*(\eta) = 0$, since there would appear to be no clear justification for making a low bet initially and then seeing. The consistency of this assumption will be established later. With $\phi_2^*(\eta) = 0$ we may write $K(\phi, \phi^*)$ as

$$\begin{aligned}
K(\phi,\phi^*) = {} & b \int_0^1 \int_0^1 [\phi_1(\xi) + \phi_2(\xi)]\phi_1^*(\eta)L(\xi,\eta)d\xi d\eta \\
& -b \int_0^1 \int_0^1 \phi_1(\xi)\phi_3^*(\eta)d\xi d\eta + b \int_0^1 \int_0^1 \phi_3(\xi)\phi_1^*(\eta)d\xi d\eta \\
& +a \int_0^1 \int_0^1 [1 - \phi_1(\xi) - \phi_3(\xi)]\phi_3^*(\eta)L(\xi,\eta)d\xi d\eta \\
& +a \int_0^1 \int_0^1 \phi_3(\xi)\phi_3^*(\eta)L(\xi,\eta)d\xi d\eta \\
= {} & \int_0^1 \phi_1(\xi)\left[b \int_0^\xi \phi_1^*(\eta)d\eta - b \int_\xi^1 \phi_1^*(\eta)d\eta - b \int_0^1 \phi_3^*(\eta)d\eta\right. \\
& \left. -a \int_0^\xi \phi_3^*(\eta)d\eta + a \int_\xi^1 \phi_3^*(\eta)d\eta\right] d\xi \\
& + \int_0^1 \phi_2(\xi)\left[b \int_0^\xi \phi_1^*(\eta)d\eta - b \int_\xi^1 \phi_1^*(\eta)d\eta\right] d\xi \\
& + \int_0^1 \phi_3(\xi)\left[b \int_0^1 \phi_1^*(\eta)d\eta\right] d\xi + \int_0^1 \int_0^1 \phi_3^*(\eta)L(\xi,\eta)d\xi d\eta,
\end{aligned} \tag{2.9.44}$$

or

$$K(\phi, \phi^*) = \int_0^1 \sum \phi_i(\xi) T_i(\xi) d\xi + Z,$$

where Z is a term independent of ϕ_i. The ϕ maximizing $K(\phi, \phi^*)$ is evaluated by choosing the component ϕ_i as large as possible whenever $T_i(\xi) = \max_j T_j(\xi)$. If the maximum of $T_j(\xi)$ is attained simultaneously by two of the T_i, then the corresponding ϕ_i may share any positive values provided their sum is 1.

Some bluffing is anticipated on low hands. This suggests that for $\xi < \xi_0$ we should have $T_1(\xi) = T_3(\xi) > T_2(\xi)$. Differentiating the identity $T_1(\xi) = T_3(\xi)$, and remembering that $\phi_1^*(\xi) + \phi_3^*(\xi) = 1$ on the interval $[0, \xi_0]$, we deduce that

$$\phi_1^*(\xi) = \frac{a}{a+b} \text{a.e. for } \xi < \xi_0.$$

With this choice of ϕ_1^*, and where $\phi_3^*(\xi) = 1$ for $\xi > \xi_0$, we obtain that $T_1(\xi) = T_3(\xi)$ is possible only if

$$\xi_0 = \frac{a-b}{a}.$$

The proposed solution is as follows:

$$\phi^* = \begin{cases} \phi_1^*(\xi) = \dfrac{a}{a+b}, \ \phi_3^*(\xi) = \dfrac{b}{a+b}, & \xi < \xi_0, \\ \phi_3^*(\xi) = 1, & \xi > \xi_0. \end{cases} \tag{2.9.45}$$

It is now routine to verify that ϕ^* as exhibited is indeed optimal. It is clear that $T_1(\xi) = T_3(\xi) > T_2(\xi)$ for $\xi < \xi_0$, and hence the maximum is achieved provided only that $\phi_1 + \phi_3 = 1$, which is certainly satisfied for ϕ^* of (2.9.45). Moreover, we have seen that $T_1(\xi) = T_3(\xi)$ uniquely determines ϕ^* to be as in (2.9.45) for $\xi < \xi_0$.

For $\xi > \xi_0$, by examining (2.9.44) we find that $T_2(\xi) = T_3(\xi) > T_1(\xi)$. Hence the maximization of $K(\phi, \phi^*)$ requires ϕ to be such that $\phi_2 + \phi_3 = 1$. But, if $\phi_2^* > 0$ in this interval, a simple calculation shows that $T_1(\xi) < T_2(\xi)$ for $\xi \geq \xi_1$ where $\xi_1 < \xi_0$. All these inferences in conjunction prove that the ϕ^* of (2.9.45) is the unique optimal strategy of the game.

2.10 Exercises and Problems

1. *Attack-defense game.* Player 1 wishes to attack with A units one of the targets $C_1, \ldots, C_n$ whose value is determined by the numbers $\tau_1 > 0, \tau_2 > 0, \ldots, \tau_n > 0$, with $\tau_1 \geq \tau_2 \geq \ldots \geq \tau_n$. A pure strategy x for Player 1 is the vector $x = (\xi_1, \ldots, \xi_n)$, $\sum_{i=1}^n \xi_i = A$, where ξ_i is the part of the units of attack allocated to the target C_i. The Defender (Player 2) has a total of B units. A pure strategy for Player 2 is the choice of a collection of y non-negative numbers $y = (\eta_1, \ldots, \eta_n)$ satisfying the condition $\sum_{i=1}^n \eta_i = B$ where η_i is a part of the units of defense assigned to the target C_i. The result of an attack on the target C_i is proportional to the difference $\xi_i - \eta_i$ if the Attacker's forces outnumber the Defender's forces, otherwise it is zero. Construct the payoff function.
2. A game on a unit square has the payoff function
$$H(x, y) = xy - \frac{1}{3}x - \frac{1}{2}y.$$
Show that $(1/2, 1/3)$ is the equilibrium point in this game.
3. Show that the game on a unit square with the payoff function
$$H(x, y) = sign(x - y)$$
has a saddle point.
4. Show that the duel type game on a unit square with the payoff function
$$H(x, y) = \begin{cases} -1/x^2, & x > y, \\ 0, & x = y, \\ 1/y^2, & x < y \end{cases}$$
has the saddle point $(0, 0)$.

5. Show that the game on a unit square with the payoff function $H(x,y) = (x-y)^2$ does not have the saddle point in pure strategies.

6. Show that in the game on a unit square with the payoff function

$$H(x,y) = \begin{cases} x+y, & x \neq 1, y \neq 0, \\ 1/2+y, & x = 1, y \neq 0, \\ 1/2+x, & x \neq 1, y = 0, \\ 2, & x = 1, y = 0 \end{cases}$$

the pair (x_ϵ, y_ϵ), where $x_\epsilon = 1-\epsilon$, $y_\epsilon = \epsilon$, is an ϵ-saddle point. Does the game have a value?

7. Compute the payoff to Player 1 in the game on a unit square with the payoff function $H(x,y)$ in the situation $(F(x), G(y))$ (F and G are distribution functions), if

(a) $H(x,y) = (x+y)/(4xy)$, $F(x) = x^2$, $G(y) = y^2$;
(b) $H(x,y) = |x-y|(1-|x-y|)$, $F(x) = x$, $G(y) = y$;
(c) $H(x,y) = (x-y)^2$, $F(x) = 1/2I_0(x) + 1/2I_1(x)$, $G(y) = I_{1/2}(x)$, where $I_k(x)$ is a step function.

8. *Game of discrete search.* Consider the following infinite game. A strategy for Player 2 is the choice of the point uniformly distributed over the circle of radius y, where y can take values from the interval $[0,1]$. Player 1 may survey in a unit circle the simply connected region whose area $a(Q) = a = const$, where $a < A$, $A = \pi$ is the area of the unit circle. His strategy x is the choice of a shape of the region Q which has the area a and lies entirely within the unit circle. The payoff $H(x,y)$ to Player 1 is the probability of being discovered, i.e. $H(x,y) = \Pr(y \in Q)$. The mixed strategy $g(y)$ for Player 2 means the density of the distribution function of a random variable $y \in [0,1]$. Find a solution to the game.

9. Prove Helly theorem, 2.5.4.

10. Consider a continuous analog of the "town defense" game (see 1.1.3). Player 1 has x units to attack the first post and $1 - x$ to attack the second post, $x \in [0, 1]$. Player 2 has y units of defense, where $y \in [0, 1]$, to allocate to the first post and $1 - y$ units of defense to the second post at which the permanent forces of defense of power $1/2$, are located. A player pays 1 to the opponent for every post at which he has less units than his opponent, and pays nothing if the players' forces are equal in number.

Construct the payoff function $H(x, y)$ for the game on a unit square. Show that this game has no solution in mixed strategies.

Hint. Make use of the result of Example 10, 2.4.12.

11. Show that in the continuous game with the payoff function

$$H(x, y) = [1 + (x + y)^2]^{-1}$$

the strategies $F^*(x) = I_{1/2}(x)$, $G^*(y) = 1/2I_0(y) + 1/2I_1(y)$ are optimal for Players 1 and 2, respectively.

12. Prove that the value of the continuous symmetric game on a unit square is zero, and optimal mixed strategies coincide (the game is symmetric if the payoff function is skew-symmetric, i.e. $H(x, y) = -H(y, x)$).

13. Define the optimal strategies and the value of the game on a unit square with the payoff function $H(x, y) = y^3 - 3xy + x^3$.

14. Show that in the game with the payoff function

$$H(x, y) = e^{\gamma y - x}\sqrt{1 - x^2/y^2}, \quad x \in [x_0, x_1], \; y \in [y_0, y_1], \; \gamma > 0.$$

Player 2 has an optimal pure strategy; clarify the form of this strategy depending on the parameter $\gamma > 0$. What can be said about Player 1's optimal strategy?

15. Verify that the payoff function from Example 10, 2.5.5, $H(x, y) = \rho(x, y)$, $x \in S(0, l)$, $y \in S(0, l)$, where $S(0, l)$ is the circle with its center at 0 and radius l, $\rho(\cdot)$ being a distance in R^2, is strictly convex in y for any x fixed.

16. Show that the sum of two convex functions is convex.

17. Prove that, if the convex function is bounded $\varphi : [\alpha, \beta] \to R^1$ then it is continuous in any point $x \in (\alpha, \beta)$. At the ends α and β of the closed interval (α, β), however, the convex function φ is upper semicontinuous, i.e.

$$\overline{\lim_{x \to \alpha}} \varphi(x) \leq \varphi(\alpha)$$

(in much the same way as $x \to \beta$).

18. Let there be given the game $\Gamma = (X, Y, H)$, $X = Y = [0, 1]$ with the bounded convex payoff function $H(x, \cdot) : [0, 1] \to R^1$. Show that Player 2 in this game has either an optimal strategy or an ϵ-optimal pure strategy for every $\epsilon > 0$. The result of the theorem given in 2.5.6 applies to Player 1.
Hint. Make use of the result of Exercise 17 and consider the auxiliary game $\Gamma_0 = (X, Y, H_0)$, where

$$H_0(x, y) = \begin{cases} H(x, y), & \text{if } y \in (0, 1), \\ \lim_{y_n \to y} H(x, y_n), & \text{if } y = 0 \text{ or } y = 1. \end{cases}$$

19. Solve the "attack-defense" game formulated in Exercise 1.

20. Consider the simultaneous game of pursuit on a plane (see Example 1 in 2.1.2) in which the strategy sets $S_1 = S_2 = S$, where S is bounded and closed convex set.

(a) Show that the value of the game discussed is R, where R is the radius of a minimal circle $S(O, R)$ containing S, and an optimal strategy for Player 2 is pure and is the choice of a center O of the circle $S(O, R)$.

(b) Show that an optimal strategy for Player 1 is mixed and constitutes a mixture of two diametrically opposite points of tangency of the set S to the circle $S(O, R)$ (if such points x_1 and x_2 exist), or, alternatively a mixture of three points of tangency x_1', x_2', x_3' such that the point O is inside a triangle whose vertices are these points.

21. Solve the simultaneous game of pursuit on a plane discussed in Example 20 assuming that Player 2 chooses not one point $y \in S$, but m points $y_1, \ldots, y_m \in S$. The payoff function of the game is

$$H(x, y) = \frac{1}{m} \sum_{i=1}^{m} \rho^2(x, y_i),$$

where $\rho(\cdot)$ is a distance in R^2.

22. Player 1 selects a system x of m points in the interval $[-1, 1]$, i.e. $x = (\xi_1, \ldots, \xi_m)$, $\xi_i \in [-1, 1]$, $i = 1, \ldots, m$. Player 2 selects independently and simultaneously a system y of n points in the same interval $[-1, 1]$, i.e. $y = (\eta_1, \ldots, \eta_n)$, $\eta_j \in [-1, 1]$, $j = 1, 2, \ldots, n$. The payoff function $H(x, y)$ is of the form

$$H(x, y) = \frac{1}{2}(\max_i \min_j |\xi_i - \eta_j| + \max_j \min_i |\xi_i - \eta_j|).$$

Find a solution to the game.

23. Consider an extension of the problem given in 2.8.3, namely, the game of search in which Player 2 selects a system of k points $y = (y_1, \ldots, y_k)$ on a sphere C and Player 1 selects, as before, a system x of s points $x = (x_1, \ldots, x_s)$ on the sphere C. The payoff function is

$$H(x, y) = \{M \mid M = |\{y_i\}| : y_i \in S(x_j, r);\ j = 1, \ldots, s\},$$

where $S(x_j, r)$ is a spherical segment (cup) with its apex at the point x_j and with r as a base radius; $|\{y_i\}|$ means the number of points of the set $\{y_i\}$. The point y_i is considered to be discovered if $y_i \in S(x_j, r)$ for at least one x_j. Thus, the payoff function is the number of the points discovered in the situation (x, y). Find a solution to the game.

Chapter 3

Nonzero-Sum Games

3.1 Definition of Noncooperative Game in Normal Form

3.1.1. The preceding chapters concentrated on zero-sum two-person games, i.e. the games in which the interests of the parties are strictly contradictory. However, a special feature of the actual problems of decision making in a conflict context is that there are too many persons involved, with the result that the conflict situation is far from being strictly contradictory. As for a two-person conflict and its models, it may be said that such a conflict is not confined to the antagonistic case alone. Although the players' interests may intersect, they are not necessarily contradictory. This, in particular, can involve situations that are of mutual benefit to both players (which is not possible in the antagonistic conflict). Cooperation (selection of an agreed decision) is thus made meaningful and tends to increase a payoff to both players. At the same time, there are conflicts for which the rules of a game do not specify any agreement or cooperation. For this reason, in nonzero-sum games, a distinction is made between noncooperative behavior, where the rules do not allow any cooperation and cooperative behavior, where the rules allow cooperation in the joint selection of strategies and side payments making. We shall consider the former case.

3.1.2. Definition. The system

$$\Gamma = (N, \{X_i\}_{i\in N}, \{H_i\}_{i\in N}),$$

where $N = \{1, 2, \ldots, n\}$ is the set of players, X_i is the strategy set for Player i, H_i is the payoff function for Player i defined on Cartesian product of the players' strategy sets $X = \prod_{i=1}^{n} X_i$ (the set of situations in the game), is called a noncooperative game.

A noncooperative n-person game is played as follows. Players choose simultaneously and independently their strategies x_i from the strategy sets X_i, $i = 1, 2, \ldots, n$, thereby generating a situation $x = (x_1, \ldots, x_n)$, $x_i \in X_i$. Each Player i receives the amount $H_i(x)$, whereupon the game ends.

If the players' pure strategy sets X_i are finite, the game is called a *finite noncooperative n-person game.*

3.1.3. The noncooperative game Γ played by two players is called a *two-person game.* The noncooperative two-person game Γ is then defined by the system $\Gamma = (X_1, X_2, H_1, H_2)$, where X_1 is the strategy set of one player, X_2 is the strategy set of the other player, $X_1 \times X_2$ is the set of situations, while $H_1 : X_1 \times X_2 \to R^1$, $H_2 : X_1 \times X_2 \to R^1$ are the payoff functions to Players 1 and 2, respectively. The finite noncooperative two-person game is called *bimatrix game.* This is due to the fact that, once the pure strategy sets of players have been designated by the numbers $1, 2, \ldots, \overline{m}$ and $1, 2, \ldots, \overline{n}$, the payoff functions can be written in the form of two matrices

$$H_1 \triangleq A \triangleq \begin{bmatrix} \alpha_{11} & \ldots & \alpha_{1\overline{n}} \\ \ldots & \ldots & \ldots \\ \alpha_{\overline{m}1} & \ldots & \alpha_{\overline{m}\overline{n}} \end{bmatrix} \text{ and } H_2 \triangleq B \triangleq \begin{bmatrix} \beta_{11} & \ldots & \beta_{1\overline{n}} \\ \ldots & \ldots & \ldots \\ \beta_{\overline{m}1} & \ldots & \beta_{\overline{m}\overline{n}} \end{bmatrix}.$$

Here the elements α_{ij} and β_{ij} of the matrices A, B are respectively the payoffs to Players 1 and 2 in the situation (i, j), $i \in \overline{M}$, $j \in \overline{N}$, $\overline{M} = \{1, \ldots, \overline{m}\}$, $\overline{N} = \{1, \ldots, \overline{n}\}$.

In line with the foregoing, the bimatrix game is played as follows. Player 1 chooses number i (the row) and Player 2 (simultaneously and independently) chooses number j (the column). Then Player 1 receives the amount $\alpha_{ij} \triangleq H_1(x_i, y_j)$ and Player 2 receives the amount $\beta_{ij} \triangleq H_2(x_i, y_j)$.

Note that the bimatrix game with matrices A and B can also be described by the $(\overline{m} \times \overline{n})$ matrix (A, B), where each element is a pair $(\alpha_{ij}, \beta_{ij})$, $i = 1, 2, \ldots, \overline{m}$, $j = 1, 2, \ldots, \overline{n}$. The game determined by the matrix A and B will be denoted as $\Gamma(A, B)$.

If the noncooperative two-person game Γ is such that $H_1(x, y) = -H_2(x, y)$ for all $x \in X_1$, $y \in X_2$, then Γ appears to be a zero-sum two-person game discussed in the preceding chapters. In the special bimatrix game, where there is $\alpha_{ij} = -\beta_{ij}$, we have a matrix game examined in Chapter 1.

3.1.4. *Example 1. Battle of the sexes.* Consider the bimatrix game determined by

$$(A, B) = \begin{array}{c} \\ \alpha_1 \\ \alpha_2 \end{array} \begin{array}{c} \begin{array}{cc} \beta_1 & \quad \beta_2 \end{array} \\ \begin{bmatrix} (4, 1) & (0, 0) \\ (0, 0) & (1, 4) \end{bmatrix} \end{array}.$$

Although this game has a variety of interpretations, the best known seems to be the following [Luce and Raiffa (1957)]. Husband (Player 1) and his wife (Player 2) may choose one of two evening entertainments: football match (α_1, β_1) or theater (α_2, β_2). If they have different desires, (α_1, β_2) or (α_2, β_1), they stay at home. The husband shows preference to the football match, while his wife prefers to go to the theater. However, it is more important for them to spend the evening together than to be alone at the entertainment (though preferable).

Example 2. "Crossroads" game [Moulin (1981)]. Two drivers move along two mutually perpendicular routes and simultaneously meet each other at a crossroad. Each driver may make a stop (1st strategy, α_1 or β_1) or continue on his way (2nd strategy, α_2 or β_2).

It is assumed that each player prefers to make a stop in order to avoid an accident, or to continue on his way if the other player has made a stop. This conflict can be formalized by the bimatrix game with the matrix

$$(A,B) = \begin{array}{c} \\ \alpha_1 \\ \alpha_2 \end{array} \begin{array}{c} \begin{array}{cc} \beta_1 & \beta_2 \end{array} \\ \begin{bmatrix} (1,1) & (1-\epsilon,2) \\ (2,1-\epsilon) & (0,0) \end{bmatrix} \end{array}$$

(the non-negative number ϵ corresponds to the feeling of dissatisfaction that one player has to make a stop and let the other go).

Example 3. Selection of a vehicle for a city tour [Moulin (1981)]. Suppose the number of players is large and each of the sets X_i consists of two elements: $X_i = \{0, 1\}$ (for definiteness: 0 is the use of a private vehicle and 1 is the use of a public vehicle). The payoff function is defined as follows:

$$H_i(x_1, \dots, x_n) = \begin{cases} a(t), & \text{with } x_i = 1, \\ b(t), & \text{with } x_i = 0, \end{cases}$$

where $t = 1/n \sum_{j=1}^{n} x_j$.

Let a and b be of the form shown in Fig. 3.1. From the form of the functions $a(t)$ and $b(t)$ it follows that if the number of players choosing 1 is greater than t_1, then the street traffic is light enough

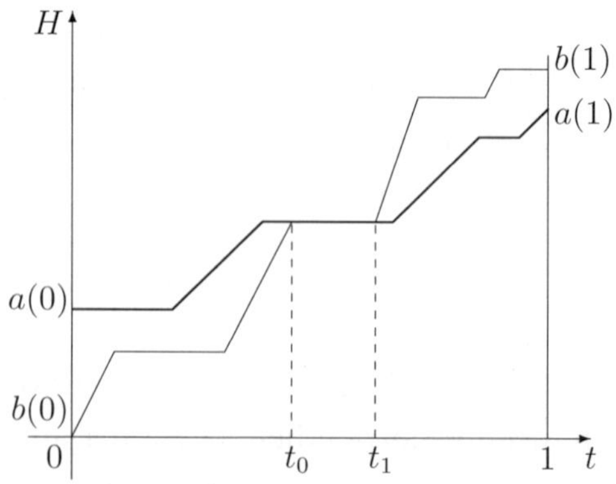

Figure 3.1

to make the driver of a private vehicle more comfortable than the passenger in a public vehicle. However, if the number of motorists is greater than $1 - t_0$, then the traffic becomes so heavy (with the natural priority for public vehicles) that the passenger in a public vehicle compares favorably with the driver of a private vehicle.

Example 4. Allocation of a limited resource taking into account the users' interests. Suppose n users have a good chance of using (accumulating) some resource whose volume is bounded by $A > 0$. Denote by x_i the volume of the resource to be used (accumulated) by the ith user. The users receive a payoff depending on the values of vector $x = (x_1, x_2, \dots, x_n)$. The payoff for the ith user is evaluated by the function $h_i(x_1, x_2, \dots, x_n)$, if the total volume of the used (accumulated) resource does not exceed a given positive value $\Theta < A$, i.e.

$$\sum_{i=1}^{n} x_i \leq \Theta, \;\; x_i \geq 0.$$

If the inverse inequality is satisfied, the payoff to the ith user is calculated by the function $g_i(x_1, x_2, \dots, x_n)$. Here the resource utility shows a sharp decrease if $\sum_{i=1}^{n} x_i > \Theta$, i.e.

$$g_i(x_1, x_2, \dots, x_n) < h_i(x_1, x_2, \dots, x_n).$$

Consider a nonzero-sum game in normal form

$$\Gamma = (N, \{X_i\}_{i \in N}, \{H_i\}_{i \in N}),$$

where the players' payoff functions is

$$H_i(x_1, x_2, \dots, x_n) = \begin{cases} h_i(x_1, \dots, x_n), & \sum_{i=1}^{n} x_i \leq \Theta, \\ g_i(x_1, \dots, x_n), & \sum_{i=1}^{n} x_i > \Theta, \end{cases}$$

$$X_i = [0, a_i], \;\; 0 \leq a_i \leq A, \;\; \sum_{i=1}^{n} a_i = A, \;\; N = \{1, 2, \dots, n\}.$$

The players in this game are the users of the resource.

Example 5. Game-theoretic model for air pollution control. In an industrial area there are n enterprises, each having an emission source. Also, in this area there is an ecologically significant zone Ω whose air pollution must not exceed a maximum permissible level. The time and area-averaged emission from n emitters can be approximately calculated by the formula

$$q = \sum_{i=1}^{n} c_i x_i, \quad 0 \leq x_i \leq a_i, \ i = 1, 2, \ldots, n.$$

Let $\Theta < \sum_{i=1}^{n} c_i a_i$ be a maximum emission concentration level.

We shall consider the enterprises to be players and construct the game, modeling an air pollution conflict situation. Suppose each enterprise can reduce its operating expenses by increasing an emission x_i. However, if the air pollution in the area Ω exceeds the maximum emission concentration level, the enterprise incurs a penalty $s_i > 0$.

Suppose player i (enterprise) has an opportunity of choosing the values x_i from the set $X_i = [0, a_i]$. The players' payoff functions are

$$H_i(x_1, \ldots, x_n) = \begin{cases} h_i(x_1, x_2, \ldots, x_n), & q \leq \Theta, \\ h_i(x_1, x_2, \ldots, x_n) - s_i, & q > \Theta, \end{cases}$$

where $h_i(x_1, x_2, \ldots, x_n)$ are the functions that are continuous and increasing in the variables x_i.

Example 6. Game-theoretic model for bargaining of divisible good [Zenkevich and Voznyuk (1994a)]. Two players take part in an auction where q units of good with minimal price p_0 are offered. Assumed that players 1, 2 have budgets M_1, M_2 respectively. The players demand their quantities of good q_1, q_2 (q_1, q_2, q –integers) and bid their prices p_1, p_2 for unit of the good simultaneously and independently in such a way that

$$q_1 + q_2 > q, \ 0 < q_1 < q, \ 0 < q_2 < q, \ p_1 \in [p_0, \overline{p_1}], \ p_2 \in [p_0, \overline{p_2}],$$

where $\overline{p_1} = M_1/(q-1)$, $\overline{p_2} = M_2/(q-1)$.

According to the bargaining process rules, a player who bids the higher price buys demanded quantity of good at this price. The other buys the rest of good at his own price. If bidden players' prices are equal then Player 1 has an advantage over Player 2. Each player objective is to maximize his profit.

This bargaining process can be described as a nonzero-sum two-person game in normal form $\Gamma = (X, Y, H_1, H_2)$, where sets of the players' strategies are

$$X = \{p_1 | p_1 \in [p_0, \overline{p_1}]\}, \; Y = \{p_2 | p_2 \in [p_0, \overline{p_2}]\}$$

and payoff functions are

$$H_1(p_1, p_2) = \begin{cases} (\overline{p_1} - p_1)q_1, & p_1 \geq p_2, \\ (\overline{p_1} - p_1)(q - q_2), & p_1 < p_2, \end{cases}$$

$$H_2(p_1, p_2) = \begin{cases} (\overline{p_2} - p_2)q_2, & p_1 < p_2, \\ (\overline{p_2} - p_2)(q - q_1), & p_1 \geq p_2. \end{cases}$$

3.2 Optimality Principles in Noncooperative Games

3.2.1. It is well known that for zero-sum games the principles of minimax, maximin and equilibrium coincide (if they are realizable, i.e. there exists an equilibrium, while maximin and minimax are reached and equal to each other). In such a case, they define a unified notion of optimality and game solutions. The theory of nonzero-sum games does not have a unified approach to optimality principles. Although, there are actually many such principles, each of them is based on some additional assumptions of players' behavior and a structure of a game.

It appears natural that each player in the game Γ seeks to reach a situation x in which his payoff function has a maximum value. The payoff function H_i, however, depends not only on the strategy of the ith player, but also on the strategies chosen by the other players. Because of this, the situations $\{x^i\}$ determining a maximum payoff to the ith player may not do the same thing for the other players. As in the case of a zero-sum game, the quest for a maximum payoff involves a conflict, and even formulation of a "good" or optimal behavior in the game becomes highly conjectural. There are many approaches to this problem. One of these is the Nash equilibrium and its various extensions and refinements. When the game Γ is zero-sum, the Nash equilibrium coincides with the notion of optimality (saddle point — equilibrium) that is the basic principle of optimality in a zero-sum game.

Suppose $x = (x_1, \dots, x_{i-1}, x_i, x_{i+1}, \dots, x_n)$ is an arbitrary situation in the game Γ and x_i is a strategy of Player i. We construct a situation that is different from x only in that the strategy x_i of Player i has been replaced by a strategy x_i'. As a result we have a situation $(x_1, \dots, x_{i-1}, x_i', x_{i+1}, \dots, x_n)$ denoted by $(x\|x_i')$. Evidently, if x_i and x_i' coincide, then $(x\|x_i') = x$.

Definition. The situation $x^* = (x_1^*, \dots, x_i^*, \dots, x_n^*)$ is called the Nash (Nash, 1950a) equilibrium if for all $x_i \in X_i$ and $i = 1, \dots, n$ there is

$$H_i(x^*) \geq H_i(x^*\|x_i). \tag{3.2.1}$$

Example 7. Consider the game from Example 3, 3.1.4. Here the Nash equilibrium is the situation for which following condition holds

$$t_0 \leq t^* - 1/n, \quad t^* + 1/n \leq t_1, \tag{3.2.2}$$

where $t^* = 1/n \sum_{j=1}^n x_j^*$. It follows from (3.2.2) that a payoff to a player remains unaffected when he shifts from one pure strategy to another provided the other players do not change their strategies.

Suppose a play of the game realizes the situation x corresponding to $t = 1/n \sum_{j=1}^n x_j$, $t \in [t_0, t_1]$, and the quantity δ is the share of the players who wish to shift from strategy 0 to strategy 1. Note that if δ is such that $b(t) = a(t) < a(t+\delta)$, then the payoffs to these players tend to increase (with such a strategy shift) provided the strategies of the other players remain unchanged. However, if this shift is actually effected, then the same players may wish to shift from strategy 1 to strategy 0, because the condition $a(t+\delta) < b(t+\delta)$ is satisfied. If this wish is realized, then the share of players, $1/n\sum_{j=1}^{n} x_j$, decreases and again falls in the interval $[t_0, t_1]$.

Similarly, let δ be the share of players, who decided, for some reason (e.g. because of random errors), to shift from strategy 1 to strategy 0, when $t-\delta < t_0$. Then, by the condition $b(t-\delta) < a(t-\delta)$, the players may wish to shift back to strategy 1. When this wish is realized, the share of the players, $1/n\sum_{j=1}^{n} x_j$, increases and again comes back to the interval $[t_0, t_1]$.

Example 8. Bertrand Paradox [Tirole (2003)]. Assume that two firms produce homogeneous goods and consumers buy from the producer who charges the lowest price. Each firm incurs a cost c per unit of production and always supplies the demand it faces. If the firms charge the same price we assume that each firm faces a demand schedule equal to half of the market demand at the common price. The market demand function is decreasing function $q = d(p)$, $p \geq 0$ and profit function of firm has the form

$$H_1(p_1, p_2) = (p_1 - c)d_1(p_1, p_2),$$

$$H_2(p_1, p_2) = (p_2 - c)d_2(p_1, p_2),$$

where the demand d_i for the output of firm i is given by $(i \neq j)$

$$d_i(p_i, p_j) = \begin{cases} d(p_i), & \text{if } p_i < p_j, \\ d(p_i)/2, & \text{if } p_i = p_j, \\ 0, & \text{if } p_i > p_j. \end{cases}$$

Therefore, the Bertrand duopoly can be described as two-person game $\Gamma_B = (X, Y, H_1, H_2)$, where $X = Y = \{p | p \geq 0\}$.

Bertrand (1883) paradox states that the unique equilibrium has the two firms charge the equilibrium price: $p_1^* = p_2^* = c$.

The proof is as follows. It is clear, that (c, c) is Nash equilibrium:

$$0 = H_1(c, c) \geq H_1(p_1, c) \quad \text{and}$$

$$0 = H_2(c, c) \geq H_2(c, p_2) \quad \text{for all } p_1 \geq 0, \quad p_2 \geq 0.$$

To prove the uniqueness of equilibrium suppose, for example, that in equilibrium $p_1^* > p_2^* > c$. Then

$$0 = H_1(p_1^*, p_2^*) < H_1(p_2^*, p_2^*) = (p_2^* - c)\frac{d(p_2^*)}{2}$$

and inequality (3.2.1) is not satisfied.

Therefore, firm 1 cannot be acting in its own best interests if it charges p_1^*. Now suppose that $p_1^* = p_2^* > c$. If firm 1 reduces its price slightly to $p_1^* - \epsilon$, we have

$$(p_1^* - c)\frac{d(p_1^*)}{2} = H_1(p_1^*, p_1^*) < H_1(p_1^* - \epsilon, p_1^*) = (p_1^* - \epsilon - c)d(p_1^* - \epsilon),$$

which is true for small $\epsilon > 0$.

Suppose now that $p_1^* > p_2^* = c$. Then

$$H_2(p_1^*, p_2^*) < H_2(p_1^*, p_2^* + \epsilon) = (p_2^* + \epsilon - c)d(p_2^*)$$

for $0 < \epsilon < p_1^* - p_2^*$ and inequality (3.2.1) is also not satisfied.

3.2.2. It follows from the definition of Nash equilibrium situation that none of Players i is interested to deviate from the strategy x_i^* appearing in this situation (by (3.2.1), when such a player uses strategy x_i instead of x_i^*, his payoff may decrease provided the other players follow the strategies generating an equilibrium x^*). Thus, if the players agree on the strategies appearing in the equilibrium x^*, then any individual non-observance of this agreement is disadvantageous to such a player.

For the noncooperative two-person game $\Gamma = (X_1, X_2, H_1, H_2)$, the situation (x^*, y^*) is equilibrium if the inequalities

$$H_1(x, y^*) \leq H_1(x^*, y^*), \quad H_2(x^*, y) \leq H_2(x^*, y^*) \tag{3.2.3}$$

hold for all $x \in X_1$ and $y \in X_2$

In particular, for the bimatrix $(m \times n)$ game $\Gamma(A, B)$ the pair (i^*, j^*) is the Nash equilibrium if the inequalities

$$\alpha_{ij^*} \leq \alpha_{i^*j^*}, \quad \beta_{i^*j} \leq \beta_{i^*j^*} \tag{3.2.4}$$

hold for all the rows $i \in M$ and columns $j \in N$. Thus, Example 1 has two equilibria at (α_1, β_1) and (α_2, β_2), whereas Example 2 has equilibria at (α_1, β_2) and (α_2, β_1).

Recall that for the zero-sum game $\Gamma = (X_1, X_2, H)$ the pair $(x^*, y^*) \in X_1 \times X_2$ is an equilibrium if

$$H(x, y^*) \leq H(x^*, y^*) \leq H(x^*, y), \quad x \in X_1,\ y \in X_2.$$

Equilibria in zero-sum games have the following properties:

1^0. A player is not interested to inform the opponent of the strategy (pure or mixed) he wishes to use. Of course, if the player announces in advance of the play, the optimal strategy to be employed, then a payoff to him will not be reduced by the announcement, though he will not win anything.

2^0. Denote by Z(Γ) the set of saddle-points (equilibria) in zero-sum game Γ, then if $(x, y) \in Z(\Gamma)$, $(x', y') \in Z(\Gamma)$ are equilibria in the game Γ, and v is the value of the game, then

$$(x', y) \in Z(\Gamma), \quad (x, y') \in Z(\Gamma), \tag{3.2.5}$$

$$v = H(x, y) = H(x', y') = H(x, y') = H(x', y). \tag{3.2.6}$$

3^0. Players are not interested in any intercourse for the purposes of developing joint actions before the game starts.

4^0. If the game Γ has an equilibrium, with x as a maximin and y as a minimax strategy for Players 1 and 2, respectively, then $(x, y) \in Z(\Gamma)$ is an equilibrium, and vice versa.

We shall investigate whether these properties hold for bimatrix games.

Example 9. Consider a "battle of the sexes" game (see Example 1 and 3.1.4). As already noted, this game has two equilibria: (α_1, β_1) and (α_2, β_2). The former is advantageous to Player 1, while the latter is advantageous to Player 2. This contradicts (3.2.6), since the payoffs to the players in these situations are different. Although the situations (α_1, β_1), (α_2, β_2) are equilibria, the pairs (α_1, β_2) and (α_2, β_1) are not Nash equilibria, i.e. property 2^0 (see (3.2.5)) is not satisfied.

If Player 1 informs his partner of the strategy α_1 to be employed, and if Player 2 is convinced that he is sure to do it, then he cannot do better than to announce the first strategy β_1. Similar reasoning applies to Player 2. Thus, it is advantageous to each player to announce his strategy, which contradicts property 1^0 for zero-sum games.

Suppose the players establish no contact with each other and make their choices simultaneously and independently (as specified by the rules of a noncooperative game). Let us do the reasoning for Player 1. He is interested in realization of the situation (α_1, β_1), whereas the situation (α_2, β_2) is advantageous to Player 2. Therefore, if Player 1 chooses strategy α_1, then Player 2 can choose strategy β_2, with the result that both players become losers (the payoff vector $(0,0)$). Then it may be wise of Player 1 to choose strategy α_2, since in the situation (α_2, β_2) he would receive a payoff 1. Player 2, however, may follow a similar line of reasoning and choose β_1, then, in the situation (α_2, β_1) both players again become losers.

Thus, this is the case where the situation is advantageous (but at the same time unstable) to Player 1. Similarly, we may examine the situation (α_2, β_2) (from Player 2's point of view). For this reason, it may be wise of the players to make, in advance of the play, contact and agree on a joint course of action, which contradicts property 3^0. Note that some difficulties may arise when the pairs of maximin strategies do not form an equilibrium.

Thus, we have an illustrative example, where none of the properties $1^0 - 4^0$ of a zero-sum game is satisfied.

Payoffs to players may vary with Nash equilibria. Furthermore, unlike the equilibrium set in a zero-sum game, the Nash equilibrium set is not rectangular. If $x = (x_1, \ldots, x_i, \ldots, x_n)$ and $x' = (x'_1, \ldots, x'_i, \ldots, x'_n)$ are two different equilibria, then the situation x'' composed of the strategies, which form the situations x and x' and coincides with none of these situations, may not be equilibrium. The Nash equilibrium is a multiple optimality principle in that various equilibria may be preferable to different players to a variable extent.

It now remains for us to answer the question: which of the equilibria can be taken as an optimality principle convenient to all players? In what follows it will be shown that the multiplicity of the optimality principle is characteristically and essential feature of an optimal behavior in the controlled conflict processes, with many participants.

Note that, unlike a zero-sum case, the equilibrium strategy x_i^* of the ith player may not always ensure at least the payoff $H_i(x^*)$ in the Nash equilibrium, since this essentially depends on whether the other players choose the strategies appearing in the given Nash equilibrium. For this reason, the equilibrium strategy should not be interpreted as an optimal strategy for the ith player. This interpretation makes sense only for the n-tuples of players' strategies, i.e. for situations.

3.2.3. An important feature of the Nash equilibrium is that any deviation from it made by two or more players may increase a payoff to one of deviating players. Let $S \subset N$ be a subset of the set of players (coalition) and let $x = (x_1, \dots, x_n)$ be a situation in the game Γ. Denote by $(x\|x'_S)$ the situation which is obtained from the situation x by replacing therein the strategies x_i, $i \in S$, with the strategies $x'_i \in X_i$, $i \in S$. In other words, the players appearing in the coalition S replace their strategies x_i by the strategies x'_i. If x^* is the Nash equilibrium, then (3.2.1) does not necessary imply

$$H_i(x^*) \geq H_i(x^*\|x_S) \quad \text{for all} \quad i \in S. \tag{3.2.7}$$

In what follows, this will be established by some simple examples.

But we may strengthen the notion of a Nash equilibrium by requiring the condition (3.2.7) or the relaxed condition (3.2.7) to hold for at least one of the players $i \in S$. Then we arrive at the following definition.

Definition. The situation x^* is called a strong equilibrium if for any coalition $S \subset N$ and $x_S \in \prod_{i \in S} X_i$ there is a player $i_0 \in S$ such that the following inequality is satisfied:

$$H_{i_0}(x^*) > H_{i_0}(x^*\|x_S). \tag{3.2.8}$$

Condition (3.2.8) guarantees that the players' agreement to enter a coalition S is inexpedient because any coalition has a player i_0 who is not interested in this agreement. Any strongly equilibrium situation is a Nash equilibrium.

If the strong equilibrium existed in a broad class of games, then it could be an acceptable principle of optimality in a noncooperative games. However, it happens extremely rare.

Example 10. Prisoners' dilemma. Consider the bimatrix game with payoffs

$$\begin{array}{c} \\ \alpha_1 \\ \alpha_2 \end{array} \begin{array}{c} \begin{array}{cc} \quad\beta_1 & \qquad\quad \beta_2 \end{array} \\ \begin{bmatrix} (5,5) & (0,10) \\ (10,0) & (1,1) \end{bmatrix} \end{array}.$$

Here we have one equilibrium situation (α_2, β_2) (though not strong equilibrium), which yields the payoff vector $(1, 1)$. However, if both players play (α_1, β_1), they obtain the payoffs $(5, 5)$, which is better to both of them. Zero-sum games have no such paradoxes. As for this particular case, the result is due to the fact that a simultaneous deviation from the equilibrium strategies may further increase the payoff to each player.

Example 11. Braess's paradox [Braess (1968)]. The model was proposed by D. Braess (1968). The road network is shown on Fig. 3.2.

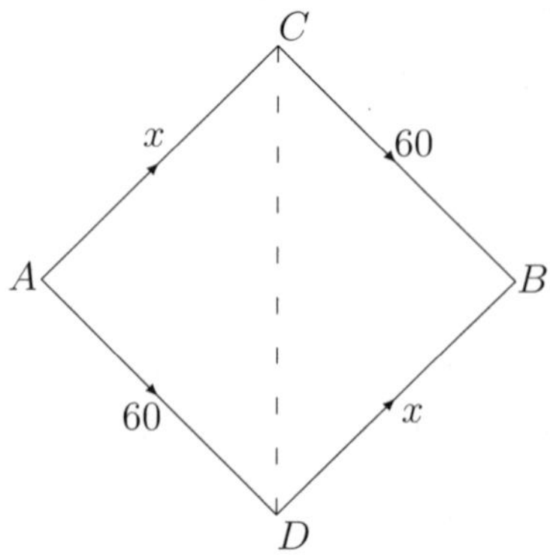

Figure 3.2

Suppose that 60 cars (players) move from point A to point B. The time for passing from C to B and from A to D equals 60 min. (and does not depend on number of cars on each of arcs AD and CB. On arcs AC and DB the passing time is equal to the number of cars using this arcs. Each player (car) from the set of 60 players (cars) has to go from A to B, and has the possibility to choose one of two roads (strategies) ACB or ADB. It is clear that Nash equilibrium ujhikjh is such allocation of cars in which the time of passing along ACB is equal to the time passing along ADB. If x is the number of cars using ACB and y is the number of cars using ADB we must have in Nash equilibrium

$$60 + x = 60 + y, \quad x + y = 60,$$

which gives us $x = y = 30$ (the proof is clear, since if one car changes his mind and switches, for instance, from road ACB to ADB he will need passing time $60+30+1 = 91$, but in Nash equilibrium his time is $60 + 30 = 90$).

Suppose now that we connect points C and D by speed way, which each car can pass in time 0. Then we see, that any car, which chooses ACB or ADB will benefit by moving along ACDB spending 60 instead of 90 along ADB or ACB. This means that allocation of cars along ACB and ADB will not be Nash equilibrium after opening a new road CD. It is easily seen, that the new Nash equilibrium in this case will be all cars use ACDB with passing time 120, since if one car deviates she will get the same amount $60 + 60 = 120$.

We observe a paradoxical case — the time passing from A to B increases from 90 to 120 after a new road construction.

3.2.4. Example 10 suggests the possibility of applying other optimality principles to a noncooperative game which may bring about situations that are more advantageous to both players than in the case of equilibrium situations. Such an optimality principle is *pareto-optimality* (optimality Pareto).

Consider a set of vectors $\{H(x)\} = \{H_1(x), \dots, H_n(x)\}$, $x \in X$, $X = \prod_{i=1}^n X_i$, i.e. the set of possible values of vector payoffs in all possible situations $x \in X$.

Definition. The situation $\overline{x}$ in the noncooperative game Γ is called pareto-optimal if there is no situation $x \in X$ for which the following inequalities hold:

$$H_i(x) \geq H_i(\overline{x}) \quad \text{for all } i \in N \text{ and}$$
$$H_{i_0}(x) > H_{i_0}(\overline{x}) \quad \text{for at least one } i_0 \in N.$$

The set of all pareto-optimal situations will be denoted by X^P.

The belonging of the situation $\overline{x}$ to the set X^P means that there is no other situation x which might be more preferable to all players than the situation $\overline{x}$.

Following Vorobjev (1977), we conceptually distinguish the notion of an equilibrium situation from that of a pareto-optimal situation. In the first case, neither player may individually increase his payoff, while in the second, all the players cannot increase acting as one player (even not strictly) a payoff to each of them.

To be noted also is that the agreement on a fixed equilibrium does not allow each individual player to deviate from it. In the pareto-optimal situation, the deviating player can occasionally obtain an essentially greater payoff. Of course, a strong equilibrium situation is also pareto-optimal. Thus, Example 9 provides a situation (α_2, β_2) which is equilibrium, but is not pareto-optimal. Conversely, the situation (α_1, β_1) is pareto-optimal, but not an equilibrium. In the game "battle of the sexes", both equilibrium situations (α_1, β_1), (α_2, β_2) are strong equilibria and pareto-optimal, but, as already noted in Example 8, they are not interchangeable. Similar reasoning also applies to the following example.

Example 12. Consider the "crossroads" game (see Example 2, 3.1.4). The situations $(\alpha_2, \beta_1), (\alpha_1, \beta_2)$ form Nash equilibria and are pareto-optimal (the situation (α_1, β_1) is pareto-optimal, but not an equilibrium). For each player the "stop" strategy α_1, β_1 is equilibrium if the other player decides to pass the crossroads and, conversely, it is advantageous for him to choose the "continue" strategy α_2, β_2 if the other player decides to pass the crossroads and, conversely, it is advantageous for him to choose the "continue" strategy α_2, β_2

if the other player makes a stop. However, each player receives a payoff of 2 units only if he chooses the "continue" strategy $\alpha_2(\beta_2)$. This necessarily involves competition for leadership, i.e. each player is interested to be the first to announce the "continue" strategy.

Note that we have reached the same conclusion from examination of the "battle of the sexes" game (see example 9).

3.2.5. We shall now consider behavior of a "leader–follower" type in a two-person game $\Gamma = (X_1, X_2, H_1, H_2)$. Denote by Z^1, Z^2 the sets of best responses for players 1 and 2, respectively, here

$$Z^1 \triangleq \{(x_1, x_2) | H_1(x_1, x_2) = \max_{y_1} H_1(y_1, x_2)\}, \tag{3.2.9}$$

$$Z^2 \triangleq \{(x_1, x_2) | H_2(x_1, x_2) = \max_{y_2} H_2(x_1, y_2)\} \tag{3.2.10}$$

(maximum in (3.2.9) and (3.2.10) are supposed to be reached, for example, if functions H_i are concave on x_i).

Definition. We call the situation $(x_1, x_2) \in X_1 \times X_2$ the Stakelberg i-equilibrium in the two-person game Γ if

$$\overline{H}_i \triangleq H_i(x_1, x_2) = \max_{(y_1, y_2) \in Z^j} H_i(y_1, y_2), \tag{3.2.11}$$

where $i = 1, 2$, $i \neq j$.

The notion of i-equilibrium may be interpreted as follows. Player 1 (Leader) knows the payoff functions of both players H_1, H_2, and hence he learns Player 2's (Follower) set of best responses Z^2 to any strategy x_1 of Player 1. Having this information he then maximizes his payoff by selecting strategy x_1 from condition (3.2.11). Thus, $\overline{H}_i$ is a payoff to the ith player acting as a "leader" in the game Γ.

Lemma. *Let $Z(\Gamma)$ be a set of Nash equilibria in the two-person game Γ. Then*

$$Z(\Gamma) = Z^1 \cap Z^2, \tag{3.2.12}$$

where Z^1, Z^2 are the sets of the best responses (3.2.9), (3.2.10) *given by the players 1,2 in the game Γ.*

Proof. Let $(x_1, x_2) \in Z(\Gamma)$ be the Nash equilibrium. Then the inequalities

$$H_1(x_1', x_2) \leq H_1(x_1, x_2), \quad H_2(x_1, x_2') \leq H_2(x_1, x_2)$$

hold for all $x_1' \in X_1$ and $x_2' \in X_2$; whence it follows that

$$H_1(x_1, x_2) = \max_{x_1'} H_1(x_1', x_2), \tag{3.2.13}$$

$$H_2(x_1, x_2) = \max_{x_2'} H_2(x_1, x_2'). \tag{3.2.14}$$

Thus, $(x_1, x_2) \in Z^1$ and $(x_1, x_2) \in Z^2$, i.e. $(x_1, x_2) \in Z^1 \cap Z^2$.

The inverse inclusion follows immediately from (3.2.13), (3.2.14).

If the maximum in (3.2.9) and (3.2.10) to be reached in unique point for each x_2 and x_1 respectively, then we can say about reaction functions $R_1(x_2)$, $R_2(x_1)$ and the sets of best responses can be rewritten as

$$Z^1 = \{(R_1(x_2), x_2) | x_2 \in X_2\},$$

$$Z^1 = \{(x_1, R_2(x_1)) | x_1 \in X_1\}.$$

This case take place, for example, if payoff functions H_i are strictly concave on x_i.

Example 13. Entry game [Tirole (2003)]. Consider a two-firm industry. Firm 1 (the existing firm, leader) chooses a level of capital x_1, which is fixed. Firm 2 (the potential entrant, follower) observes x_1 and then chooses its level of capital x_2.

Assume that payoff functions of the two firms are specified by

$$H_1(x_1, x_2) = x_1(1 - x_1 - x_2),$$

$$H_2(x_1, x_2) = x_2(1 - x_1 - x_2).$$

The game between the two firms is a two-stage game. Firm 1 must foresee the reaction of firm 2 to capital level x_1. Reaction functions $x_2 = R_2(x_1)$, $x_1 = R_1(x_2)$ exists because functions H_i are strictly concave on x_i. Payoff maximization H_2 on x_2 give us first-order condition

$$\frac{\partial H_2}{\partial x_2} = 1 - x_1 - 2x_2 = 0$$

and reaction function

$$x_2 = R_2(x_1) = \frac{1 - x_1}{2}.$$

Therefore, firm 1 maximizes function

$$H_1\left(x_1, \frac{1 - x_1}{2}\right) = x_1\left(1 - x_1 - \frac{1 - x_1}{2}\right)$$

from which we can determine Stakelberg 1-equilibrium:

$$x_1^1 = \frac{1}{2}, \quad x_2^1 = \frac{1}{4}, \quad H_1^1 = \frac{1}{8}, \quad H_2^1 = \frac{1}{16}.$$

If the two firms choose their strategies simultaneously, each would react to other optimally, so that $x_2 = R_2(x_1)$ and $x_1 = R_1(x_2)$, where

$$x_2 = R_2(x_1) = \frac{1 - x_1}{2}, \quad x_1 = R_1(x_2) = \frac{1 - x_2}{2}.$$

Solving the equations we receive the Nash equilibrium $x_1^* = x_2^* = \frac{1}{3}$, $H_1^* = H_2^* = \frac{1}{9}$.

Comparing the Nash and Stakelberg equilibrium we see the leader's advantage.

Definition [Moulin (1981)]. We say that the two-person game $\Gamma = (X_1, X_2, H_1, H_2)$ involves competition for leadership if there exists a situation $(x_1, x_2) \in X_1 \times X_2$ such that

$$\overline{H}_i \leq H_i(x_1, x_2), \quad i = 1, 2. \tag{3.2.15}$$

Theorem [Moulin (1981)]. *If the two-person game* $\Gamma = (X_1, X_2, H_1, H_2)$ *has at least two pareto-optimal and Nash equilibrium situations* (x_1, x_2), (y_1, y_2) *with different payoff vectors*

$$(H_1(x_1, x_2), H_2(x_1, x_2)) \neq (H_1(y_1, y_2), H_2(y_1, y_2)), \tag{3.2.16}$$

then the game Γ *involves competition for leadership.*

Proof. By (3.2.12), for any Nash equilibrium $(z_1, z_2) \in Z(\Gamma)$ we have

$$H_i(z_1, z_2) \leq \overline{H}_i, \quad i = 1, 2.$$

Suppose the opposite is true, i.e. the game Γ does not involve competition for leadership. Then there is a situation $(z_1, z_2) \in X_1 \times X_2$ for which

$$H_i(x_1, x_2) \leq \overline{H}_i \leq H_i(z_1, z_2), \tag{3.2.17}$$

$$H_i(y_1, y_2) \leq \overline{H}_i \leq H_i(z_1, z_2), \tag{3.2.18}$$

$i = 1, 2$. But (x_1, x_2), (y_1, y_2) are pareto-optimal situations, and hence the inequalities (3.2.17), (3.2.18) are satisfied as equalities, which contradicts (3.2.16). This completes the proof of the theorem.

In conclusion we may say that the games "battle of the sexes" and "crossroads" (as in 3.1.4) satisfy the condition of the theorem (as in 3.2.5) and hence involve competition for leadership.

Example 14. Auction of an indivisible goods [Moulin (1986)]. A seller has one indivisible unit of an object for sale with reservation price c. There are n potential bidders with valuations

$$c \leq v_n \leq \ldots \leq v_2 \leq v_1$$

and these valuations and the reservation price are common knowledge. The bidders simultaneously submit bids $x_i \geq c$. The highest bidder wins the object.

a) *First-price auction.* In this case, the winner pays the highest bid. Set of strategies for each bidder is $X_i = X = [c, +\infty)$. Suppose that profile $x = (x_1, \ldots, x_n)$ is realized. Denote the set of highest bidders as

$$w(x) = \{i | x_i = \max_j x_j\}.$$

Then the payoff function for bidder i have the following form

$$H_i(x_1, \dots, x_i, \dots, x_n) = \begin{cases} v_i - x_i, & i = \min_{j \in w(x)} j, \\ 0, & i \neq \min_{j \in w(x)} j. \end{cases}$$

b) *Second-price auction.* In second-price auction winner i pays the second bid, i.e.

$$x_i^+ = \max_{j \neq i} x_j.$$

Then the payoff function for bidder i is

$$H_i(x_1, \dots, x_i, \dots, x_n) = \begin{cases} v_i - x_i^+, & i = \min_{j \in w(x)} j, \\ 0, & i \neq \min_{j \in w(x)} j. \end{cases}$$

There are many Nash equilibria in the models. The structure of Nash equilibrium is the following: $x^* = (v_2, v_2, x_3, \dots, x_n)$, where $x_i \in [c, v_i]$, $i = 3, 4, \dots, n$.

But if bidders' valuations are not common knowledge and bidder i knows its own valuation, then the first-price auction does not have Nash equilibrium in pure strategies. In this case second-price auction has Nash equilibrium in truthful strategies: $x^* = (v_1, v_2, \dots, v_n)$. However, the Nash equilibrium is pareto-dominated in both cases.

3.3 Mixed Extension of Noncooperative Game

3.3.1. We shall examine a noncooperative two-person game $\Gamma = (X_1, X_2, H_1, H_2)$. In the nonzero-sum case, we have already seen that an equilibrium in pure strategies generally does not exist. The matrix games have an equilibrium in mixed strategies. For this reason, it would appear natural that the Nash equilibrium in a noncooperative game would be sought in the class of mixed strategies.

As in the case of zero-sum games, we identify the player's mixed strategy with the probability distribution over the set of pure strategies. For simplicity, we assume that the sets of strategies X_i are finite, and introduce the notion of a mixed extension of the game. Let

$$\Gamma = (N, \{X_i\}_{i \in N}, \{H_i\}_{i \in N}) \tag{3.3.1}$$

be an arbitrary finite noncooperative game. For definitiveness, suppose the player i in the game Γ has m_i strategies.

Denote by μ_i an arbitrary mixed strategy of Player i, i.e. some probability distribution over the set of strategies X_i to be referred to as pure strategies. Also, denote by $\mu_i(x_i)$ the probability prescribed by strategy μ_i to the particular pure strategy $x_i \in X_i$. The set of all mixed strategies of player i will be denoted by $\overline{X}_i$.

Suppose each player $i \in N$ uses his mixed strategy μ_i, i.e. he chooses pure strategies with probabilities $\mu_i(x_i)$. The probability that a situation $x = (x_1, \ldots, x_n)$ may arise is equal to the product of the probabilities of choosing its component strategies, i.e.

$$\mu(x) = \mu_1(x_1) \times \mu_2(x_2) \times \ldots \times \mu_n(x_n). \tag{3.3.2}$$

Formula (3.3.2) defines the probability distribution over the set of all situations $X = \prod_{i=1}^{n} X_i$ determined by mixed strategies $\mu_1, \mu_2, \ldots, \mu_n$. The n-tuple $\mu = (\mu_1, \ldots, \mu_n)$ is called a situation in mixed strategies. The situation in mixed strategies μ realizes various situations in pure strategies with some probabilities; hence the value of the payoff function for each player turns out to be a random variable. The value of the payoff function for the ith player in the situation μ is taken to be the mathematical expectation of this random variable:

$$\begin{aligned} K_i(\mu) &\triangleq \sum_{x \in X} H_i(x)\mu(x) \\ &= \sum_{x_1 \in X_1} \cdots \sum_{x_n \in X_n} H_i(x_1, \ldots, x_n) \times \mu_1(x_1) \times \ldots \times \mu_n(x_n), \end{aligned}$$

$$i \in N,\ x = (x_1, \ldots, x_n) \in X. \tag{3.3.3}$$

We introduce the notation

$$K_i(\mu\|x'_j) \triangleq \sum_{x_1\in X_1} \cdots \sum_{x_{j-1}\in X_{j-1}} \sum_{x_{j+1}\in X_{j+1}} \cdots \times \sum_{x_n\in X_n} H_i(x\|x'_j) \prod_{k\neq j} \mu_k(x_k). \tag{3.3.4}$$

Let μ'_j be an arbitrary mixed strategy for player j in the game Γ. Multiplying (3.3.4) by $\mu'(x'_j)$ and summing over all $x'_j \in X_j$, we obtain

$$\sum_{x'_j\in X_j} K_i(\mu\|x'_j)\mu'_j(x'_j) \triangleq K_i(\mu\|\mu'_j).$$

Definition. The game $\overline{\Gamma} = (N, \{\overline{X}_i\}_{i\in N}\{K_i\}_{i\in N})$, in which N is the set of players, $\overline{X}_i$ is the set of mixed strategies of each player i, and the payoff function is defined by (3.3.3), is called a mixed extension of the game Γ.

If the inequality $K_j(\mu\|x_i) \leq a$ holds for any pure strategy x_i of player i, then the inequality $K_j(\mu\|\mu^*_i) \leq a$ holds for any mixed strategy μ^*_i. The truth of this assertion follows from (3.3.3) and (3.3.4) by a standard shift to mixed strategies.

3.3.2. For the bimatrix $(m \times n)$ game $\Gamma(A, B)$ we may define the respective sets of mixed strategies X_1, X_2 for Players 1 and 2 as

$$X_1 = \{x \mid xu = 1,\ x \geq 0,\ x \in R^m\},$$

$$X_2 = \{y \mid yw = 1,\ y \geq 0,\ y \in R^n\},$$

where $u = (1, \ldots, 1) \in R^m$, $w = (1, \ldots, 1) \in R^n$. We also define the players' payoffs K_1 and K_2 at (x, y) in mixed strategies as payoff expectations

$$K_1(x, y) \triangleq xAy, \quad K_2(x, y) \triangleq xBy, \quad x \in X_1,\ y \in X_2.$$

Thus, we have constructed formally a mixed extension $\overline{\Gamma}(A, B)$ of the game $\Gamma(A, B)$, i.e. the noncooperative two-person game $\overline{\Gamma}(A, B) = (X_1, X_2, K_1, K_2)$.

For the bimatrix game (just as for the matrix game) the set $M_x = \{i|\xi_i > 0\}$ will be called Player 1's *spectrum of mixed strategy* $x = (\xi_1, \ldots, \xi_m)$, while the strategy x, for which $M_x = M$, $M = \{1, 2, \ldots, m\}$, will be referred to as *completely mixed*. Similarly, $N_y = \{j|\eta_j > 0\}$ will be Player 2's spectrum of mixed strategy $y = \{\eta_1, \ldots, \eta_n\}$ in the bimatrix $(m \times n)$ game $\Gamma(A, B)$. The situation (x, y), in which both strategies x and y are completely mixed, will be referred to as *completely mixed*.

We shall now use the "battle of the sexes" game to demonstrate that the difficulties encountered in examination of a noncooperative game (Example 9, 3.2.2) are not resolved through introduction of mixed strategies.

Example 15. Suppose Player 1 in the "battle of the sexes" game wishes to maximize his guaranteed payoff. This means that he is going to choose a mixed strategy $x^0 = (\xi^0, 1 - \xi^0)$, $0 \leq \xi^0 \leq 1$ so as to maximize the least of the two quantities $K_1(x, \beta_1)$ and $K_1(x, \beta_2)$, i.e.

$$\max_x \min\{K_1(x, \beta_1), K_1(x, \beta_2)\} = \min\{K_1(x^0, \beta_1), K_1(x^0, \beta_2)\}.$$

The maximin strategy x^0 of Player 1 is of the form $x^0 = (1/5, 4/5)$ and guarantees him, on the average, a payoff of 4/5. If Player 2 chooses strategy β_1, then the players' payoffs are $(4/5, 1/5)$. However, if he uses strategy β_2, then the players' payoffs are $(4/5, 16/5)$.

Thus, if Player 2 suspects that his partner will use the strategy x^0, then he will choose β_2 and receive a payoff of 16/5. (If Player 1 can justify the choice of β_2 for Player 2, then he may also improve his own choice.) Similarly, suppose Player 2 uses a maximin strategy that is $y^0 = (4/5, 1/5)$. If Player 1 chooses strategy α_1 then the players' payoffs are $(16/5, 4/5)$. However, if he chooses α_2, then the players' payoffs are $(1/5, 4/5)$. Therefore, it is advantageous for him to use his strategy α_1 against the maximin strategy y^0.

If both players follow this line of reasoning, they will arrive at a situation (α_1, β_2), in which the payoff vector is $(0, 0)$. Hence the situation (x^0, y^0) in maximin mixed strategies is not a Nash equilibrium.

3.3.3. Definition. The situation μ^* is called a Nash equilibrium in mixed strategies in the game Γ if for any player i, and for any mixed strategies μ_i the following inequality holds:

$$K_i(\mu^* \| \mu_i) \leq K_i(\mu^*), \ i = 1, \dots, n.$$

Example 15 shows that a situation in maximin mixed strategies is not necessarily a Nash equilibrium in mixed strategies.

Example 16. The game of "crossroads" (see Example 12, 3.2.4) has two Nash equilibria in pure strategies: (α_1, β_2) and (α_2, β_1). These situations are also pareto-optimal. The mixed extension of the game gives rise to one more equilibrium situation, namely the pair (x^*, y^*):

$$x^* = y^* = \frac{1-\epsilon}{2-\epsilon} u_1 + \frac{1}{2-\epsilon} u_2,$$

where $u_1 = (1, 0)$, $u_2 = (0, 1)$ or $x^* = y^* = ((1-\epsilon)/(2-\epsilon), 1/(2-\epsilon))$.

Indeed, we have

$$K_1(\alpha_1, y^*) = \frac{1-\epsilon}{2-\epsilon} + \frac{1-\epsilon}{2-\epsilon} = 1 - \frac{\epsilon}{2-\epsilon},$$

$$K_1(\alpha_2, y^*) = 2\frac{1-\epsilon}{2-\epsilon} = 1 - \frac{\epsilon}{2-\epsilon}.$$

Furthermore, since for any pair of mixed strategies $x = (\xi, 1-\xi)$ and $y = (\eta, 1-\eta)$, we have

$$K_1(x, y^*) = \xi K_1(\alpha_1, y^*) + (1-\xi) K_1(\alpha_2, y^*) = 1 - \frac{\epsilon}{2-\epsilon},$$

$$K_2(x^*, y) = \eta K_2(x^*, \beta_1) + (1-\eta) K_2(x^*, \beta_2) = 1 - \frac{\epsilon}{2-\epsilon},$$

then we get

$$K_1(x, y^*) = K_1(x^*, y^*), \ K_2(x^*, y) = K_2(x^*, y^*)$$

for all mixed strategies $x \in X_1$ and $y \in X_2$. Therefore, (x^*, y^*) is a Nash equilibrium. Furthermore, it is a completely mixed equilibrium. But the situation (x^*, y^*) is not pareto-optimal, since the vector $K(x^*, y^*) = (1 - \epsilon/(2-\epsilon), 1 - \epsilon/(2-\epsilon))$ is strictly (component-wise) smaller than the payoff vector $(1, 1)$ in the situation (α_1, β_1).

Let $K(\mu^*) = \{K_i(\mu^*)\}$ be a payoff vector in some Nash equilibrium. Denote $v_i = K_i(\mu^*)$ and $v = \{v_i\}$. While the zero-sum games have the same value v of the payoff function in all equilibrium points and hence this value was uniquely defined for each zero-sum game, which had such an equilibrium, in the nonzero-sum games there is a whole set of vectors v. Thus, every vector v is connected with a special equilibrium point μ^*, $v_i = K_i(\mu^*)$, $\mu^* \in \overline{X}$, $\overline{X} = \prod_{i=1}^n \overline{X}_i$.

In the game of "crossroads", the equilibrium payoff vector (v_1, v_2) at the equilibrium point (α_1, β_2) is of the form $(1-\epsilon, 2)$, whereas at (x^*, y^*) it is equal to $(1 - \epsilon/(2-\epsilon), 1 - \epsilon/(2-\epsilon))$ (see Example 12).

3.3.4. If the strategy spaces in the noncooperative game $\Gamma = (X_1, X_2, H_1, H_2)$ are infinite, e.g. $X_1 \subset R^m$, $X_2 \subset R^n$, then as in the case of zero-sum infinite game, the mixed strategies of the players are identified with the probability measures given on Borel σ-algebras of the sets X_1 and X_2. If μ and ν are respectively the mixed strategies of Players 1 and 2, then a payoff to player i in this situation $K_i(\mu, \nu)$ is the mathematical expectation of payoff, i.e.

$$K_i(\mu, \nu) = \int_{X_1}\int_{X_2} H_i(x, y) d\mu(x) d\nu(y), \tag{3.3.5}$$

where the integrals are taken to be Lebesgue–Stieltjes integrals. Note that, the payoffs to the players at (x, ν) and (μ, y) are

$$K_i(x, \nu) = \int_{X_2} H_i(x, y) d\nu(y),$$

$$K_i(\mu, y) = \int_{X_1} H_i(x, y) d\mu(x), \;\; i = 1, 2.$$

(All integrals are assumed to exist.)

Formally, the mixed extension of the noncooperative two-person game Γ can be defined as a system $\overline{\Gamma} = (\overline{X}_1, \overline{X}_2, K_1, K_2)$, where

$\overline{X}_1 = \{\mu\}$, $\overline{X}_2 = \{\nu\}$ with K_1 and K_2 determined by (3.3.5). The game $\overline{\Gamma}$ is a noncooperative two-person game, and hence the situation (μ^*, ν^*) is equilibrium if and only if the inequalities (as in (3.2.3)) are satisfied.

3.4 Existence of Nash Equilibrium

3.4.1. In the theory of zero-sum games, the continuity of 1 payoff function and the compactness of strategy sets (see 2.4.4) is sufficient for the existence of an equilibrium in mixed strategies. It turns out that these conditions also suffice for the existence of a Nash equilibrium in mixed strategies where a noncooperative two-person game is concerned.

First we prove the existence of an equilibrium in mixed strategies for a bimatrix game. This proof is based on the familiar Kakutani's fixed point theorem. This theorem will be given without proof (see 3.5.5).

Theorem. *Let S be a convex compact set in R^n and ψ be a multi-valued map which corresponds to each point of S the convex compact subsets of S and satisfies the condition: if $x_n \in S$, $x_n \to x$, $y_n \in \psi(x_n)$, $y_n \to y$ and $y \in \psi(x)$. Then there exists $x^* \in S$ such that $x^* \in \psi(x^*)$.*

Theorem. *Let $\Gamma(A, B)$ be a bimatrix $(m \times n)$ game. Then there are mixed strategies $x^* \in X_1$ and $y^* \in X_2$ for Players 1 and 2, respectively, such that the pair (x^*, y^*) is a Nash equilibrium.*

Proof. The mixed strategy sets X_1 and X_2 of Players 1 and 2 are convex polyhedra. Hence the set of situations $X_1 \times X_2$ is a convex compact set.

Let ψ be a multi-valued map,

$$\psi : X_1 \times X_2 \to X_1 \times X_2,$$

determined by the relationship

$$\psi : (x_0, y_0) \to \left\{ (x', y') \;\middle|\; \begin{array}{l} K_1(x', y_0) = \max_{X_1} K_1(x, y_0), \\ K_2(x_0, y') = \max_{X_2} K_2(x_0, y), \end{array} \right.$$

i.e. the image of the map ψ consists of the pairs of the players' best responses to the strategies y_0 and x_0, respectively.

The functions K_1 and K_2 as the mathematical expectations of the payoffs in the situation (x, y) are bilinear in x and y, and hence the image $\psi(x_0, y_0)$ of the situation (x_0, y_0) with ψ as a map represents a convex compact subset in $X_1 \times X_2$. Furthermore, if the sequence of pairs $\{(x_0^n, y_0^n)\}$, $(x_0^n, y_0^n) \in X_1 \times X_2$ and $\{(x'_n, y'_n)\}$, $(x'_n, y'_n) \in \psi(x_0^n, y_0^n)$ have limit points, i.e.

$$\lim_{n\to\infty} (x_0^n, y_0^n) = (x_0, y_0), \quad \lim_{n\to\infty} (x'_n, y'_n) = (x', y'),$$

then by the bilinearity of the functions K_1 and K_2, and because of the compactness of the sets X_1 and X_2, we have $(x', y') \in \psi(x_0, y_0)$. Then, by the Kakutani's theorem, there exists a situation $(x^*, y^*) \in X_1 \times X_2$ for which $(x^*, y^*) \in \psi(x^*, y^*)$, i.e.

$$K_1(x^*, y^*) \geq K_1(x, y^*), \quad K_2(x^*, y^*) \geq K_2(x^*, y),$$

for all $x \in X_1$ and $y \in X_2$. This completes the proof of the theorem.

3.4.2. The preceding theorem can be extended to the case of continuous payoff functions H_1 and H_2. To prove this result, we have to use the well-known Brouwer fixed point theorem [Parthasarathy and Raghavan (1971)].

Theorem. *Let S be a convex compact set in R^n which has an interior. If φ is a continuous self-map of S, then there exists a fixed point x^* of the map φ, i.e. $x^* \in S$ and $x^* = \varphi(x^*)$.*

Theorem. *Let $\Gamma = (X_1, X_2, H_1, H_2)$ be a noncooperative two-person game, where the strategy spaces $X_1 \subset R^m$, $X_2 \subset R^n$ are convex compact subsets and the set $X_1 \times X_2$ has an interior. Also, let the payoff functions $H_1(x, y)$ and $H_2(x, y)$ be continuous in $X_1 \times X_2$, with $H_1(x, y)$ being concave in x at every fixed y and $H_2(x, y)$ being concave in y at every fixed x.*

Then the game Γ has the Nash equilibrium (x^*, y^*).

Proof. Let $p \triangleq (x, y) \in X_1 \times X_2$ and $q \triangleq (\overline{x}, \overline{y}) \in X_1 \times X_2$ be two situations in the game Γ. Consider the function

$$\theta(p, q) = H_1(x, \overline{y}) + H_2(\overline{x}, y).$$

First we show that there exists a situation $q^* = (x^*, y^*)$ for which

$$\max_{p \in X_1 \times X_2} \theta(p, q^*) = \theta(q^*, q^*).$$

Suppose this is not the case. Then for each $q \in X_1 \times X_2$ there is a $p \in X_1 \times X_2$, $p \neq q$, such that $\theta(p, q) > \theta(q, q)$. Introduce the set

$$G_p = \{q | \theta(p, q) > \theta(q, q)\}.$$

Since the function θ is continuous (H_1 and H_2 are continuous in all their variables) and $X_1 \times X_2$ is a convex compact set, then the sets G_p are open. Furthermore, by the assumptions, $X_1 \times X_2$ is covered by the sets from the family G_p.

It follows from the compactness of $X_1 \times X_2$ that there is a finite collection of these sets which covers $X_1 \times X_2$. Suppose these are the sets $G_{p_1}, \dots, G_{p_k}$. Denote

$$\varphi_j(q) = \max\{\theta(p_j, q) - \theta(q, q), 0\}.$$

The functions $\varphi_j(q)$ are non-negative, and, by the definition of G_{p_j}, at least one of the functions φ_j takes a positive value at every point q.

We shall now define the self-map ψ of the set $X_1 \times X_2$ as follows:

$$\psi(q) \triangleq \frac{1}{\varphi(q)} \sum_j \varphi_j(q) p_j,$$

where $\varphi(q) = \sum_j \varphi_j(q)$. The functions φ_j are continuous and hence ψ is a continuous self-map of $X_1 \times X_2$. By the Brouwer's fixed point theorem, there is a point $\overline{q} \in X_1 \times X_2$ such that $\psi(\overline{q}) = \overline{q}$, i.e.

$$\overline{q} = (1/\varphi(\overline{q})) \sum_j \varphi_j(\overline{q}) p_j.$$

Consequently,

$$\theta(\overline{q}, \overline{q}) = \theta\left(\frac{1}{\varphi(\overline{q})} \sum_j \varphi_j(\overline{q}) p_j, \overline{q}\right).$$

But the function $\theta(p, q)$ is concave in p, with q fixed, and hence

$$\theta(\overline{q}, \overline{q}) \geq \frac{1}{\varphi(\overline{q})} \sum_j \varphi_j(\overline{q})\theta(p_j, \overline{q}). \tag{3.4.1}$$

On the other hand, if $\varphi_j(\overline{q}) > 0$, then $\theta(\overline{q}, \overline{q}) < \theta(p_j, \overline{q})$, and if $\varphi_j(\overline{q}) = 0$, then $\varphi_j(\overline{q})\theta(p_j, \overline{q}) = \varphi_j(\overline{q})\theta(\overline{q}, \overline{q})$. Since $\varphi_j(\overline{q}) > 0$ for some j, we get the inequality

$$\theta(\overline{q}, \overline{q}) < \frac{1}{\varphi(\overline{q})} \sum_j \varphi_j(\overline{q})\theta(p_j, \overline{q}),$$

which contradicts (3.4.1).

Thus, there always exists q^* for which

$$\max_{p \in X_1 \times X_2} \theta(p, q^*) = \theta(q^*, q^*).$$

Which means that

$$H_1(x, y^*) + H_2(x^*, y) \leq H_1(x^*, y^*) + H_2(x^*, y^*)$$

for all $x \in X_1$ and $y \in X_2$. Setting successively $x = x^*$ and $y = y^*$ in the last inequality, we obtain the inequalities

$$H_2(x^*, y) \leq H_2(x^*, y^*), \quad H_1(x, y^*) \leq H_1(x^*, y^*),$$

which hold for all $x \in X_1$ and $y \in X_2$. This completes the proof of the theorem.

The result given below holds for the noncooperative two-person games played on compact sets (specifically, on a unit square) with a continuous payoff function.

Example 17. Cournot duopoly [Tirole (2003)]. Consider the Cournot model of duopoly producing a homogeneous good. The strategies are quantities. Firms 1 and 2 simultaneously choose their strategies q_i from feasible sets $X = Y = \{q | q \geq 0\}$. Suppose that market demand function $q = d(c)$, $p \geq 0$ is strictly decreasing. Then exist strictly decreasing inverse function

$$p = p(q) = d^{-1}(q), \quad q \geq 0.$$

Firms sell their good at the market-clearing price $p(q)$, where $q = q_1 + q_2$. Firm i's cost of production is increasing function $c_i(q_i)$ and firm i's total profit is

$$H_1(q_1, q_2) = q_1 p(q) - c_1(q_1),$$
$$H_2(q_1, q_2) = q_2 p(q) - c_2(q_2).$$

The Cournot duopoly can be described as nonzero-sum two person game in normal form $\Gamma_C = (X, Y, H_1, H_2)$.

The existence of a pure-strategy Nash equilibrium in Cournot duopoly follows from concavity profit functions H_i on q_i. Calculate partial derivatives

$$\frac{\partial H_i}{\partial q_i} = p(q_i + q_j) - c_i'(q_i) + q_i p'(q_i + q_j)$$

and

$$\frac{\partial^2 H_i}{\partial q_i^2} = 2p' + q_i p'' - c_i''.$$

Recall that $p' < 0$. For the payoff function to be concave $\left(\frac{\partial^2 H_i}{\partial q_i^2} < 0\right)$, it suffices that the firm's cost function be convex $(c'' \geq 0)$ and that the inverse-demand function be concave $(p'' \geq 0)$.

The Nash equilibrium is easily derived in case of linear demand and cost. Suppose that $d(p) = a - p$ (or $p(q) = a - q$) and $c_i(q_i) = c_i q_i$, $c_i > 0$. Let us consider the first-order condition to find Nash equilibrium

$$\begin{cases} \dfrac{\partial H_1}{\partial q_1} = a - 2q_1 - q_2 - c_1 = 0, \\ \dfrac{\partial H_2}{\partial q_2} = a - q_1 - 2q_2 - c_2 = 0. \end{cases}$$

Hence, the Nash equilibrium is given by

$$q_1^* = \frac{a - 2c_1 + c_2}{3}, \quad q_2^* = \frac{a + c_1 - 2c_2}{3}$$

and the profits are

$$\Pi_1^* = \frac{(a - 2c_1 + c_2)^2}{9}, \quad \Pi_2^* = \frac{(a + c_1 - 2c_2)^2}{9}.$$

Example 18. Hotelling competition [Fudenberg and Tirole (1992)]. A city of length 1 lies on the line, and consumers are uniformly distributed with density 1 along the interval $[0, 1]$ on this line. There are two stores located at the two extremes of city, which sell the same product. p_1, p_2 are prices of the unit of good in stores 1 and 2 correspondingly. Suppose that store 1 is at $x = 0$ and store 2 at $x = 1$.

Consumers incur a transportation cost t per unit of distance and have unit demands.

A consumer who is indifferent between the two stores is located at x, where x can be calculated from the equation

$$p_1 + tx = p_2 + t(1 - x).$$

The demand for store 1 is equal to the number of consumers who find it cheaper to buy from store 1 and is given by

$$d_1(p_1, p_2) = x = \frac{p_2 - p_1 + t}{2t}$$

and

$$d_2(p_1, p_2) = 1 - x = \frac{p_1 - p_2 + t}{2t}.$$

Firm's profits are

$$H_1(p_1, p_2) = (p_1 - c)\frac{(p_2 - p_1 + t)}{2t},$$

$$H_2(p_1, p_2) = (p_2 - c)\frac{(p_1 - p_2 + t)}{2t}.$$

This game is a strictly concave, that is why we can calculate thee Nash equilibrium from first-order conditions

$$\begin{cases} p_2 + c + t - 2p_1 = 0, \\ p_1 + c + t - 2p_2 = 0. \end{cases}$$

Using the symmetry of the problem, we obtain the solution:

$$p_1^* = p_2^* = c + t, \quad \Pi_1^* = \Pi_2^* = \frac{t}{2} > 0.$$

Theorem. *Let $\Gamma = (X_1, X_2, H_1, H_2)$ be a noncooperative two-person game, where H_1 and H_2 are continuous functions on $X_1 \times X_2$; X_1, X_2 are compact subsets in finite-dimensional Euclidean spaces. Then the game Γ has an equilibrium (μ, ν) in mixed strategies.*

This theorem is given without proof, since it is based on the continuity and bilinearity of the function

$$K_i(\mu, \nu) = \int_{X_1}\int_{X_2} H_i(x, y) d\mu(x) d\nu(y), \quad i = 1, 2.$$

over the set $\overline{X}_1 \times \overline{X}_2$ and almost exactly repeats the proof of the preceding theorem.

We shall discuss in more detail the construction of mixed strategies in noncooperative n-person games with an infinite number of strategies. Note that if the players' payoff functions $H_i(x)$ are continuous on the Cartesian product $X = \prod_{i=1}^n X_i$ of the compact sets of pure strategies, then in such a noncooperative game there always exists a Nash equilibrium in mixed strategies. As for the existence of pareto-optimal situations, it suffices to ensure the compactness of the set $\{H(x)\}$, $x \in X$, which in turn can be ensured by the compactness in some topology of the set of all situations X and the continuity in this topology of all the payoff functions K_i, $i = 1, 2, \ldots, n$. It is evident that this is always true for finite noncooperative games.

3.5 Kakutani Fixed-Point Theorem and Proof of Existence of an Equilibrium in n-Person Games

3.5.1. The reader can read Sec. 3.5 without referring to Sec. 3.4. Given any game $\Gamma = \langle N, \{X_i\}_{i \in N}, \{H_i\}_{i \in N}\rangle$ with finite sets of strategies X_i in normal form ($|N| = n$), a mixed strategy for any player i is a probability distribution over X_i. We let $\overline{X}_i$ denote the set of all possible mixed strategies for player i. To underline the distinction from mixed strategies, the strategies in X_i will be called pure strategies.

A mixed strategy profile is any vector that specifies one mixed strategy for each player, so the set of all mixed strategy profiles (situations in mixed strategies) is a Cartesian product $\overline{X} = \prod_{i=1}^{n} \overline{X}_i$. $\mu = (\mu_1, \ldots, \mu_n)$ is a mixed-strategy profile in $\prod_{i=1}^{n} \overline{X}_i$ if and only if, for each Player i and each pure strategy $x_i \in X_i$, μ prescribes a non-negative real number $\mu_i(x_i)$, representing the probability that Player i would choose x_i, such that

$$\sum_{x_i \in X_i} \mu_i(x_i) = 1, \text{ for all } i \in N.$$

If the players choose their pure strategies independently, according to the mixed strategy profile μ, then the probability, that they will choose the pure strategy profile $x = (x_1, \ldots, x_i, \ldots, x_n)$ is $\prod_{i=1}^{n} \mu_i(x_i)$, the multiplicative product of the individual strategy probabilities.

For any mixed strategy profile μ, let $K_i(\mu)$ denote the mathematical expectation of payoff that player i would get when the players independently choose their pure strategies according to μ. Denote $X = \prod_{i=1}^{n} X_i$ (X_i is the set of all possible situation in pure strategies), then

$$K_i(\mu) = \sum_{x \in X} \left(\prod_{j \in N} \mu_j(x_j) \right) H_i(x), \text{ for all } i \in N.$$

For any $\tau_i \in \overline{X}_i$, we denote by $(\mu \| \tau_i)$ the mixed strategy profile in which the i-component is τ_i and all other components are as in μ. Thus

$$K_i(\mu \| \tau_i) = \sum_{x \in X} \left(\prod_{j \in N \setminus \{i\}} \mu_j(x_j) \right) \tau_i(x_i) H_i(x).$$

We shall not use any special notation for the mixed strategy $\mu_i \in \sum_i$ that puts probability 1 on the pure strategy x_i, denoting this mixed strategy by x_i (in the same manner as the corresponding pure strategy).

If player i used the pure strategy x_i, while all other players behaved independently according to the mixed-strategy profile μ, then Player i's mathematical expectation of payoff would be

$$K_i(\mu\|x_i) = \sum_{x_j \in X_j,\ j \neq i} \left(\prod_{j \in N \setminus \{i\}} \mu_j(x_j) \right) H_i(\mu\|x_i).$$

3.5.2. Definition. The mixed strategy profile μ is Nash equilibrium in mixed strategies if

$$K_i(\mu\|\tau_i) \leq K_i(\mu), \text{ for all } \tau_i \in \overline{X}_i,\ i \in N.$$

3.5.3. Lemma. *For any $\mu \in \prod_{i=1}^{n} \overline{X}_i$ and any player $i \in N$,*

$$\max_{x_i \in X_i} K_i(\mu\|x_i) = \max_{\tau_i \in \overline{X}_i} K_i(\mu_i\|\tau_i).$$

Furthermore, $p_i \in \arg\max_{\tau_i \in \overline{X}_i} K_i(\mu\|\tau_i)$ if and only if $p_i(x_i) = 0$ for every x_i such that $x_i \notin \arg\max_{x_i \in X_i} K_i(\mu\|x_i)$.

Proof. Notice that for any $\tau_i \in \overline{X}_i$

$$K_i(\mu\|\tau_i) = \sum_{x_i \in X_i} \tau_i(x_i) K_i(\mu\|x_i).$$

$K_i(\mu\|\tau_i)$ is a mathematical expectation of terms $K_i(\mu\|x_i)$. This mathematical expectation cannot be greater than the maximum value of the random variable $K_i(\mu\|x_i)$, and is strictly less than this maximum value, whenever any nonmaximal value of $K_i(\mu\|x_i)$ gets a positive probability ($\tau_i(x_i) \geq 0$, $\sum_{x_i \in X_i} \tau_i(x_i) = 1$).

So the highest expected payoff that player i can get against any combination of other player's mixed strategies is the same whether he uses a mixed strategy or not.

3.5.4. As we have seen in the two-person case, the Kakutani fixed-point theorem is a useful mathematical tool for proving existence of solution concepts in game theory including Nash equilibrium. To state the Kakutani fixed-point theorem we first develop some terminology.

The set S of a finite dimensional vector space R^m is closed if for every convergent sequence of vectors $\{x^j\}$, $j = 1, \ldots, \infty$, if $x^j \in S$ for every j, then $\lim_{j\to\infty} x^j \in S$.

The set S is bounded if there exists some positive number K such that for every $x \in S$, $\sum_{i=1}^{m} \xi_i^2 \leq K$ (here $x = \{\xi_i\}$, ξ_i are the components of x).

A point-to-set correspondence $F : X \to Y$ is any mapping that specifies, for every point x in X, a set $F(x)$ that is subset of Y. Suppose that X and Y are any metric spaces, so the notion of convergence and limit are defined for sequences of points in X and in Y. A correspondence $F : X \to Y$ is upper-semicontinuous if, for every sequence x^j, y^j, $j = 1, \ldots, \infty$, if $x^j \in S$ and $y^j \in F(x^j)$ for every j, and the sequence $\{x^j\}$ converges to some point $\overline{x}$, and the sequence $\{y^j\}$ converges to some point $\overline{y}$, then $\overline{y} \in F(\overline{x})$. Thus $F : X \to Y$ is upper-semicontinuous, if the set $\{(x, y) : x \in X, y \in F(x)\}$ is a closed subset of $X \times Y$.

A fixed point of a correspondence $F : S \to S$ is any x in S such that $x \in F(x)$.

3.5.5. Theorem (Kakutani). *Let S be any nonempty, convex, bounded, and closed subset of a finite-dimensional vector space R^m. Let $F : S \to S$ be any upper-semicontinuous point-to-set correspondence such that, for every x in S, $F(x)$ is a nonempty convex subset of S. Then there exists some $\overline{x}$ in S such that $\overline{x} \in F(\overline{x})$.*

Proofs of the Kakutani fixed-point Theorem can be found in Burger (1963).

With the help of the previous theorem we shall prove the following fundamental result.

3.5.6. Theorem. *Given any finite n-person game* Γ *in normal form, there exists at least one equilibrium in mixed strategies (in* $\prod_{i\in N}\overline{X}_i$*).*

Proof. Let Γ be any finite game in normal form

$$\Gamma = (N, \{X_i\}_{i\in N}, \{H_i\}_{i\in N}).$$

The set of mixed-strategy profiles $\prod_{i=1}^{n}\overline{X}_i$ is a nonempty, convex, closed, and bounded subset of a finite dimensional vector space. This set satisfies the above definition of boundedness with $K = |N|$ and it is a subset of R^m, where $m = \sum_{i=1}^{n}|X_i|$ (here $|A|$ means the number of elements in a finite set A).

For any $\mu \in \prod_{i=1}^{n}\overline{X}_i$ and any player $j \in N$, let

$$R_j(\mu) = \arg\max_{\tau_j\in\overline{X}_j} K_i(\mu\|\tau_j).$$

That is, $R_j(\mu)$ is the set of best responses in $\overline{X}_j$ to the combination of independently mixed strategies $(\mu_1, \dots, \mu_{j-1}, \mu_{j+1}, \dots, \mu_N)$ of other players. By previous lemma $R_j(\mu)$ is the set of all probability distributions ρ_j over X_j such that

$$\rho_j(x_j) = 0 \text{ for every } x_j \text{ such that } x_j \notin \arg\max_{y_j\in\overline{X}_j} K_j(\mu\|y_j).$$

Thus, $R_j(\mu)$ is convex, because it is a subset of $\overline{X}_j$ that is defined by a collection of linear equalities. Furthermore, $R_j(\mu)$ is nonempty, because it includes x_j from the set $\arg\max_{y_j\in\overline{X}_j} K_j(\mu\|y_j)$, which is nonempty.

Let $R : \prod_{i=1}^{n}\overline{X}_i \to \prod_{i=1}^{n}\overline{X}_i$ be the point-to-set correspondence such that

$$R(\mu) = \prod_{i=1}^{n} R_j(\mu), \text{ for all } \mu \in \prod_{i=1}^{n}\overline{X}_i.$$

That is, $\tau \in R(\mu)$ if and only if $\tau_j \in R_j(\mu)$ for every $j \in N$. For each μ, $R(\mu)$ is nonempty and convex, because it is the Cartesian product of nonempty convex sets.

To show that R is upper-semicontinuous, suppose that $\{\mu^k\}$ and $\{\tau^k\}$, $k = 1, \ldots, \infty$ are convergent sequences, $\mu^k \in \prod_{i \in N} \overline{X}_i$, $k = 1, 2, \ldots$; $\tau_k \in R(\mu_k)$, $k = 1, 2, \ldots$; $\overline{\mu} = \lim_{k \to \infty} \mu^k$, $\overline{\tau} = \lim_{k \to \infty} \tau^k$.

We have to show that $\overline{\tau} \in R(\overline{\mu})$. For every player $j \in N$ and every $\rho_j \in \overline{X}_j$

$$K_j(\mu^k \| \tau_j^k) \geq K_j(\mu^k \| \rho_j), \quad k = 1, 2, \ldots.$$

By continuity of the mathematical expectation $K_j(\mu)$ on $\prod_{i=1}^n \overline{X}_i$, this in turn implies that, for every $j \in N$ and $\rho_j \in \overline{X}_j$,

$$K_j(\overline{\mu} \| \overline{\tau}_j) \geq K_j(\overline{\mu} \| \rho_j).$$

Thus $\overline{\tau}_j \in R_j(\overline{\mu})$ for every $j \in N$, and by the definition of $R(\overline{\mu})$, $\overline{\tau} \in R(\overline{\mu})$. And we have proved that $R : \prod_{i \in N} \overline{X}_i \to \prod_{i \in N} \overline{X}_i$ is an upper-semicontinuous correspondence.

By the Kakutani fixed-point theorem, there exists some mixed strategy profile μ in $\prod_{i \in N} \overline{X}_i$ such that $\mu \in R(\mu)$. That is $\mu_j \in R_j(\mu)$ for every $j \in N$ thus $K_j(\mu) \geq K_j(\mu \| \tau_j)$ for all $j \in N$, $\tau_j \in \overline{X}_j$, and so μ is a Nash equilibrium of Γ.

3.6 Refinements of Nash Equilibria

3.6.1. Many attempts were made to choose a particular Nash equilibrium from the set of all possible Nash equilibria profiles. There are some approaches, but today it is very difficult to distinguish among them to find out the most perspective ones. We shall introduce only some of them, but for the better understanding of this topic we refer to the book of Eric van Damme (1991).

3.6.2. One of the ideas is that each player with a small probability makes mistakes, and as a consequence every pure strategy is chosen with a positive (although small) probability. This idea is modelled through perturbed games, i.e. games in which players have to use only completely mixed strategies.

Let $\Gamma = \langle N, X_1, \ldots, X_n, H_1, \ldots, H_n \rangle$ be n-person game in normal form. Denote as before $\overline{X}_i$ — the set of mixed strategies of player i,

K_i — mathematical expectation of the payoff of player i in mixed strategies. For $i \in N$, let $\eta_i(x_i)$, $x_i \in X_i$ and $\overline{X}_i(\eta_i)$ be defined by

$$\overline{X}_i(\eta_i) = \Bigg\{ \mu_i \in \overline{X}_i : \mu_i(x_i) \geq \eta_i(x_i), \quad \text{for all } x_i \in X_i, \text{ where}$$

$$\eta_i(x_i) > 0, \quad \sum_{x_i \in X_i} \eta_i(x_i) < 1 \Bigg\}.$$

Let $\eta(x) = (\eta_1(x_1), \ldots, \eta_n(x_n))$, $x_i \in X_i$, $i = 1, \ldots, n$ and $\overline{X}[\eta(x)] = \prod_{i=1}^{n} \overline{X}_i(\eta_i(x_i))$. The perturbed game (Γ, η) is the infinite game in normal form

$$\Gamma = \big\langle N, \overline{X}_1(\eta_1(x_1)), \ldots, \overline{X}_n(\eta_n(x_n)),$$
$$K_1(\mu_1, \ldots, \mu_n), \ldots, K_n(\mu_1, \ldots, \mu_n)\big\rangle$$

defined over the strategy sets $\overline{X}_i(\eta_i(x_i))$ with payoffs $K_i(\mu_1, \ldots, \mu_n)$, $\mu_i \in \overline{X}_i(\eta_i(x_i))$, $i = 1, \ldots, n$.

3.6.3. It is easily seen that a perturbed game (Γ, η) satisfies the conditions under which the Kakutani fixed point theorem can be used and so such a game possesses at least one equilibrium. It is clear that in such an equilibrium a pure strategy which is not a best reply has to be chosen with a minimum probability. And we have the following lemma.

Lemma. *A strategy profile $\mu \in \overline{X}(\eta)$ is an equilibrium of (Γ, η) if and only if the following condition is satisfied:*

$$\text{if } K_i(\mu \| x_k) < K_i(\mu \| x_l), \text{ then } \mu_i(x_k) = \eta_i(x_k), \text{ for all } i, x_k, x_l.$$

3.6.4. Definition. Let Γ be a game in normal form. An equilibrium μ of Γ is a perfect equilibrium of Γ if μ is a limit point of a sequence $\{\mu(\eta)\}_{\eta \to \infty}$ with $\mu(\eta)$ being Nash equilibrium in a perturbed game (Γ, η) for all η.

For an equilibrium μ of Γ to be perfect it is sufficient that some perturbed games (Γ, η) with η close to zero possess an equilibrium close to μ and that it is not required that all perturbed games (Γ, η) with η close to zero possess such an equilibrium. Let $\{(\Gamma, \eta^k)\}$, $k = 1, \ldots, \infty$ be a sequence of perturbed games for which $\eta^k \to 0$ as $k \to \infty$. Since every game (Γ, η^k) possesses at least one equilibrium μ^k, and since μ is an element of compact set $\overline{X} = \prod_{j=1}^n \overline{X}_i$, there exists one limit point of $\{\mu^k\}$. It can be easily seen that this limit point is an equilibrium of Γ and this will be a perfect equilibrium. Thus the following theorem holds.

Theorem [Selten (1975)]. *Every game in normal form possesses at least one perfect equilibrium.*

3.6.5. *Example 19.* Consider a bimatrix game Γ

$$\begin{array}{c} \\ L_1 \\ R_1 \end{array} \begin{array}{c} \begin{array}{cc} L_2 & \quad R_2 \end{array} \\ \begin{bmatrix} (1,1) & (0,0) \\ (0,0) & (0,0) \end{bmatrix} \end{array}.$$

This game has two equilibria (L_1, L_2) and (R_1, R_2). Consider a perturbed game (Γ, η). In the situation (R_1, R_2) in the perturbed game the strategies R_1 and R_2 will be chosen with probabilities $1 - \eta_1(L_1)$ and $1 - \eta_2(L_2)$ respectively and the strategies L_1 and L_2 will be chosen with probabilities $\eta_1(L_1)$ and $\eta_2(L_2)$. Thus the payoff $K_1^\eta(R_1 R_2)$ in (Γ, η) will be equal

$$K_1^\eta(R_1, R_2) = \eta_1(L_1) \cdot \eta_2(L_2).$$

In the situation (R_1, L_2) the strategies R_1 and L_2 will be chosen with probabilities $(1 - \eta_1(L_1))$ and $(1 - \eta_2(R_2))$, and

$$K_1^\eta(L_1, R_2) = \eta_1(L_1)(1 - \eta_2(L_2)).$$

Since η is small we get

$$K_1^\eta(L_1, R_2) > K_1^\eta(R_1, R_2).$$

Then we see that in perturbed game (R_1, R_2) is not an equilibrium, from this it follows that (R_1, R_2) is not a perfect equilibrium in the original game. It is easily seen that (L_1, L_2) is a perfect equilibrium.

Consider now the game with matrix

$$\begin{array}{c} \\ L_1 \\ R_1 \end{array}\begin{array}{c} \begin{array}{cc} L_2 & \quad R_2 \end{array} \\ \begin{bmatrix} (1,1) & (10,0) \\ (0,10) & (10,10) \end{bmatrix}. \end{array}$$

In this game we can see that a perfect equilibrium (L_1, L_2) is payoff dominated by a non-perfect one. This game also has two different equilibria, (L_1, L_2) and (R_1, R_2). Consider the perturbed game (Γ, η). Show that (L_1, L_2) is a perfect equilibrium in (Γ, η)

$$\begin{aligned} K_1(L_1, L_2) &= (1 - \eta_1(R_1))(1 - \eta_2(R_2)) + 10(1 - \eta_1(R_1))\eta_2(R_2) \\ &\quad + 10\eta_1(R_1)\eta_2(R_2), \\ K_1(R_1, L_2) &= 10(1 - \eta_1(L_1))\eta_2(R_2) + 1 \cdot \eta_1(L_1)(1 - \eta_2(R_2)) \\ &\quad +10\eta_1(L_1)\eta_2(R_2). \end{aligned}$$

For η_1, η_2 small we have

$$K_1(L_1, L_2) > K_2(R_1, L_2).$$

In the similar way we can show that

$$K_2(L_1, L_2) > K_2(L_1, R_2).$$

Consider now (R_1, R_2) in (Γ, η)

$$\begin{aligned} K_1(R_1, R_2) &= 10(1 - \eta_1(L_1))(1 - \eta_2(L_2)) + 10(1 - \eta_2(L_2))\eta_1(L_1) \\ &\quad + \eta_1(L_1)\eta_2(L_2) \\ &= 10(1 - \eta_2(L_2)) + \eta_1(L_1)\eta_2(L_2), \\ K_1(L_1, R_2) &= 10(1 - \eta_1(R_1))(1 - \eta_2(L_2)) + 10\eta_1(R_1)(1 - \eta_2(L_2)) \\ &\quad + (1 - \eta_1(R_1))\eta_2(L_2) \\ &= 10(1 - \eta_2(L_2)) + (1 - \eta_1(R_1))\eta_2(L_2). \end{aligned}$$

For small η we have $K_1^\eta(L_1, R_2) > K_1^\eta(R_1, R_2)$. Thus (R_1, R_2) is not an equilibrium in (Γ, η) and it cannot be a perfect equilibrium in Γ.

It can be seen that (L_1, L_2) equilibrium in (Γ, η), and the only perfect equilibrium in Γ, but this equilibrium is payoff dominated by (R_1, R_2). We see that the perfectness refinement eliminates equilibria with attractive payoffs. At the same time the perfectness concept does not eliminate all intuitively unreasonable equilibria.

As it is seen from the example of [Myerson (1978)]

$$\begin{array}{c} \\ L_1 \\ R_1 \\ A_1 \end{array} \begin{array}{ccc} L_2 & R_2 & A_2 \\ \left[\begin{array}{c} (1,1) \\ (0,0) \\ (-2,-1) \end{array}\right. & \begin{array}{c} (0,0) \\ (0,0) \\ (-2,0) \end{array} & \left.\begin{array}{r} (-1,-2) \\ (0,-2) \\ (-2,-2) \end{array}\right] \end{array}.$$

It can be seen that an equilibrium (R_1, R_2) in this game is also perfect. Namely if the players have agreed to play (R_1, R_2) and if each player expects, that the mistake A will occur with a larger probability than the mistake L, then it is optimal for each player to play R. Hence adding strictly dominated strategies may change the set of perfect equilibria.

3.6.6. There is another refinement of equilibria concept introduced by Myerson (1978), which exclude some "unreasonable" perfect equilibria like (R_1, R_2) in the last example.

This is the so-called *proper equilibrium.* The basic idea underlying the properness concept is that a player when making mistakes, will try much harder to prevent more costly mistakes than he will try to prevent the less costly ones, i.e. that there is some rationality in the mechanism of making mistakes. As a result, a more costly mistake will occur with a probability which is of smaller order than the probability of a less costly one.

3.6.7. Definition. Let $\langle N, \overline{X}_1, \dots, \overline{X}_n, K_1, \dots, K_n \rangle$ be an n-person normal form game in mixed strategies. Let $\epsilon > 0$, and $\mu^\epsilon \in \prod_{i=1}^n \overline{X}_i$. We say that the strategy profile μ^ϵ is an ϵ-proper equilibrium of Γ if μ^ϵ is completely mixed and satisfies

$$\text{if } K_i(\mu^\epsilon \| x_k) < K_i(\mu^\epsilon \| x_l), \text{ then } \mu_i^\epsilon(x_k) < \epsilon \mu_i^\epsilon(x_l) \text{ for all } i, k, l.$$

$\mu \in \prod_{i=1}^{n} \overline{X}_i$ is a proper equilibrium of Γ if μ is a limit point of a sequence $\mu^\epsilon(\epsilon \to 0)$, where μ^ϵ is an ϵ-proper equilibrium of Γ.

The following theorem holds.

Theorem [Myerson (1978)]. *Every normal form game possesses at least one proper equilibrium.*

3.6.8. When introducing perfectness and properness concepts we considered refinements of the Nash equilibrium which are based on the idea that a reasonable equilibrium should be stable against slight perturbations in the equilibrium strategies. There are refinements based on the idea that a reasonable equilibrium should be stable against perturbations in the payoffs of the game. But we do not cover all possible refinements. We recommend the readers to the book of Eric van Damme (1991) for a complete investigation of the problem, see also Berge (1957).

3.7 Properties of Optimal Solutions

3.7.1. We shall now present some of the equilibrium properties which may be helpful in finding a solution of a noncooperative two-person game.

Theorem. *In order for a mixed strategy situation (μ^*, ν^*) in the game $\Gamma = (X_1, X_2, H_1, H_2)$ to be an equilibrium, it is necessary and sufficient that for all the players' pure strategies $x \in X_1$ and $y \in X_2$ the following inequalities be satisfied:*

$$K_1(x, \nu^*) \leq K_1(\mu^*, \nu^*), \tag{3.7.1}$$

$$K_2(\mu^*, y) \leq K_2(\mu^*, \nu^*). \tag{3.7.2}$$

Proof. The necessity is evident, since every pure strategy is a special case of a mixed strategy, and hence inequalities (3.7.1), (3.7.2) must be satisfied. To prove the sufficiency, we need to shift to the mixed strategies of Players 1 and 2, respectively, in inequalities (3.7.1), (3.7.2).

This theorem (as in the case of zero-sum games) shows that, for the proof that the situation forms an equilibrium in mixed strategies it only suffices to verify inequalities (3.7.1), (3.7.2) for opponent's pure strategies. For the bimatrix $(m \times n)$ game $\Gamma(A, B)$ these inequalities become

$$K_1(i, y^*) \triangleq a_i y^* \leq x^* A y^* = K_1(x^*, y^*), \tag{3.7.3}$$

$$K_2(x^*, j) \triangleq x^* b^j \leq x^* B y^* = K_2(x^*, y^*), \tag{3.7.4}$$

where $a_i(b^j)$ are rows (columns) of the matrix $A(B)$, $i = 1, \ldots, m$, $j = 1, \ldots, n$.

3.7.2. Recall that, for matrix games, each essentially pure strategy equalizes any optimal strategy of the opponent (see 1.7.6). A similar result is also true for bimatrix games.

Theorem. *Let $\Gamma(A, B)$ be a bimatrix $(m \times n)$ game and let $(x, y) \in Z(\Gamma)$ be a Nash equilibrium in mixed strategies. Then the equations*

$$K_1(i, y) = K_1(x, y), \tag{3.7.5}$$

$$K_2(x, j) = K_2(x, y) \tag{3.7.6}$$

hold for all $i \in M_x$ and $j \in N_y$, where $M_x(N_y)$ is the spectrum of a mixed strategy $x(y)$.

Proof. By the Theorem 3.7.1, we have

$$K_1(i, y) \leq K_1(x, y) \tag{3.7.7}$$

for all $i \in M_x$. Suppose that at least one strict inequality in (3.7.7) is satisfied. That is

$$K_1(i_0, y) < K_1(x, y), \tag{3.7.8}$$

where $i_0 \in M_x$. Denote by ξ_i the components of the vector $x = (\xi_1, \ldots, \xi_m)$. Then $\xi_{i_0} > 0$ and

$$\begin{aligned} K_1(x, y) &= \sum_{i=1}^{m} \xi_i K_1(i, y) \\ &= \sum_{i \in M_x} \xi_i K_1(i, y) < K_1(x, y) \sum_{i \in M_x} \xi_i = K_1(x, y). \end{aligned}$$

The contradiction proves the validity of (3.7.5). Equation (3.7.6) can be proved in the same way.

This theorem provides a means of finding equilibrium strategies of players in the game $\Gamma(A, B)$. Indeed, suppose we are looking for an equilibrium (x, y), with the strategy spectra M_x, N_y being given. The optimal strategies must then satisfy a system of linear equations

$$ya_i = v_1, \quad xb^j = v_2, \tag{3.7.9}$$

where $i \in M_x$, $j \in N_y$, v_1, v_2 are some numbers. If, however, the equilibrium (x, y) is completely mixed, then the system (3.7.9) becomes

$$Ay = v_1 u, \quad xB = v_2 w, \tag{3.7.10}$$

where $u = (1, \dots, 1)$, $w = (1, \dots, 1)$ are the vectors of suitable dimensions composed of unit elements, and the numbers $v_1 = xAy$, $v_2 = xBy$ are the players' payoffs in the situation (x, y).

3.7.3. Theorem. *Let $\Gamma(A, B)$ be a bimatrix $(m \times m)$ game, where A, B are nonsingular matrices. If the game Γ has a completely mixed equilibrium, then it is unique and is defined by formulas*

$$x = v_2 u B^{-1}, \tag{3.7.11}$$

$$y = v_1 A^{-1} u, \tag{3.7.12}$$

where

$$v_1 = 1/(uA^{-1}u), \quad v_2 = 1/(uB^{-1}u). \tag{3.7.13}$$

Conversely, if $x \geq 0$, $y \geq 0$ hold for the vectors $x, y \in R^m$ defined by (3.7.11)–(3.7.13), *then the pair (x, y) forms an equilibrium in mixed strategies in the game $\Gamma(A, B)$ with the equilibrium payoff vector (v_1, v_2).*

Proof. If (x, y) is a completely mixed equilibrium, then x and y necessarily satisfy system (3.7.10). Multiplying the first of the equations (3.7.10) by A^{-1}, and the second by B^{-1}, we obtain (3.7.11), (3.7.12). On the other hand, since $xu = 1$ and $yu = 1$, we find values for v_1 and v_2. The uniqueness of the completely mixed situation

(x, y) follows from the uniqueness of the solution of system (3.7.10) in terms of the theorem.

We shall now show that the reverse is also true. By the construction of the vectors x, y in terms of (3.7.11)–(3.7.13), we have $xu = yu = 1$. From this, and from the conditions $x \geq 0,\ y \geq 0$, it follows that (x, y) is a situation in mixed strategies in the game Γ.

By Theorem 3.7.1, for the situation (x, y) to be an equilibrium in mixed strategies in the game $\Gamma(A, B)$, it suffices to satisfy the conditions

$$a_i y = K_1(i, y) \leq xAy,\ i = \overline{1, m},$$

$$xb^j = K_2(x, j) \leq xBy,\ j = \overline{1, m},$$

or

$$Ay \leq (xAy)u, \quad xB \leq (xBy)u.$$

Let us check the validity of these relations for $x = \dfrac{uB^{-1}}{uB^{-1}u}$ and $y = \dfrac{A^{-1}u}{uA^{-1}u}$. We have

$$Ay = \frac{u}{uA^{-1}u} = \frac{(uB^{-1}AA^{-1}u)u}{(uB^{-1}u)(uA^{-1}u)} = (xAy)u,$$

$$xB = \frac{u}{uB^{-1}u} = \frac{(uB^{-1}BA^{-1}u)u}{(uB^{-1}u)(uA^{-1}u)} = (xBy)u,$$

which proves the statement.

We shall now demonstrate an application of the theorem with the example of a "battle of the sexes" game as in 3.1.4. Consider a mixed extension of the game. The set of points representing the payoff vectors in mixed strategies can be represented graphically (Fig. 3.3, Exercise 6). It can be easily seen that the game satisfies the conditions of the theorem; therefore, it has a unique, completely mixed equilibrium (x, y) which can be computed by the formulas (3.7.11)–(3.7.13): $x = (4/5, 1/5)$, $y = (1/5, 4/5)$, $(v_1, v_2) = (4/5, 4/5)$.

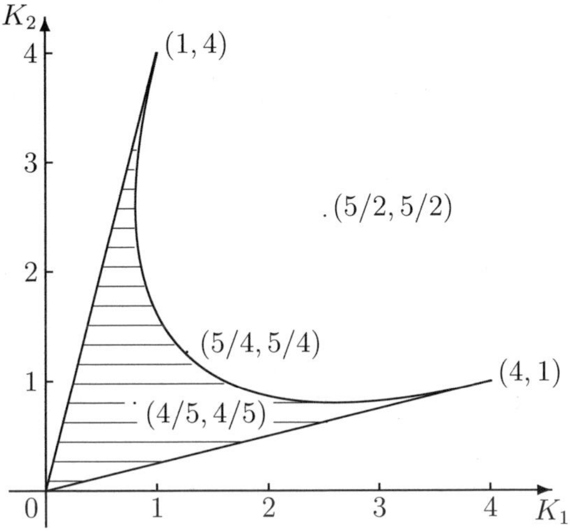

Figure 3.3

3.7.4. We shall now consider the properties of various optimality principles. Note that the definitions given in Sec. 3.2 of Nash equilibria and pareto-optimal situations apply (in particular) to an arbitrary noncooperative game; therefore, they are also true for the mixed extension $\overline{\Gamma}$. For this reason, the theorem of competition for leadership (see 3.2.12) holds for the two-person game:

$$Z(\overline{\Gamma}) = \overline{Z}^1 \cup \overline{Z}^2,$$

where $Z(\overline{\Gamma})$ is the set of Nash equilibria, $\overline{Z}^1$ and $\overline{Z}^2$ are the sets of the best responses to be given by Players 1 and 2, respectively, in the game $\overline{\Gamma}$.

Things become more complicated where the Nash equilibria and pareto-optimal situations are concerned. The examples given in Sec. 3.2 suggest the possibility of the cases where the situation is Nash equilibrium, but not pareto-optimal, and vice versa. However, the same situation can be optimal in both senses (see 3.2.4).

Example 16 in 3.3.3 shows that an additional equilibrium arising in the mixed extension of the game Γ is not pareto-optimal in the mixed extension of Γ. This appears to be a fairly common property of bimatrix games.

Theorem. *Let $\Gamma(A, B)$ be a bimatrix $(m \times n)$ game. Then the following assertion is true for almost all $(m \times n)$ games (except for no more than a countable set of games).*

Nash equilibrium situations in mixed strategies, which are not equilibrium in the original game, are not pareto-optimal in the mixed extension.

For the proof of this theorem, see Moulin (1981).

3.7.5. In conclusion of this section, we examine an example of the solution of bimatrix games.

Example 20. Bimatrix (2×2) *games* [Moulin (1981)]. Consider the game $\Gamma(A, B)$, in which each player has two pure strategies. Let

$$(A, B) = \begin{array}{c} \\ \delta_1 \\ \delta_2 \end{array} \begin{array}{c} \begin{array}{cc} \quad\tau_1 & \qquad\qquad\tau_2 \end{array} \\ \begin{bmatrix} (\alpha_{11}, \beta_{11}) & (\alpha_{12}, \beta_{12}) \\ (\alpha_{21}, \beta_{21}) & (\alpha_{22}, \beta_{22}) \end{bmatrix} \end{array}.$$

Here the indices $\delta_1, \delta_2, \tau_1, \tau_2$ denote pure strategies of Players 1 and 2, respectively.

For simplicity, assume that the numbers α_{11}, α_{12}, α_{21}, α_{22}, (β_{11}, β_{12}, β_{21}, β_{22}) are different.

Case 1. In the original game Γ, at least one player, say Player 1, has a strictly dominant strategy, say δ_1 (see Sec. 1.8). Then the game Γ and its mixed extension $\overline{\Gamma}$ have a unique Nash equilibrium. In fact, inequalities $\alpha_{11} > \alpha_{21}$, $\alpha_{12} > \alpha_{22}$ cause the pure strategy δ_1 in the game $\overline{\Gamma}$ to dominate strictly all the other mixed strategies of Player 1. Therefore, an equilibrium is represented by the pair (δ_1, τ_1) if $\beta_{11} > \beta_{12}$, or by the pair (δ_1, τ_2) if $\beta_{11} < \beta_{12}$.

Case 2. The game Γ does not have a Nash equilibrium in pure strategies. Here two mutually exclusive cases *a*) and *b*) are possible:

$$a)\ \alpha_{21} < \alpha_{11},\ \alpha_{12} < \alpha_{22},\ \beta_{11} < \beta_{12},\ \beta_{22} < \beta_{21},$$

$$b)\ \alpha_{11} < \alpha_{21},\ \alpha_{22} < \alpha_{12},\ \beta_{12} < \beta_{11},\ \beta_{21} < \beta_{22},$$

where $\det A \neq 0$, $\det B \neq 0$ and hence the conditions of Theorem 3.7.3 are satisfied. The game, therefore, has the equilibrium (x^*, y^*), where

$$x^* = \left(\frac{\beta_{22} - \beta_{21}}{\beta_{11} + \beta_{22} - \beta_{21} - \beta_{12}}, \frac{\beta_{11} - \beta_{12}}{\beta_{11} + \beta_{22} - \beta_{21} - \beta_{12}}\right), \quad (3.7.14)$$

$$y^* = \left(\frac{\alpha_{22} - \alpha_{12}}{\alpha_{11} + \alpha_{22} - \alpha_{21} - \alpha_{12}}, \frac{\alpha_{11} - \alpha_{21}}{\alpha_{11} + \alpha_{22} - \alpha_{21} - \alpha_{12}}\right) \quad (3.7.15)$$

while the corresponding equilibrium payoffs v_1 and v_2 are determined by

$$v_1 = \frac{\alpha_{11}\alpha_{22} - \alpha_{12}\alpha_{21}}{\alpha_{11} + \alpha_{22} - \alpha_{21} - \alpha_{12}}, \quad v_2 = \frac{\beta_{11}\beta_{22} - \beta_{12}\beta_{21}}{\beta_{11} + \beta_{22} - \beta_{12} - \beta_{21}}.$$

Case 3. The game Γ has two Nash equilibria. This occurs when one of the following conditions is satisfied:

$$a)\ \alpha_{21} < \alpha_{11},\ \alpha_{12} < \alpha_{22},\ \beta_{12} < \beta_{11},\ \beta_{21} < \beta_{22},$$
$$b)\ \alpha_{11} < \alpha_{21},\ \alpha_{22} < \alpha_{12},\ \beta_{11} < \beta_{22},\ \beta_{12} < \beta_{21}.$$

In case a), the situations (δ_1, τ_1), (δ_2, τ_2) are found to be equilibrium, whereas in case b), the situations (δ_1, τ_2), (δ_2, τ_1) form an equilibrium. The mixed extension, however, has one more completely mixed equilibrium (x^*, y^*) determined by (3.7.14), (3.7.15).

The above cases provide an exhaustive examination of a (2×2) game with the matrices having different elements.

3.8 Symmetric Bimatrix Games and Evolutionary Stable Strategies

3.8.1. Let $\Gamma = (X, Y, A, B)$ be a bimatrix game. Γ is said to be symmetric if the sets X and Y coincide $X = Y$ and $\alpha_{ij} = \beta_{ji}$ for all i, j.

This definition of symmetry is not invariant with respect to permutations of strategy sets. Suppose $|X| = |Y| = m$, and the pure strategies will be denoted by i or j. In the evolutionary game theory the payoff matrix A of Player 1 is also called the fitness matrix of the game, and in symmetric case the matrix A determines the game completely, thus we shall identify the game Γ with A and speak of the game A. Mixed strategies x, y are defined in the usual way. And the mathematical expectation of the payoff to Player 1 in the situation (x, y) is equal

$$E(x,y) = xAy = \sum_{i=1}^{m}\sum_{j=1}^{m} \alpha_{ij}\xi_i\eta_j.$$

For any mixed strategy $p = \{\xi\}$ define $C(p)$ as a carrier of p and $B(p)$ the set of pure best replies against p in the game A

$$C(p) = \{i : \xi_i > 0\}, \quad B(p) = \{i : \ E(i,p) = \max_j E(j,p)\}.$$

3.8.2. Consider now an example of Hawk–Dove game of Maynard Smith and Price (1973) which leads us to the notion of the *evolutionary stable strategy* (*ESS*).

Example 21. The Hawk–Dove game is 2×2 symmetric bimatrix game with the following matrices:

$$A = \begin{array}{c} \\ H \\ D \end{array}\begin{array}{c} \begin{array}{cc} H & D \end{array} \\ \begin{bmatrix} 1/2(V-C) & V \\ 0 & 1/2V \end{bmatrix} \end{array}, \quad B = \begin{array}{c} \\ H \\ D \end{array}\begin{array}{c} \begin{array}{cc} H & D \end{array} \\ \begin{bmatrix} 1/2(V-C) & 0 \\ V & 1/2V \end{bmatrix} \end{array}. \tag{3.8.1}$$

Suppose two animals are contesting a resource (such as territory in a favourable place) of value V, i.e. by obtaining the resource, an animal increases the expected number of offspring (fitness) by V. For simplicity assume that only two pure strategies, hawk and dove, are possible. An animal adopting the hawk strategy always fights as hard as it can, retreated only when seriously injured. A dove merely threats in a conventional way and quitely retreats when seriously challenged, without ever being wounded. Two doves can share the resource peacefully, but two hawks go on fighting until one is wounded and forced to retreat. It is assumed that a wound reduces the fitness by an amount C. If we furthermore assume that there are no differences in size or age that influence the probability of winning then the conflict can be represented by the symmetric bimatrix game (3.8.1).

If $V > C$, the Hawk–Dove game has a unique Nash equilibrium, (H, H) hence, it is always reasonable to fight. In a population of hawks and doves, the hawks have greater reproductive success the dove will gradually die out and in the long run only hawks will exist.

If $V < C$, then (H, H) is not an equilibrium. Consequently, a monomorphic population of hawks is not stable. In such a population, a mutant dove has greater reproductive success and, therefore, doves will spread through the population. Similarly, a population of doves can also be successfully invaded by hawks, because (D, D) is not a Nash equilibrium.

If $V < C$ the game has the unique symmetric Nash equilibrium in mixed strategies

$$\overline{x} = (\xi, 1 - \xi), \quad \overline{y} = (\xi, 1 - \xi),$$

where $\xi = V/C$.

There are also two asymmetric equilibria (H, D) and (D, H).

3.8.3. Assume now that a monomorphic population is playing the mixed strategy p in the game with fitness matrix A and suppose that a mutant playing q arises. Then we may suppose that the population will be in perturbed state in which a small fraction ϵ of the individuals is playing q. The population will return to its original position if the mutant is selected again, i.e. the fitness of a q-individuals is smaller than that of an individual playing p. Suppose that (p, p) is a symmetric Nash equilibrium in a symmetric bimatrix game, and suppose that the second player in the game instead of playing the strategy p decides to play a mixture of the two mixed strategies: p and q with the probabilities $1 - \epsilon$, ϵ, where ϵ is small enough. Then in general for the new mixed strategy $y = (1 - \epsilon)p + \epsilon q$ the set of Player's 1 best replies against $y = (1 - \epsilon)p + \epsilon q$ will not necessarily contain the starting p. And also it may happen that q is better reply against y than p. But if for any q there exists such an $\epsilon > 0$, that p is better reply against $y = (1-\epsilon)p+\epsilon q$ than q, then we see that the use of p by Player 1 is in some sense stable against small perturbations of the opponents strategies, i.e. for any q there exists $\epsilon > 0$ such that

$$qA((1-\epsilon)p+\epsilon q) < pA((1-\epsilon)p+\epsilon q), \tag{3.8.2}$$

If (p, p) is a strict equilibrium ($pAp > qAp$ for all q), then (3.8.2) always holds. There is also an evolutionary interpretation of (3.8.2), based on the example of Hawk–Dove game.

If (3.8.2) holds, we have

$$(1-\epsilon)qAp + \epsilon qAq < (1-\epsilon)pAp + \epsilon pAq. \tag{3.8.3}$$

From (3.8.3) we see that $qAp > pAp$ is impossible because in this case (3.8.2) will not hold for small $\epsilon > 0$. Then from (3.8.3) it uniquely follows

$$qAp < pAp, \tag{3.8.4}$$

or

$$\text{if } qAp = pAp, \text{ then } qAq < pAq. \tag{3.8.5}$$

From (3.8.4), (3.8.5) the condition (3.8.3) trivially follows for sufficiently small $\epsilon > 0$ (this ϵ depends upon q).

3.8.4. Definition. A mixed strategy p is an ESS if (p, p) is a Nash equilibrium, and the following stability condition is satisfied:

$$\text{if } q \neq p \text{ and } qAp = pAp, \text{ then } qAq < pAq.$$

3.8.5. Consider the correspondence

$$p \to \{q \in Y,\ C(q) \subset B(p)\}.$$

This correspondence satisfies the conditions of the Kakutani fixed point theorem, and hence there exists a point p^* for which

$$p^* \in \{q \in Y,\ C(q) \subset B(p)\},$$

and thus

$$C(p^*) \subset B(p^*). \tag{3.8.6}$$

From (3.8.6) it follows that

$$p^*Ap^* \geq qAp^*$$

for all $q \in Y$, and (p^*, p^*) is a symmetric Nash equilibrium. We proved a theorem.

Theorem [Nash (1951)]. *Every symmetric bimatrix game has a symmetric Nash equilibrium.*

3.8.6. If (p, p) is a strict Nash equilibrium, then p is an ESS (this follows also directly from the definition of the ESS, since in this case there are no such $q \in Y$ that $qAp = pAp$).

Not every bimatrix games possess an ESS. For example if in A all $\alpha_{ij} = \alpha$ and do not depend upon i, j then it is impossible to satisfy (3.8.6).

3.8.7. Theorem. *If A is 2×2 matrix with $\alpha_{11} \neq \alpha_{21}$ and $\alpha_{12} \neq \alpha_{22}$, then A has an ESS. If $\alpha_{11} > \alpha_{21}$ and $\alpha_{22} > \alpha_{12}$, then A has two strict equilibria $(1, 1), (2, 2)$ and they are ESS. If $\alpha_{11} < \alpha_{21}$, $\alpha_{22} < \alpha_{12}$ then A has a unique symmetric equilibrium (p, p), which is completely mixed ($C(p) = B(p) = \{1, 2\}$).*

Proof. For $q \neq p$, we have $qAq - pAq = (q - p)A(q - p)$. If $q = (\eta_1, \eta_2)$, $p = (\xi_1, \xi_2)$, then

$$(q - p)A(q - p) = (\eta_1 - \xi_1)^2(\alpha_{11} - \alpha_{21} + \alpha_{22} - \alpha_{12}) < 0.$$

Hence (3.8.3) is satisfied, and p is ESS.

3.8.8. Consider the game, where the matrix A has the form

$$A = \begin{bmatrix} b & a & -a & a & -a \\ -a & b & a & -a & a \\ a & -a & b & a & -a \\ -a & a & -a & b & a \\ a & -a & a & -a & b \end{bmatrix}. \tag{3.8.7}$$

If $0 < b < a$ this game does not have an ESS. The game has a unique symmetric equilibrium $p = (1/5, 1/5, 1/5, 1/5, 1/5)$, $pAp = b/5$, and every stratetgy is a best reply against p and for any i, $e_i A e_i = \alpha_{ii} = b > b/5 = pAp = pAe_i$ (where $e_i = (0, \ldots, 0, 1_i, 0, \ldots, 0)$), and the condition (3.8.5) in ESS is violated.

Thus for the games with more than two pure strategies the theorem does not hold.

3.8.9. It is interesting that the number of ESS in the game is always finite (although may be equal to zero).

If (p, p) and (q, q) are Nash equilibria of A with $q \neq p$, and $C(q) \subset B(p)$, then p cannot be an ESS, since q is the best reply against p and q.

Theorem. *If p is an ESS of A and (q, q) is a symmetric Nash equilibrium of A with $C(q) \subset B(p)$, then $p = q$.*

3.8.10. Let (p^n, p^n) be the sequence of symmetric Nash equilibrium's of A such that $\lim_{n\to\infty} p^n = p$. Then from the definition of the limit we get, that there exists such N that for all $n \geq N$

$$C(p) \subset C(p^n) \subset B(p^n) \subset B(p).$$

From the previous theorem we have that $p^n = p$ for $n \geq N$. It follows, that every ESS is isolated within the set of symmetric equilibrium strategies. From the compactness of the set of situations in mixed strategies we have that if there would be infinitely many ESS, there would be a cluster point, but the previous discussion shows that this is impossible. Thus the following theorem holds:

Theorem [Haigh (1975)]. *The number ESS is finite (but possibly zero).*

3.9 Equilibrium in Joint Mixed Strategies

3.9.1. We shall continue discussion of two-person games. As already noted in Sec. 3.2, even though an equilibrium is not dominated (pareto-optimal), we may have the cases where one equilibrium is advantageous to Player 1, while the other is advantageous to Player 2. This presents certain problems in finding a mutually acceptable solution to a nonantagonistic conflict which arises where a noncooperative game is to be formalized. For this reason, we have to examine a nonantagonistic conflict in the formalization which allows the players to make joint decisions. This approach can be illustrated with an example of the "battle of the sexes" game (see Example 1, 3.1.4).

Example 22. Consider a mixed extension of the "battle of the sexes" game. The set of points corresponding to the payoff vectors in mixed strategies in the game can be represented graphically (see Fig. 3.3, and 3.7.3). Figure 3.3 shows two Nash equilibria in pure strategies with the payoff vectors $(1, 4)$, $(4, 1)$ and one completely mixed equilibrium with the payoff vectors $(4/5, 4/5)$ (this may be found by employing Theorem 3.7.3), which is less preferable to players than every equilibrium in pure strategies. Thus here the following situations form an equilibrium: (α_1, β_1), (α_2, β_2), (x^*, y^*), where $x^* = (4/5, 1/5)$, $y^* = (1/5, 4/5)$, and the situations $(\alpha_1, \beta_1), (\alpha_2, \beta_2)$ are also pareto-optimal.

If the game is repeated, then it may be wise for the players to make their choice jointly, i.e. to choose with probability $1/2$ the situation (α_1, β_1) or (α_2, β_2). Then the expected payoff to the players is, on the average, $(5/2, 5/2)$. This point, however, is not lying in the set of payoff vectors corresponding to possible situations in a noncooperative game (Fig. 3.4), i.e. it cannot be realized if the players choose mixed strategies independently.

A *joint mixed strategy* of players is the probability distribution over the set of all possible pairs (i, j) (a situation in pure strategies), which is not necessary generated by the players' independent random choices of pure strategies. Such strategies can be realized by a mediator before the game starts.

Denote by M a joint mixed strategy in the game $\Gamma(A, B)$. If this strategy is played by Players 1 and 2, their expected payoffs $K_1(M)$,

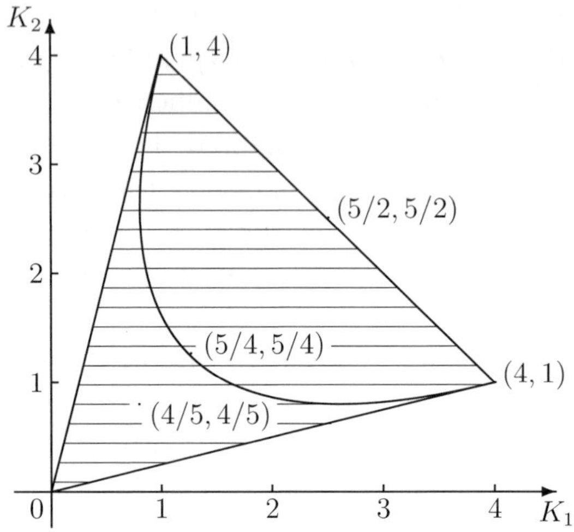

Figure 3.4

$K_2(M)$ respectively are

$$K_1(M) = \sum_{i,j} \alpha_{ij}\mu_{ij}, \quad K_2(M) = \sum_{i,j} \beta_{ij}\mu_{ij},$$

where $A = \{\alpha_{ij}\}$, $B = \{\beta_{ij}\}$ are the players' payoff matrices, $M = \{\mu_{ij}\}$, and $uMw = 1$, $M \geq 0$, $u = (1, \ldots, 1) \in R^m$, $w = (1, \ldots, 1) \in R^n$. Geometrically, the set of vector payoffs corresponding to the joint mixed strategies is the convex hull of the set of possible vector payoffs in pure strategies. For the game in Example 22 it is of the form as shown in Fig. 3.4.

Note that the joint mixed strategy $M^* = \begin{bmatrix} 1/2 & 0 \\ 0 & 1/2 \end{bmatrix}$ is pareto-optimal and corresponds to the payoff vector $(5/2, 5/2)$. Thus, M^* can be suggested as a solution to the game "battle of the sexes".

Definition. For the bimatrix $(m \times n)$ game $\Gamma(A, B)$, denote by $M = \{\mu_{ij}\}$ the joint probability distribution over the pairs (i, j), $i = 1, \ldots, m$, $j = 1, \ldots, n$. Denote by $\mu_i(j)$ the conditional probability of realizing strategy j provided strategy i has been realized. Similarly, denote by $\nu_j(i)$ the conditional probability of realizing strategy i provided strategy j has been realized. Then

$$\mu_i(j) \triangleq \begin{cases} \mu_{ij}/\sum_{j=1}^{n}\mu_{ij}, & \text{if } \sum_{j=1}^{n}\mu_{ij} \neq 0, \\ 0, & \text{if } \mu_{ij} = 0,\ j = 1,\dots,n, \end{cases}$$

$$\nu_j(i) \triangleq \begin{cases} \mu_{ij}/\sum_{i=1}^{m}\mu_{ij}, & \text{if } \sum_{i=1}^{m}\mu_{ij} \neq 0, \\ 0, & \text{if } \mu_{ij} = 0,\ i = 1,\dots,m. \end{cases}$$

We say that $M^* = \{\mu^*_{ij}\}$ is an *equilibrium in joint mixed strategies* in the game $\Gamma(A,B)$ if the inequalities

$$\sum_{j=1}^{n}\alpha_{ij}\mu^*_i(j) \geq \sum_{j=1}^{n}\alpha_{i'j}\mu^*_i(j),$$

$$\sum_{i=1}^{m}\beta_{ij}\nu^*_j(i) \geq \sum_{i=1}^{m}\beta_{ij'}\nu^*_j(i) \tag{3.9.1}$$

hold for all $i,i' \in \{1,2,\dots,m\}$ and $j,j' \in \{1,2,\dots,n\}$.

3.9.2. The game $\Gamma(A,B)$ in joint mixed strategies can be interpreted as follows. Suppose the players have reached an agreement on joint strategy $M^* = \{\mu^*_{ij}\}$ and a chance device has yielded the pair (i,j), i.e. Player 1(2) has received the strategy number $i(j)$. Note that each player knows only his own course of action. In general, he may not agree on the realization i (j, respectively) of the joint strategy and choose the strategy $i'(j')$. If M^* is an equilibrium, then it is disadvantageous for each player to deviate from the proposed realization i (j, respectively), which follows from (3.9.1), where the left-hand sides of the inequalities coincide with the expected payoff to Player 1(2) provided he agrees on the realization $i(j)$.

Suppose the strategy i of Player 1 is such that $\mu_{ij} = 0$ for all $j = 1,2,\dots,n$. Then the first of the inequalities (3.9.1) seems to be satisfied. Similarly, if $\mu_{ij} = 0$ for all $i = 1,\dots,m$, then the second inequality in (3.9.1) is satisfied. We substitute the expressions for $\mu_i(j)$ and $\nu_j(i)$ in terms of μ_{ij} into (3.9.1). Then it follows that the necessary and sufficient condition for the situation $M^* = \{\mu^*_{ij}\}$ to be equilibrium is that the inequalities

$$\sum_{j=1}^{n} \alpha_{ij}\mu^*_{ij} \geq \sum_{j=1}^{n} \alpha_{i'j}\mu^*_{ij}, \ \sum_{i=1}^{m}\sum_{j=1}^{n} \mu^*_{ij} = 1,$$

$$\sum_{i=1}^{m} \beta_{ij}\mu^*_{ij} \geq \sum_{i=1}^{m} \beta_{ij'}\mu^*_{ij}, \ \mu^*_{ij} \geq 0 \tag{3.9.2}$$

hold for all $i, i' \in \{1, 2, \dots, m\}$ and $j, j' \in \{1, 2, \dots, n\}$.

Denote by $Z_c(\Gamma)$ the set of equilibria in joint mixed strategies.

Theorem. *The following assertions are true*:

1. *The set $Z_c(\Gamma)$ of equilibria in joint mixed strategies in the bimatrix $(m \times n)$ game $\Gamma(A, B)$ is a nonempty convex compact set in the space $R^{m\times n}$.*
2. *If (x, y) is a situation in mixed strategies in the game $\Gamma(A, B)$, then the joint mixed strategy situation $M = \{\mu_{ij}\}$ generated by the situation (x, y), is equilibrium if and only if (x, y) is the Nash equilibrium in mixed strategies in the game $\Gamma(A, B)$.*

Proof. Suppose that (x, y), $x = (\xi_1, \dots, \xi_m)$, $y = (\eta_1, \dots, \eta_n)$ is the situation in mixed strategies in the game $\Gamma(A, B)$, while $M = \{\mu_{ij}\}$ is the corresponding situation in joint strategies, i.e. $\mu_{ij} = \xi_i \cdot \eta_j$, $i = 1, \dots, m$, $j = 1, \dots, n$. The necessary and sufficient condition for M to be equilibrium is provided by the system of inequalities (3.9.2), i.e.

$$\xi_i K_1(i, y) \geq \xi_i K_1(i', y), \quad \eta_j K_2(x, j) \geq \eta_j K_2(x, j'), \tag{3.9.3}$$

where $i, i' \in \{1, 2, \dots, m\}$, $j, j' \in \{1, \dots, n\}$. If $\xi_i = 0$ ($\eta_j = 0$), then the inequalities (3.9.3) are trivially satisfied. Therefore, the system of inequalities (3.9.3) is equivalent to the following:

$$K_1(i, y) \geq K_1(i', y), \ K_2(x, j) \geq K_2(x, j'), \tag{3.9.4}$$

$i, i' \in \{1, \dots, m\}$, $j, j' \in \{1, \dots, n\}$, where i and j belong to the spectra of strategies x and y. Let (x, y) be the Nash equilibrium in mixed strategies in the game $\Gamma(A, B)$. Then, by the Theorem 3.5.2,

$$K_1(i, y) = K_1(x, y), \quad K_2(x, j) = K_2(x, y)$$

for all i and j from the spectra of optimal strategies. Therefore, inequalities (3.9.4) are satisfied and $M \in Z_c(\Gamma)$.

Conversely, if (3.9.3) is satisfied, then summing the inequalities (3.9.3) over i and j, respectively, and applying Theorem 3.7.2, we have that the situation (x, y) is a Nash equilibrium.

The convexity and compactness of the set $Z_c(\Gamma)$ follow from the fact that $Z_c(\Gamma)$ is the set of solutions to the system of linear inequalities (3.9.2) which is bounded, whereas its nonemptiness follows from the existence of the Nash equilibrium in mixed strategies (see 3.4.1). This completes the proof of the theorem.

Note that the joint mixed strategy $M^* = \begin{bmatrix} 1/2 & 0 \\ 0 & 1/2 \end{bmatrix}$ is equilibrium in the game "battle of the sexes" (see Example 1, 3.1.4), which may be established by mere verification of inequalities (3.9.2).

3.10 The Bargaining Problem

3.10.1. This section deals with the question: how rational players can come to an agreement on a joint choice by negotiations. Before stating the problem, we return to the game "battle of the sexes" once again.

Example 23. Consider the set R corresponding to possible payoff vectors in joint mixed strategies for the game "battle of the sexes" (this region is shaded in Fig. 3.5). Acting together, the players can ensure any payoff in mixed strategies in the region R. However, this does not mean that they can agree on any outcome of the game. Thus, the point $(4, 1)$ is preferable to Player 1 whereas the point $(1, 4)$ is preferable to Player 2. Neither of the two players can agree with the results of negotiations if his payoff is less than the maximin value, since he can receive this payoff independently of his partner. Maximin mixed strategies for the players in this game are respectively

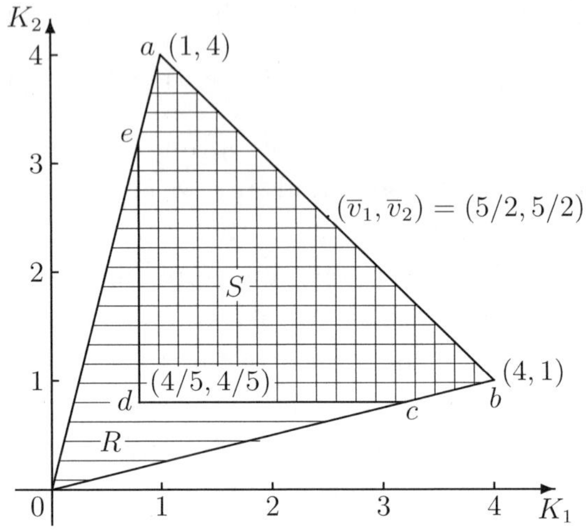

Figure 3.5

$x^0 = (1/5, 4/5)$ and $y^0 = (4/5, 1/5)$, while the payoff vector in maximin strategies (v_1^0, v_2^0) is $(4/5, 4/5)$. Therefore, the set S, which can be used in negotiations, is bounded by the points a, b, c, d, e (see Fig. 3.5). This set will be called a *bargaining set of the game.* Furthermore, acting jointly, the players can always agree to choose points on the line segment $\overline{ab}$, since this is advantageous to both of them (the line segment $\overline{ab}$ corresponds to pareto-optimal situations).

3.10.2. The problem of choosing the points $(\overline{v}_1, \overline{v}_2)$ from S by bargaining will be a *bargaining problem.* (Nash, J., 1950b). This brings us to the following consideration. Let the bargaining set S and the maximin payoff vector (v_1^0, v_2^0) be given for the bimatrix game $\Gamma(A, B)$. We need to find the device capable for solving bargaing problem, i.e. to find a function φ such that

$$\varphi(S, v_1^0, v_2^0) = (\overline{v}_1, \overline{v}_2), \tag{3.10.1}$$

where $(\overline{v}_1, \overline{v}_2) \in S$ is the solution. The point (v_1^0, v_2^0) is called a disagreement point.

It appears that, under some reasonable assumptions, it is possible to construct such a function $\varphi(S, v_1^0, v_2^0)$.

Theorem. *Let S be a convex compact set in R^2, and let (v_1^0, v_2^0) be a maximin payoff vector in the game $\Gamma(A, B)$. The set S, the pair $(\overline{v}_1, \overline{v}_2)$ and the function φ satisfy the following conditions:*

1. $(\overline{v}_1, \overline{v}_2) \geq (v_1^0, v_2^0)$.
2. $(\overline{v}_1, \overline{v}_2) \in S$.
3. *If* $(v_1, v_2) \in S$ *and* $(v_1, v_2) \geq (\overline{v}_1, \overline{v}_2)$, *then* $(v_1, v_2) = (\overline{v}_1, \overline{v}_2)$.
4. *If* $(\overline{v}_1, \overline{v}_2) \in \overline{S} \subset S$ *and* $(\overline{v}_1, \overline{v}_2) = \varphi(S, v_1^0, v_2^0)$, *then* $(\overline{v}_1, \overline{v}_2) = \varphi(\overline{S}, v_1^0, v_2^0)$.
5. *Let T be obtained from S by linear transformation* $v_1' = \alpha_1 v_1 + \beta_1$, $v_2' = \alpha_2 v_2 + \beta_2$; $\alpha_1 > 0$, $\alpha_2 > 0$. *If* $\varphi(S, v_1^0, v_2^0) = (\overline{v}_1, \overline{v}_2)$, *then* $\varphi(T, \alpha_1 v_1^0 + \beta_1, \alpha_2 v_2^0 + \beta_2) = (\alpha_1 \overline{v}_1 + \beta_1, \alpha_2 \overline{v}_2 + \beta_2)$.
6. *If for any* $(v_1, v_2) \in S$, *also* $(v_2, v_1) \in S$, $v_1^0 = v_2^0$ *and* $\varphi(S, v_1^0, v_2^0) = (\overline{v}_1, \overline{v}_2)$, *then* $\overline{v}_1 = \overline{v}_2$.

There exists a unique function φ, satisfying 1–6 *such that*

$$\varphi(S, v_1^0, v_2^0) = (\overline{v}_1, \overline{v}_2).$$

The function φ, which maps the bargaining game (S, v_1^0, v_2^0) into the payoff vector set $(\overline{v}_1, \overline{v}_2)$ and satisfies conditions 1–6, is called a Nash bargaining scheme [Owen (1968)], conditions 1–6 are called *Nash axioms*, and the vector $(\overline{v}_1, \overline{v}_2)$ is called a *bargaining solution vector*. Thus, the bargaining scheme is a realizable optimality principle in the bargaining game.

Before going to prove the theorem we will discuss its conditions using the game "battle of the sexes" as an example (see Fig. 3.5). Axioms 1 and 2 imply that the payoff vector $(\overline{v}_1, \overline{v}_2)$ is contained in the set bounded by the points a, b, c, d, e. The axiom 3 implies that $(\overline{v}_1, \overline{v}_2)$ is pareto-optimal. Axiom 4 shows that the function φ is independent of irrelevant alternatives. This says that if the solution outcome of a given problem remains feasible for a new problem obtained from it by contraction, then it should also be the solution outcome of this new problem. Axiom 5 is the scale invariance axiom and axiom 6 shows that the two players possess equal rights.

The proof of the Theorem 3.10.2 is based on the following auxiliary results.

3.10.3. Lemma. *If there are points* $(v_1, v_2) \in S$ *such that* $v_1 > v_1^0$ *and* $v_2 > v_2^0$, *then there exists a unique point* $(\overline{v}_1, \overline{v}_2)$ *which maximizes the function*

$$\theta(v_1, v_2) = (v_1 - v_1^0)(v_2 - v_2^0)$$

over a subset $S_1 \subset S$, $S_1 = \{\, (v_1, v_2) \mid (v_1, v_2) \in S,\ v_1 \geq v_1^0\}$.

Proof. By condition, S_1 is a nonempty compact set while θ is a continuous function, and hence achieves its maximum $\overline{\theta}$ on this set. By assumption, $\overline{\theta}$ is positive.

Suppose there are two different points of maximum (v_1', v_2') and (v_1'', v_2'') for the function θ on S_1. Note that $v_1' \neq v_1''$; otherwise the form of the function θ would imply $v_2' = v_2''$.

If $v_1' < v_1''$, then $v_2' > v_2''$. Since the set S_1 is convex, then $(\overline{v}_1, \overline{v}_2) \in S_1$, where $\overline{v}_1 = (v_1' + v_1'')/2$, $\overline{v}_2 = (v_2' + v_2'')/2$. We have

$$\begin{aligned}\theta(\overline{v}_1, \overline{v}_2) &= \frac{(v_1' - v_1^0) + (v_1'' - v_1^0)}{2}\,\frac{(v_2' - v_2^0) + (v_2'' - v_2^0)}{2} \\ &= \frac{(v_1' - v_1^0)(v_2' - v_2^0)}{2} + \frac{(v_1'' - v_1^0)(v_2'' - v_2^0)}{2} \\ &\quad + \frac{(v_1' - v_1'')(v_2'' - v_2')}{4}.\end{aligned}$$

Each of the first two summands in the last sum is equal to $\overline{\theta}/2$, while the third summand is positive, which is impossible, because $\overline{\theta}$ is the maximum of the function θ. Thus, the point $(\overline{v}_1, \overline{v}_2)$, which maximizes the function θ over the set S_1, is unique.

3.10.4. Lemma. *Suppose that* S *satisfies the conditions of Lemma* 3.10.3, *while* $(\overline{v}_1, \overline{v}_2)$ *is the point of maximum for the function* $\theta(v_1, v_2)$. *Define*

$$\delta(v_1, v_2) \triangleq (\overline{v}_2 - v_1^0)v_1 + (\overline{v}_1 - v_1^0)v_2.$$

If $(v_1, v_2) \in S$, *then the following inequality holds*:

$$\delta(v_1, v_2) \leq \delta(\overline{v}_1, \overline{v}_2).$$

Proof. Suppose there exists a point $(v_1, v_2) \in S$ such that $\delta(v_1, v_2) > \delta(\overline{v}_1, \overline{v}_2)$. From the convexity of S we have: $(v_1', v_2') \in S$, where $v_1' = \overline{v}_1 + \epsilon(v_1 - \overline{v}_1)$ and $v_2' = \overline{v}_2 + \epsilon(v_2 - \overline{v}_2)$, $0 < \epsilon < 1$. By linearity, $\delta(v_1 - \overline{v}_1, v_2 - \overline{v}_2) > 0$. We have

$$\theta(v_1', v_2') = \theta(\overline{v}_1, \overline{v}_2) + \epsilon\delta(v_1 - \overline{v}_1, v_2 - \overline{v}_2) + \epsilon^2(v_1 - \overline{v}_1)(v_2 - \overline{v}_2).$$

For a sufficiently small $\epsilon > 0$ we obtain the inequality $\theta(v_1', v_2') > \theta(\overline{v}_1, \overline{v}_2)$, but this contradicts the maximality of $\theta(\overline{v}_1, \overline{v}_2)$.

3.10.5. We shall now prove Theorem 3.10.2. To do this, we shall show that the point $(\overline{v}_1, \overline{v}_2)$, which maximizes $\theta(v_1, v_2)$, is a solution of the bargaining problem.

Proof. Suppose the conditions of Lemma 3.10.3 are satisfied. Then the point $(\overline{v}_1, \overline{v}_2)$ maximizing $\theta(v_1, v_2)$ is defined. It is easy to verify that $(\overline{v}_1, \overline{v}_2)$ satisfies conditions 1–4 of Theorem 3.10.2. This point also satisfies Condition 5 of this theorem, because if $v_1' = \alpha_1 v_1 + \beta_1$ and $v_2' = \alpha_2 v_2 + \beta_2$, then

$$\theta'(v_1', v_2') = [v_1' - (\alpha_1 v_1^0 + \beta_1)][v_2' - (\alpha_2 v_2^0 + \beta_2)] = \alpha_1\alpha_2\theta(v_1, v_2),$$

and if $(\overline{v}_1, \overline{v}_2)$ maximizes $\theta(v_1, v_2)$, then $(\overline{v}_1', \overline{v}_2')$ maximizes $\theta'(v_1', v_2')$. Suppose that the set S is symmetric in the sense of Condition 6 and $v_1^0 = v_2^0$. Then $(\overline{v}_2, \overline{v}_1) \in S$ and $\theta(\overline{v}_1, \overline{v}_2) = \theta(\overline{v}_2, \overline{v}_1)$. Since $(\overline{v}_1, \overline{v}_2)$ is a unique point, which maximizes $\theta(v_1, v_2)$ over S_1, then $(\overline{v}_1, \overline{v}_2) = (\overline{v}_2, \overline{v}_1)$, i.e. $\overline{v}_1 = \overline{v}_2$.

Thus, the point $(\overline{v}_1, \overline{v}_2)$ satisfies conditions 1–6. Show that this is a unique solution to the bargaining problem. Consider the set

$$R \stackrel{\triangle}{=} \{\ (v_1, v_2) \mid \delta(v_1, v_2) \leq \delta(\overline{v}_1, \overline{v}_2)\}. \qquad (3.10.2)$$

By Lemma 3.10.4, the inclusion $S \subset R$ holds. Suppose T is obtained from R by transformation

$$v_1' = \frac{v_1 - v_1^0}{\overline{v}_1 - v_1^0}, \quad v_2' = \frac{v_2 - v_2^0}{\overline{v}_2 - v_2^0}. \tag{3.10.3}$$

Expressing v_1 and v_2 in terms of (3.10.3) and substituting into (3.10.2), we obtain

$$T = \{\ (v_1', v_2') \mid v_1' + v_2' \leq 2\}$$

and $v_1'^0 = v_2'^0 = 0$. Since T is symmetric, it follows from property 6 that a solution (if any) must lie on a straight line $v_1' = v_2'$, and, by Condition 3, it must coincide with the point (1,1), i.e. $(1, 1) = \varphi(T, 0, 0)$. Reversing the transform (3.10.3) and using property 5, we obtain $(\overline{v}_1, \overline{v}_2) = \varphi(R, v_1^0, v_2^0)$. Since $(\overline{v}_1, \overline{v}_2) \in S$ and $S \subset R$, then by property 4, the pair $(\overline{v}_1, \overline{v}_2)$ is a solution of (S, v_1^0, v_2^0).

Now suppose that the conditions of Lemma 3.10.3 are not satisfied, i.e. there are no points $(v_1, v_2) \in S$ for which $v_1 > v_1^0$ and $v_2 > v_2^0$. Then the following cases are possible:

a) There are points, at which $v_1 > v_1^0$ and $v_2 = v_2^0$. Then $(\overline{v}_1, \overline{v}_2)$ is taken to be the point in S, which maximizes v_1 under constraint $v_2 = v_2^0$.
b) There are points, at which $v_1 = v_1^0$ and $v_2 > v_2^0$. In this case, $(\overline{v}_1, \overline{v}_2)$ is taken to be the point in S, which maximizes v_2 under constraint $v_1 = v_1^0$.
c) The bargaining set S degenerates into the point (v_1^0, v_2^0) of maximin payoffs (e.g. the case of matrix games). Set $\overline{v}_1 = v_1^0$, $\overline{v}_2 = v_2^0$.

It can be immediately verified that these solutions satisfy properties 1–6, and properties 1–3 imply uniqueness. This completes the proof of the theorem.

In the game "battle of the sexes", the Nash scheme yields bargaining payoff $(\overline{v}_1, \overline{v}_2) = (5/2, 5/2)$ (see Fig. 3.5).

3.10.6. In this section, we survey the axiomatic theory of bargaining for n players. Although alternatives to the Nash solution were proposed soon after the publication of Nash's paper, it is fair to say until the mid-1970s, the Nash solution was often seen by economists and game theorists as the main, if not the only, solution to the bagaining problem. Since all existing solutions are indeed invariant under parallel transformation of the origin and since our own formulation will also assume this invariance, it is convenient to take as admissible only problems that have already been subjected to parallel transformation bringing their disagreement point to the origin. Consequently, $v^0 = (v_1^0, \ldots, v_n^0) = (0, \ldots, 0) \in R^n$ always, and a typical problem is simply denoted by S instead of $(S, 0)$. Finally, all problems are taken to be subsets of R_+^n (instead of R^n). This means that all alternatives that would give any player less than what he gets at the disagreement point $v^0 = 0$ are disregarded.

Definition. The Nash solution N is defined by setting, for all convex, compact, comprehensive subsets $S \subset R_+^n$ containing at least one vector with all positive coordinates (denote $S \in \sum^n$), $N(S)$ equal to the maximizer in $v \in S$ of the "Nash product" $\prod_{i=1}^n v_i$.

Nash's theorem is based on the following axioms:

1^0. *Pareto-optimality.* For all $S \in \sum^n$, for all $v \in R^n$, if $v \geq \varphi(S)$ and $v \neq \varphi(S)$, then $v \notin S$ [denote $\varphi(S) \in PO(S)$].

A slightly weaker condition is:

2^0. *Weak pareto-optimality.* For all $S \in \sum^n$, for all $v \in R^n$, if $v > \varphi(S)$, then $v \notin S$.

Let $\Pi^n : \{1, \ldots, n\} \to \{1, \ldots, n\}$ be the *class of permutations of order n*. Given $\Pi \in \Pi^n$, and $v \in R^n$, let $\pi(v) \triangleq (v_{\pi(1)}, \ldots, v_{\pi(n)})$. Also, given $S \subset R^n$, let $\pi(S) \triangleq \{v' \in R^n \mid \exists v \in S \text{ with } v' = \pi(v)\}$.

3^0. *Symmetry.* For all $S \in \sum^n$, if for all $\pi \in \Pi^n, \pi(S) = S$, then $\varphi_i(S) = \varphi_j(S)$ for all i, j (note that $\pi(S) \in \sum^n$).

Let $L^n : R^n \to R^n$ be the class of *positive, independent person-by-person, and linear transformations of order n*. Each $l \in L^n$ is characterised by n positive numbers α_i such that given $v \in R^n, l(v) = (\alpha_1 v_1, \ldots, \alpha_n v_n)$. Now, given $S \subset R^n$, let $l(S) \triangleq \{v' \in R^n | \exists v \in S$ with $v' = l(v)\}$.

4^0. *Scale invariance.* For all $S \in \sum^n$, for all $l \in L^n$, $\varphi(l(S)) = l(\varphi(S))$ [note that $l(S) \in \sum^n$].

5^0. *Independence of irrelevant alternatives.* For all $S, S' \subset \sum^n$, if $S' \subset S$ and $\varphi(S) \in S'$ then $\varphi(S') = \varphi(S)$.

In previous section we showed the Nash theorem for $n = 2$, i.e. only one solution satisfies these axioms. This result extends directly to arbitrary n.

Theorem. *A solution $\varphi(S)$, $S \in \sum^n$ satisfies 1^0, 3^0, 4^0, 5^0 if and only if it is the Nash solution.*

This theorem constitutes the foundation of the axiomatic theory of bargaining. It shows that a *unique* point can be identified for each problem, representing an equitable compromise.

In the mid-1970s, Nash's result become the object of a considerable amount of renewed attention, and the role played by each axiom in the characterization was scrutinized by several authors.

6^0. *Strong individual rationality.* For all $S \in \sum^n$, $\varphi(S) > 0$.

Theorem [Roth (1977)]. *A solution $\varphi(S)$, $S \in \sum^n$ satisfies $3^0, 4^0, 5^0, 6^0$ if and only if it is the Nash solution.*

If 3^0 is dropped from the list of axioms in Theorem 3.10.6, a somewhat wider but still small family of additional solutions become admissible.

3.10.7. Definition. Given $a = (\alpha_1, \dots, \alpha_n)$, $\alpha_i > 0$, $i = 1, \dots, n$, $\sum_{i=1}^n \alpha_i = 1$, the asymmetric Nash solution with weights a, N^a, is defined by setting, for all $S \in \sum^n$, $N^a(S) \triangleq \arg\max \prod_{i=1}^n v_i^{\alpha_i}$, $v \in S$.

These solutions were introduced by Harsanyi and Selten (1972).

Theorem. *A solution $\varphi(S)$, $S \in \sum^n$ satisfies $4^0, 5^0, 6^0$ if and only if it is an asymmetric Nash solution.*

If 6^0 is not used, a few other solutions became available.

3.10.8. Definition. Given $i \in \{1, \dots, n\}$ the ith Dictatorial solution D^i is defined by setting, for all $S \in \sum^n, D^i(S)$ equals to the maximal point of S in the direction of the ith unit vector.

Note that all D^i satisfy $4^0, 5^0$, and 2^0 (but not 1^0). To recover full optimality, one may proceed as follows. First, select an ordering

π of the n players. Then given $S \in \sum^n$, pick $D^{\pi(1)}(S)$ if the point belongs to pareto-optimal subset of S; otherwise, among the points whose $\pi(1)$th coordinate is equal to $D^{\pi(1)}(S)$, find the maximal point in the direction of the unit vector pertaining to player $\pi(2)$. Pick this point if it belongs to pareto-optimal subset of S; otherwise, repeat the operation with $\pi(3), \ldots$. This algorithm is summarized in the following definition.

3.10.9. Definition. Given an ordering π of $\{1, \ldots, n\}$, the lexicographic Dictatorial solution relative to π, D^π, if defined by setting, for all $S \in \sum^n$, $D^\pi(S)$ to be the lexicographic maximizer over $v \in S$ of $v_{\pi(1)}, v_{\pi(2)}, \ldots, v_{\pi(n)}$.

All of these solutions satisfy $1^0, 4^0$, and 5^0, and there are no others if $n = 2$.

3.10.10. *The Kalai–Smorodinsky solution.* A new impulse was given to the axiomatic theory of bargaining when Kalai and Smorodinsky (1975) provided a characterization of the following solution (see Fig. 3.6).

Definition. The Kalai–Smorodinsky solution K is defined by setting, for all $S \in \sum^n, K(S)$ to be the maximal point of S on the segment connecting the origin to $a(S)$, the ideal point of S, defined by $v_i(S) \triangleq \max\{v_i \mid v \in S\}$ for each i.

An important distinguishing feature between the Nash solution and the Kalai–Smorodinsky solution is that the latter responds much

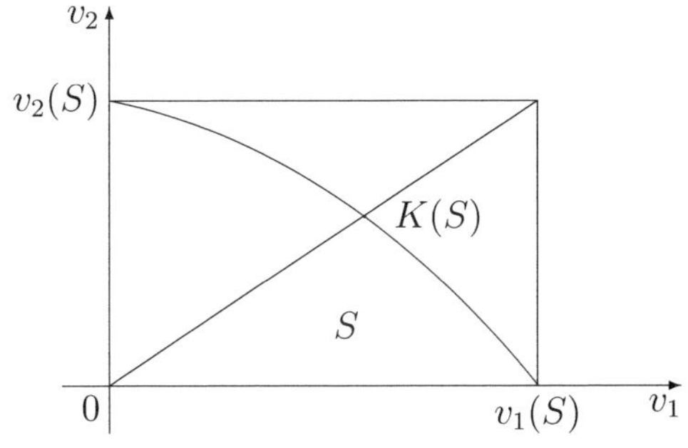

Figure 3.6

more satisfactorily to expansions and contractions of the feasible set. In particular, it satisfies the following axiom.

7^0. *Individual monotonicity*: For all $S, S' \in \sum^2$, for all i, if $v_j(S) = v_j(S')$ and $S' \supset S$, then $\varphi_i(S') \geq \varphi_i(S)$.

Theorem. *A solution* $\varphi(S)$, $S \in \sum^2$ *satisfies* $1^0, 3^0, 7^0$ *if and only if it is the Kalai–Smorodinsky solution.*

Although the extension of the definition of the Kalai–Smorodinsky solution to the n-person case itself causes no problem, the generalization of the preceding results to the n-person case is not as straightforward as was the case of the extensions of the results concerning the Nash solution from $n = 2$ to arbitrary n. First of all, for $n > 2$, the n-person Kalai–Smorodinsky solution satisfies 2^0 only. This is not a serious limitations since, for most problems S, $K(S)$ in fact is (fully) pareto-optimal. But it is not the only change that has to be made in the axioms of Theorem to extend the characterization of the Kalai–Smorodinsky solution to the case $n > 2$.

3.10.11. *The Egalitarian solution.* We now turn to a third solution, which is the main distinguishing feature from the previous two.

Definition. The Egalitarian solution E is defined by setting, for all $S \in \sum^n$, $E(S)$ to be the maximal point of S of equal coordinates (see Fig. 3.7).

The most striking feature of this solution is that it satisfies the following monotonicity condition, which is very strong, since no

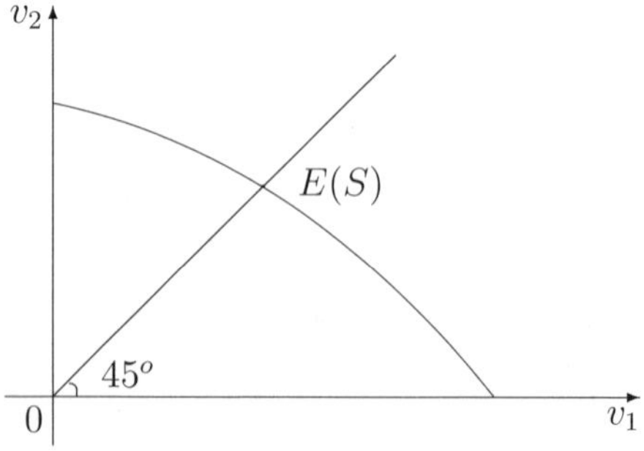

Figure 3.7

restriction are imposed in its hypotheses on the sort of expansions that take S into S'. In fact, this axiom can serve to provide an easy characterization of the solution.

8^0. *Strong monotonicity.* For all $S, S' \in \sum^n$, if $S \subset S'$, then $\varphi(S) \leq \varphi(S')$.

The following characterization result is a variant of a theorem due to Kalai and Smorodinsky (1975).

Theorem. *A solution* $\varphi(S)$, $S \in \sum^n$ *satisfies* $2^0, 3^0, 8^0$ *if and only if it is the Egalitarian solution.*

3.10.12. *The Utilitarian solution.* We close this review with a short discussion of the Utilitarian solution.

Definition. A Utilitarian solution U is defined by choosing, for each $S \in \sum^n$ among the maximizers of $\sum_{i=1}^n v_i$ for $v \in S$ (see Fig. 3.8).

Obviously, all Utilitarian solutions satisfy 1^0. They also satisfy 3^0 if appropriate selections are made. However, no Utilitarian solution satisfies 4^0. Also, no Utilitarian solution satisfies 5^0, because of the impossibility of performing appropriate selections. The Utilitarian solution has been characterized by Myerson (1981).

Other solutions have been discussed in the literature by Luce and Raiffa (1957), and Perles and Mashler (1981). In this section we follow Thomson and Lensberg (1989), where the reader can find proofs of the theorems.

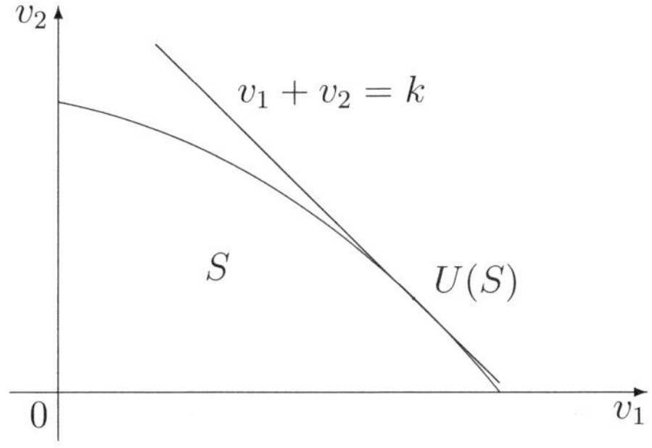

Figure 3.8

3.11 Exercises and Problems

1. Two companies are engaged in exploration of n mineral deposits. The exploration funds allocated by the companies 1 and 2 are α and β, respectively. The profit from mining the ith field is $\gamma_i > 0$. It is distributed between companies in proportion with their contributions to the ith mining field. If they make no investment in the ith field, then their profits from this field are zero.

(a) Describe this conflict as a two-person game, taking a payoff to each company to be the total profit from mining of all fields.

(b) Find a Nash equilibrium.

Hint. Use the convexity of the function H_1 in x and that of H_2 in y.

2. In the ecologically significant region there are n industrial enterprises, each having one pollution source. Concentration q_i of emission from the ith enterprise is proportional to the value $0 \leq x_i \leq a_i$, $i = 1, \dots, n$, of emission from this enterprise. The losses incurred by the ith enterprise and made up of the waste utilization expenses ($f_i(x_i)$) and the pollution tax which is proportional to the total concentration q of emission from all enterprises. The quantity q should not exceed Θ that is the maximum level of emission concentration; otherwise each ith enterprise has to pay an extra penalty s_i.

Describe this conflict, as a noncooperative n-person game, taking the losses incurred by each enterprise to be the total environmental projection expenses.

Hint. Use the result of Example 5, 3.1.4.

3. Find the sets of all Nash equilibria (in pure strategies) in the following ($m \times n$) bimatrix games with the matrices $A = \{\alpha_{ij}\}$ and $B = \{\beta_{ij}\}$.

(a) The matrices A and B are diagonal and positive, i.e. $m = n$, $\alpha_{ij} = \beta_{ij} = 0$, $i \neq j$ and $\alpha_{ii} > 0$, $\beta_{ii} > 0$, $i = 1, \dots, m$, $j = 1, \dots, n$.

(b)

$$A = \begin{bmatrix} 2 & 0 & 5 \\ 2 & 2 & 3 \end{bmatrix}, \quad B = \begin{bmatrix} 2 & 2 & 1 \\ 0 & 7 & 8 \end{bmatrix}.$$

(c)

$$A = \begin{bmatrix} 3 & 8 & -1 \\ 4 & 0 & 2 \\ 1 & 2 & 3 \end{bmatrix}, \quad B = \begin{bmatrix} 1 & 3 & 4 \\ 2 & 1 & 8 \\ 2 & 3 & 0 \end{bmatrix}.$$

4. Show that in the bimatrix game with the matrices

$$A = \begin{bmatrix} 1 & 2 & 0 \\ 1 & 3 & 1 \\ 2 & 2 & 1 \end{bmatrix}, \quad B = \begin{bmatrix} 3 & 4 & 0 \\ 1 & 3 & 2 \\ 1 & 3 & 0 \end{bmatrix}$$

the situation (2,2) is equilibrium. Is it strongly equilibrium?

5. Find all pareto-optimal situations in pure strategies in the bimatrix game with the matrices

$$A = \begin{bmatrix} 4 & 1 & 0 \\ 2 & 7 & 5 \\ 6 & 0 & 1 \end{bmatrix}, \quad B = \begin{bmatrix} 0 & 5 & 6 \\ 7 & 0 & 2 \\ 2 & 6 & 1 \end{bmatrix}.$$

Does this game have pure strategy equilibria?

6. Show graphically in coordinates (K_1, K_2) the set of all possible mixed strategy payoff vectors in the game "battle of the sexes" (see 3.1.4).

Hint. Arbitrary mixed strategies x and y for Players 1 and 2, respectively, can be written as $x = (\xi, 1-\xi)$, $y = (\eta, 1-\eta)$, $\xi, \eta \in [0,1]$. Writing the mixed strategy payoff functions K_1 and K_2 and eliminating one of the parameters we obtain a single-parameter family of line segments the union of which is the required set (see Fig. 3.2). The curvilinear part of the boundary represents an envelope for this family of line segments and is a part of the parabola $5K_1^2+5K_2^2-10K_1K_2-18(K_1+K_2)+45 = 0$.

7. Find a completely mixed Nash equilibrium in the bimatrix game with the matrices

$$A = \begin{bmatrix} 6 & 0 & 2 \\ 0 & 4 & 3 \\ 7 & 0 & 0 \end{bmatrix}, \quad B = \begin{bmatrix} 6 & 0 & 7 \\ 0 & 4 & 0 \\ 2 & 3 & 0 \end{bmatrix}.$$

Does this game also have equilibria in mixed strategies?

Hint. First find a completely mixed equilibrium (x, y), $x = (\xi_1, \xi_2, \xi_3)$, $y = (\eta_1, \eta_2, \eta_3)$, then an equilibrium for which $\xi_1 = 0$, etc.

8. *"Originality game"* [Vorobjev (1984)]. Consider a noncooperative n-person game $\Gamma = (N, \{X_i\}_{i \in N}, \{H_i\}_{i \in N})$, where $X_i = \{0, 1\}$, $H_i(0, \ldots, 0\|_i 1) = g_i > 0$, $H_i(1, \ldots, 1\|_i 0) = h_i > 0$, $H_i(x) = 0$ in the remaining cases where $\|_i$ means that a replacement is made in the ith position.

(a) Interpret the game in terms of advertising.
(b) Find a completely mixed equilibrium.

9. As is shown in 1.10.1, zero-sum two-person games can be solved by the "fictious play" method. Examining the bimatrix game with the matrices

$$A = \begin{bmatrix} 2 & 0 & 1 \\ 1 & 2 & 0 \\ 0 & 1 & 2 \end{bmatrix}, \quad B = \begin{bmatrix} 1 & 0 & 2 \\ 2 & 1 & 0 \\ 0 & 2 & 1 \end{bmatrix},$$

show that this method cannot be used in finding an equilibrium in bimatrix games.

10. *"Musical chairs" game* [Moulin (1981)]. There are two players and three chairs designated by numbers 1,2,3. A strategy of a player is to choose a chair number. Both players may suffer losses due to a choice of the same chair. If, however, their choices are different, then the player, say i, whose chair is located immediately after Player j's chair, wins twice as much as Player j (it is assumed that chair 1 is located after chair 3). We have the bimatrix game $\Gamma(A, B)$,

$$(A, B) = \begin{bmatrix} (0,0) & (1,2) & (2,1) \\ (2,1) & (0,0) & (1,2) \\ (1,2) & (2,1) & (0,0) \end{bmatrix}.$$

(a) Show that the unique completely mixed Nash equilibrium is an equiprobable choice of chairs to be made by each player.

(b) Show that an equilibrium in joint mixed strategies is of the form

$$L(i, j) = \begin{cases} 1/6, & \text{if } i \neq j, \\ 0, & \text{if } i = j. \end{cases}$$

(c) Show that the payoffs in Nash equilibrium are not pareto-optimal, while a joint mixed strategy equilibrium may result in pareto-optimal payoffs $(3/2, 3/2)$.

11. The equilbrium in joint mixed strategies does not imply that the players must necessarily follow the pure strategies resulting from the adopted joint mixed strategy (see definition in 3.6.1). However, if we must adhere to the results of a particular realization of the joint mixed strategy, then it is possible to extend the concept an "equilibrium in joint mixed strategies". For all $i \in N$, denote by $\mu(N \setminus \{i\})$ the restriction of distribution μ to the set $X_{N\setminus\{i\}} = \prod_{i \in N\setminus\{i\}} X_i$, namely

$$\mu(N \setminus \{i\}) = \sum_{x_i \in X_i} \mu(x \| x_i),$$

for all $x \in \prod_{i \in N} X_i$. We say that μ is the weak equilibrium in joint mixed strategies if the following inequalities hold for all $i \in N$ and $y_i \in X_i$:

$$\sum_{x \in \prod_{i \in N} X_i} H_i(x)\mu(x) \geq \sum_{x \in \prod_{i \in N} X_i} H(x \| y_i)\mu(N \setminus \{i\}).$$

(a) Prove that any equilibrium in joint mixed strategies is the weak equilibrium in joint mixed strategies.

(b) Let $\mu = (\mu_1, \ldots, \mu_n)$ be a mixed strategy situation in the game Γ. Show that the probability measure $\overline{\mu} = \prod_{i \in N} \mu_i$ on the set $X = \prod_{i \in N} X_i$ is a weak equilbrium in joint mixed strategies and an equilibrium in joint strategies if and only if the situation $\mu = (\mu_1, \ldots, \mu_n)$ is Nash equilibrium.

12. (a) Prove that in the game formulated in Example 12, 3.2.3 the set of Nash equilibria, the set of joint strategy equilibria and the set of weak equilibria in joint mixed strategies do not coincide.

(b) Show that the interval $[(5/3, 4/3), (4/3, 5/3)]$ is covered by the set of vector payoffs that are pareto-optimal among the payoffs in joint mixed strategy equilibria, while the interval $[(2, 1), (1, 2)]$ is covered by the payoffs that are pareto-optimal among the weak equilibrium payoffs in joint mixed strategies.

13. Find an arbitration solution to the bimatrix game with the matrices $A = \begin{bmatrix} 2 & -1 \\ -1 & 1 \end{bmatrix}$, $B = \begin{bmatrix} 1 & -1 \\ -1 & 2 \end{bmatrix}$ by employing the Nash bargaining procedure.

14. Consider the bimatrix (2×2) game with the matrix

$$(A, B) = \begin{array}{c} \\ \alpha_1 \\ \alpha_2 \end{array} \begin{array}{c} \begin{array}{cc} \beta_1 & \beta_2 \end{array} \\ \begin{bmatrix} (1, 1) & (1, 2) \\ (2, 1) & (-5, 0) \end{bmatrix} \end{array}.$$

This is a modification of the "crossroads" game (see Example 2 in 3.1.4) with the following distinction. A car driver (Player 1) and a truck driver (Player 2) make different assessments of an accident (situation (α_2, β_2)). Show that an analysis of the game in threat strategies prescribes a situation (α_1, β_2), i.e. the car must "go" and the truck must "make a stop".

15. Consider Braess's paradox for the case when $N = 6$, time for passing BC and DE $10x$ (if x is the number of cars on the arc), $50 + x$ for BD and CE and $10 + x$ for CD.

16. *Majority voting* [Fudenberg and Tirole (1992)]. There are three Players 1, 2 and 3, and three alternatives a, b, c. Players vote simultaneously for the alternative; abstaining is not allowed. Thus, the strategy spaces are $X_i = \{a, b, c\}$. The alternative with the most votes wins; if no alternative receives the majority, then alternative a is selected. Player's utility functions are

$$u_1(c) < u_1(b) < u_1(a),$$

$$u_2(b) < u_2(a) < u_2(c),$$

$$u_3(a) < u_3(c) < u_3(b).$$

Then the payoff functions are

$$H_i(x_1, x_2, x_3) = u_i(s(x_1, x_2, x_3)),$$

where winning function s is

$$s(x_1, x_2, x_3) = \begin{cases} x_1 & \text{if } x_2 \neq x_3, \\ x_2 & \text{if } x_2 = x_3. \end{cases}$$

Find all Nash equilibria in the game.

17. *Bertrand duopoly* [Tirole (2003)]. We consider the case if differentiated products (see example 8 for the case of homogeneous products). If firms 1 and 2 choose prices p_1 and p_2, respectively, the consumers demand from firm i is

$$q_i(p_i, p_j) = a - p_i + bq_j,$$

where $b > 0$ reflects the extent to which firm i's product is a substitute for firm j's product. Suppose that $b < 2$, that marginal costs are constant at c, where $c < a$, and that the firms choose their prices simultaneously. As before we need to translate the problem into a normal-form game.

Find Nash equilibrium in the game.

18. *Cournot duopoly.* Consider Cournot duopoly with n firms, which produce a homogeneous good. All firms simultaneously choose their strategies q_i from feasible set $X = \{q|q \geq 0\}$ (see Example 17, 3.4.2 for Cournot duopoly) and have linear demand and costs. Suppose that inverse-demand function $p(q) = a - q$ and cost function $c_i(q_i) = c_i q_i$, $c_i > 0$, $i = 1, 2, \ldots, n$.

In the case firm i's total profit

$$H_i(q_1, \ldots, q_n) = q_i(a - q) - c_i q_i,$$

where $q = \sum_{i=1}^{n} q_i$.

(a) Suppose, that $c_i = c$, $i = 1, 2, \ldots, n$. Find Nash equilibrium in the game.
(b) Find Nash equilibrium for arbitrary $c_i > 0$, $i = 1, 2, \ldots, n$.

19. [Tirole (2003)]. Consider a duopoly producing a homogeneous product. Firm 1 produces one unit of output with one unit of labor and one unit of raw material. Firm 2 produces one unit of output with two unit of labor one unit of raw material. The unit costs of labor and raw material are w and r. The demand is $p = a - q_1 - q_2$, and firms compete in quantities.

(a) Compute Nash equilibrium.
(b) Show that firm 1's profit is not affected by the price of labor.

20. [Tirole (2003)]. There are two firms in a market. They produce a homogeneous product at cost $c(q) = q^2/2$. The demand is $p = a - (q + q_2)$.

(a) Compute Nash equilibrium in Cournot model.
(b) Suppose, that firm 1 has the opportunity to sell the same output on another market as well. The demand on the second market is $p = b - x_1$. Consider the Cournot game in which firm 1 chooses q_1 and x_1 and firm 2 chooses q_2 simultaneously. Compute the Nash equilibrium.

Chapter 4

Cooperative Games

4.1 Games in Characteristic Function Form

4.1.1. Chapter 3 demonstrates how the players can arrive at a mutually acceptable decisions on the arising conflict by an agreed choice of strategies (strategic cooperation). We now suppose that the conditions of a game admit the players' joint actions and redistribution of a payoffs. This implies that the utilities of various players can be evaluated by a single scale (transferable payoffs), and hence the mutual redistribution of payoffs does not affect the conceptual statement of the original problem. It appears natural that, from the point of view of each player, the best results may also be produced by uniting players into a maximal coalition (the coalition composed of all players). In this case, we are interested not only in the ways the coalition of players ensures its total payoff, but also in the ways it is distributed among the members of this coalition (cooperative approach).

The chapter deals with the cooperative theory of n-person games. This theory is focused on the conditions under which integration of players into a maximal coalition is advisable and individual players are not interested in forming smaller groups or act individually.

4.1.2. Let $N = \{1, \ldots, n\}$ be a set of all players. Any nonempty subset $S \subset N$ is called a *coalition.*

Definition. The real-valued function v defined on coalitions $S \subset N$ is called a characteristic function of the n-person game. Here the inequality

$$v(T) + v(S) \leq v(T \cup S), \quad v(\oslash) = 0 \tag{4.1.1}$$

holds for any nonintersecting coalitions T, S ($T \subset N$, $S \subset N$).

Property (4.1.1) is called a *superadditivity property.* This property is necessary for the number $v(T)$ to be conceptually interpreted as a guaranteed payoff to a coalition T when this coalition is acting independently of other players. This interpretation of inequality (4.1.1) implies that the coalition $S \cup T$ has no less opportunities than the two nonintersecting coalitions S and T when they act independently.

From the superadditivity of v it follows that for any system of nonintersecting coalitions $S_1, \ldots, S_k$ there is

$$\sum_{i=1}^{n} v(S_i) \leq v(N).$$

This, in particular, implies that there is no decomposition of the set N into coalitions such that the guaranteed total payoff to these coalitions exceeds the maximum payoff to all players $v(N)$.

4.1.3. *Example 1. Investment fund* [Lemaire (1991)]. Three investment fund managers consider investment possibilities for a year. Fund manager 1 has \$3,000,000 to invest, manager 2 has \$1,000,000 and manager 3 has \$2,000,000. There is an investment scheme (Table 4.1)

Table 4.1 Interest rates

	Deposit	Interest rate
1	less than \$2000000	8%
2	\$2000000 up to \$5000000	9%
3	\$5000000 and more	10%

This situation can be readily translated into a characteristic function v for a three-player game, $N = \{1, 2, 3\}$ (in \$10,000 units):

$$v(N) = 60, \quad v(\{1,2\}) = 36, \quad v(\{1,3\}) = 50, \quad v(\{2,3\}) = 27,$$
$$v(\{1\}) = 27, \quad v(\{2\}) = 8, \quad v(\{3\}) = 18.$$

It's simply to test that characteristic function $v(S)$, $S \subset N$ has superadditivity property.

Example 2. Farmer, manufacturer, subdivider [Shubik (1982)]. A farmers land is worth \$100,000 to him for agricultural use; to a manufacturer it is worth \$200,000 as a plant site; a subdivider would pay up to \$300,000.

Denoting the farmer as player 1, the manufacture as player 2 and the subdivider as player 3, one obtains the player set $N = \{1, 2, 3\}$. The characteristic function v (in \$100,000 units) follows immediately from the description of the situation:

$$v(N) = 3, \quad v(\{1,2\}) = 2, \quad v(\{1,3\}) = 3, \quad v(\{2,3\}) = 0,$$
$$v(\{1\}) = 1, \quad v(\{i\}) = 0, \quad \text{for } i = 2, 3.$$

Example 3. Weighted majority voting [Eichberger (1993)]. Consider a group of n shareholders of a company that has to decide on some investment project. Denote by w_i the number of shares shareholder i owns. The company charter assigns one vote to each share and requires a minimum of q votes for the adoption of a project. Suppose the investment project yields a return R to be distributed equally per share.

This problem can be modeled as a game in characteristic function form with player set $N = \{1, 2, \ldots, n\}$ and characteristic function v:

$$v(S) = \begin{cases} \left(\sum_{i \in S} w_i\right) \frac{R}{w}, & \text{if } \sum_{i \in S} w_i \geq q, \\ 0, & \text{if } \sum_{i \in S} w_i < q, \end{cases}$$

where $w = \sum_{i \in N} w_i$.

Example 4. Water treatment system [Moulin (1986)]. Three communities sharing a large city consider developing a water treatment system. So far these communities dispose off their sewage by sending it to a central treatment plant operated by city authorities at a monthly fee.

A cost–benefit study estimates the present value of these payments over the usual lifetime of water treatment plant at \$100 per household. To build and operate a water treatment plant for the same period is estimated at present value cost of \$500000 for up to 5,000 households; \$600000 for up to 10,000 households, and \$700000 for up to 15,000 households.

Community 1 is estimated to serve 5,000 households on average during the period under consideration, community 2 has to serve 3,000 households and community 3 has 4,000 households.

The decision problem of the three communities can be modeled as a game in characteristic function form with set of players $N = \{1, 2, 3\}$ and characteristic function v (measured in \$100000 units):

$$v(N) = 5, \quad v(\{1,2\}) = 2, \quad v(\{1,3\}) = 3, \quad v(\{2,3\}) = 1,$$
$$v(\{1\}) = v(\{2\}) = v(\{3\}) = 0.$$

The net benefit of a coalition is calculated as difference between the joint benefit (number of households times present value of fees saved) minus the cost of a water treatment system of appropriate size. A single community can of couse still send its sewage to the central treatment plant. This guarantees it a net benefit of zero.

Example 5. "Jazz band" game [Moulin (1981)]. Manager of a club promises singer S, pianist P, and drummer D to pay \$100 for a joint performance. He values a singer-pianist duet at \$80, a drummer-pianist duet at \$65 and a pianist at \$30.

A singer-drummer duet may earn \$50 and a singer, on the average, \$20 for doing an evening performance. A drummer may not earn anything by playing alone.

Designating players S, P, and D by numbers 1,2,3, respectively, we are facing a cooperative game (N, v), where $N = \{1, 2, 3\}$,

$v(1,2,3) = 100$, $v(1,3) = 50$, $v(1) = 20$, $v(1,2) = 80$, $v(2,3) = 65$, $v(2) = 30$, $v(3) = 0$.

4.1.4. We shall now consider a noncooperative game $\Gamma = (N, \{X_i\}_{i\in N}, \{H_i\}_{i\in N})$.

Suppose the players appearing in a coalition $S \subset N$ unite their efforts for the purpose of increasing their total payoff. Let us find the largest payoff they can guarantee themselves. The joint actions of the players from a coalition S imply that this coalition S acting for all its members as one player (call him Player 1) takes the set of pure strategies to be the set of all possible combinations of strategies for its constituent players from S, i.e. the elements of the Cartesian product

$$X_S = \prod_{i\in S} X_i.$$

The community of interests of the players from S means that a payoff to the coalition S (Player 1) is the sum of payoffs to the players from S, i.e.

$$H_S(x) \triangleq \sum_{i\in S} H_i(x),$$

where $x \in X_N$, $x = (x_1, \ldots, x_n)$ is a situation in pure strategies.

We are interested in the largest payoff the players from S can guarantee themselves. In the worst case for Player 1, the remaining players from $N\backslash S$ may also unite into a collective Player 2 with the set of strategies $X_{N\backslash S} = \prod_{i\in N\backslash S} X_i$, where interests are diametrically opposite to those of Player 1 (i.e. Player 2's payoff at x is $-H_S(x)$). As a result of this reasoning, the question of the largest guaranteed payoff to the coalition S becomes the issue of the largest guaranteed payoff to Player 1 in the zero-sum game $\Gamma_S = (X_S, X_{N\backslash S}, H_S)$. In the mixed extension $\overline{\Gamma}_S = (\overline{X}_S, \overline{X}_{N\backslash S}, K_S)$ of the game Γ_S, the guaranteed payoff $v(S)$ to Player 1 can merely be increased in comparison with that in the game Γ_S. For this reason, the following discussion concentrates on the mixed extension of Γ_S. In particular, it should be noted that, according to this interpretation, $v(S)$ coincides with the value of the game $\overline{\Gamma}_S$ (if any), while $v(N)$ is the maximum total payoff to the players. Evidently, $v(S)$ only depends on the coalition S (and on the original noncooperative game itself, which remains unaffected in our reasoning) and is a function of S. We shall verify that this function is a characteristic function of a noncooperative game. To do this, it suffices to show that the conditions (4.1.1) is satisfied.

Note that $v(\oslash) = 0$ for every noncooperative game constructed above.

Lemma (on superadditivity). *For the noncooperative game that* $\Gamma = (N, \{X_i\}_{i\in N}, \{H_i\}_{i\in N})$, *we shall construct the function* $v(S)$ *as*

$$v(S) = \sup_{\mu_S} \inf_{\nu_{N\setminus S}} K_S(\mu_S, \nu_{N\setminus S}), \quad S \subset N, \tag{4.1.2}$$

where $\mu_S \in \overline{X}_S$, $\nu_{N\setminus S} \in \overline{X}_{N\setminus S}$ *and* $\overline{\Gamma}_S = (\overline{X}_S, \overline{X}_{N\setminus S}, K_S)$ *is a mixed extension of the zero-sum game* Γ_S. *Then for all* $S, T \subset N$ *for which* $S \cap T = \oslash$, *the following inequality holds:*

$$v(S \cap T) \geq v(S) + v(T). \tag{4.1.3}$$

Proof. Note that

$$v(S \cup T) = \sup_{\mu_{S\cup T}} \inf_{\nu_{N\setminus(S\cup T)}} \sum_{i\in S\cup T} K_i(\mu_{S\cup T}, \nu_{N\setminus(S\cup T)}),$$

where $\mu_{S\cup T}$ is the mixed strategy of coalition $S \cup T$, i.e. arbitrary probability measures on $X_{S\cup T}$, $\nu_{N\setminus(S\cup T)}$ is probability measure on $X_{N\setminus(S\cup T)}$, K_i is a payoff to player i in mixed strategies. If we restrict ourselves to those probability measures on $X_{S\cup T}$, which are the products of independent distributions μ_S and ν_T over the Cartesian product $X_S \times X_T$, then the range of the variable, in terms of which maximization is taken, shrinks and supremum merely decreases. Thus we have

$$v(S \cup T) \geq \sup_{\mu_S} \sup_{\mu_T} \inf_{\nu_{N\setminus(S\cup T)}} \sum_{i\in S\cup T} K_i(\mu_S \times \mu_T, \nu_{N\setminus(S\cup T)}).$$

Hence

$$v(S \cup T) \geq \inf_{\nu_{N\setminus(S\cup T)}} \sum_{i\in S\cup T} K_i(\mu_S \times \mu_T, \nu_{N\setminus(S\cup T)})$$
$$= \inf_{\nu_{N\setminus(S\cup T)}} \Bigg(\sum_{i\in S} K_i(\mu_S \times \mu_T, \nu_{N\setminus(S\cup T)})$$
$$+ \sum_{i\in T} K_i(\mu_S \times \mu_T, \nu_{N\setminus(S\cup T)})\Bigg).$$

Since the sum of infimums does not exceed the infimum of the sum, we have

$$v(S \cup T) \geq \inf_{\nu_{N\setminus(S\cup T)}} \sum_{i\in S} K_i(\mu_S \times \mu_T, \nu_{N\setminus(S\cup T)})$$
$$+ \inf_{\nu_{N\setminus(S\cup T)}} \sum_{i\in T} K_i(\mu_S \times \mu_T, \nu_{N\setminus(S\cup T)}).$$

Minimizing the first addend on the right-hand side of the inequality over μ_T, and the second addend over μ_S (for uniformity, these will be renamed as ν_T and ν_S), we obtain

$$v(S \cup T) \geq \inf_{\nu_T} \inf_{\nu_{N\setminus(S\cup T)}} \sum_{i\in S} K_i(\mu_S \times \nu_T, \nu_{N\setminus(S\cup T)})$$
$$+ \inf_{\nu_S} \inf_{\nu_{N\setminus(S\cup T)}} \sum_{i\in T} K_i(\nu_S \times \mu_T, \nu_{N\setminus(S\cup T)})$$
$$\geq \inf_{\nu_{N\setminus S}} \sum_{i\in S} K_i(\mu_S, \nu_{N\setminus S}) + \inf_{\nu_{N\setminus T}} \sum_{i\in T} K_i(\mu_T, \nu_{N\setminus T}).$$

The last inequality holds for any values of measures μ_S and μ_T. Consequently, these make possible the passage to suprema

$$v(S \cup T) \geq \sup_{\mu_S} \inf_{\nu_{N\setminus S}} \sum_{i\in S} K_i(\mu_S, \nu_{N\setminus S}) + \sup_{\mu_T} \inf_{\nu_{N\setminus T}} \sum_{i\in T} K_i(\mu_T, \nu_{N\setminus T}),$$

whence, using (4.1.2), we obtain

$$v(S \cup T) \geq v(S) + v(T).$$

The superadditivity is proved.

Note that inequality (4.1.3) also holds if the function $v(S)$ is constructed by the rule

$$v(S) = \sup_{x_S} \inf_{x_{N\setminus S}} H_S(x_S, x_{N\setminus S}), \quad S \subset N,$$

where $x_S \in X_S$, $x_{N\setminus S} \in X_{N\setminus S}$, $\Gamma_S = (X_S, X_{N\setminus S}, H_S)$. In this case, the proof literarily repeats the one given above.

4.1.5. Definition. The noncooperative game $\Gamma = (N, \{X_i\}_{i\in N}, \{H_i\}_{i\in N})$ is called a constant sum game if

$$\sum_{i\in N} H_i(x) = c = const$$

for all $x \in X_N$, $X_N = \prod_{i\in N} X_i$.

Lemma. *Let* $\Gamma = (N, \{X_i\}_{i\in N}, \{H_i\}_{i\in N})$ *be a noncooperative constant sum game, the function* $v(S)$, $S \subset N$, *be defined as in Lemma* 4.1.4, *and the games* Γ_S, $S \subset N$, *have the values in mixed strategies. Then*

$$v(N) = v(S) + v(N\setminus S), \quad S \subset N.$$

Proof. The definition of the constant sum game implies that

$$v(N) = \sum_{i\in N} H_i(x) = \sum_{i\in N} K_i(\mu) = c$$

for all situations x in pure strategies and all situations μ in mixed strategies.

On the other hand,

$$\begin{aligned} v(S) &= \sup_{\mu_S} \inf_{\nu_{N\setminus S}} \sum_{i\in S} K_i(\mu_S, \nu_{N\setminus S}) \\ &= \sup_{\mu_S} \inf_{\nu_{N\setminus S}} \Big(c - \sum_{i\in N\setminus S} K_i(\mu_S, \nu_{N\setminus S})\Big) \\ &= c - \inf_{\nu_{N\setminus S}} \sup_{\mu_S} \sum_{i\in N\setminus S} K_i(\mu_S, \nu_{N\setminus S}) = c - v(N\setminus S), \end{aligned}$$

which is what we set out to prove.

4.1.6. In what follows, by the cooperative game is meant a pair (N, v), where v is the characteristic function satisfying inequality (4.1.1). The conceptual interpretation of the characteristic function justifying property (4.1.1) is not essential for what follows (see Examples 1–5).

The main problem in the cooperative theory of n-person games is to construct realizable principles for optimal distribution of a maximum total payoff $v(N)$ among players Aumann (1959).

Let α_i be an amount the player i receives by distribution of maximum total payoff $v(N)$, $N = \{1, 2, \dots, n\}$.

Definition. The vector $\alpha = (\alpha_1, \dots, \alpha_n)$, which satisfies the conditions

$$\alpha_i \geq v(\{i\}), \quad i \in N, \tag{4.1.4}$$

$$\sum_{i=1}^{n} \alpha_i = v(N), \tag{4.1.5}$$

where $v(\{i\})$ is the value of characteristic function for a single-element coalition $S = \{i\}$ is called an imputation.

Condition (4.1.4) is called an individual rationality condition and implies that every member of coalition received at least the same amount he could ensure by acting alone, without any support of other players. Furthermore, condition (4.1.5) must be satisfied, since in the case $\sum_{i \in N} \alpha_i < v(N)$ there is a distribution α', on which every player $i \in N$ receives more than his share α_i. However, if $\sum_{i \in N} \alpha_i > v(N)$, then players from N distribute among themselves an unrealized payoff. For this reason, the vector α can be taken to be admissible only if condition (4.1.5) is satisfied. This condition is called a collective (or group) rationality condition.

By (4.1.4), (4.1.5), for the vector $\alpha = (\alpha_1, \dots, \alpha_n)$ to be an imputation in the cooperative game (N, v), it is necessary and sufficient that it could be represented as

$$\alpha_i = v(\{i\}) + \gamma_i, \quad i \in N,$$

and

$$\gamma_i \geq 0, \quad i \in N, \quad \sum_{i \in N} \gamma_i = v(N) - \sum_{i \in N} v(\{i\}).$$

Definition. The game (N, v) is called essential if

$$\sum_{i \in N} v(\{i\}) < v(N), \tag{4.1.6}$$

otherwise it is called nonessential.

For any imputation α, we denote the quantity $\sum_{i \in S} \alpha_i$ by $\alpha(S)$ and the set of all imputations by D. The nonessential game has a unique imputation $\alpha = (v(\{1\}), v(\{2\}), \ldots, v(\{n\}))$.

In any essential game with more than one player, the imputation set is infinite. We shall examine such games by using a dominance relation.

Definition. Imputation α dominates imputation β in coalition S (denoted as $\alpha \overset{S}{\succ} \beta$) if

$$\alpha_i > \beta_i, \quad i \in S, \quad \alpha(S) \leq v(S). \tag{4.1.7}$$

The first condition in (4.1.7) implies that imputation α is more advantageous to all members of coalition S than imputation β, while the second condition accounts for the fact that imputation α can be realized by coalition S (that is, coalition S can actually offer an amount α_i to every player $i \in S$).

Definition. Imputation α is said to dominanate imputation β if there is a coalition S for which $\alpha \overset{S}{\succ} \beta$. Dominance of imputation β by imputation α is denoted as $\alpha \succ \beta$.

Dominance is not possible in a single element coalition and in the set of all players N. Indeed, $\alpha \overset{i}{\succ} \beta$ had to imply $\beta_i < \alpha_i \leq v(\{i\})$ which contradicts condition (4.1.5).

4.1.7. Combining cooperative games into one or another class may substantially simplify their subsequent examination. We may examine equivalency classes of games.

Definition. The cooperative game (N, v) is called equivalent to the game (N, v') if there exists a positive number k and n arbitrary real numbers c_i, $i \in N$, such that for any coalition $S \subset N$ there is

$$v'(S) = kv(S) + \sum_{i \in S} c_i. \tag{4.1.8}$$

The equivalence of the game (N, v) to (N, v') will be denoted as $(N, v) \sim (N, v')$ or $v \sim v'$.

It is obvious that $v \sim v$. This can be verified by setting $c_i = 0$, $k = 1$, $v' = v$ in (4.1.8). This property is called reflexity.

We shall prove the symmetry of the relation, i.e. that the condition $v \sim v'$ implies $v' \sim v$. In fact, setting $k' = 1/k$, $c'_i = -c_i/k$ we obtain

$$v(S) = k'v'(S) + \sum_{i \in S} c'_i,$$

i.e. $v' \sim v$.

Finally, if $v \sim v'$ and $v' \sim v''$, then $v \sim v''$. This property is called transitivity. This can be verified by successively applying (4.1.8).

Since the equivalence relation is reflexive, symmetric and transitive, it decomposes the set of all n-person games into mutually nonintersecting classes of equivalent games.

Theorem. *If two games v and v' are equivalent, then the map $\alpha \to \alpha'$, where*

$$\alpha'_i = k\alpha_i + c_i, \quad i \in N,$$

establishes the one-to-one mapping of the set of all imputations in the game v onto the imputation set in the game v', so that $\alpha \overset{S}{\succ} \beta$ implies $\alpha' \overset{S}{\prec} \beta'$.

Proof. Let us verify that α' is an imputation in the game (N, v'). Indeed,

$$\alpha'_i = k\alpha_i + c_i \geq kv(\{i\}) + c_i = v(\{i\}),$$

$$\sum_{i \in N} \alpha'_i = \sum_{i \in N} (k\alpha_i + c_i) = kv(N) + \sum_{i \in N} c_i = v'(N).$$

It follows that conditions (4.1.4), (4.1.5) hold for α'. Furthermore, if $\alpha \overset{S}{\succ} \beta$, then

$$\alpha_i > \beta_i, \quad i \in S, \quad \sum_{i \in S} \alpha_i \leq v(S),$$

and hence

$$\alpha_i' = k\alpha_i + c_i > k\beta_i + c_i = \beta_i' \ (k > 0),$$

$$\sum_{i \in S} \alpha_i' = k \sum_{i \in S} \alpha_i + \sum_{i \in S} c_i \leq kv(S) + \sum_{i \in S} c_i = v'(S),$$

i.e. $\alpha' \overset{S}{\succ} \beta'$. The one-to-one correspondence follows from the existence of the inverse mapping (it was used in the proof of the symmetry of the equivance relation). This completes the proof of the theorem.

4.1.8. When decomposing the set of cooperative games into mutually disjoint classes of equivalence, we are faced with the problem of choosing the simplest representative from each class.

Definition. The game (N, v) is called the game in (0–1)-reduced form, if for all $i \in N$

$$v(\{i\}) = 0, \quad v(N) = 1.$$

Theorem. *Every essential cooperative game is equivalent to some game in (0–1)-reduced form.*

Proof. Let

$$k = \frac{1}{v(N) - \sum_{i \in N} v(\{i\})} > 0,$$

$$c_i = -\frac{v(\{i\})}{v(N) - \sum_{i \in N} v(\{i\})}, \quad v'(S) = kv(S) + \sum_{i \in S} c_i.$$

Then $v'(\{i\}) = 0$, $v'(N) = 1$. This completes the proof of the theorem.

This theorem implies that the game theoretic properties involving the notion of dominance can be examined on the games in (0–1)-reduced form. If v is the characteristic function of an arbitrary essential game (N, v), then

$$v'(S) = \frac{v(S) - \sum_{i \in S} v(\{i\})}{v(N) - \sum_{i \in N} v(\{i\})}, \tag{4.1.9}$$

is (0–1)-normalization corresponding to the function v. In this case, an imputation is found to be any vector $\alpha = (\alpha_1, \ldots, \alpha_n)$ whose components satisfy the conditions

$$\alpha_i \geq 0, \quad i \in N, \quad \sum_{i \in N} \alpha_i = 1, \tag{4.1.10}$$

i.e. imputations can be regarded as the points of the $(n-1)$-dimensional simplex generated by the unit vectors $w_j = (0, \ldots, 0, 1, 0, \ldots, 0)$, $j = 1, \ldots, n$ in the space R^n.

4.2 The Core and NM-Solution

We shall now turn to the principles of optimal behavior in cooperative games. As already noted in 4.1.4, we are dealing with the principles of optimal distribution of a maximum total payoff among players.

4.2.1. The following approach is possible. Suppose the players in the cooperative game (N, v) have come to an agreement on distribution of a payoff to the whole coalition N (imputation α^*), under which none of the imputations dominates α^*. Then such a distribution is stable in that it is disadvantageous for any coalition S to separate from other players and distribute a payoff $v(S)$ among its members. This suggests that it may be wise to examine the set of nondominant imputations.

Definition. The set of nondominant imputations in the cooperative game (N, v) is called core.

Then we have the theorem which characterizes core.

Theorem. *For the imputation α to belong to core, it is necessary and sufficient that*

$$v(S) \leq \alpha(S) = \sum_{i \in S} \alpha_i \tag{4.2.1}$$

hold for all $S \subset N$.

Proof. This theorem is straightforward for nonessential games, and, by Theorem 4.1.8, it suffices to prove it for the games in (0–1)-reduced form.

Prove first that the statement of the theorem is sufficient. Suppose that condition (4.2.1) holds for the imputation α. Show that the imputation α belongs to the core. Suppose this is not so. Then there is an imputation β such that $\beta \overset{S}{\succ} \alpha$, i.e. $\beta(S) > \alpha(S)$ and $\beta(S) \leq v(S)$ which contradicts (4.2.1).

We shall now prove the necessity of (4.2.1). For any imputation α, which does not satisfy (4.2.1), there exists a coalition S for which $\alpha(S) < v(S)$. Let

$$\beta_i = \alpha_i + \frac{v(S) - \alpha(S)}{|S|}, \quad i \in S,$$

$$\beta_i = \frac{1 - v(S)}{|N| - |S|}, \quad i \notin S,$$

where $|S|$ is the number of elements of the set S. It can be easily seen that $\beta(N) = 1$, $\beta_i \geq 0$ and $\beta \overset{S}{\succ} \alpha$. Then it follows that α does not belong to the core.

Theorem 4.2.1 implies that core is a closed convex subset of the set of all imputations (core may also be an empty set).

4.2.2. Suppose the players are negotiating the choice of a cooperative agreement. It follows from the superadditivity of v that such an agreement brings about the formation of the coalition N of all players. The question is tackled as to the way of distributing the total payoff $v(N)$, i.e. the way of choosing a vector $\alpha \in R^n$ for which $\sum_{i \in N} \alpha_i = v(N)$.

The minimum requirement for obtaining the players' consent to choose a vector α is the individual rationality of this vector, i.e. the condition $\alpha_i \geq v(\{i\})$, $i \in N$. Suppose the players are negotiating the choice of the particular imputation α. Some coalition S demanding a more advantageous imputation may raise an objection against the choice of this imputation. The coalition S lays down this demand, threatening to break up general cooperation (this threat is quite real, since the payoff $v(N)$ can only be ensured by unanimous consent on the part of all players). Suppose the other players $N\backslash S$ respond to this threat by uniting their efforts against the coalition S. The maximum guaranteed payoff to the coalition S is evaluated by the number $v(S)$. Condition (4.2.1) implies that there exists a stabilizing threat to the coalition S from the coalition N. Thus, a core of the game (N, v) is the set of distributions of the maximum total payoff $v(N)$ which is immune to cooperative threats.

We shall bring forward one more criterion to judge whether an imputation belongs to the core.

Lemma. *Let α be an imputation in the game (N, v). Then α belongs to the core if and only if the inequality*

$$\sum_{i \in S} \alpha_i \leq v(N) - v(N\backslash S) \tag{4.2.2}$$

holds for all coalitions $S \subset N$.

Proof. Since $\sum_{i \in N} \alpha_i = v(N)$, the above inequality can be written as

$$v(N\backslash S) \leq \sum_{i \in N\backslash S} \alpha_i.$$

Now the assertion of the lemma follows from (4.2.1).

Condition (4.2.1) shows that if the imputation α belongs to the core, then no coalition S can guarantee itself the amount exceeding $\sum_{i \in S} \alpha_i = \alpha(S)$, i.e. the total payoff ensured by the coalition members using the imputation α. This makes unreasonable the existence of coalitions S other than the maximal coalition N.

Theorem 4.2.1 provides enough reason to use core as an important optimality principle in the cooperative theory. However, in many cases the core appears to be empty, whereas in the other cases it represents a multiple optimality principle and the question as to which of the imputation are to be chosen from the core in the particular case is still open.

Example 6. Consider the "jazz band" game (see Example 5, 4.1.3). The total receipts of three musicians is maximum (\$100) when they do performance jointly. If the singer does performance separately from the pianist and drummer, they receive \$65 + \$20 all together. If the pianist does performance separately from the singer and drummer, they receive \$30 + \$50 all together. Finally, if the pianist and singer do performance without the drummer, their total receipts amount to \$80. What is the distribution of the maximum total receipts to be considered rational in terms of the above-mentioned partial cooperation and individual behavior?

The vector $\alpha = (\alpha_1, \alpha_2, \alpha_3)$ in the "jazz band" game belongs to the core if and only if

$$\begin{cases} \alpha_1 \geq 20, \ \ \alpha_2 \geq 30, \ \ \alpha_3 \geq 0, \\ \alpha_1 + \alpha_2 + \alpha_3 = 100, \\ \alpha_1 + \alpha_2 \geq 80, \ \ \alpha_2 + \alpha_3 \geq 65, \ \ \alpha_1 + \alpha_3 \geq 50. \end{cases}$$

This set is a convex hull of the following three imputations: $(35, 45, 20)$, $(35, 50, 15)$, $(30, 50, 20)$. Thus, the payoffs of each player in different imputations differs on the amount not more than five rubles. The typical representative of the core is the arithmetical mean of extreme points of core, namely $\alpha^* = (33.3, 48.3, 18.3)$. The characteristic feature of the imputation a^* is that all two-component coalitions have the same additional receipts: $\alpha_i + \alpha_j - v(\{i, j\}) = 1.6$. The imputation α^* is a fair compromise from the interior of the core.

Example 7. Recall the characteristic function of example 2, 4.1.3:

$$v(N) = 3, \quad v(\{1, 2\}) = 2, \quad v(\{1, 3\}) = 3, \quad v(\{2, 3\}) = 0,$$
$$v(\{1\}) = 1, \quad v(\{i\}) = 0, \quad \text{for } i = 2, 3.$$

After substituting the $\alpha_3 = 3 - \alpha_1 - \alpha_2$, the core of this game is given by the following inequalities:

$$\alpha_1 + \alpha_2 \geq 2, \quad \alpha_2 \leq 0, \quad \alpha_1 \leq 3, \quad \alpha_1 \geq 1, \quad \alpha_2 \geq 0, \quad \alpha_1 + \alpha_2 \leq 3.$$

Obviously, Player 2 has no power in this example.

Each imputation in the core will assign zero to Player 2. Player 1, however, plays a major role, since the coalition of Player 3 with Player 1 guarantees Player 1 at least 2. Hence, $C(v) = \{(\alpha, 0, 3 - \alpha) | \alpha \in [2, 3]\}$ is the set of core imputations.

Example 8. In the problem of cost sharing in Example 4, 4.1.3 the following characteristic function v can be established:

$$v(N) = 5, \quad v(\{1,2\}) = 2, \quad v(\{1,3\}) = 3, \quad v(\{2,3\}) = 1,$$
$$v(\{1\}) = v(\{2\}) = v(\{3\}) = 0.$$

Substituting $\alpha_3 = 5 - \alpha_1 - \alpha_2$ into inequalities (4.2.1) defining the core, one derives the following inequalities:

$$\alpha_1 + \alpha_2 \geq 2, \quad \alpha_2 \leq 2, \quad \alpha_1 \leq 4, \quad \alpha_1 \geq 0, \quad \alpha_2 \geq 0, \quad \alpha_1 + \alpha_2 \leq 5.$$

It's clear, that the core of the problem is not empty.

4.2.3. The fact that the core is empty does not mean that the cooperation of all players N is impossible. This simply means that no imputation can be stabilized with the help of simple threats as above. The kernel is empty when intermediate coalitions are too strong. This assertion can be explained as follows.

Example 9. Symmetric games [Moulin (1981)]. In a symmetric game, coalitions with the same number of players have the same payoffs. The characteristic function v is

$$v(S) = f(|S|)$$

for all $S \subset N$, where $|S|$ is the number of elements of the set S.

We may assume, without loss of generality, that $f(1) = 0$ and $N = \{1, \ldots, n\}$. Then the imputation set in the game (N, v) is the following simplex in R^n

$$\sum_{i=1}^{n} \alpha_i = f(n) = v(N), \quad \alpha_i \geq 0, \; i = 1, \ldots, n.$$

The core is a subset of the imputation set defined by linear inequalities (4.2.1), i.e. a convex polyhedron. By the symmetry of $v(S)$, the

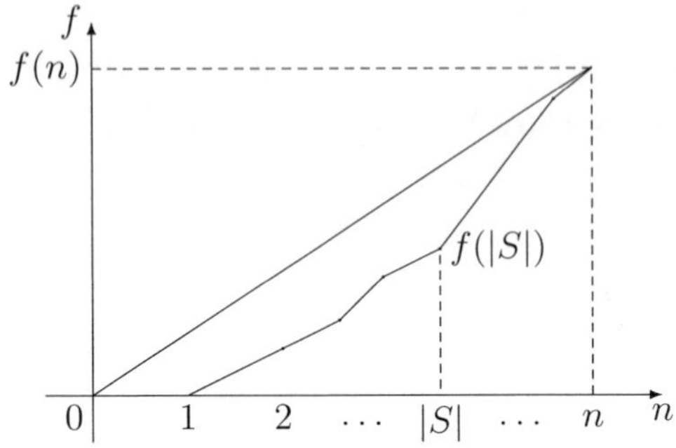

Figure 4.1

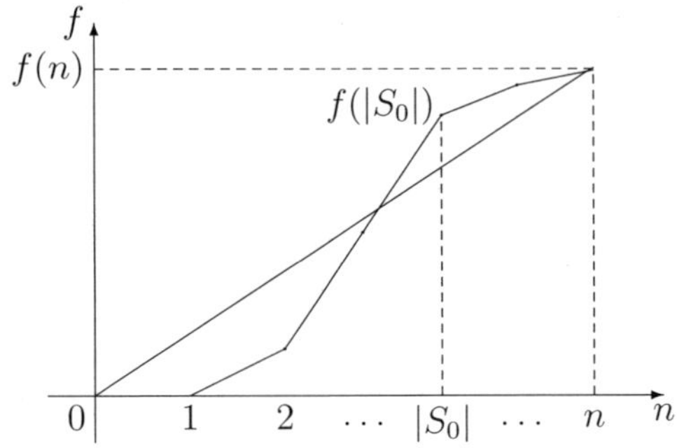

Figure 4.2

core is also symmetric, i.e. invariant under any permutation of components $\alpha_1, \ldots, \alpha_n$. Furthermore, by the convexity of the core, it can be shown that the core is nonempty if and only if it contains the center α^* of the set of all distributions ($\alpha_i^* = f(n)/n$, $i = 1, \ldots, n$). Returning to system (4.2.1), we see that the core is nonempty if and only if the inequality $(1/|S|)f(|S|) \leq (1/n)f(n)$ holds for all $|S| = 1, \ldots, n$. Thus, the core is nonempty if and only if there is no intermediate coalition S, in which the average payment to each player exceeds the corresponding amount in the coalition N. Figure 4.1(4.2) corresponds to the case where the core is nonempty (empty).

4.2.4. *Example 10* [Vorobjev (1977)]. Consider a general three-person game in (0–1)-reduced form. For its characteristic function we have $v(\oslash) = v(1) = v(2) = v(3) = 0$, $v(1,2,3) = 1$, $v(1,2) = c_3$, $v(1,3) = c_2$, $v(2,3) = c_1$, where $0 \leq c_i \leq 1$, $i = 1,2,3$. By the Theorem 4.2.1, for the imputation α to belong to the core, it is necessary and sufficient that

$$\alpha_1 + \alpha_2 \geq c_3, \quad \alpha_1 + \alpha_3 \geq c_2, \quad \alpha_2 + \alpha_3 \geq c_1$$

or

$$\alpha_3 \leq 1 - c_3, \quad \alpha_2 \leq 1 - c_2, \quad \alpha_1 \leq 1 - c_1. \tag{4.2.3}$$

Summing inequalities (4.2.3) we obtain

$$\alpha_1 + \alpha_2 + \alpha_3 \leq 3 - (c_1 + c_2 + c_3),$$

or, since the sum of all α_i, $i = 1,2,3$, is identically equal to 1,

$$c_1 + c_2 + c_3 \leq 2. \tag{4.2.4}$$

The last inequality is the necessary condition for the existence of a nonempty core in the game of interest. On the other hand, if (4.2.4) is satisfied, then there are non-negative ξ_1, ξ_2, ξ_3 such that

$$\sum_{i=1}^{3}(c_i + \xi_i) = 2, \quad c_i + \xi_i \leq 1, \; i = 1,2,3.$$

Let $\beta_i = 1 - c_i - \xi_i$, $i = 1,2,3$. The numbers β_i satisfy inequalities (4.2.3) in such a way that the imputation $\beta = (\beta_1, \beta_2, \beta_3)$ belongs to the core of the game ($\sum_{i=1}^{3} \beta_i = 1$); hence relation (4.2.4) is also sufficient for a nonempty core to exist.

Geometrically, the imputation set in the game involved is the simplex: $\alpha_1 + \alpha_2 + \alpha_3 = 1$, $\alpha_i \geq 0$, $i = 1,2,3$ (triangle ABC shown in Fig. 4.3). The nonempty core is an intersection of the imputation set ($\triangle ABC$) and a convex polyhedron (parallelepiped) $0 \leq \alpha_i \leq 1 - c_i$, $i = 1,2,3$. It is the part of triangle ABC cut out by the lines of intersections of the planes

$$\alpha_i = 1 - c_i, \quad i = 1,2,3 \tag{4.2.5}$$

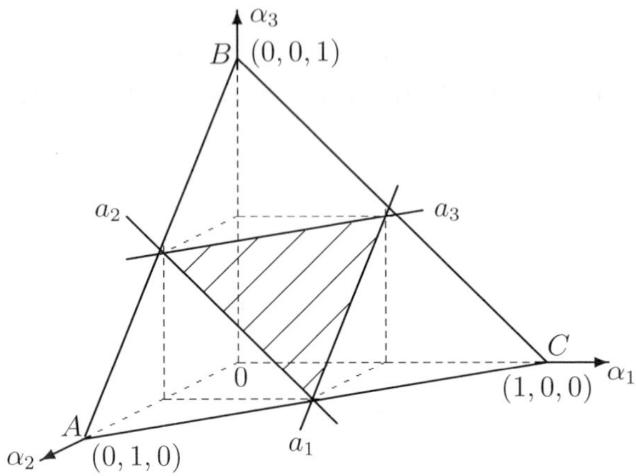

Figure 4.3

with the plane $\triangle ABC$. Referring to Fig. 3.10, we have α_i, $i = 1, 2, 3$ standing for the line formed by intersection of the planes $\alpha_i = 1 - c_i$ and $\alpha_1 + \alpha_2 + \alpha_3 = 1$. The intersection point of two lines, α_i and α_j, belongs to triangle ABC if the kth coordinate of this point, with $(k \neq i,\ k \neq j)$, is non-negative; otherwise it is outside of $\triangle ABC$ (Figs. 4.4(a) and 4.4(b)). Thus, the core has the form of a triangle if a joint solution to any pair of equations (4.2.5) and equation $\alpha_1 + \alpha_2 + \alpha_3 = 1$ is non-negative. This requirement holds for

$$c_1 + c_2 \geq 1, \quad c_1 + c_3 \geq 1, \quad c_2 + c_3 \geq 1. \tag{4.2.6}$$

The core can take one or another form, as the case requires (whereas a total of eight is possible here). For example, if none of the three inequalities (4.2.6) is satisfied, then the core appears to be a hexagon (Fig. 4.4(b)).

4.2.5. Another optimality principle in cooperative games is NM-solution, which is actually a multiple optimality principle in the set of all imputations, as also is the core. Although the elements of the core are not dominated by other imputations, but we cannot say that for any previously given imputation α in the core there is its associated dominating imputation. For this reason, it seems to be

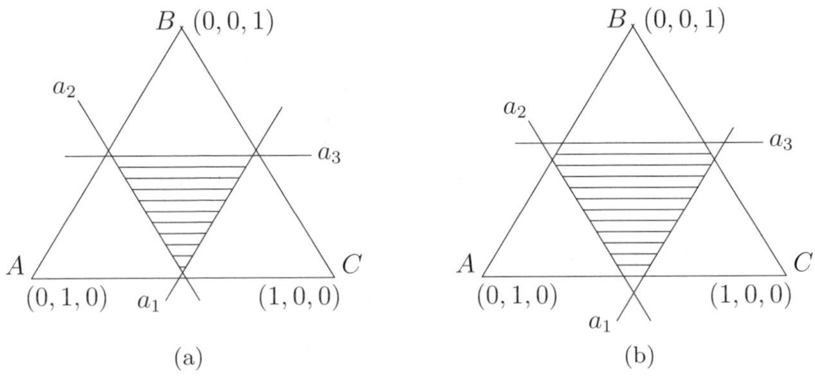

Figure 4.4

wise to formulate an optimality principle by taking into account this situation.

Definition. The imputation set L in the cooperative game (N, v) is called NM-solution if:

(1) $\alpha \succ \beta$ implies that either $\alpha \notin L$ or $\beta \notin L$ (interior stability);

(2) for any $\alpha \notin L$ there is an imputation $\beta \in L$ such that $\beta \succ \alpha$ (exterior stability).

Unfortunately, the definition is not constructive and thus the notion of NM-solution cannot find practical use, and has more philosophical significance rather than a practical meaning.

There is a particular relation between the core in a cooperative game and its NM-solution. For example, if the core is nonempty and NM-solution exists, then it contains the core. Suppose the imputation α belongs to the core. In fact, if it did not belong to NM-solution L, then, by property 2, there would be an imputation α' such that $\alpha' \succ \alpha$. This, however, contradicts the fact that α belongs to the core as a set of nondominant imputations.

Theorem. *If the inequalities*

$$v(S) \leq \frac{1}{n - |S| + 1},$$

where $|S|$ is the number of players in coalition S, hold for the characteristic function of the game (N, v) in (0–1)-reduced form ($|N| = n$), then the core of this game is nonempty, and is its NM-solution.

Proof. Take an arbitrary imputation α which is exterior to the core. Then there exists a nonempty set of those coalitions S in which it is possible to dominate α, i.e. these are the coalitions for which $\alpha(S) < v(S)$. The set $\{S\}$ is partially ordered in the inclusion, i.e. $S_1 > S_2$ if $S_2 \subset S_1$. Take in it a minimal element S_0 which apparently exists. Let k be the number of players in the coalition S_0. Evidently, $2 \le k \le n-1$. Let us construct the imputation β as follows:

$$\beta_i = \begin{cases} \alpha_i + \dfrac{v(S_0) - \alpha(S_0)}{k}, & i \in S_0, \\ \dfrac{1 - v(S_0)}{n-k}, & i \notin S_0. \end{cases}$$

Since $\beta(S_0) = v(S_0)$, $\beta_i > \alpha_i$, $i \in S_0$, then β dominates α in the coalition S_0. Show that β is contained in the core. To do this, it suffices to show that $\beta(S) \ge v(S)$ for an arbitrary S. At first, let $|S| \le k$. Note that β is not dominated for any coalition $S \subset S_0$, since $\beta_i > \alpha_i$ ($i \in S_0$), while S_0 is a minimal coalition for which it is possible to dominate α. If, however, at least one player from S is not contained in S_0, then

$$\beta(S) \ge \frac{1 - v(S_0)}{n-k} \ge \frac{1 - \frac{1}{n-k+1}}{n-k} = \frac{1}{n-k+1} \ge \frac{1}{n-|S|+1} \ge v(S).$$

Thus, β is not dominated for any coalition containing at most k players.

Now, let $|S| > k$. If $S_0 \subset S$, then

$$\beta(S) = \frac{(|S|-k)(1 - v(S_0))}{n-k} + v(S_0) \ge \frac{|S|-k}{n-k}$$
$$\ge \frac{|S|-k+k-|S|+1}{n-k+k-|S|+1} = \frac{1}{n-|S|+1} \ge v(S).$$

However, if S does not contain S_0, then the number of players of the set S, not contained in S_0, is at least $|S| - k + 1$; hence

$$\beta(S) \geq \frac{(|S| - k + 1)(1 - v(S_0))}{n - k}$$
$$\geq \frac{|S| - k + 1}{n - k + 1} \geq \frac{1}{n - |S| + 1} \geq v(S).$$

Thus, β is not dominated for any one of the coalitions S. Therefore, β is contained in the core. Furthermore, β dominates α. We have thus proved that the core is nonempty and satisfies property 2 which characterizes the set of NM-solutions. By definition, the core satisfies property 1 automatically. This completes the proof of the theorem.

4.2.6. Definition. The game (N, v) in (0–1)-reduced form is called simple if for any $S \subset N$ $v(S)$ takes only one of the two values, 0 or 1. A cooperative game is called simple if its (0–1)-reduced form is simple.

Example 11 [Vorobjev (1977)]. Consider a three-person simple game in (0–1)-reduced form, in which the coalition composed of two or three players wins ($v(S) = 1$), while the one-player coalition loses ($v(\{i\}) = 0$). For this game, we consider three imputations

$$\alpha_{12} = (1/2, 1/2, 0), \quad \alpha_{13} = (1/2, 0, 1/2), \quad \alpha_{23} = (0, 1/2, 1/2). \tag{4.2.7}$$

None of the three imputations dominates each other. The imputation set (4.2.7) also has the property as follows: any imputation (except for three imputations α_{ij}) is dominated by one of the imputations α_{ij}. This can be verified by examining some imputation $\alpha = (\alpha_1, \alpha_2, \alpha_3)$. Since we are examining the game in (0–1)-reduced form, then $\alpha_i \geq 0$ and $\alpha_1 + \alpha_2 + \alpha_3 = 1$. Therefore, no more than two components of the vector α can be at least $1/2$. If there are actually two components, then each of them is $1/2$, whereas the third component is 0. But this means that α coincides with one of α_{ij}. However, if α is some other imputation, then it has no more than one component which is at least $1/2$. We thus have at least two components, say,

α_i and α_j $(i < j)$, which are less than 1/2. But here $\alpha_{ij} \overset{ij}{\succ} \alpha$. Now three imputations (4.2.7) form NM-solution. But this is not the only NM-solution.

Let c be any number from the interval $[0, 1/2]$. It can be easily verified that the set

$$L_{3,c} = \{(a, 1 - c - a, c) \mid 0 \leq a \leq 1 - c\}$$

also is NM-solution. Indeed, this set contains imputations, on which Player 3 receives a constant c, while the players 1 and 2 divide the remaining part in all possible proportions. Internal stability follows from the fact that for any two imputations α and β from this set we have: if $\alpha_1 > \beta_1$, then $\alpha_2 < \beta_2$. But dominance for a single player coalition is not possible. To prove the external stability $L_{3,c}$, we may take any imputation $\beta \notin L_{3,c}$. This means that either $\beta_3 > c$ or $\beta_3 < c$. Let $\beta_3 > c$, e.g. $\beta_3 = c + \epsilon$. Define the imputation α as follows:

$$\alpha_1 = \beta_1 + \epsilon/2, \quad \alpha_2 = \beta_2 + \epsilon/2, \quad \alpha_3 = c.$$

Then $\alpha \in L_{3,c}$ and $\alpha \succ \beta$ for coalition $\{1, 2\}$. Now, let $\beta_3 < c$. It is clear that either $\beta_1 \leq 1/2$ or $\beta_2 \leq 1/2$ (otherwise their sum would be greater than 1). Let $\beta_1 \leq 1/2$. Set $\alpha = (1 - c, 0, c)$. Since $1 - c > 1/2 \geq \beta_1$, then $\alpha \succ \beta$ for coalition $\{1, 3\}$. Evidently $\alpha \in L_{3,c}$. However, if $\beta_2 \leq 1/2$, then we may show in a similar manner that $\gamma \succ \beta$, where $\gamma = (0, 1 - c, c)$. Now, aside from the symmetric NM-solution, the game involved has the whole family of solutions which allow Player 3 to obtain a fixed amount c from the interval $0 \leq c \leq 1/2$. These NM-solutions are called discriminant. In the case of the set $L_{3,0}$ Player 3 is said to be completely discriminated or excluded.

From symmetry it follows that there are also two families of NM-solutions, $L_{1,c}$ and $L_{2,c}$, which discriminate Players 1 and 2, respectively.

The preceding example shows that the game may have many NM-solutions. It is not clear which of them is to be chosen. If, however, one NM-solution has been chosen, it remains unclear which of the imputations is to be chosen from this particular solution.

Although the existence of NM-solutions in the general case has not been proved, some special results have been obtained. Some of them are concerned with the existence of NM-solutions, while the others are related to the existence of NM-solutions of a particular type [Diubin and Suzdal (1981)].

4.3 The Shapley Value

4.3.1. The multiplicity of the previously discussed optimality principles (core and NM-solution) in cooperative games and the rigid conditions on the existence of these principles force us to a search for the principles of optimality, existence and uniqueness of which may be ensured in every cooperative game. Among such optimality principles is Shapley value. The Shapley value is defined axiomatically.

Definition. The carrier of the game (N, v) is called a coalition T such that $v(S) = v(S \cap T)$ for any coalition $S \subset N$.

Conceptually, this definition states that any player, not a member of the carrier, is a "dummy", i.e. he has nothing to contribute to any one of the coalitions.

We shall consider an arbitrary permutation P of the ordered set of players $N = \{1, 2, \dots, n\}$. This permutation has associated with the substitution π, i.e. one-to-one function $\pi : N \to N$ such that for $i \in N$ the value $\pi(i) \in N$ is an element of N to which $i \in N$ changes in a permutation P.

Definition. Suppose that (N, v) is an n-person game, P is a permutation of the set N and π is its associated substitution. Denote by $(N, \pi v)$ a game (N, u) such that for any coalition $S \subset N$, $S = \{i_1, i_2, \dots, i_s\}$

$$u(\{\pi(i_1), \pi(i_2), \dots, \pi(i_s)\}) = v(S).$$

The game $(N, \pi v)$ and the game (N, v) differ only in that in the latter the players exchange their roles in accordance with permutation P.

The definition permit the presentation of Shapley axiomatics. First, note that since cooperative n-person games are essentially identified with real-valued (characteristic) functions, we may deal with the sum of two or more games and the product of game by number.

4.3.2. We shall set up a correspondence between every cooperative game (N, v) and the vector $\varphi(v) = (\varphi_1[v], \ldots, \varphi_n[v])$ whose components are interpreted to mean the payoffs received by players under an agreement or an arbitration award. Here, this correspondence is taken to satisfy the following axioms.

Shapley axioms.

1. *If S is any carrier of the game (N, v), then*

$$\sum_{i \in S} \varphi_i[v] = v(S).$$

2. *For any substitution of π and $i \in N$*

$$\varphi_{\pi(i)}[\pi v] = \varphi_i[v].$$

3. *If (N, u) and (N, v) are any cooperative games, then*

$$\varphi_i[u + v] = \varphi_i[u] + \varphi_i[v].$$

Definition. Suppose φ is the function which, by axioms 1–3, sets up a correspondence between every game (N, v) and the vector $\varphi[v]$. Then $\varphi[v]$ is called the vector of values or the Shapley value of the game (N, v).

It turns out that these axioms suffice to define uniquely values for all n-person games.

Theorem. *There exists a unique function φ which is defined for all games (N, v) and satisfies axioms* 1–3 (*see formula* (4.3.7)).

4.3.3. The proof of the theorem is based on the following results.

Lemma. *Let the game (N, w_S) be defined for any coalition $S \subset N$ as follows:*

$$w_S(T) = \begin{cases} 0, & S \not\subset T, \\ 1, & S \subset T. \end{cases} \tag{4.3.1}$$

Then for the game (N, w_S) the vector $\varphi[w_S]$ is uniquely defined by axioms 1, 2:

$$\varphi_i[w_S] = \begin{cases} 1/s, & i \in S, \\ 0, & i \notin S, \end{cases} \tag{4.3.2}$$

where $s = |S|$ is the number of players in S.

Proof. It is obvious that S is the carrier of w_S, as is any set T containing the set S. Now, by axiom 1, if $S \subset T$, then

$$\sum_{i \in T} \varphi_i[w_S] = 1.$$

But this means that $\varphi_i[w_S] = 0$ for $i \notin S$. Further, if π is any substitution which converts S to itself, then $\pi w_S = w_S$. Therefore, by axiom 2, for any $i, j \in S$ there is the equality $\varphi_i[w_S] = \varphi_j[w_S]$. Since there is a total of $s = |S|$ and their sum is 1, we have $\varphi_i[w_S] = 1/s$ if $i \in S$.

The game with the characteristic function w_S defined by (4.3.1) is called the *simple n-person game*. Now the lemma states that for the simple game (N, w_S) the Shapley value for the game (N, w_S) is determined in a unique manner.

Corollary. *If $c \geq 0$, then*

$$\varphi_i[cw_S] = \begin{cases} c/s, & i \in S, \\ 0, & i \notin S. \end{cases}$$

The proof is straightforward. Thus $\varphi[cw_S] = c\varphi[w_S]$ for $c \geq 0$.

We shall now show that if $\sum_S c_S w_S$ is a characteristic function, then

$$\varphi_i\left(\sum_S c_S w_S\right) = \sum_S \varphi_i(c_S w_S) = \sum_S c_S \varphi_i(w_S). \tag{4.3.3}$$

In the case of $c_S \geq 0$, the first equation in (4.3.3) is stated by axiom 3, while the second follows from the corollary. Further, if u, v and $u - v$ are characteristic functions, then, by axiom 3, $\varphi[u - v] = \varphi[u] - \varphi[v]$. Hence it follows that (4.3.3) holds for any c_S. Indeed, if $\sum_S c_S w_S$ is a characteristic function, then

$$v = \sum_S c_S w_S = \sum_{S|c_S \geq 0} c_S w_S - \sum_{S|c_S<0} (-c_S) w_S,$$

hence

$$\varphi[v] = \varphi\left[\sum_{S|c_S\geq 0} c_S w_S\right] - \varphi\left[\sum_{S|c_S<0} (-c_S) w_S\right]$$
$$= \sum_{S|c_S\geq 0} c_S \varphi[w_S] - \sum_{S|c_S<0} (-c_S)\varphi[w_S] = \sum_S c_S \varphi[w_S].$$

4.3.4. Lemma. *Let (N, v) be any game. Then there are $2^n - 1$ real numbers c_S such that*

$$v = \sum_{S \subset N} c_S w_S, \tag{4.3.4}$$

where w_S are defined by (4.3.1) and summation is made over all subsets S of the set N, exclusive of an empty set. Here, representation (4.3.4) is unique.

Proof. Set

$$c_S = \sum_{T|T\subset S} (-1)^{s-t} v(T) \tag{4.3.5}$$

(here t is the number of elements in T). Show that these numbers c_S satisfy the conditions of the lemma. Indeed, if U is an arbitrary coalition, then

$$\sum_{S|S\subset N} c_S w_S(U) = \sum_{S|S\subset U} c_S$$

$$= \sum_{S|S\subset U}\left(\sum_{T|T\subset S}(-1)^{s-t}v(T)\right) = \sum_{T|T\subset U}\left[\sum_{S|T\subset S\subset U}(-1)^{s-t}\right]v(T).$$

We shall now consider the quantity which is bracketed in the last expression. For every value s between t and u there are C_{u-t}^{u-s} of sets S with s-elements such that $T \subset S \subset U$. Therefore, the bracketed expression can be replaced by the following:

$$\sum_{s=t}^{u} C_{u-t}^{u-s}(-1)^{s-t} = \sum_{s=t}^{u} C_{u-t}^{s-t}(-1)^{s-t},$$

but this is a binomial decomposition of $(1-1)^{u-t}$; hence it is 0 for all $t < u$, and 1 for $t = u$. Therefore, for all $U \subset N$

$$\sum_{S|S\subset N} c_S w_S(U) = v(U).$$

We shall prove the uniqueness of representation (4.3.4). To every characteristic function v corresponds an element in the space R^{2^n-1}. Now, let us order coalitions $T \subset U$. Then for every nonempty coalition $T \subset U$ corresponds component of the vector equal to $v(T)$. These vectors will be denoted by v as functions. It is obvious that to the simple characteristic functions w_S correspond the vectors whose components are 0 or 1. We shall prove that the simple characteristic functions (or, more specifically, their associated vectors) are linearly independent. Indeed, let

$$\sum_{S\subset N} \lambda_S w_S(T) = 0 \quad \text{for all } T \subset N.$$

Then for all $T = \{i\}$ we have $w_S(\{i\}) = 0$ if $S \neq \{i\}$, and $w_S(\{i\}) = 1$ if $S = \{i\}$. Hence $\lambda_{\{i\}} = 0$ for all $i \subset N$. Continue the proof by using the induction method. Let $\lambda_S = 0$ for all $S \subset T$, $S \neq T$. Show that $\lambda_T = 0$. Indeed,

$$\sum_{S \subset N} \lambda_S w_S(T) = \sum_{S \subset T} \lambda_S w_S(T) = \lambda_T = 0.$$

Now, we have $2^n - 1$ linearly independent vectors in $R^{2^n - 1}$; therefore every vector, and hence every characteristic function v is uniquely expressed as a linear combination (4.3.4) of the simple characteristic functions w_S. This completes the proof of the lemma.

4.3.5. We shall now turn to the proof of Theorem 4.3.2. Lemma 4.3.4 shows that any game can be represented as a linear combination of games w_S and the representation (4.3.4) is unique. By 4.3.3, the function $\varphi[v]$ is uniquely defined by relations (4.3.3), (4.3.2).

Let (N, v) be an arbitrary game. We now obtain an expression for the vector $\varphi[v]$. By (4.3.3), (4.3.4),

$$\varphi_i[v] = \sum_{S | S \subset N} c_S \varphi_i[w_S] = \sum_{S | i \subset S \subset N} c_S (1/s),$$

but c_S are determined by (4.3.5). Substituting (4.3.5) into this expression we obtain

$$\begin{aligned} \varphi_i[v] &= \sum_{S | i \subset S \subset N} (1/s) \left[\sum_{T | T \subset S} (-1)^{s-t} v(T) \right] \\ &= \sum_{T | T \subset N} \left[\sum_{S | T \cup i \subset S \subset N} (-1)^{s-t} (1/s) v(T) \right]. \end{aligned}$$

Set

$$\gamma_i(T) \triangleq \sum_{S | T \cup i \subset S \subset N} (-1)^{s-t} (1/s). \tag{4.3.6}$$

If $i \notin T'$ and $T = T' \cup \{i\}$, then $\gamma_i(T') = -\gamma_i(T)$. In fact, all terms on the right-hand side of (4.3.6) in both cases are the same and only

$t = t' + 1$; hence they differ only in sign. Thus we have

$$\varphi_i[v] = \sum_{T|i \in T \subset N} \gamma_i(T)[v(T) - v(T \backslash \{i\})].$$

Further, if $i \in T$, then there are exactly C_{n-t}^{s-t} coalitions S with s-elements such that $T \subset S$. This brings us to the well-known definite integral

$$\gamma_i(T) = \sum_{s=t}^{n} (-1)^{s-t} C_{n-t}^{s-t} (1/s) = \sum_{s=t}^{n} (-1)^{s-t} C_{n-t}^{s-t} \int_0^1 x^{s-1} dx$$

$$= \int_0^1 \sum_{s=t}^{n} (-1)^{s-t} C_{n-t}^{s-t} x^{s-1} dx = \int_0^1 x^{t-1} \sum_{s=t}^{n} (-1)^{s-t} C_{n-t}^{s-t} x^{s-t} dx$$

$$= \int_0^1 x^{t-1} (1-x)^{n-t} dx.$$

Thus we have

$$\gamma_i(T) = (t-1)!(n-t)!/(n!)$$

and hence

$$\varphi_i[v] = \sum_{T|i \in T \subset N} \frac{(t-1)!(n-t)!}{n!} [v(T) - v(T \backslash \{i\})].$$

Equation (4.3.7) determines explicitly the components of Shapley value. This expression satisfies axioms 1–3 in 4.3.2.

Note that the vector $\varphi[v]$ is an imputation. Indeed, by the superadditivity of the function v,

$$\varphi_i[v] \geq v(\{i\}) \sum_{T|i \subset T \subset N} \frac{(t-1)!(n-t)!}{n!}$$

$$= v(\{i\}) \sum_{t=1}^{n} C_{n-1}^{t-1} \frac{(t-1)!(n-t)!}{n!} = v(\{i\}).$$

4.3.6. Axiomatic definition apart, the Shapley value expressed by (4.3.7) can be interpreted conceptually as follows. Suppose the players (elements of the set N) have decided to meet in a specified place at a specified time. It would appear natural that, because of random deviations, they would arrive at various instants of time. However, it is assumed that the players' arrival orders (i.e. their permutations) have the same probability, namely $1/(n!)$. Suppose that if, on arrival, player i finds in place (only) the members of coalition $T\backslash\{i\}$, then he receives a payoff $v(T) - v(T\backslash\{i\})$, that is, the amount he contributes to that coalition. Then the component of Shapley value $\varphi_i[v]$ represents the mathematical expectation of player i's contribution to randomly selected coalition in terms of this randomization.

4.3.7. For a simple game (as in Sec. 3.9.6), the formula for Shapley value seems to be particularly descriptive. Indeed, $v(T)-v(T\backslash\{i\})$ is always either 0 or 1, and this expression equals 1 if the coalition T wins, while 0 if the coalition $T\backslash\{i\}$ fails to win. Hence we have

$$\varphi_i[v] = \sum_T (t-1)!(n-t)!/n!,$$

where summation is extended over all such winning coalitions $T \supset i$ for which the coalition $T\backslash\{i\}$ is not the winning one.

Example 12. Game with major player [Vorobjev (1977)]. The game is played by n players. One of the players is called "major". Coalition S wins 1 if it contains either a major player and at least one more player or all its $n-1$ "ordinary" players. If n is the major player, then the characteristic function of this game can be written as

$$v(S) = \begin{cases} 1, & S \supset \{i, n\},\ i \neq n, \\ 1, & S \supset \{1, \dots, n-1\}, \\ 0, & \text{otherwise.} \end{cases}$$

It is obvious that the conditions $v(T) = 1$ and $v(T\backslash\{n\}) = 0$ hold for any coalition $T \supset \{n\}$ if and only if $2 \leq |T| \leq n-1$. Hence

$$\varphi_n[v] = \sum_{t=2}^{n-1} C_{n-1}^{t-1} \frac{(t-1)!(n-t)!}{n!} = \frac{n-2}{n}.$$

Since the game is in (0-1)-reduced form,

$$\sum_{i=1}^{n-1} \varphi_i[v] = 1 - \varphi_n[v] = 2/n.$$

All ordinary players possess equal rights; hence, by symmetry,

$$\varphi_i[v] = \frac{2}{n(n-1)}, \quad i = 1, \ldots, n-1.$$

Now, the "monopolistic" position of major player ensures him the payoff $(n-1)(n-2)/2$ times that of ordinary players.

4.3.8. *Example 13.* Reconsider the weighted majority voting game (Example 3, 4.1.3) for the case of a company with five shareholders $N = \{1, 2, 3, 4, 5\}$. Shareholder 1 owns 80 shares and the others hold 30 each. Let R, the return on the investment project under consideration, be 100. A decision is reached with 50% of the voting shares ($q = 0,5$). This yields to the following characteristic function:

$$v(S) = \begin{cases} 0,5\left(\sum_{i\in S} w_i\right), & \text{for } \sum_{i\in S} w_i \geq 0,5, \\ 0, & \text{for } \sum_{i\in S} w_i < 0,5. \end{cases}$$

The following considerations will simplify computations. Firstly, note that all players but player 1 contribute in the same way to each coalition that does not contain them. Hence, they are symmetric and must obtain equal payoff. Denote this payoff by α, then $\phi_2 = \phi_3 = \phi_4 = \phi_5 = \alpha$ must hold and $\alpha = [v(N) - \phi_1]/4$. Secondly, to compute $\phi_1(v)$ note that for $\{S|\{1\} \in S\}$

$$[v(S) - v(S\backslash\{1\})] = \begin{cases} 0 & S = 1, \\ 40 + 15(S-1) & 2 \leq S \leq 5, \\ 40 & S = 5. \end{cases}$$

Using the value formula, one computes easily $\phi(v) = (\phi_1, \phi_2, \phi_3, \phi_4, \phi_5)$, where $\phi_1 = 50$; $\phi_i = 12,5$, $i = 2, 3, 4, 5$.

4.3.9. *Example 14. Land-lord and farm labourers* [Vorobjev (1977)]. Suppose there are $n-1$ farm laborers (players $i = 1, \ldots, n-1$) and a land-lord (player n). The land-lord engages k laborers and derives from the harvest a profit $f(k)$ ($f(k)$ increases monotonically). The farm laborers cannot derive a profit for themselves. This is described by the characteristic function

$$v(S) = \begin{cases} f(|S|-1), & \{n\} \in S, \\ 0, & \text{otherwise.} \end{cases}$$

Here, for all $T \supset \{n\}$, $|T| > 1$, $v(T) - v(T \backslash \{n\}) = f(t-1)$, where $t = |T|$ and from (4.3.7) follows

$$\varphi_n[v] = \sum_{t=2}^{n} C_{n-1}^{t-1} \frac{(t-1)!(n-t)!}{n!} f(t-1) = \frac{1}{n} \sum_{t=1}^{n-1} f(t).$$

By the efficiency and symmetry of all laborers, we have

$$\varphi_i[v] = \frac{1}{n-1}\left(f(n-1) - \frac{1}{n}\sum_{t=1}^{n-1} f(t)\right), \quad i = 1, \ldots, n-1.$$

In what follows we shall denote $\varphi_i[v]$ by Sh^i.

4.4 The Potential of the Shapley Value

4.4.1. Consider as before the n-person games in characteristic function form with transferable payoffs. We studied different solution concepts or different optimality principles. Some of them constitute subsets of payoff vectors or imputations (such as core and NM-solution). Finally, the Shapley value represents an optimality principle consisting from the unique payoff vector (unique imputation). In this section, we follow Hart and Mas-Colell (1988) and introduce here one number which specifies the cooperative game. By using the "marginal contribution" principle we assign to each player his marginal contribution according to the numbers defined for the game. It happens that the only requirement, that the resulting payoff vector be "efficient" (i.e. that the payoffs add up to the worth of the grand coalition), determines this process uniquely.

4.4.2. A cooperative game with transferable payoffs is a pair (N, v), where N is a finite set of players and $v : 2^N \to R$ is the characteristic function, satisfying $v(\oslash) = 0$. A subset $S \subset N$ is called a coalition, and $v(S)$ is the worth of the coalition S. Given a game (N, v) and a coalition $S \subset N$, we write (S, v) for the subgame obtained by restricting v to (the subsets of) S; that is, the domain of the function v is restricted to 2^S.

4.4.3. Let $\hat{\Gamma}$ denote the set of all games. Given a function $P : \hat{\Gamma} \to R$ that associates a real number $P(N, v)$ to every game (N, v), the marginal contribution of player i in game (N, v) is defined as

$$D^i P(N, v) = P(N, v) - P(N \backslash \{i\}, v),$$

where $i \in N$. (The game $(N\backslash\{i\}, v)$ is restriction of (N, v) to $(N\backslash\{i\}, v)$.)

A function $P : \hat{\Gamma} \to R$ with $P(\oslash, v) = 0$ is called a *potential function* if it satisfies the following condition:

$$\sum_{i \in N} D^i P(N, v) = v(N) \tag{4.4.1}$$

for all games (N, v). Thus, a potential function is such that its marginals are always efficient; that is, they add up to the worth of the grand coalition.

4.4.4. Theorem. *There exists a unique potential function P. For every game (N, v) the resulting payoff vector $(D^i P(N, v))_{i \in N}$ of marginal contributions coincides with the Shapley value of the game. Moreover, the potential of a game (N, v) is uniquely determined by* (4.4.1) *applied only to the game and its subgames (i.e. to (S, v) for all $S \subset N$).*

Proof. Rewrite (4.4.1) as

$$P(N, v) = \frac{1}{|N|}\left(v(N) + \sum_{i \in N} P(N\backslash\{i\}, v)\right). \tag{4.4.2}$$

Starting with $P(\oslash, v) = 0$, (4.4.2) determines $P(N, v)$ recursively. This proves the existence and uniqueness of the potential function P and that $P(N, v)$ is uniquely determined by (4.4.1) (or (4.4.2)) applied just to (S, v) for all $S \subset N$.

It remains to show that $D^i P(N, v) = Sh^i(N, v)$ for all games (N, v) and all players $j \in N$, where P is the (unique) potential function and $Sh^i(N, v)$ denotes the Shapley value of player i in the game (N, v) (see (4.3.7)). We prove that all the axioms that uniquely determine the Shapley value are satisfied by $D^i P$. Efficiency is just (4.4.1); the other three axioms — dummy (null) player, symmetry, and additivity — are proved inductively using (4.4.2). Indeed, let i be a null player in the game (N, v) (i.e. $v(S) = v(S \backslash \{i\})$ for all S). We claim that this implies $P(N, v) = P(N \backslash \{i\}, v)$; hence $D^i P(N, v) = 0$. Assume the assertion holds for all games with less than $|N|$ players; in particular, $P(N \backslash \{j\}, v) = P(N \backslash \{j, i\}, v)$ for all $j \neq i$. Now subtract (4.4.2) for $N \backslash \{i\}$ from (4.4.2) for N, to obtain

$$\begin{aligned} |N|[P(N, v) - P(N \backslash \{i\}, v)] &= [v(N) - v(N \backslash \{i\})] \\ &\quad + \sum_{j \neq i} [P(N \backslash \{j\}, v) - P(N \backslash \{j, i\}, v)] = 0. \end{aligned}$$

Next, assume players i and j are substitutes in the game (N, v). This implies that $P(N \backslash \{i\}, v) = P(N \backslash \{j\}, v)$ (use (4.4.2), noting that i and j are substitutes in $(N, \{k\}, v)$ for all $k \neq i, j$); thus $D^i P(N, v) = D^j P(N, v)$. Finally, another inductive argument on (4.4.2) shows that $P(N, v + w) = P(N, v) + P(N, w)$, implying additivity.

Present another way of viewing the potential. Given a game (N, v), the allocation of marginal contributions (i.e. $v(N) - v(N \backslash \{i\})$ to player i) is, in general, not efficient. One way to resolve this difficulty is to add a new player, say player 0, and extend the game to $N_0 = N \cup \{0\}$ in such a way that the allocation of marginal contributions in the extended game becomes efficient. Formally, let (N_0, v_0) be an extension of (N, v) (i.e. $v_0(S) = v(S)$ for all $S \subset N$). Then the requirement is

$$v_0(N_0) = \sum_{i \in N_0} [v_0(N_0) - v_0(N_0 \backslash \{i\})]$$

$$= [v_0(N_0) - v(N)] + \sum_{i \in N} [v_0(N_0) - v_0(N_0 \backslash \{i\})]. \qquad (4.4.3)$$

This reduces to

$$v(N) = \sum_{i \in N} [v_0(N_0) - v_0(N_0 \backslash \{i\})], \qquad (4.4.4)$$

which, yields the following restatement of the result of Theorem 4.4.4.

4.4.5. Corollary. *There exists a unique extension v_0 of v whose marginal contributions to the grand coalition are always efficient (more precisely, (4.4.3) is satisfied for the game and all its subgames); it is given by $v_0(S \cup \{0\}) = P(S, v)$ for all $S \subset N$, where P is the potential function.*

Note that the payoffs to the original players (in N) add up correctly to $v(N)$ (4.4.4); these are the Shapley values. Player 0, whose payoff is the residual $P(N, v) - v(N)$, may be regarded as a "hidden player", similarly to the "hidden factor" introduced by McKenzie in the study of production functions in order to explain the residual profit (or loss).

In (4.4.1) and (4.4.2), the potential is only given implicitly. We now present two explicit formulas. The T-*unanimity* game u_T (where T is a nonempty finite set) is defined by $u_T(S) = 1$ if $S \supset T$, and $u_T(S) = 0$ otherwise. It is well known that these games form a linear basis for Γ: Each game (N, v) has a unique representation (see Shapley [1953] and 4.3.4)

$$v = \sum_{T \subset N} \alpha_T u_T,$$

where, for all $T \subset N$,

$$\alpha_T \equiv \alpha_T(N, v) = \sum_{S \subset T} (-1)^{|T|-|S|} v(S). \qquad (4.4.5)$$

4.4.6. Theorem. *The potential function P satisfies*

$$P(N,v) = \sum_{T\subset N} \frac{1}{|T|}\alpha_T$$

for all games (N,v), where α_T is given by (4.4.5).

Proof. Let $Q(N,v)$ denote the right-hand side in the preceding formula. Then $Q(\oslash,v) = 0$ and $Q(N,v) - Q(N\backslash\{i\},v) = \sum_{i\in T}\alpha_T/|T|$, which when summed up over i shows that Q satisfies (4.4.1). Therefore, by Theorem 4.1.4, Q coincides with the unique potential function P.

The number $\delta_T = \alpha_T/|T|$ is called the *dividend* of each member of the coalition T and $Sh^i(N,v) = \sum_{i\in T}\delta_T$ [Harsanyi (1963)].

4.4.7. Theorem. *The potential function P satisfies*

$$P(N,v) = \sum_{S\subset N} \frac{(s-1)!(n-s)!}{n!}v(S),$$

where $n = |N|$ and $s = |S|$.

Proof. The marginal contributions of the function on the right side are easily seen to yield the Shapley value.

To interpret this last formula, consider the following probabilistic model of choosing a random nonempty coalition $S\subset N$: First, choose a size $s = 1,2,\ldots,n = |N|$ uniformly (i.e. with probability $1/n$ each). Second, choose a subset S of size s, again uniformly (i.e. each of the C_n^s subsets has the same probability). Equivalently, choose a random order of the n elements of N (with probability $1/n!$ each), choose a cutting point $s(1\le s\le n)$, and let S be the first s elements in that order. The probability of choosing of a set S with $|S| = s$ is

$$\pi_S = \frac{s!(n-s)!}{n\cdot n!} = \frac{s}{n}\frac{(s-1)!(n-s)!}{n!}.$$

Therefore, the formula of Theorem 4.1.7 may be rewritten as

$$P(N,v) = \sum_{S\subset N} \pi_S\frac{n}{s}v(S) = E\left[\frac{|N|}{|S|}v(S)\right], \tag{4.4.6}$$

where E denotes expectation over S with respect to the foregoing probability model. The interpretation of (4.4.6) is that the potential is the *expected normalized worth*, or, equivalently, the *per capita potential* $P(N, v)/|N|$ equals the *average per capita worth* $v(S)/|S|$. This shows that the potential may be viewed as an appropriate "summary" of the characteristic function into one number (from which marginal contributions are then computed).

4.5 The τ-Value and Nucleolus

4.5.1. Suppose we have a cooperative game $\Gamma = (N, v)$. Define two types of vectors which represent upper and lower values of players claims.

Definition. An "utopia" (upper value) vector

$$M(v) = \{M_1(v), \ldots, M_i(v), \ldots, M_n(v)\}, \tag{4.5.1}$$

where

$$M_i(v) = v(N) - v(N \backslash \{i\}), \quad i = 1, \ldots, n. \tag{4.5.2}$$

It is seen that $M_i(v)$ is a marginal contribution of player $i \in N$ to grand coalition N.

4.5.2. Definition. The "minimum rights" vector

$$m(v) = \{m_1(v), \ldots, m_i(v), \ldots, m_n(v)\}, \tag{4.5.3}$$

where

$$m_i(v) = \max_{S: i \in S} [v(S)] - \sum_{j \in S \backslash \{i\}} M_j(v). \tag{4.5.4}$$

Getting the payoff according to "utopia" vector means that the player i is getting the whole surplus from joining the coalition $N \backslash \{i\}$, as if other players do not affect the increase of $v(N \backslash \{i\})$ to $v(N)$. Getting the payoff according to "minimum rights" means that the player i is satisfied in getting the surplus in a coalition under the condition that other coalition members are getting their "utopia" values.

4.5.3. Definition. The vector $\tau(v)$ representing the intersection of the segment $[m(v), M(v)]$ and the hyperplane $\sum_{i\in N} = v(N)$ is called τ-value of cooperative game.

4.5.4. *Nucleolus.* The concept of Nucleolus was introduced by D. Schmeidler (1969). The principal role in its definition plays excess.

Definition. For each coalition $S \subset N$ the excess of a coalition S with respect to imputation α is defined as

$$e(x, S) = v(S) - \sum_{i\in N} \alpha_i, \quad S \subset N. \tag{4.5.5}$$

The excess measures the dissatisfaction of coalition S with the imputation $\alpha = (\alpha_1, \ldots, \alpha_n)$.

For each imputation α form an $2^{|N|} - 1$ dimensional vector of excesses of nonempty coalitions placing them in descending order

$$e(x) = (e_1(x), \ldots, e_i(x), \ldots, e_m(x)),$$

where $e_i(x) = e(x, S_i)$, $S_i \subset N$, $i = 1, 2, \ldots, m = 2^{n-1}$, and $(e_1(x) \geq e_2(x) \geq \cdots \geq e_m(x)$.

4.5.5. Definition. Let $x, y \in R^n$. We shall say that vector y lexicographically exceeds vector x (and denote this by $x \leq_l y$) if $e(x) = e(y)$ or there exists k: $1 \leq k \leq m$ such that $e_i(x) = e_i(y)$ for all $i = 1, \ldots, k-1$ and $e_k(x) < e_k(y)$.

For example, vector $(5, 1, 6, 0)$ lexicographically exceeds vector $(5, 1, 4, 8)$.

4.5.6. Definition. The lexicographical minimum with respect to the preference relation $<_l$ is called nucleolus.

From the definition it follows that nucleolus is an imputation which minimizes the maximal dissatisfaction of coalitions.

4.5.7. The following theorems are true (see Mazalov, V (2014)).

Theorem. *In every cooperative game there exist a unique nucleolus.*

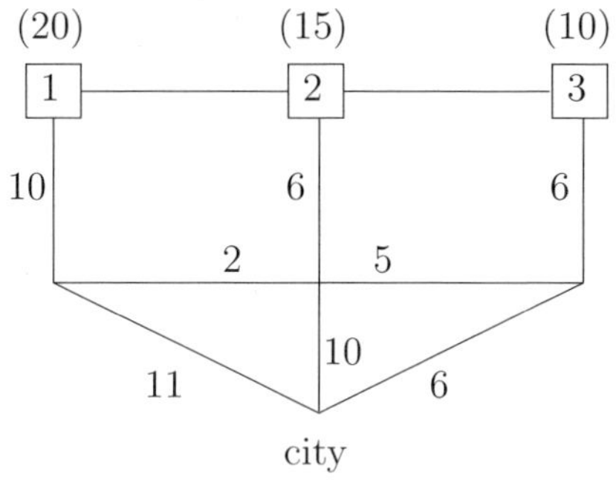

Figure 4.5

4.5.8. Theorem. *If the core of a cooperative game is not empty the nucleolus belongs to the core.*

4.5.9. *Example 15.* Nucleolus in the road construction game see Mazalov (2014). Three farmers want to build a road connecting their farms with the city (market). There are different possibilities presented on the following Fig. 4.5.

The farmers are denoted by squares and in brackets above the incomes of each farmer is represented. The road to the city is any path connecting farmer with the city in the presented network. Each road consists from arcs and the cost of construction of each arch is written over this arc.

Try to write down the characteristic function.

$$v(\{1\}) = 20 - 21 = -1,$$

since the income of farmer 1 from selling his product in the city is equal to 20 and the minimal cost to construct the road by himself is 21.

From the same argumentation we get $v(\{2\}) = 15 - 16 = -1$, $v(\{3\}) = 10 - 12 = -2$. By cooperation players get

$$v(\{1,2\}) = 35 - 23 = 12, \quad v(\{2,3\}) = 25 - 20 = 5,$$
$$v(\{1,3\}) = 30 - 27 = 3, \quad v(\{1,2,3\}) = 45 - 27 = 18.$$

Table 4.2 Nucleolus evaluation

S	v	$e(x,S)$	$8,6,4$	$8,8,2$	$7,9,2$
$\{1\}$	-1	$-1-\alpha_1$	-9	-9	-8
$\{2\}$	-1	$-1-\alpha_2$	-7	-9	-10
$\{3\}$	-2	$-2-\alpha_3$	-6	-4	-4
$\{1,2\}$	12	$12-\alpha_1-\alpha_2=\alpha_3-6$	-2	-4	-4
$\{1,3\}$	3	$3-\alpha_1-\alpha_3=\alpha_2-15$	-9	-7	-6
$\{2,3\}$	5	$5-\alpha_2-\alpha_3=\alpha_1-13$	-5	-5	-6

Now try to compute the nucleolus. Start with the imputation $\alpha = (8, 6, 4)$. Computing the excess for all coalitions we find out, that the maximal excess is attained on coalition $S = \{1, 2\}$. $e(x, S) = \alpha_3 - 6$, and can be reduced by decreasing α_3 and increasing α_2 until in coincides with $e(x, 3) = -2 + \alpha_3$ (until $x_3 = 2$). The next excess (in decreasing order) is $e(x, \{2, 3\}) = \alpha_1 - 13$; one can reduce it by decreasing x_1 and simultaneously increasing x_2 (until it coincides with the next excess $e(x, \{1, 3\}) = \alpha_2 - 15$). This happens under $\alpha_1 = 7$, $\alpha_2 = 9$. The next reduction of excess is not possible. We get the final imputation $\alpha = (7, 9, 2)$ which is nucleolus. The total cost to construct the road under imputation α is 27 with shares of players $c_1 = 20 - 7 = 13$, $c_2 = 15 - 9 = 6$, $c_3 = 10 - 2 = 8$ (see Table 4.2).

4.6 Exercises and Problems

1. Draw the core in examples "Investment fund" (Example 1, 4.1.3), "Farmer, manufacturer, subdivider" (Example 2, 4.1.3), "Jazz band" (Example 5, 4.1.3). Prove that core in this cases are nonempty.
2. Derive the NM-solution in examples "Investment fund" (Example 1, 4.1.3), "Farmer, manufacturer, subdivider" (Example 2, 4.1.3), "Water treatment system" (Example 4, 4.1.3), "Jazz band" (Example 5, 4.1.3).
3. For each of the following games compute the Shapley value. Is it in the core?

(a) "Investment fund" (4.1.3)
(b) "Farmer, manufacture, subdivider" (4.1.3)
(c) "Water treatment system" (4.1.3)
(d) "Jazz band" (4.1.3)

4. For a general 3-person game in 0–1 normalized form $v(1) = v(2) = v(3) = 0$, $v(\{1,2\}) = a$, $v(\{1,3\}) = b$, $v(\{2,3\}) = c$, $v(\{1,2,3\}) = 1$. Show that the Shapley value are given by

$$\phi_1 = \frac{a+b+2(1-c)}{6}, \quad \phi_2 = \frac{a+c+2(1-b)}{6},$$
$$\phi_3 = \frac{b+c+2(1-a)}{6}.$$

5. The 0–1 normalized form of a general superadditive 3-person game is $v(1) = v(2) = v(3) = 0$, $v(\{1,2\}) = a$, $v(\{1,3\}) = b$, $v(\{2,3\}) = c$, $v(\{1,2,3\}) = 1$ with $0 \le a,\ b,\ c \le 1$. Find a general condition on a, b, c for the core of the game to be nonempty.

6. For the game $v(1) = v(2) = v(3) = 0$, $v(\{1,2\}) = 5$, $v(\{1,3\}) = 2$, $v(\{2,3\}) = 3$, $v(\{1,2,3\}) = 6$. Draw the core in the imputation triangle. What is its shape? List the imputations at its vertices.

7. Consider three countries negotiating trade agreement. The joint benefit of an agreement for the different coalitions is as follows:

$$v(\{1,2,3\}) = R', \quad v(\{1,2\}) = v(\{1,3\}) = 0,6R',$$
$$v(\{2,3\}) = 0,3R', \quad v(\{i\}) = 0 \quad \text{for } i = 1,2,3.$$

(a) Draw a diagram of the core for this game. Does the size of the core depend on R?
(b) Derive the Shapley value for this game. Is the value imputation in the core? Does the answer to this question depend on R?

8. Consider a parliament in which three parties share the seats in the following proportions: $(s_1, s_2, s_3) = (0,4; 0,4; 0,2)$. A proposed bill will pass if it obtains at least 50% of the votes in the parliament. Suppose that the benefit from winning is 1 for winning coalition and 0 for a losing one.

(a) Derive the characteristic function of this game.
(b) Derive the Shapley value of the game.
(c) Is the Shapley value an element of the core in this game?

9. Suppose the core has a nonempty intersection with all the bounds $x_i = v(\{i\})$ of the imputation set. Show that here it is a unique NM-solution.

10. For the cooperative game (N, v) we define a semi-imputation to be the vector $\alpha = (\alpha_1, \ldots, \alpha_n)$ for which $\alpha_i \geq v(\{i\})$ and $\sum_{i=1}^{n} \alpha_i \leq v(N)$. Show that if L is an NM-solution of the game (N, v) and α is the semi-imputation which does not belong to L, then there exists an imputation $\beta \in L$ such that $\beta > \alpha$.

11. For the game (N, v) we define β_i by

$$\beta_i = \max_{S \subset N \backslash \{i\}} [v(S \cup \{i\}) - v(S)].$$

Show that if there is an i for which $\alpha_i > \beta_i$, then the imputation α cannot belong either to the core or to one of the NM-solutions.

12. Let (N, v) be a simple game in $(0, 1)$-reduced form (see 3.10.6) Player i is called a "veto" player if $v(N \backslash \{i\}) = 0$.

(a) Prove that in order for the core to be nonempty in a simple game, it is necessary and sufficient that there be at least one "veto" player in the game.
(b) Let S be the set of all "veto" players. Show that the imputation $\alpha = (\alpha_1, \ldots, \alpha_n)$ belongs to the core if $\sum_{i \in S} \alpha_i = 1$, $\alpha_i \geq 0$, for $i \in S$ and $\alpha_i = 0$, for $i \notin S$.

13. In the game (N, v) we interpret the quasi-imputation to mean a vector $\alpha = (\alpha_1, \ldots, \alpha_n)$ such that $\sum_{i \in N} \alpha_i = v(N)$. For every $\epsilon > 0$ we define a strict ϵ-core $C_\epsilon(v)$ to be the set of imputations such that for every coalition

$$\sum_{i \in S} \alpha_i \geq v(S) - \epsilon.$$

(a) Show that if $\epsilon < \epsilon'$, then $C_\epsilon(v) \subset C_{\epsilon'}(v)$.

(b) Show that there exists the smallest number $\epsilon > 0$ for which $C_\epsilon(v) \neq 0$. For such an ϵ the set $C_\epsilon(v)$ is called the minimal ϵ-core and is denoted by $MC(v)$.

(c) Find a minimal ϵ-core in the game (N, v), where $N = \{1, 2, 3\}$, $v(\{i\}) = 0$, $v(\{1, 2\}) = 50$, $v(\{1, 3\}) = 80$, $v(\{2, 3\}) = 90$, $v(\{N\}) = 100$.

(d) Let (N, v), (N, v') be two cooperative games and suppose the equality $C_\epsilon(v) = C_{\epsilon'}(v') \neq 0$, holds for some ϵ and ϵ'. Show that here $C_{\epsilon-\delta}(v) = C_{\epsilon'-\delta}(v')$ for all $\delta > 0$, $\delta < \min[\epsilon, \epsilon']$.

14. Show that if (N, v) is a constant sum game (see 4.1.5), then the Shapley value Sh is determined by

$$Sh_i(v) = 2 \sum_{S: S \subset N,\ i \in S} \left[\frac{(n-s)!(s-1)!}{n!} v(S) \right] - v(N).$$

15. The game (N, v) is called convex if for all $S, T \subset N$

$$v(S \cup T) + v(S \cap T) \geq v(S) + v(T).$$

(a) Prove that the convex game has a nonempty core and the Shapley value belongs to the core.

(b) Show that (N, v) is a convex game if

$$v(S) = \left(\sum_{i \in S} m_i \right)^2, \quad S \subset N,$$

while $m = (m_1, \ldots, m_n)$ is a non-negative vector.

16. Consider a simple game (N, v) in (0-1)-reduced form. We interpret player i's "jump" to mean the set $S \subset N$ for which $v(S) = 1$ and $v(S \backslash \{i\}) = 0$. Denote by Θ_i the number of player i's jumps in the game. Then the vector $\beta(v) = (\beta_1(v), \ldots, \beta_n(v))$, where $\beta_i(v) = \Theta_i / \sum_{j=1}^n \Theta_j$, is called a *Banzaf vector* for a simple game.

(a) Show that $\Theta_1 = 6$, $\Theta_2 = \Theta_3 = \Theta_4 = 2$, and hence $\beta(v) = (1/2, 1/6, 1/6, 1/6)$ for a simple four-person game (N, v) in which the coalition S wins if it comprises either two players and $\{1\} \in S$ or three or four players.

(b) Show that in the above game the $\beta(v)$ coincides with the Shapley value.

17. Let (N, v) be a simple three-person game in which the coalitions $(1, 2)$, $(1, 3)$, $(1, 2, 3)$ are the only winning coalitions. Show that in this game $\Theta_1 = 3$, $\Theta_2 = \Theta_3 = 1$, and hence the Banzaf vector is $\beta(v) = (3/5, 1/5, 1/5)$, while the Shapley value is $Sh[v] = (2/3, 1/6, 1/6)$.

18. Consider a non-negative vector $p = (\pi_1, \ldots, \pi_n)$ and a number $\Theta > 0$. Let $0 < \Theta \leq \sum_{i=1}^{n} \pi_i$. The weighted maj. game (Example 3, 4.1.3) is taken to be a simple game (N, v) in which the characteristic function v is determined by

$$v(S) = \begin{cases} 0, & \sum\limits_{i \in S} \pi_i < \theta, \\ 1, & \sum\limits_{i \in S} \pi_i \geq \theta. \end{cases}$$

Let $\Theta = 8$ and $p = (4, 3, 3, 2, 2, 1)$, $n = 6$. Compute the Shapley value and Banzaf vector for the simple weighted majority game.

19. Nucleolus of two-player game. Let (N, v) be an essential two-player game. $\left(v(N) > \sum_{i=1}^{n} v(\{i\})\right)$. Compute the nucleolus.

20. Consider the cooperative game $(N, v) = \Gamma$ with $v(N) > \sum_{i=1}^{n} v(\{i\})$ (essential game). Suppose that the core is not empty in Γ, prove that nucleolus always belongs to the core.

21. Compute τ-value and Shapley value for "jazz band" game (Example 5, 4.1.3).

22. Show that at least one of vectors "utopia" and "minimum rights" is not an imputation.

Chapter 5

Positional Games

5.1 Multistage Games with Perfect Information

5.1.1. The preceding chapters dealt with games in normal form. A dynamic (i.e. continued during a period of time, not instantaneous) conflictly controlled process can be reduced to a normal form by formal introduction of the notion of a pure strategy. In the few cases when the power of a strategy space is not great and the possibility exists of numerical solutions such an approach seems to be allowable. However, in the majority of the problems connected with an optimal behavior of participants in the conflictly-controlled process the passage to normal form, i.e. the reduction of the problem to a single choice of pure strategies as elements of large dimension spaces or functional spaces, does not lead to effective ways of finding solutions, though permits illustration of one or another of the optimality principles. In a number of cases the general existence theorems for games in normal form does not allow finding or even specifying the optimal behavior in the games for which they constitute their normalizations. As is shown below, in "chess" there exists a solution in pure strategies. This result, however, cannot be obtained by direct investigation of matrix games. By investigation of differential games of pursuit and evasion it is possible in a number of cases to find explicit solutions. In such cases however, the normal form of differential game is so general that, for practical purposes, it is impossible to obtain specific results.

5.1.2. Mathematical dynamic models of conflict are investigated in the theory of *positional games.* The simplest class of positional games is the class of *finite stage games with perfect information.* To define a finite stage n-person game with perfect information we need a rudimentary knowledge of graph theory.

Let X be a finite set. The rule f setting up a correspondence between every element $x \in X$ and an element $f(x) \in X$ is called a single-valued map of X into X or the function defined on X and taking values in X. The set-valued map F of the set X into X is the rule which sets up a correspondence between every element $x \in X$ and a subset $F_x \subset X$ (here $F_x = \oslash$ is not ruled out). In what follows, for simplicity, the term "map" will be interpreted to mean a "set-valued map".

Let F be the map of X into X, while $A \subset X$. By the image set A will mean the set

$$FA \triangleq \cup_{x \in A} F_x.$$

By definition, let $F(\oslash) = \oslash$. It can be seen that if $A_i \subset X$, $i = 1, \dots, n$, then

$$F(\cup_{i=1}^n A_i) = \cup_{i=1}^n FA_i, \quad F(\cap_{i=1}^n A_i) \subset \cap_{i=1}^n FA_i.$$

Define the maps $F^2, F^3, \dots, F^k, \dots$ as follows:

$$F_x^2 \triangleq F(F_x),\ F_x^3 = F(F_x^2),\ \dots,\ F_x^k = F(F_x^{k-1}), \dots. \tag{5.1.1}$$

The map $\hat{F}$ of the set X into X is called a *transitive closure* of the map F, if

$$\hat{F}_x \triangleq \{x\} \cup F_x \cup F_x^2 \cup \dots \cup F_x^k \cup \dots. \tag{5.1.2}$$

The map F^{-1} that is inverse to the map F is defined as

$$F_y^{-1} \triangleq \{x | y \in F_x\},$$

i.e. this is the set of points x whose image contains the point y. The map $(F^{-1})_y^k$ is defined in much the same way as the map F_x^k, i.e.

$$\begin{aligned}&(F^{-1})_y^2 = F^{-1}((F^{-1})_y), \\ &(F^{-1})_y^3 = F^{-1}((F^{-1})_y^2), \ldots, (F^{-1})_y^k = F^{-1}((F^{-1})_y^{k-1}).\end{aligned} \tag{5.1.3}$$

If $B \subset X$, then let

$$F^{-1}(B) \triangleq \{x | F_x \cap B \neq \oslash\}. \tag{5.1.4}$$

Example 1. (*Chess.*) Every position on a chess-board is defined by the number and composition of chess pieces for each player as well as by the arrangement of chess pieces at a given moment and the indication as to whose move it is. Suppose X is the set of positions, F_x, $x \in X$ is the set of those positions which can be realized immediately after the position x has been realized. If in the position x the number of black or white pieces is zero, then $F_x \triangleq \oslash$. Now F_x^k defined by (5.1.1) is the set of positions which can be obtained from x in k moves, $\hat{F}_x$ is the set of all positions which can be obtained from x, $F^{-1}(A)$ $(A \subset X)$ is the set of all positions from which it is possible to make, in one move, the transition to positions of the set A (see (5.1.2) and (5.1.4)).

Depicting positions by dots and connecting by an arrow two positions x and y, $y \in F_x$, it is possible to construct the graph of a game emanating from the original position. However, because of a very large number of positions it is impossible to draw such a graph in reality.

The use of set-valued maps over finite sets makes it possible to represent the structure of many multistage games: chess, draughts, go, and other games.

Definition. The pair (X, F) is called a graph if X is a finite set and F is a map of X into X.

The graph (X, F) is denoted by G. In what follows, the elements of the set X are represented by points on a plane, and the pairs of points x and y, for which $y \in F_x$, are connected by the solid line with the arrow pointing from x to y. Then every element of the set X is called a vertex or a node of the graph, and the pair of elements (x, y), where $y \in F_x$ is called the *arc* of the graph. For the arc $p = (x, y)$ the nodes x and y are called the boundary nodes of the arc with x as the origin and y as the end point of the arc. Two arcs p and q are called contingent if they are distinct and have a boundary point in common.

The set of arcs in the graph is denoted by P. The set of arcs in the graph $G = (X, F)$ determines the map F, and vice versa, the map F determines the set P. Therefore, the graph G can be represented as $G = (X, F)$ or $G = (X, P)$.

The *path* in the graph $G = (X, F)$ is called a sequence of arcs, $p = (p_1, p_2, \dots, p_k, \dots)$, such that the end of the preceding arc coincides with the origin of the next one. The length of the path $p = (p_1, \dots, p_k)$ is the number $l(p) = k$ of arcs in the sequence; in the case of an endless path p we set $l(p) \triangleq \infty$.

The edge of the graph $G = (X, P)$ is called the set made up of two elements $x, y \in X$, for which either $(x, y) \in P$ or $(y, x) \in P$. The orientation is of no importance in the edge as opposed to the arc. The edges are denoted by p, q, and the set of edges by P. By the chain is meant a sequence of edges $(p_1, p_2, \dots)$, where one of the boundary nodes of each edge p_k is also boundary for p_{k-1}, while the other is boundary for p_{k+1}.

The cycle is a finite chain starting in some node and terminating in the same node. The graph is called connected if its any two nodes can be connected by a chain.

By definition, the tree or the graph tree is a finite connected graph without cycles which has at least two nodes and has a unique node x_0 such that $\hat{F}_{x_0} = X$. The node x_0 is called the initial node of the graph G.

Example 2. Figure 5.1 shows the graph or the *graph tree* with its origin at x_0. The nodes $x \in X$ or the vertices of the graph are marked

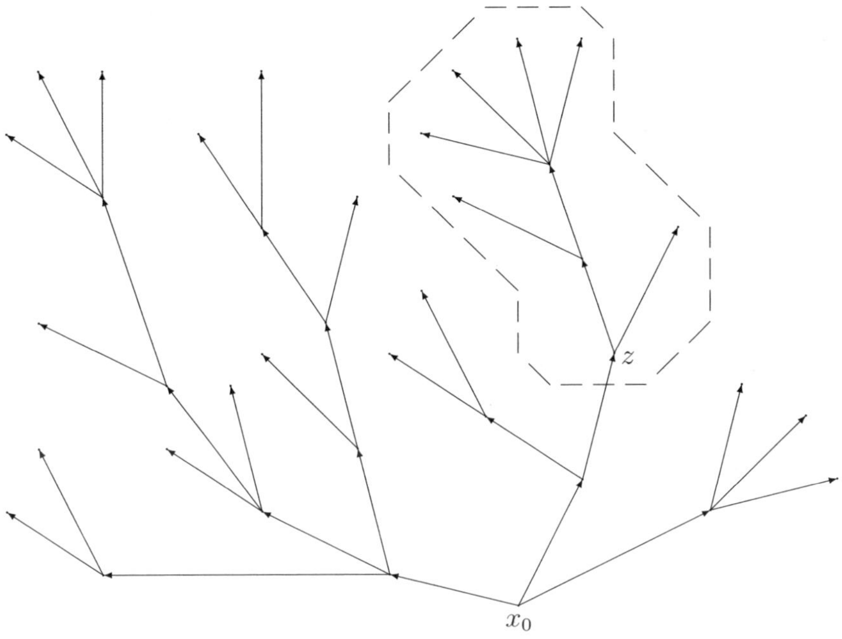

Figure 5.1

by dots. The arcs are depicted as the arrowed segments emphasizing the origin and the end point of the arc.

Example 3. Generally speaking, draughts or chess cannot be represented by a graph tree if by the node of the graph is meant an arrangement of draughtsmen or chess pieces on the board at a given time and an indication of a move, since the same arrangement of pieces can be obtained in a variety of ways. However, if the node of the graph representing a structure of draughtsmen or chess pieces at a given time is taken to mean an arrangement of pieces on the board at a given time, an indication of a move and the past course of the game (all successive positions of pieces on the earlier moves), then each node is reached from the original one in a unique way (i.e. there exists the only chain passing from the original node to any given node); hence the corresponding graph of the game, contains no cycles and is the tree.

5.1.3. Let $z \in X$. The *subgraph* G_z of the tree graph $G = (X, F)$ is called a graph of the form (X_z, F_z), where $X_z = \hat{F}_z$ and $F_z x = F_x \cap X_z$. In Fig. 5.1, the dashed line encircles the subgraph starting in the node z. On the tree graph for all $x \in X_z$ the set F_x and the set $F_z x$ coincide, i.e. the map F_z is a restriction of the map F to the set X_z. Therefore, for the subgraphs of the tree graph we use the notation $G_z = (X_z, F)$.

5.1.4. We shall now define the *multistage game with perfect information on a finite tree graph.*

Let $G = (X, F)$ be a tree graph. Consider the partition of the node set X into $n+1$ sets $X_1, \dots, X_n, X_{n+1}$, $\cup_{i=1}^{n+1} X_i = X$, $X_k \cap X_l = \oslash$, $k \neq l$, where $F_x = \oslash$ for $x \in X_{n+1}$. The set X_i, $i = 1, \dots, n$ is called the *set of personal positions* of the ith player, while the set X_{n+1} is called the *set of final* or *terminal positions*. The real-valued functions $H_1(x), \dots, H_n(x)$, $x \in X_{n+1}$ are defined on the set of final positions X_{n+1}. The function $H_i(x)$, $i = 1, \dots, n$ is called a *payoff* to the ith player.

The game proceeds as follows. Let there be given the set N of players designated by natural numbers $1, \dots, i, \dots, n$ (hereafter denoted as $N \triangleq \{1, 2, \dots, n\}$). Let $x_0 \in X_{i_1}$, then in the node (position) x_0 player i_1 "makes a move" and chooses the next node (position) $x_1 \in F_{x_0}$. If $x_1 \in X_{i_2}$, then in the node x_1 Player i_2 "makes a move" and chooses the next node (position) $x_2 \in F_{x_1}$ and so on. Thus, if the node (position) $x_{k-1} \in X_{i_k}$ is realized at the kth step, then in this node Player i_k "makes a move" and selects the next node (position) from the set $F_{x_{k-1}}$. The game terminates as soon as the terminal node (position) $x_l \in X_{n+1}$, (i.e. the node for which $F_{x_l} = \oslash$) is reached.

Such a step-by-step selection implies a unique realization of some sequence $x_0, \dots, x_k, \dots, x_l$ determining the path in the tree graph G which emanates from the initial position and reaches one of the final positions of the game. In what follows, such a path is called a *play* or *path of the game*. Because of the tree-like structure of the graph G, each play uniquely determines the final position x_l to be reached and, conversely, the final position x_l uniquely determines the play. In the position x_l each of the players i, $i = 1, \dots, n$, receives a payoff $H_i(x_l)$.

We assume that Player i making his choice in position $x \in X_i$ knows this position and hence, because of the tree-like structure of the graph G, can restore all previous positions. In this case, the players are said to have perfect information. Chess and draughts provide a good example of the game with perfect information, because players record their moves, and hence they know the past course of the game when making each move in turn.

Definition. The single-valued map u_i which sets up a correspondence between each node (position) $x \in X_i$ and some unique node (position) $y \in F_x$ is called a strategy for player i.

The set of all possible strategies for player i is denoted by U_i. Now the strategy of ith player prescribes him, in any position x from his personal positions X_i, a unique choice of the next position.

The ordered set $u = (u_1, \dots, u_i, \dots, u_n)$, where $u_i \in U_i$, is called a *situation in the game*, while the Cartesian product $U = \prod_{i=1}^{n} U_i$ is called the *set of situations*. Each situation $u = (u_1, \dots, u_i, \dots, u_n)$ uniquely determines a play in the game, and hence payoffs of the players. Indeed, let $x_0 \in X_{i_1}$. In the situation $u = (u_1, \dots, u_i, \dots, u_n)$ the next position x_1 is then uniquely determined by the rule $u_{i_1}(x_0) = x_1$. Now let $x_1 \in X_{i_2}$. Then x_2 is uniquely determined by the rule $u_{i_2}(x_1) = x_2$. If the position $x_{k-1} \in X_{i_k}$ is realized at the k-th step, then x_k is uniquely determined by the rule $x_k = u_{i_k}(x_{k-1})$ and so on.

Suppose that the situation $u = (u_1, \dots, u_i, \dots, u_n)$ in the above sense determines a play $x_0, x_1, \dots, x_l$. Then we may introduce the notion of the *payoff function* K_i of player i by equating its value in each situation u to the value of the payoff H_i in the final position of the play $x_0, \dots, x_l$ corresponding to the situation $u = (u_1, \dots, u_n)$, that is

$$K_i(u) = K_i(u_1, \dots, u_i, \dots, u_n) \triangleq H_i(x_l), \quad i = 1, \dots, n.$$

Functions K_i, $i = 1, \ldots, n$, are defined on the set of situations $U = \prod_{i=1}^{n} U_i$. Thus, constructing the players' strategy sets U_i and defining the payoff functions K_i, $i=1, \ldots, n$, on the Cartesian product of strategy sets of players we obtain a game in normal form

$$\Gamma = (N, \{U_i\}_{i \in N}, \{K_i\}_{i \in N}),$$

where $N = \{1, \ldots, i, \ldots, n\}$ is the set of players, U_i is the strategy set for player i, and K_i is the payoff function for player i, $i = 1, \ldots, n$.

5.1.5. For the purposes of further investigation of the game Γ we need to introduce the notion of a *subgame*, i.e. the game on a subgraph of the graph G in the main game (see 1.1.1).

Let $z \in X$. Consider a subgraph $G_z = (X_z, F)$ which is associated with the subgame Γ_z as follows. The players personal positions in the subgame Γ_z are determined by the rule $Y_i^z \triangleq X_i \cap X_z$, $i = 1, \ldots, n$, the set of final positions $Y_{n+1}^z = X_{n+1} \cap X_z$, player i's payoff $H_i^z(x)$ in the subgame is taken to be

$$H_i^z(x) \triangleq H_i(x), \ x \in Y_{n+1}^z, \ i = 1, \ldots, n.$$

Accordingly, player i's strategy u_i^z in the subgame Γ_z is defined to be the truncation of player i's strategy u_i in the game Γ to the set Y_i^z, i.e.

$$u_i^z(x) = u_i(x), \ x \in Y_i^z = X_i \cap X_z, \ i = 1, \ldots, n.$$

The set of all strategies for player i in the subgame is denoted by U_i^z. Then each subgraph G_z is associated with the subgame in normal form

$$\Gamma_z = (N, \{U_i^z\}, \{K_i^z\}),$$

where the payoff function K_i^z, $i = 1, \ldots, n$ are defined on the Cartesian product $U^z = \prod_{i=1}^{n} U_i^z$.

5.2 Absolute Equilibrium (Subgame-Perfect)

In Chapter 3, we introduced the notion of a Nash equilibrium for the n-person game in normal form. It turns out that for multistage games it is possible to strengthen the notion of equilibrium by introducing the notion of an absolute equilibrium.

5.2.1. Definition. The Nash equilibrium $u^* = (u_1^*, \dots, u_n^*)$ is called an absolute Nash equilibrium in the game Γ if for any $z \in X$ the situation $(u^*)^z = ((u_1^*)^z, \dots, (u_n^*)^z)$, where $(u_i^*)^z$ is the truncation of strategy u_i^* to the subgame Γ_z, is Nash equilibrium in this subgame.

Then the following fundamental theorem is valid.

Theorem. *In any multistage game with perfect information on a finite tree graph there exists an absolute Nash equilibrium.*

Preparatory to proving this theorem we will first introduce the notion of a game length. By definition, the length of the game Γ means the length of the longest path in the corresponding graph $G = (X, F)$.

Proof will be carried out by induction along the length of the game. If the length of the game Γ is 1, then a move can be made by only one of the players who, by choosing the next node from the maximization condition of his payoff, will act in accordance with the strategy constituting an absolute Nash equilibrium.

Now, suppose the game Γ has the length k and $x_0 \in X_{i_1}$ (i.e. in the initial position x_0 player i_1 makes his move). Consider the family of subgames Γ_z, $z \in F_{x_0}$, where the length of each subgame does not exceed $k-1$. Suppose this theorem holds for all games whose length does not exceed $k-1$, and prove it for the game of length k. Since the subgame Γ_z, $z \in F_{x_0}$ has the length $k-1$ at most, under the assumption of induction the theorem holds for them and thus there exists an absolute Nash equilibrium situation. For each subgame Γ_z, $z \in F_{x_0}$ this situation will be denoted by

$$(u^*)^z = [(u_1^*)^z, \dots, (u_n^*)^z]. \tag{5.2.1}$$

Using absolute equilibria in the subgames Γ_z we construct an absolute equilibrium in the game Γ. Let $u_i^*(x) = (u_i^*(x))^z$, for $x \in X_i \cap X_z$, $z \in F_{x_0}$, $i = 1, \ldots, n$, $u_{i_1}^*(x_0) = z^*$, where z^* is obtained from the condition

$$K_{i_1}^{z^*}\left[(u^*)^{z^*}\right] = \max_{z \in F_{x_0}} K_{i_1}^z\left[(u^*)^z\right]. \qquad (5.2.2)$$

The function u_i^* is defined on player i's set of personal positions X_i, $i = 1, \ldots, n$, and for every fixed $x \in X_i$ the value $u_i^*(x) \in F_x$. Thus, u_i^*, $i = 1, \ldots, n$, is a strategy for player i in the game Γ, i.e. $u_i^* \in U_i$. By construction, the truncation $(u_i^*)^z$ of the strategy u_i^* to the set $X_i \cap X_z$ is the strategy appearing in the absolute Nash equilibrium of the game Γ_z, $z \in F_{x_0}$. Therefore, to complete the proof of the theorem, it suffices to show that the strategies u_i^*, $i = 1, \ldots, n$ constructed by formulas (5.2.2) constitute a Nash equilibrium in the game Γ. Let $i \neq i_1$. By the construction of the strategy $u_{i_1}^*$ after a position z^* has been chosen by player i_1 at the first step, the game Γ becomes the subgame Γ_{z^*}. Therefore,

$$K_i(u^*) = K_i^{z^*}\left\{(u^*)^{z^*}\right\} \geq K_i^{z^*}\left\{(u^*\|u_i)^{z^*}\right\}$$
$$= K_i(u^*\|u_i), \; u_i \in U_i, \; i = 1, \ldots, n, \; i \neq i_1, \qquad (5.2.3)$$

since $(u^*)^{z^*}$ is an absolute equilibrium in the subgame Γ_{z^*}. Let $u_{i_1} \in U_{i_1}$ be an arbitrary strategy for player i_1 in the game Γ. Denote $z_0 = u_{i_1}(x_0)$. Then

$$K_{i_1}(u^*) = K_{i_1}^{z^*}\left\{(u^*)^{z^*}\right\} = \max_{z \in F_{x_0}} K_{i_1}^z\{(u^*)^z\} \geq K_{i_1}^{z_0}\{(u^*)^{z_0}\}$$
$$\geq K_{i_1}^{z_0}\{(u^*\|u_{i_1})^{z_0}\} = K_{i_1}(u^*\|u_{i_1}). \qquad (5.2.4)$$

The assertion of this theorem now follows from (5.2.3), (5.2.4).

5.2.2. *Example 4.* Suppose the game Γ is played on the graph depicted in Fig. 5.2, and the set N is composed of two players: $N = \{1, 2\}$. Referring to Fig. 5.2, we determine sets of personal positions. The nodes of the set X_1 are represented by circles and those of the set X_2 by blocks. Players' payoffs are written in final

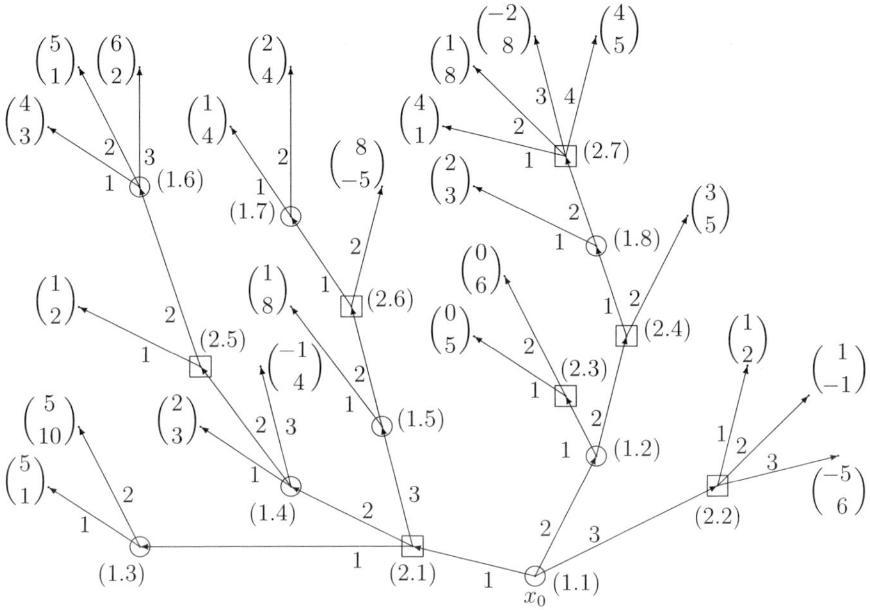

Figure 5.2

positions. Designate by double indices the positions appearing in the sets X_1 and X_2, and by one index the arcs emanating from each node. The choice in the node x is equivalent to the choice of the next node $x' \in F_x$; therefore we assume that the strategies indicate in each node the index of the arc along which it is necessary to move further. For example, Player 1's strategy $u_1 = (2, 1, 2, 3, 1, 2, 1, 1)$ tells him to choose arc 2 in node 1, arc 1 in node 2, arc 2 in node 3, arc 3 in node 4, and so on. Since the set of personal positions of the first player is composed of eight nodes, the strategy for him is an eight-dimensional vector. Similarly, any strategy for Player 2 is a seven-dimensional vector. Altogether there are 864 strategies for Player 1 and 576 strategies for Player 2. Thus the corresponding normal form appears to be an 864×576 bimatrix game. It appears natural that the solution of such bimatrix games by the methods proposed in Chapter 3 is not only difficult, but also impossible. However, the game involved is simple enough to be solved by the backward construction of an absolute Nash equilibrium as proposed in the proof of Theorem 1 in 5.2.1.

Indeed, denote by $v_1(x), v_2(x)$ the payoffs in the subgame Γ_x in a fixed absolute equilibrium. First we solve subgames $\Gamma_{1.6}$, $\Gamma_{1.7}$, $\Gamma_{2.7}$. It can be easily seen that $v_1(1.6) = 6$, $v_2(1.6) = 2$, $v_1(1.7) = 2$, $v_2(1.7) = 4$, $v_1(2.7) = 1$, $v_2(2.7) = 8$. Further, solve subgames $\Gamma_{2.5}$, $\Gamma_{2.6}$, $\Gamma_{1.8}$. Subgame $\Gamma_{2.5}$ has two Nash equilibria, because Player 2 does not care which of the alternative to choose. At the same time, his choice appears to be essential for Player 1, because with Player 2's choice of left-hand arc Player 1 scores +1 or, with Player 2's choice of right-hand arc, Player 1 scores +6. We point out this feature and suppose Player 2 "favors" and chooses the right-hand arc in position (2.5). Then $v_1(2.5) = v_1(1.6) = 6$, $v_2(2.5) = v_2(1.6) = 2$, $v_1(2.6) = v_1(1.7) = 2$, $v_2(2.6) = v_2(1.7) = 4$, $v_1(1.8) = 2$, $v_2(1.8) = 3$. Further, solve games $\Gamma_{1.3}$, $\Gamma_{1.4}$, $\Gamma_{2.3}$, $\Gamma_{1.5}$, $\Gamma_{2.4}$. Subgame $\Gamma_{1.3}$ has two Nash equilibria, because Player 1 does not care which of the alternative to choose. At the same time, his choice, appears to be essential for Player 2, because with Player 1's choice of the left-hand alternative he scores +1, whereas with the choice of the right-hand alternative he scores +10. Suppose Player 1 "favors" and chooses in position (1.3) the right-hand alternative. Then $v_1(1.3) = 5$, $v_2(1.3) = 10$, $v_1(1.4) = v_1(2.5) = 6$, $v_2(1.4) = v_2(2.5) = 2$, $v_1(1.5) = v_1(2.6) = 2$, $v_2(1.5) = v_2(2.6) = 4$, $v_1(2.3) = 0$, $v_2(2.3) = 6$, $v_1(2.4) = 3$, $v_2(2.4) = 5$. Further, solve games $\Gamma_{2.1}$, $\Gamma_{1.2}$, $\Gamma_{2.2}$; $v_1(2.1) = v_1(1.3) = 5$, $v_2(2.1) = v_2(1.3) = 10$, $v_1(1.2) = v_1(2.4) = 3$, $v_2(1.2) = v_2(2.4) = 5$, $v_1(2.2) = -5$, $v_2(2.2) = 6$. Now solve the game $\Gamma = \Gamma_{1.1}$. Here $v_1(1.1) = v_1(2.1) = 5$, $v_2(1.1) = v_2(2.1) = 10$.

As a result we have an absolute Nash equilibrium (u_1^*, u_2^*), where

$$u_1^* = (1, 2, 2, 2, 2, 3, 2, 1), \;\; u_2^* = (1, 3, 2, 2, 2, 1, 2). \tag{5.2.5}$$

In the situation (u_1^*, u_2^*) the game follows the path (1.1), (2.1), (1.3). It is apparent from the construction that the strategies u_i^*, $i = 1, 2$, are "favorable" in that the player i making his move and being equally interested in the choice of the subsequent alternatives, chooses that alternative which is favorable for player $3 - i$.

The game Γ has absolute equilibria in which the payoffs to players are different. To construct such equilibria, it suffices to replace the players' "favorableness" condition by the inverse condition, i.e. the "unfavorableness" condition. Denote by $\overline{v}_1(x)$, $\overline{v}_2(x)$ the payoffs to players in subgame Γ_x when players use an "unfavorable" equilibrium. Then we have: $v_1(1.6) = \overline{v}_1(1.6) = 6$, $v_2(1.6) = \overline{v}_2(1.6) = 2$, $v_1(1.7) = \overline{v}_1(1.7) = 2$, $v_2(1.7) = \overline{v}_2(1.7) = 4$, $v_1(2.7) = \overline{v}_1(2.7) = -2$, $v_2(2.7) = \overline{v}_2(2.7) = 8$. As noted before, subgame $\Gamma_{2.5}$ has two Nash equilibria. Contrary to the preceding case, we assume that Player 2 "does not favor" and chooses the node which ensures a maximum payoff to him and a minimum payoff to Player 1. Then $\overline{v}_1(2.5) = 1$, $\overline{v}_2(2.5) = 2$, $\overline{v}_1(2.6) = v_1(1.7) = 2$, $\overline{v}_2(2.6) = v_2(1.7) = 4$, $\overline{v}_1(1.8) = v_1(1.8) = 2$, $\overline{v}_2(1.8) = v_2(1.8) = 3$. Further, we seek a solution to the games $\Gamma_{1.3}$, $\Gamma_{1.4}$, $\Gamma_{1.5}$, $\Gamma_{2.3}$, $\Gamma_{2.4}$. Subgame $\Gamma_{1.3}$ has two Nash equilibria. As in the preceding case, we choose "unfavorable" actions for Player 1. Then we have $\overline{v}_1(1.3) = v_1(1.3) = 5$, $\overline{v}_2(1.3) = 1$, $\overline{v}_1(1.4) = 2$, $\overline{v}_2(1.4) = 3$, $\overline{v}_1(1.5) = \overline{v}_1(2.6) = v_1(1.5) = 2$, $\overline{v}_2(1.5) = \overline{v}_2(2.6) = v_2(2.6) = 4$, $\overline{v}_1(2.3) = v_1(2.3) = 0$, $\overline{v}_2(2.3) = v_2(2.3) = 6$, $\overline{v}_1(2.4) = v_1(2.4) = 3$, $\overline{v}_2(2.4) = v_2(2.4) = 5$. Further, we solve games $\Gamma_{2.1}$, $\Gamma_{1.2}$, $\Gamma_{2.2}$. We have: $\overline{v}_1(2.1) = \overline{v}_1(1.5) = 2$, $\overline{v}_2(2.1) = \overline{v}_2(1.5) = 4$, $\overline{v}_1(1.2) = \overline{v}_1(2.4) = 3$, $\overline{v}_2(1.2) = \overline{v}_2(2.4) = 5$, $\overline{v}_2(2.2) = v_2(2.2) = 6$, $\overline{v}_1(2.2) = v_1(2.2) = -5$. Now solve the game $\Gamma = \Gamma_{1.1}$. Here $\overline{v}_1(1.1) = \overline{v}_1(1.2) = 3$, $\overline{v}_2(1.1) = \overline{v}_2(1.2) = 5$.

We have thus obtained a new Nash equilibrium

$$\overline{u}_1^*(\cdot) = (2, 2, 1, 1, 2, 3, 2, 1),\ \ \overline{u}_2^*(\cdot) = (3, 3, 2, 2, 1, 1, 3). \qquad (5.2.6)$$

Payoff to both players in situation (5.2.6) are less than those in situation (5.2.5). Just as situation (5.2.5), situation (5.2.6) is an absolute equilibrium.

5.2.3. It is apparent that in parallel with "favorable" and "unfavorable" absolute Nash equilibria there exists the whole family of intermediate absolute equilibria. Of interest is the question concerning the absence of two distinct absolute equilibria differing by payoffs to players.

Theorem [Rochet (1980)]. *Let the players' payoffs* $H_i(x)$, $i = 1, \ldots, n$, *in the game* Γ *be such that if there exists an* i_0 *and there are* x, y *such that* $H_{i_0}(x) = H_{i_0}(y)$, *then* $H_i(x) = H_i(y)$ *for all* $i \in N$. *Then in the game* Γ, *the players' payoffs coincide in all absolute equilibria.*

Proof. Consider the family of subgames Γ_x of the game Γ and prove the theorem by induction over their length $l(x)$. Let $l(x) = 1$ and suppose player i_1 makes a move in a unique nonterminal position x. Then in the equilibrium he makes his choice from the condition

$$H_{i_1}(\overline{x}) = \max_{x' \in F_x} H_{i_1}(x').$$

If the point $\overline{x}$ is unique, then so is the payoff vector in the equilibrium which is here equal to $H(\overline{x}) = \{H_1(\overline{x}), \ldots, H_n(\overline{x})\}$. If there exists a point $\overline{\overline{x}} \neq \overline{x}$ such that $H_{i_1}(\overline{\overline{x}}) = H_{i_1}(\overline{x})$, then there is one more equilibrium with payoffs $H(\overline{\overline{x}}) = \{H_1(\overline{\overline{x}}), \ldots, H_{i_1}(\overline{\overline{x}}), \ldots, H_n(\overline{\overline{x}})\}$. From the condition of the theorem, however, it follows that if $H_{i_1}(\overline{\overline{x}}) = H_{i_1}(\overline{x})$, then $H_i(\overline{\overline{x}}) = H_i(\overline{x})$ for all $i \in N$.

Let $v(x) = \{v_i(x)\}$ be the payoff vector in the equilibrium in a single-stage subgame Γ_x which, as is shown above, is determined in a unique way. Show that if the equality $v_{i_0}(x') = v_{i_0}(x'')$ holds for some i_0 (x', x'' are such that the lengths of the subgames $\Gamma_{x'}$, $\Gamma_{x''}$ are 1), then $v_i(x') = v_i(x'')$ for all $i \in N$. Indeed, let $x' \in X_{i_1}$, $x'' \in X_{i_2}$, then

$$v_{i_1}(x') = H_{i_1}(\overline{x}') = \max_{y \in F_{x'}} H_{i_1}(y),$$

$$v_{i_2}(x'') = H_{i_2}(\overline{x}'') = \max_{y \in F_{x''}} H_{i_2}(y)$$

and $v_i(x') = H_i(\overline{x}')$, $v_i(x'') = H_i(\overline{x}'')$ for all $i \in N$. From the equality $v_{i_0}(x') = v_{i_0}(x'')$ it follows that $H_{i_0}(\overline{x}') = H_{i_0}(\overline{x}'')$. But, under the condition of the theorem, $H_i(\overline{x}') = H_i(\overline{x}'')$ for all $i \in N$. Hence $v_i(x') = v_i(x'')$ for all $i \in N$.

We now assume that in all subgames Γ_x of length $l(x) \leq k-1$ the payoff vector in equilibria is determined uniquely and if for some two subgames $\Gamma_{x'}, \Gamma_{x''}$ whose length does not exceed $k-1$, $v_{i_0}(x') = v_{i_0}(x'')$ for some i_0, then $v_i(x') = v_i(x'')$ for all $i \in N$.

Suppose the game Γ_{x_0} is of length k and player i_1 makes his move in the initial position x_0. By the induction hypothesis, for all $z \in F_{x_0}$ in the game Γ_z the payoffs in Nash equilibria are determined uniquely. Let the payoff vector in Nash equilibria in the game Γ_z be $\{v_i(z)\}$. Then as follows from (5.2.2), in the node x_0 player i chooses the next node $\overline{z} \in F_{x_0}$ from the condition

$$v_{i_1}(\overline{z}) = \max_{z \in F_{x_0}} v_{i_1}(z). \tag{5.2.7}$$

If the point $\overline{z}$ determined by (5.2.7) is unique, then the vector with components $v_i(x_0) = v_i(\overline{z})$, $i = 1, \ldots, n$, is a unique payoff vector in Nash equilibria in the game Γ_{x_0}. If, however, there exist two nodes $\overline{z}$, $\overline{\overline{z}}$ for which $v_{i_1}(\overline{z}) = v_{i_1}(\overline{\overline{z}})$, then, by the induction hypothesis, since the lengths of subgames $\Gamma_{\overline{z}}$ and Γ_z do not exceed $k-1$, the equality $v_{i_1}(\overline{z}) = v_{i_1}(z)$ implies the equality $v_i(\overline{z}) = v_i(z)$ for all $i \in N$. Thus, in this case the payoffs in equilibria $v_i(x_0)$, $i \in N$ are also determined uniquely.

5.2.4. *Example 5.* We have seen in the previous example that "favorableness" of the players give them higher payoffs in the corresponding Nash equilibria, than the "unfavorable" behavior. But it is not always the case. Sometimes the "unfavorable" Nash equilibrium gives higher payoffs to all the players than "favorable" one. We shall illustrate this rather nontrivial fact on example. Consider the two-person game on the Fig. 5.3. The nodes from the personal positions X_1 are represented by circles and those from X_2 by blocks, with players payoffs written in final position. On the figure positions from the sets X_i $(i = 1, 2)$ are numbered by double indexes (i, j) where i is the index of the player and j the index of the node x in the set X_i. One can easily see that the "favorable" equilibrium has the form $((2, 2, 1, 1, 1), (2, 1))$ with payoffs $(2, 1)$. The

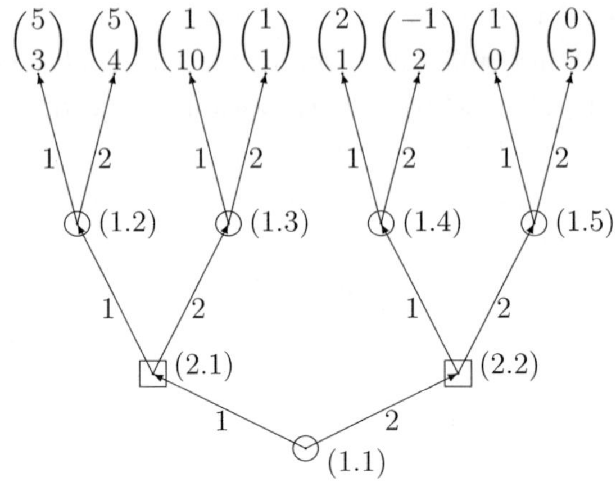

Figure 5.3

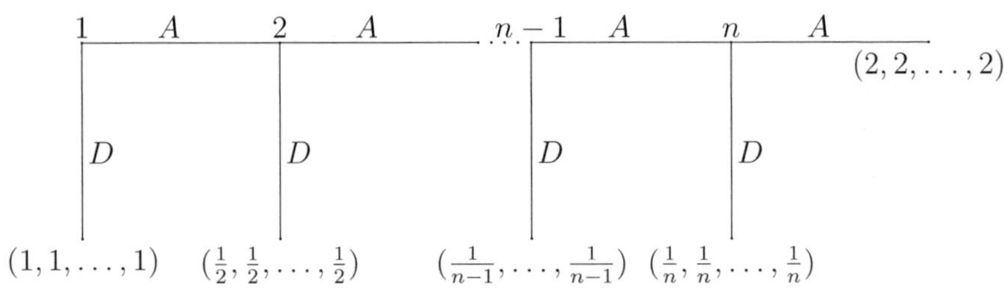

Figure 5.4

"unfavorable" equilibrium has the form $((1, 1, 2, 1, 1), (1, 1))$ with payoffs $(5, 3)$.

5.2.5. [Fudenberg and Tirole (1992)]. Consider the n-person game with perfect information, where each player $i \leq n$ can either end the game by playing D or play A and give the move to player $i + 1$ (see Fig. 5.4).

If player i selects D, each player gets $1/i$, if all players select A each gets 2. The backward induction algorithm for computing the subgame perfect (absolute) equilibria predicts that all players should play A. Thus the situation $(A, A, \ldots, A)$ is a subgame perfect Nash equilibrium. (Note that in the game under consideration each player is moving only once and has two alternatives, wich are also his strategies.) But there are also other equilibria. One class of Nash equilibria has the form $(D, A, A, D, \ldots)$, where the first player selects D and at least one of the others selects D. The payoffs in the first case are $(2, 2, \ldots, 2)$ and in the second $(1, 1, \ldots, 1)$. On the basis of robustness argument it seems that the equilibrium $(A, A, \ldots, A)$ is inefficient if n is very large. The equilibrium $(D, A, A, D, \ldots)$ is such because Player 4 uses the punishment strategy to enforce Player 1 to play D. This equilibrium is not subgame perfect, because it is not an equilibrium in any subgame starting from the positions 2, 3.

5.2.6. ***Cuban missile crises.*** Consider now the example which can in a very simplified version be modeled by a game on the tree with perfect information. Namely the Cuban missile crises between the United States under John Kennedy and Soviet Union under Nikita Khrushchev in 1963 [see Dixit and Skeath (1999)].

Khrushchev got information from his secret service that USA is planning a nuclear air over USSR planning to destroy 60 main USSR cities. To prevent this attack he starts the game by deciding whether or not to place intermediate range ballistic missiles in Cuba. If he places the missiles, his opponent player, Kennedy, will have three options: not react, blockade Cuba or eliminate the missiles by special airstrike. If Kennedy chooses the aggressive action of a blockade or an airstrike, Khrushchev may acquiesce or he may go by way of escalation with possible nuclear war at the end.

Consider the game tree on Fig. 5.5.

In this game Khrushchev — the first player — moves in circled vertexes and Kennedy — the second player — in squared vertexes. Payoffs are written in the final vertexes and on the first place is the payoff of Khrushchev.

Interpret the payoffs. If Player 1 (Khrushchev) decides not to place missiles his payoff is (-5), since there will be the probability of nuclear strike against USSR, in this case Player 2 (Kennedy) will get (5) since at that time USSR did not have the opportunity to strike back with nuclear bombs on long distances. If Player 1

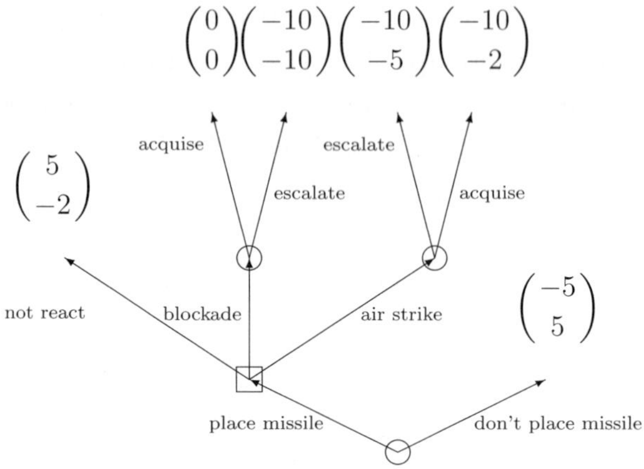

Figure 5.5

chooses on first stage place missiles, then the second player after some time will be informed and will have the possibility of taking one of following alternatives: not react (n), blockade Cuba (b), air strike to eliminate the missiles (a). If Player 2 chooses (n) then the game is over and the payoff of the Player 1 will be (5) and payoff of the Player 2 will be (-2). Since in this case the nuclear war will have a very small probability but strategic position of USSR would be much better than of USA. If Player 2 decides to blockade, then Player 1 has two alternatives: to acquire (a) or to escalate (e). In the case Player 1 chooses (a) the game is draw with payoffs (0, 0), which means that USSR will not keep missiles in Cuba and USA will forget the idea of nuclear air strike against USSR after Cuba lesson. If Player 1 decides to escalate the nuclear war is possible and the payoffs will be (-10, -10) the same for both, since compared with the case after 1 player alternative do not place missiles the nuclear war will make symmetric damage on both countries. Suppose now that Player 2 chooses air strike (alternative a), then if Player 1 chooses (a) the payoffs will be (-10, -2) since in this case Player 2 will have incentive to start nuclear war (as he wanted before the crises) but maybe some missiles in Cuba will remain untouched by air strike and few USA cities may be destroyed by nuclear attack (-2).

In the case if Player 1 choused (e), the payoff will be $(-10, -5)$ with less damage for USA since air strike can decrease the power of USSR missiles in Cuba (but not too much). Nash equilibrium can be found by backward induction and has payoff (0, 0), which really happens.

5.2.7. Indifferent equilibrium. As we have seen from 5.2.4 Example 5 a subgame perfect equilibrium may appear non-unique in an extensive form game. This happens when the payoffs of some players coincides in terminal positions. Then the behavior of the player depends on his attitude to his opponents and the behavior of the player type naturally arises. In two person cases one way destinguish with between two types of players "favorable" and "unfavorable". The resulting two different subgame perfect Nash equilibrium where demonstrated in Example 5.

There is another approach to avoid ambiguity in the behavior of players when any continuation of the game gives the same payoffs. Such an approach was proposed in Petrosyan *et al.* (2012). It realizes the following idea: in a given personal position the player moving in this position randomizes the alternatives yielding the same payoffs with equal probabilities. It can be easily proved that such behavior will also form a subgame perfect Nash equilibrium (not only in the case the randomization is made with equal probabilities, but also if it is made with arbitrary probability distribution over the alternatives yielding the same payoff).

For instance let us evaluate an indifferent equilibrium in the game from Example 5. In this example the alternatives 1 and 2 in position (1.2) will be chosen with probabilities $(\frac{1}{2}, \frac{1}{2})$, and in the same manner the alternatives 1 and 2 in position (1.3). In this case the Nash equilibrium (indifferent) will give the same payoffs (2,1) as favorable equilibrium.

Consider another example (see Fig. 5.6).

Here Player I moves in positions $\{(1.1), (1.2), (1.3)\}$, and Player II in positions $\{(2.1), (2.2)\}$. In this game in "favorable" equilibrium Player I chooses alternative 2 in position (1.2), and Nash

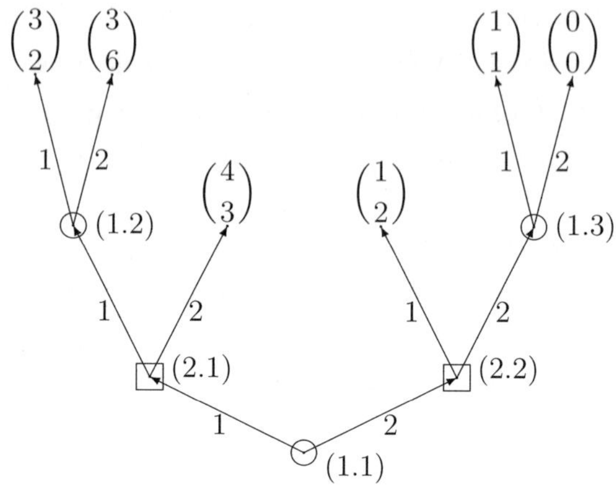

Figure 5.6

equilibrium will give the payoffs (3, 6). In unfavorable equilibrium Player I chooses in position (1.2) alternative 1, and Nash equilibrium will give the payoffs (4, 3). In indifferent equilibrium Player I will choose alternatives 1, 2 with probabilities $(\frac{1}{2}, \frac{1}{2})$ in position (1.2). The payoffs in indifferent equilibrium will be (3, 4). In this example as well as in Example 5 there are infinite many subgame perfect Nash equilibrium, since Player I in position (1.2) can choose the alternatives 1 and 2 with any probability $p = (p_1, p_2)$, $p_1 \geq 0$, $p_2 \geq 0$, $p_1 + p_2 = 1$, and all this mixed behavior in position (1.2) will be part of some subgame perfect Nash equilibrium.

5.3 Fundamental Functional Equations

5.3.1. We shall consider *multistage zero-sum games with perfect information.* If in conditions of 5.1.4 the set of players is composed of two elements $N = \{1, 2\}$ and $H_2(x) = -H_1(x)$ for all $x \in X_3$ (X_3 is the set of final positions in the game Γ), then

$$\Gamma = \langle N, U_i, K_i \rangle$$

appears to be a multistage zero-sum game with perfect information. It is apparent that this property is possessed by all subgames Γ_z of the game Γ.

Since an immediate consequence of the condition $H_2(x) = -H_1(x)$ is that $K_2(u_1, u_2) = -K_1(u_1, u_2)$ for all $u_1 \in U_1$, $u_2 \in U_2$, in the Nash equilibrium (u_1^*, u_2^*) the inequalities $K_1(u_1, u_2^*) \leq K_1(u_1^*, u_2^*) \leq K_1(u_1^*, u_2)$ hold for all $u_1 \in U_1$, $u_2 \in U_2$. The pair (u_1^*, u_2^*) is now called an *equilibrium* or a *saddle point* and the strategies forming an equilibrium are called *optimal.* The value of the payoff function in an equilibrium is denoted by v and is called the value of the game Γ.

5.3.2. From the Theorem in 5.2.1 it follows that in a multistage zero-sum game with perfect information on a tree graph there exists an absolute equilibrium, i.e. the equilibrium (u_1^*, u_2^*) whose truncation to any subgame Γ_z of the game Γ forms an equilibrium in Γ_z. For any subgame Γ_y it is possible to determine the number $v(y)$ which represents the value of the payoff function in the equilibrium of this subgame and is called the *value of the subgame* Γ_y. As shown in 1.3.2, the value of a zero-sum game (i.e. the value of Player 1's payoff function in the equilibrium) is determined uniquely, therefore the function $v(y)$ is defined for all $y \in X_1$, $y \in X_2$ and is unique.

5.3.3. Let us derive a functional equation to compute the function $v(y)$. From the definition of $v(y)$ it follows that

$$v(y) = K_1^y((u_1^*)^y, (u_2^*)^y) = -K_2^y((u_1^*)^y, (u_2^*)^y),$$

where $((u_1^*)^y, (u_2^*)^y)$ is an equilibrium in the subgame Γ_y that is the truncation of the absolute equilibrium (u_1^*, u_2^*).

Let $y \in X_1$ and $z \in F_y$. Then, as follows from (5.2.2), we have

$$v(y) = \max_{z \in F_y} K_1^z((u_1^*)^z, (u_2^*)^z) = \max_{z \in F_y} v(z). \tag{5.3.1}$$

Similarly, for $y \in X_2$ we have

$$\begin{aligned} v(y) &= -K_2^y((u_1^*)^y, (u_2^*)^y) = -\max_{z \in F_y} K_2^z((u_1^*)^z, (u_2^*)^z) \\ &= -\max_{z \in F_y}(-v(z)) = \min_{z \in F_y} v(z). \end{aligned} \tag{5.3.2}$$

From (5.3.1) and (5.3.2) we finally get

$$v(y) = \max_{z \in F_y} v(z), \;\; y \in X_1, \tag{5.3.3}$$

$$v(y) = \min_{z \in F_y} v(z), \;\; y \in X_2. \tag{5.3.4}$$

Equations (5.3.3), (5.3.4) are solved under the boundary condition

$$v(y)|_{y \in X_3} = H_1(y). \tag{5.3.5}$$

The system of equations (5.3.3), (5.3.4) with the boundary condition (5.3.5) makes possible the backward recursion for finding the value of the game and optimal strategies for players. Indeed, suppose the values of all subgames Γ_z of length $l(z) \leq k - 1$ are known and equal to $v(z)$. Let Γ_y be a subgame of length $l(y) = k$. Now, if $y \in X_1$, then $v(y)$ is determined from (5.3.3), if $y \in X_2$, then $v(y)$ is determined from (5.3.4). Here the values $v(z)$ in formulas (5.3.3), (5.3.4) are known, since the corresponding subgames have the length not exceeding $k - 1$. The same formulas indicate the way of constructing optimal strategies for players. Indeed, if $y \in X_1$, then Player 1 (maximizer) has to choose at the point y the node $z \in F_y$ for which the value of the next subgame is maximum. However, if $y \in X_2$, then Player 2 (minimizer) has to choose the position $z \in F_y$ for which the value of the next subgame is minimum.

When the players' choices in a multistage zero-sum game (an alternating game) alternate, equations (5.3.3), (5.3.4) can be written as one equation. Indeed, consider the subgame Γ_x and, for definiteness, let $x \in X_1$. Then in the next position Player 2 makes his move or this position is final (an alternating game), i.e. $F_x \subset X_2 \cup X_3$. Therefore, we may write

$$v(x) = \max_{y \in F_x} v(y), \;\; x \in X_1, \tag{5.3.6}$$

$$v(y) = \min_{z \in F_y} v(z), \;\; y \subset F_x \subset X_2 \cup X_3. \tag{5.3.7}$$

Subtracting (5.3.7) into (5.3.6) we obtain

$$v(x) = \max_{y \in F_x}[\min_{z \in F_y} v(z)], \;\; x \in X_1. \tag{5.3.8}$$

If $x \in X_2$, then in a similar way we get

$$v(x) = \min_{y \in F_x}[\max_{z \in F_y} v(z)]. \tag{5.3.9}$$

Equations (5.3.8), (5.3.9) are equivalent and must be considered with the initial condition $v(x)|_{x \in X_3} = H_1(x)$.

5.3.4. The Theorem in 5.2.1, considered for multistage zero-sum alternating games, shows the existence of an equilibrium in the game of chess, draughts in the class of pure strategies, while equations (5.3.8), (5.3.9) show a way of finding the value of the game. At the same time, it is apparent that for the foreseeable future no computer implementation is possible for solving these functional equations in order to find the value of the game and optimal strategies. It is highly improbable that we will know whether a player, "black" or "white", can guarantee the winning in any play of chess or there can always be a draw. In chess and draughts, however, successful attempts are made to construct approximately optimal solutions by creating programs capable to foresee several steps. Use is also made of various (obtained empirically) estimations of current positions. Such an approach is possible in the investigation of general multistage zero-sum games with perfect information. Successive iteration of estimations (for several steps ahead) may lead to desirable results.

5.4 Penalty Strategies

5.4.1. In 5.2.1, we proved the existence of absolute (Nash) equilibria in multistage games with perfect information on a finite graph tree. However, the investigation of particular games of this class may reveal the whole family of equilibria whose truncations are not necessarily equilibria in all subgames of the original game. Among such

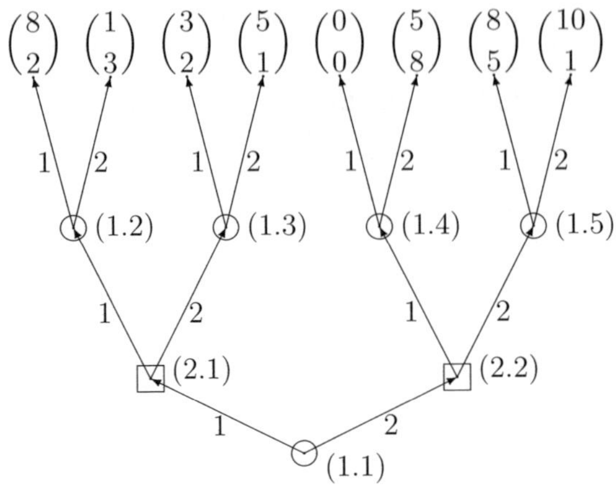

Figure 5.7

equilibria are equilibria in penalty strategies. We shall demonstrate this with the examples below.

Example 6. Suppose the game Γ proceeds on the graph depicted in Fig. 5.7. The set $N = \{1, 2\}$ is made up of two players. In Fig. 5.7, as an Example 5, the circles represent the nodes making up the set X_1 and the blocks represent the set X_2. The nodes of the graph are designated by double indices and the arcs by single indices.

It can be easily seen that the situation $u_1^* = (1, 1, 2, 2, 2)$, $u_2^* = (1, 1)$ is an absolute equilibrium in the game Γ. In this case, the payoffs to players are 8 and 2 units, respectively. Now consider the situation $\overline{u}_1 = (2, 1, 2, 1, 2)$, $\overline{u}_2 = (2, 2)$. In this situation the payoffs to players respectively are 10 and 1, and thus Player 1 receives a greater amount than in the situation (u_1^*, u_2^*). The situation $(\overline{u}_1, \overline{u}_2)$ is equilibrium in the game Γ but not absolute equilibrium. In fact, in the subgame $\Gamma_{1.4}$ the truncation of the strategy $\overline{u}_1$ tells Player 1 to choose the left-hand arc, which is not optimal for him in position 1.4. Such an action taken by Player 1 in position 1.4 can be interpreted as a "penalty" threat to Player 2 if he avoids Player 1's desirable choice of arc 2 in position 2.2, thereby depriving Player 1 of the maximum payoff of 10 units. But this "penalty" threat is unlikely to be treated as valid, because the penalizer (Player 1) may lose in case 5 units (acting nonoptimally).

5.4.2. We shall now give a strict definition of penalty strategies. For simplicity, we shall restrict ourselves to the case of a nonzero-sum two-person game. Let there be a multistage nonzero sum two-person game

$$\Gamma = \langle U_1, U_2, K_1, K_2 \rangle.$$

The game Γ will be associated with two zero-sum games Γ_1 and Γ_2 as follows. The game Γ_1 is a zero-sum game constructed in terms of the game Γ, where Player 2 plays against Player 1, i.e. $K_2 = -K_1$. The game Γ_2 is a zero-sum game constructed in terms of the game Γ, where Player 1 plays against Player 2, i.e. $K_1 = -K_2$. The graphs of the games $\Gamma_1, \Gamma_2, \Gamma$ and the sets therein coincide. Denote by (u_{11}^*, u_{21}^*) and (u_{12}^*, u_{22}^*) absolute equilibria in the games Γ_1, Γ_2 respectively. Let Γ_{1x}, Γ_{2x} be subgames of the games Γ_1, Γ_2; $v_1(x), v_2(x)$ are the values of these subgames. Then the situations $\{(u_{11}^*)^x, (u_{21}^*)^x\}$ and $\{(u_{12}^*)^x, (u_{22}^*)^x\}$ are equilibria in the games Γ_{1x}, Γ_{2x}, respectively, and $v_1(x) = K_1^x((u_{11}^*)^x, (u_{21}^*)^x)$, $v_2(x) = K_2^x((u_{12}^*)^x, (u_{22}^*)^x)$.

Consider an arbitrary pair (u_1, u_2) of strategies in the game Γ. Of course, this pair is the same in the games Γ_1, Γ_2. Let $Z = (x_0 = z_0, z_1, \ldots, z_l)$ be the path to be realized in the situation (u_1, u_2).

Definition. The strategy $\tilde{u}_1(\cdot)$ is called a penalty strategy of Player 1 if

$$\tilde{u}_1(z_k) = z_{k+1} \text{ for } z_k \in Z \cap X_1, \tag{5.4.1}$$
$$\tilde{u}_1(y) = u_{12}^*(y) \text{ for } y \in X_1,\ y \notin Z.$$

The strategy $\tilde{u}_2(\cdot)$ is called a penalty strategy for Player 2 if

$$\tilde{u}_2(z_k) = z_{k+1} \text{ for } z_k \in Z \cap X_2, \tag{5.4.2}$$
$$\tilde{u}_2(y) = u_{21}^*(y) \text{ for } y \in X_2,\ y \notin Z.$$

5.4.3. From the definition of penalty strategies we immediately obtain the following properties:

1^0. $K_1(\tilde{u}_1(\cdot), \tilde{u}_2(\cdot)) = H_1(z_l)$, $K_2(\tilde{u}_1(\cdot), \tilde{u}_2(\cdot)) = H_2(z_l)$.

2^0. Suppose one of the players, say, Player 1, uses strategy $u_1(\cdot)$ for which the position $z_k \in Z \cap X_1$ is the first in the path Z, where $u_1(\cdot)$ dictates the choice of the next position z'_{k+1} that is different from the choice dictated by the strategy $\tilde{u}_1(\cdot)$, i.e. $z'_{k+1} \neq z_{k+1}$. Hence from the definition of the penalty strategy $\tilde{u}_2(\cdot)$ it follows that

$$K_1(u_1(\cdot), \tilde{u}_2(\cdot)) \leq v_1(z_k). \tag{5.4.3}$$

Similarly, if Player 2 uses strategy $u_2(\cdot)$ for which the position $z_k \in Z \cap X_2$ is the first in the path Z, where $u_2(\cdot)$ dictates the choice of the next position z'_{k+1} that is different from the choice dictated by $\tilde{u}_2(\cdot)$ i.e. $z'_{k+1} \neq z_{k+1}$, then from the definition of the penalty strategy $\tilde{u}_1(\cdot)$ it follows that

$$K_2(\tilde{u}_1(\cdot), u_2(\cdot)) \leq v_2(z_k). \tag{5.4.4}$$

Hence, in particular, we obtain the following theorem.

Theorem. *Let $(\tilde{u}_1(\cdot), \tilde{u}_2(\cdot))$ be a situation in penalty strategies. For the situation $(\tilde{u}_1(\cdot), \tilde{u}_2(\cdot))$ to be equilibrium, it is sufficient that for all $k = 0, 1, \ldots, l-1$ there be the inequalities*

$$K_1(\tilde{u}_1(\cdot), \tilde{u}_2(\cdot)) \geq v_1(z_k), \tag{5.4.5}$$

$$K_2(\tilde{u}_1(\cdot), \tilde{u}_2(\cdot)) \geq v_2(z_k),$$

where $z_0, z_1, \ldots, z_l$ is the path realized in the situation $(\tilde{u}_1(\cdot), \tilde{u}_2(\cdot))$.

5.4.4. Suppose that $u^*_{11}(\cdot)$ and $u^*_{22}(\cdot)$ are optimal strategies for Players 1 and 2, respectively, in the auxiliary zero-sum games Γ_1 and Γ_2 and $\overline{Z} = \{\overline{z}_0, \overline{z}_1, \ldots, \overline{z}_l\}$ is the path corresponding to the situation $(u^*_{11}(\cdot), u^*_{22}(\cdot))$. Also, suppose the penalty strategies $\tilde{\tilde{u}}_1(\cdot)$ and $\tilde{\tilde{u}}_2(\cdot)$ are such that $\tilde{\tilde{u}}_1(\overline{z}_k) = u^*_{11}(z_k)$ for $\overline{z}_k \in \overline{Z} \cap X_1$ and $\tilde{\tilde{u}}_2(\overline{z}_k) = u^*_{21}(\overline{z}_k)$ for $\overline{z}_k \in \overline{Z} \cap X_2$. Then the situation $(\tilde{\tilde{u}}_1(\cdot), \tilde{\tilde{u}}_2(\cdot))$ forms a Nash equilibrium in penalty strategies. To prove this assertion it suffices to show that

$$K_1(u_{11}^*(\cdot), u_{22}^*(\cdot)) = K_1(\tilde{\tilde{u}}_1(\cdot), \tilde{\tilde{u}}_2(\cdot)) \geq v_1(\overline{z}_k), \tag{5.4.6}$$
$$K_2(u_{11}^*(\cdot), u_{22}^*(\cdot)) = K_2(\tilde{\tilde{u}}_1(\cdot), \tilde{\tilde{u}}_2(\cdot)) \geq v_2(\overline{z}_k),\ k = 0, 1, \ldots, l-1,$$

and use the Theorem in 5.4.3. Inequalities (5.4.6) follow from the optimality of strategies $u_{11}^*(\cdot)$ and $u_{22}^*(\cdot)$ in the games Γ_1 and Γ_2, respectively. The proof is offered as an exercise. We have thus obtained the following theorem.

Theorem. *In the game* Γ *there always exists an equilibrium in penalty strategies. In the special case described above (subsection* 5.4.4), *the payoffs in this situation are equal to* $K_i(u_{11}^*(\cdot), u_{22}^*(\cdot))$, *where* $u_{11}^*(\cdot)$ *and* $u_{22}^*(\cdot)$ *are optimal strategies for Players* 1 *and* 2 *in the auxiliary zero-sum games* Γ_1 *and* Γ_2, *respectively.*

The meaning of penalty strategies is that a player causes his partner to follow the particular path in the game (the particular choices) by constantly threatening to shift to a strategy that is optimal in a zero-sum game against the partner. Although the set of equilibria in the class of penalty strategies is sufficiently representative, these strategies should not be regarded as very "good", because by penalizing the partner the player can penalize himself to a greater extent.

5.5 Repeated Games and Equilibrium in Punishment (Penalty) Strategies

Consider the "Prisoners dilemma" game G from 3.2.1 (Example 5).

$$G = \begin{array}{c} \\ \alpha_1 \\ \alpha_2 \end{array} \begin{array}{c} \begin{array}{cc} \beta_1 & \beta_2 \end{array} \\ \begin{bmatrix} (5,5) & (0,10) \\ (10,0) & (1,1) \end{bmatrix} \end{array}.$$

Here the unique Nash equilibrium is (α_2, β_2) with payoff (1,1) which is dominated by pareto-optimal outcome (α_1, β_1) with payoff (5,5).

Assume that the game G is played infinitely many times [see Peters (2008)], and after each play (stage) players have the information about previous choices of each other. As a result we get infinite steam of payoffs, and there is a common discount factor $0 < \delta < 1$, such that the payoff of each player in the infinitely repeated game is defined as

$$\sum_{t=0}^{\infty} \delta^t \times (\text{payoff from } t\text{th play of the stage game}).$$

The notation stage game is used for one shot game G in order to distinguish this game from the infinitely repeated game. As it was explained earlier a strategy of player prescribes at each moment an action (behavior) of player or mixed behavior for each sequence of length t of stage strategies $\{(\alpha_1, \beta_1), (\alpha_1, \beta_2), (\alpha_2, \beta_1), (\alpha_2, \beta_2)\}$.

Denote the just defined game by G_δ^∞. As solution we shall try to find the subgame perfect Nash equilibrium. It is clear that each subgame starting from the stage t $G_\delta^\infty(t)$ coincides with the initial game G_δ^∞.

There are many subgame perfect Nash equilibrium (NE) in G_δ^∞. A. One is trivial: play in each stage game NE (α_2, β_2). If both players use (α_2, β_2), the payoff of each player will be

$$\sum_{t=0}^{\infty} \delta^t = \frac{1}{1-\delta}.$$

B. In stage game G an each stage t such that on the previous stages only (α_1, β_1) has occurred play (α_1, β_1). Otherwise play (α_2, β_2).

If this strategies are used by both players, then they will always play (α_1, β_1) and the payoffs of both players will be

$$\sum_{t=0}^{\infty} 5\delta^t = \frac{5}{1-\delta}.$$

There is also a subgame perfect Nash equilibrium for δ large enough. It is easy to derive the conditions on δ. Suppose one of the players, say Player II deviates, this means that there exist the first time instant (first stage game) $\bar{t}$, when he chooses β_2 instead of β_1.

Then since on the next stage $\bar{t}+1$ first player will choose α_2 till the end of the game, his payoff (of deviation) will be equal at most to

$$\sum_{t=0}^{\bar{t}-1} 5\delta^t + 10\delta^{\bar{t}} + \sum_{t=\bar{t}+1}^{\infty} 1\delta^t.$$

If player two is not deviating his payoff will be

$$\sum_{t=0}^{\infty} 5\delta^t = \frac{5}{1-\delta}.$$

The deviation will be not preferable if

$$\sum_{t=0}^{\infty} 5\delta^t > \sum_{t=0}^{\bar{t}-1} 5\delta^t + 10\delta^{\bar{t}} + \sum_{t=\bar{t}+1}^{\infty} 1\delta^t,$$

or if $\delta > 0,6$.

Thus we proved that for $\delta > 0,6$ the strategy pair constructed in B is Nash equilibrium (it can be proved that it is also a subgame perfect Nash equilibrium).

5.6 Hierarchical Games

There exists an important subclass of multistage nonzero-sum games, referred to as *hierarchical games.* Hierarchical games model conflict-controlled systems with a hierarchical structure. This structure is determined by a sequence of control levels ranking in a particular priority order. Mathematically, it is convenient to classify hierarchical games according to the number of levels and the nature of vertical relations. The simplest of them is a two-level system as depicted in Fig. 5.8.

5.6.1. The functioning of a *two-level conflict-controlled system* is as follows. The control (coordinating) center A_0 is at the first level of hierarchy, selects a vector $u = (u_1, \dots, u_n)$ from a given control

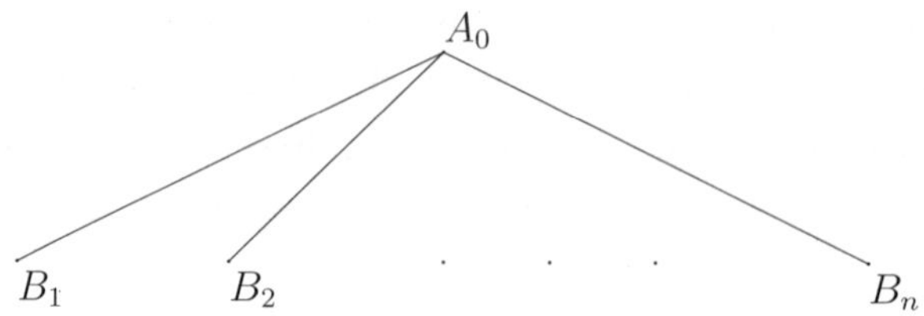

Figure 5.8

set U, where u_i is a control influence of the center on its subordinate divisions B_i, $i = 1, \ldots, n$ standing at the second level of the hierarchy. B_i, $i = 1, \ldots, n$ in its turn, selects controls $v_i \in V_i(u_i)$, where $V_i(u_i)$ is the set of controls for division B_i predetermined by the control u of center A_0. Now, the control center has the priority right to make the first move and may restrict the possibilities of its subordinate divisions by channeling their actions as desired. The aim of center A_0 is to maximize the functional $K_0(u, v_1, \ldots, v_n)$ over u, whereas divisions B_i, $i = 1, \ldots, n$, which have their own goals to pursue, seek to maximize the functionals $K_i(u_i, v_i)$ over v_i.

5.6.2. We shall formalize this problem as a noncooperative $(n+1)$-person game Γ (an administrative center A_0 and production divisions $B_1, \ldots, B_n$) in normal form.

Suppose Player A_0 selects a vector $u \in U$, where

$$U = \Bigg\{ u = (u_1, \ldots, u_n) : \; u_i \geq 0,$$

$$u \in R^l, \; i = 1, \ldots, n, \sum_{i=1}^{n} u_i \leq b \Bigg\}, \; b \geq 0$$

is the set of strategies for player A_0 in the game Γ. The vector u_i will be interpreted to mean the vector of resources of l items allocated by center A_0 to the ith production division.

Suppose each of the players B_i in the original problem (see 5.6.1) knows the choice by A_0 and selects the vector $v_i \in V_i(u_i)$ where

$$V_i(u_i) = \{ v_i \in R^m : \; v_i A_i \leq u_i + \alpha_i, \; v_i \geq 0 \} . \tag{5.6.1}$$

The vector v_i is interpreted as a production program of the ith division for various products; A_i is the production or the technological matrix of the ith production division ($A_i \geq 0$); α_i is the vector of available resources for the ith production division ($\alpha_i \geq 0$).

By definition, the strategies of Player B_i in the game Γ mean the set of functions $v_i(\cdot)$ setting up a correspondence among the elements $u_i : (u_1, \ldots, u_i, \ldots, u_n) \in U$ and the vector $v_i(u_i) \in V_i(u_i)$. The set of such functions is denoted by V_i, $i = 1, \ldots, n$.

Let us define the players' payoff functions in the game Γ. The payoff function for Player A_0 is

$$K_0(u, v_1(\cdot), \ldots, v_n(\cdot)) = \sum_{i=1}^{n} a_i v_i(u_i),$$

where $a_i \geq 0$, $a_i \in R^m$ is a fixed vector, $i = 1, \ldots, n$; $a_i v_i(u_i)$ is the scalar product of vectors a_i and $v_i(u_i)$. The payoff function for player B_i is

$$K_i(u, v_1(\cdot), \ldots, v_n(\cdot)) = c_i v_i(u_i),$$

where $c_i \geq 0$, $c_i \in R^m$ is a fixed vector, $i = 1, \ldots, n$.

Now the game Γ becomes $\Gamma = (U, V_1, \ldots, V_n, K_0, K_1, \ldots, K_n)$.

5.6.3. We shall construct a Nash equilibrium in the game Γ. Let $v_i^*(u_i) \in V_i(u_i)$ be a solution to a linear parametric programming problem (with the vector u_i as parameter)

$$\max_{v_i \in V_i(u_i)} c_i v_i = c_i v_i^*(u_i), \; i = 1, \ldots, n, \tag{5.6.2}$$

and let $u^* \in U$ be a solution to the problem

$$\max_{u \in U} K_0(u, v_1^*(u_1), \ldots, v_n^*(u_n)). \tag{5.6.3}$$

For simplicity assume that the maxima in (5.6.2) and (5.6.3) are achieved. Note that (5.6.3) is a nonlinear programming problem with an essentially discontinuous objective function (maximization is taken over u, and $v_i^*(u_i)$ are generally discontinuous functions of the parameter u_i). Show that the point $(u^*, v_1^*(\cdot), \dots, v_n^*(\cdot))$ is an equilibrium in the game Γ. Indeed,

$$K_0(u^*, v_1^*(u_1), \dots, v_n^*(u_n)) \geq K_0(u, v_1^*(u_1), \dots, v_n^*(u_n)), \ u \in U.$$

Further, with all $i = 1, \dots, n$ the inequality

$$\begin{aligned} K_i(u^*, v_1^*(u_1), \dots, v_n^*(u_n)) &= c_i v_i^*(u_i^*) \geq c_i v_i(u_i^*) \\ &= K_i(u^*, v_1^*(u_1), \dots, v_{i-1}^*(u_{i-1}), \\ &\quad \times v_i(u_i), v_{i+1}^*(u_{i+1}), \dots, v_n^*(u_n)) \end{aligned}$$

holds for any $v_i(\cdot) \in V_i$. Thus, it is not advantageous for every player A_0, $B_1, \dots, B_n$ to depart individually from the situation $(u^*, v_1^*(u_1), \dots, v_n^*(u_n))$, i.e. it is an equilibrium. Note that this situation is also stable against departures from it of any coalition $S \subset \{B_1, \dots, B_n\}$, since the payoff K_i to the ith player does not depend on strategies $v_j(\cdot)$, $j \in \{1, \dots, n\}$, $j \neq i$.

5.6.4. If in the solution (5.6.2) $v_i^*(u_i)$ is unique for each u_i, $i \in N$ (which is a very rare event), then the constructed Nash equilibrium coincides with Stackelberg solution. But if this is not the case the Stackelberg solution will be different.

Denote the set of all possible $v_i^*(u_i)$ on which the maximum in (5.6.2) is achieved by $V_i^*(u_i)$. Suppose that there exist

$$\min_{v_i \in V_i^*(u_i)} (a_i, v_i(u_i)) = (a_i, \bar{v}_i(u_i)). \tag{5.6.4}$$

Then the Stackelberg solution for the game for player A_0 will be to select such $\bar{u} = (\bar{u}_1, \dots, \bar{u}_n)$, that

$$\sum_{i=1}^{n} (a_i, \bar{v}_i(\bar{u}_i)) = \max_{u \in U} \sum_{i=1}^{n} (a_i \bar{v}_i(u_i)). \tag{5.6.5}$$

It is clear that the situation $(\bar{u}, \bar{v}_i(u_i))$ is not Nash equilibrium. But $v_i^*(u_i)$ only if it is uniquely defined from (5.6.2), then $v_i^*(u_i) = \bar{v}_i(u_i)$ and also $\bar{u} = u^*$.

5.7 Hierarchical Games (Cooperative Version)

This section deals with a cooperative version of the simplest hierarhical games (including the games defined in 5.6.1, 5.6.2). Characteristic functions are constructed and existence conditions for a nonempty core are studied.

5.7.1. Starting from the conceptual basis of problem 5.6.1, 5.6.2 and using the strategies which form a Nash equilibrium, we define for every coalition $S \subset N = \{A_0, B_1, \dots, B_n\}$ its guaranteed profit $v(S)$ as follows:

$$v(S) = \begin{cases} 0, & \text{if } S = \{A_0\}, \quad (5.7.1) \\ \sum_{i:B_i \in S} c_i v_i^*(0), & \text{if } A_0 \notin S, \quad (5.7.2) \\ \max_{\{u \in U:\ \sum_{i:B_i \in S} u_i = b\}} & \\ \quad \sum_{i:B_i \in S}(a_i + c_i) v_i^*(u_i), & \text{if } A_0 \in S, \quad (5.7.3) \end{cases}$$

where $v_i^*(u_i)$, $i = 1, \dots, n$ is a solution to the linear parametric programming problem (5.6.2).

Equality (5.7.1) holds, since the coalition $\{B_1, \dots, B_n\}$ can ensure a zero payoff to Player A_0 by selecting all $v_i = 0$, $i = 1, \dots, n$; equality (5.7.2) holds, since Player A_0 can always guarantee for S the payoff at most (5.7.2) by allocating to every $B_i \in S$ a zero resource; equality (5.7.3) holds, since coalition S incorporating A_0 can always ensure distribution of the whole resource only among its members.

Let S be an arbitrary coalition containing A_0. Denote by $u^S = (u_1^S, \dots, u_n^S)$ the maximizing vector in the nonlinear programming problem (5.7.3) (condition $u_i^S = 0$ holds for $i : B_i \notin S$). The following expression holds for any coalition $\overline{S} \subset S$, $\overline{S} \neq A_0$, $A_0 \in S$:

$$\sum_{i:B_i \in S} (a_i + c_i) v_i^* \left(u_i^S\right) \geq \sum_{i:B_i \in S} (a_i + c_i) v_i^* \left(u_i^{\overline{S}}\right)$$

$$= \sum_{i:B_i \in \overline{S}} (a_i + c_i) v_i^* \left(u_i^{\overline{S}}\right) + \sum_{i:B_i \in S \setminus \overline{S}} (a_i + c_i) v_i^*(0).$$

Let $S, R \subset N$, $S \cap R = \oslash$ and $A_0 \in S \neq A_0$. Then $A_0 \notin R$. In view of the condition $a_i \geq 0$, $c_i \geq 0$, $v_i \geq 0$, $i = 1, \ldots, n$, we have

$$v(S \cup R) = \sum_{i:B_i \in S \cup R} (a_i + c_i) v_i^* \left(u_i^{S \cup R}\right) \geq \sum_{i:B_i \in S \cup R} (a_i + c_i) v_i^* \left(u_i^S\right)$$

$$= \sum_{i:B_i \in S} (a_i + c_i) v_i^* \left(u_i^S\right) + \sum_{i:B_i \in R} (a_i + c_i) v_i^*(0)$$

$$= v(S) + v(R) + \sum_{i:B_i \in R} a_i v_i^*(0) \geq v(S) + v(R),$$

where $\sum_{i:B_i \in R} a_i v_i^*(0) \geq 0$ is the profit of center A_0 from "self-supporting" enterprises. When $A_0 \notin S \cup R$ or $S = A_0 \notin R$ the inequality $v(S \cup R) \geq v(S) + v(R)$ is obvious.

The function $v(S)$ defined by (5.7.1)–(5.7.3) is super-additive and we may consider the cooperative game $(\{A_0, B_1, \ldots, B_n\}, v)$ in a characteristic function form.

5.7.2. Consider an $(n+1)$ vector

$$\xi = \left(\sum_{i=1}^{n} a_i v_i^*(\overline{u}_i), c_1 v_1^*(\overline{u}_1), \ldots, c_n v_n^*(\overline{u}_n)\right), \tag{5.7.4}$$

where $\overline{u} = u^N$. The vector ξ is an imputation since the following conditions are satisfied:

1) $\displaystyle\sum_{k=0}^{n} \xi_k = \sum_{i=1}^{n} (a_i + c_i) v_i^*(\overline{u}_i) = v(N);$

2) $\displaystyle\xi_0 = \sum_{i=1}^{n} a_i v_i^*(\overline{u}_i) \geq 0 = v(A_0),$

$$\xi_i = c_i v_i^*(\overline{u}_i) \geq c_i v_i^*(0) = v(B_i), \ i = 1, \ldots, n.$$

By the Theorem 3.10.1, the necessary and sufficient condition for the imputation $(\xi_0, \xi_1, \ldots, \xi_n)$ to belong to core is that the inequality

$$\sum_{i \in S} \xi_i \geq v(S) \tag{5.7.5}$$

holds for all coalitions $S \subset \{A_0, B_1, \ldots, B_n\}$.

Let us introduce the condition under which the imputation ξ belongs to the core. If $S = \{A_0\}$ or $S \subset \{B_1, \ldots, B_n\}$, then condition (5.6.5) is satisfied, since

$$\xi_0 = \sum_{i=1}^{n} a_i v_i^*(\overline{u}_i) \geq 0 = v(\{A_0\}),$$

$$\sum_{i \in S} \xi_i = \sum_{i:B_i \in S} c_i v_i^*(\overline{u}_i) \geq \sum_{i:B_i \in S} c_i v_i^*(0) = v(S).$$

If $A_0 \in S \neq A_0$, then condition (5.7.5) can be written as

$$\begin{aligned}\sum_{i=1}^{n} a_i v_i^*(\overline{u}_i) + \sum_{i:B_i \in S} c_i v_i^*(\overline{u}_i) &= \sum_{i:B_i \in S} a_i v_i^*(\overline{u}_i) + \sum_{i:B_i \in S} c_i v_i^*(\overline{u}_i) \\ &\quad + \sum_{i:B_i \notin S} a_i v_i^*(\overline{u}_i) \\ &\geq \sum_{i:B_i \in S} (a_i + c_i) v_i^*(u_i^S).\end{aligned}$$

Therefore, the imputation (5.7.4) belongs to the core if the inequality

$$\sum_{i:B_i \notin S} a_i v_i^*(\overline{u}_i) \geq \sum_{i:B_i \in S} (a_i + c_i)[v_i^*(u_i^S) - v_i^*(\overline{u}_i)]$$

holds for all $S : A_0 \in S$.

Note that in this case we have defined the characteristic function of the game by using the payoff in Nash equilibrium and the quantity $v(N) = \max_u \sum_{i=1}^{n} (a_i + c_i) v_i^*(u_i)$ is generally less than the total maximum payoff to all players that is

$$\max_{u \in U} \max_{v_k \in V_k(u_k)} \left[\sum_{k=1}^{n} (a_k + c_k) v_k(u_k) \right],$$

(different from the definition of a characteristic function adopted in Ch. 3).

5.7.3. The characteristic function of the game can be constructed in the ordinary way, that is: it can be defined for every coalition S as the value of a zero-sum game between this coalition and the coalition of the other players $N \setminus S$. We shall now construct the characteristic function exactly in this way. In this case we shall slightly generalize the preceding problem by introducing arbitrary payoff functions for players.

As in the previous case, we assume that center A_0 distributes resources among divisions $B_1, \dots, B_n$ which use these resources to manufacture products. The payoffs to the control center A_0 and "production" divisions $B_1, \dots, B_n$ depend on the output of products by $B_1, \dots, B_n$. The vector of resources available to center A_0 is denoted by b. Center (Player) A_0 selects a system of n vectors $u = (u_1, \dots, u_n)$ from the set

$$U = \left\{ u = (u_1, \dots, u_n) : u_k \geq 0, u_k \in R^l, \sum_{k=1}^{n} u_k \leq b, k = 1, \dots, n \right\}.$$

Here u_k is interpreted as the vector of resources allocated by center A_0 to the production division B_k. The capacities of enterprise (Player) B_k are determined by the resource u_k obtained from A_0, i.e. enterprise B_k selects its production program x_k from the set $B_k(u_k) \subset R^m$ of non-negative vectors. We assume that the sets $B_k(u_k)$ for all u_k contain a zero vector and increase monotonically in the inclusion, i.e. from $u'_k > u_k$ follows $B_k(u'_k) \supset B_k(u_k)$, and the condition $B_k(0) = \oslash$ (impossibility of production because of the lack of resources) is satisfied.

Let $x = (x_1, \dots, x_n)$. The payoff to Player A_0 is determined by the function $l_0(x) \geq 0$, whereas the payoffs to the players B_k are taken to be $l_k(x_k) \geq 0$, $k = 1, \dots, n$ (the payoff to Player B_k depends only upon his production program). For simplicity, assume that the payoff to center A_0 satisfies the condition

$$l_0(x) = \sum_{k=1}^{n} l(x_k),$$

where the term $l(x_k)$ is interpreted to mean the payoff to Player A_0 due from Player B_k. We also assume that $l(x_k) \geq 0$ for all $x_k \in B_k(u_k)$ and $l_k(0) = 0$, $l(0) = 0$, $k = 1, \ldots, n$.

Just as in Sec. 5.6, so can the hierarchical game 5.7.3 be represented as a noncooperative $(n+1)$ person game in normal form, where the strategies for Player A_0 are the vectors $u \in U$, while the strategies for players B_k are the functions from the corresponding sets. Let us construct the characteristic function $\overline{v}(\cdot)$ for this game following 3.9.2. For each players' subset of S, we take $\overline{v}(S)$ to be the value (if it exists in conditions of the subsection) of a zero-sum game between coalitions S and $N \setminus S$, in which the payoff to coalition S is determined as the sum of payoffs belonging to the players set S.

Let $N = \{A_0, B_1, \ldots, B_n\}$. Then

$$\overline{v}(N) = \sup_{\{u \in U : \sum_{k=1}^{n} u_k = b\}} \sup_{x_k \in B_k(u_k),\ k=1,\ldots,n} \left\{ \sum_{k=1}^{n} [l(x_k) + l_k(x_k)] \right\}.$$

Note that for all $S \subset \{B_1, \ldots, B_n\}$, $\overline{v}(S) = 0$, since Player A_0 can always distribute the whole resource b among the members of coalition $N \setminus S$, to which he also belongs, thereby depriving coalition S of resources (i.e. A_0 can always set $u_k \equiv 0$ for $k : B_k \in S$, which results in $B_k(0) = \oslash$ for all $B_k \in S$). Using this line of reasoning we get $\overline{v}(A_0) = 0$, since the players $B_1, \ldots, B_n$ can always nullify a payoff to center A_0 by setting $x_k = 0$ for $k = 1, \ldots, n$ (without turning out products). It is apparent that A_0 will distribute the whole resource among the members of the coalition when coalition S contains center A_0. This reasoning leads to the following formula:

$$\overline{v}(S) = \sup_{\{u \in U : \sum_{k:B_k \in S} u_k = b\}} \sup_{x_k \in B_k(u_k),\ k:B_k \in S} \left\{ \sum_{k:B_k \in S} [l(x_k) + l_k(x_k)] \right\}$$

for $S : A_0 \in S$.

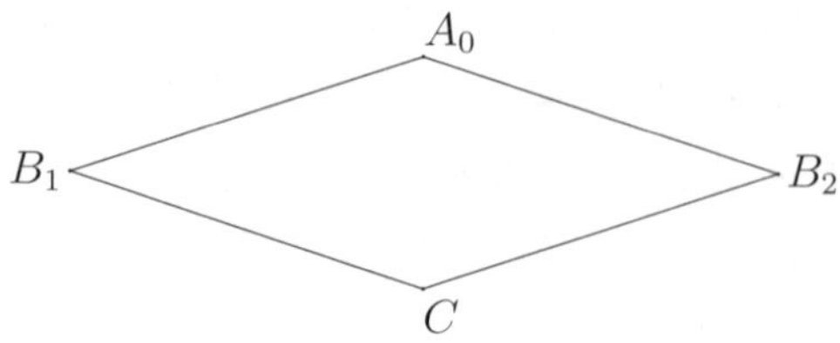

Figure 5.9

It can be shown that, under such definition of the characteristic function, the core of the imputation set

$$\alpha = (\alpha_0, \alpha_1, \ldots, \alpha_n) : \alpha_i \geq 0, i = 0, 1, \ldots, n, \sum_{i=0}^{n} \alpha_i = \overline{v}(N)$$

is always nonempty.

5.7.4. Hierarchical systems with double subordination are called diamond-shaped (Fig. 5.9). Control of a double subordination division C depends on control B_1 and control B_2.

One can envision a situation in which center B_1 represents the interests of an industry, while B_2 represents regional interests, including the issues of environment protection. A simple diamond-shaped system is an example of a hierarchical two-level decision making system. At the upper level there is an administrative center which is in charge of material and labor resources. It brings an influence to bear upon activities of its two subordinate centers belonging to the next level. The decisions made by these centers determine an output of the enterprise standing at a lower level of the hierarchical system.

We shall consider this decision making process as a four-person game. Denote this game by Γ. Going to the game setting, we assume that at the first step Player A_0 moves and selects an element (strategy) $u = (u_1, u_2)$ from a certain set U, where U is a strategy set for Player A_0. The element $u \in U$ restricts the possibilities for players B_1 and B_2 to make their choices at the next step. In other words, the set of choices for Player B_1 is function of the parameter u_1 (denoted by $B_1(u_1)$). Similarly, the set of choices for Player B_2 is function of parameter u_2 (denoted by $B_2(u_2)$). Denote by $\omega_1 \in B_1(u_1)$ and $\omega_2 \in B_2(u_2)$ the elements of the sets of choices for players B_1 and B_2 respectively. The parameters ω_1 and ω_2 selected by players B_1 and B_2 specify restrictions on the set of choices for Player C at the third step of the game, i.e. this set turns out to be the function of parameters ω_1 and ω_2. Denote it by $C(\omega_1, \omega_2)$, and the elements of this set (production programs) by v.

Suppose the payoffs of all players A_0, B_1, B_2, C depend only on the production program v selected by Player C and are respectively equal to $l_1(v), l_2(v), l_3(v), l_4(v)$, where $l_i(v) \geq 0$.

This hierarchical game can be represented as a noncooperative four-person game in normal form if the strategies for Player A_0 are taken to be the elements $u = (u_1, u_2) \in U$, while the strategies for players B_1, B_2, and C are taken to be the functions $\omega_1(u_1), \omega_2(u_2)$ and $v(\omega_1, \omega_2)$ with values in the sets $B_1(u_1), B_2(u_2), C(\omega_1, \omega_2)$, respectively, (the sets of such functions will be denoted by B_1, B_2, C) which set up a correspondence between every possible choice by the player (or the players) standing at a higher level and the choice made by this player. Setting

$$K_i(u, \omega_1(\cdot), \omega_2(\cdot), v(\cdot)) = l_i(v(\omega_1(u_1), \omega_2(u_2)), \; i = \overline{1,4}$$

we obtain the normal form of the game Γ

$$\Gamma = (U, B_1, B_2, C, K_1, K_2, K_3, K_4).$$

5.7.5. We shall now seek a Nash equilibrium in the game Γ. To this end, we shall perform additional constructions.

For every fixed pair (ω_1, ω_2), $(\omega_1, \omega_2) \in \cup_{u \in U} B_1(u_1) \times B_2(u_2)$ we denote by $v^*(\omega_1, \omega_2)$ a solution to the parametric extremal problem.

$$\max_{v \in C(\omega_1, \omega_2)} l_4(v) = l_4(v^*(\omega_1, \omega_2)). \tag{5.7.6}$$

(The maximum in (5.7.6) is supposed to be achieved.) The solution $v^*(\cdot) = v^*(\omega_1, \omega_2)$ of problem (5.7.6) is function of the parameters ω_1, ω_2 and $v^*(\cdot) \in C$.

Consider an auxiliary parametric (with parameters u_1, u_2) nonzero-sum two-person (B_1 and B_2) game $\Gamma'(u_1, u_2) = \{B_1(u_1), B_2(u_2), l_2, l_3\}$, where $l_2 = l_2(v^*(\omega_1, \omega_2))$, $l_3 = l_3(v^*(\omega_1, \omega_2))$. The elements $\omega_1 \in B_1(u_1)$ are strategies for Player B_1 in $\Gamma'(u_1, u_2)$, while the elements $\omega_2 \in B_2(u_2)$ are strategies for Player B_2. Suppose the game $\Gamma'(u_1, u_2)$ has the Nash equilibrium denoted as $(\omega_1^*(u_1), \omega_2^*(u_2))$. Note that $\omega_i^*(\cdot)$ is function of the parameter u_i and $\omega_i^*(\cdot) \in B_i$, $i = 1, 2$.

Further, let $u^* = (u_1, u_2)$ be a solution to the extremal problem

$$\max_{u \in U} l_1(v^*(\omega_1^*(u_1), \omega_2^*(u_2))). \tag{5.7.7}$$

Lemma. *The situation* $(u^*, \omega_1^*(\cdot), \omega_2^*(\cdot), v^*(\cdot))$ *is a Nash equilibrium in the game* Γ.

Proof. By the definition of u^*, from (5.7.7) follows

$$\begin{aligned} K_1(u^*, \omega_1^*(\cdot), \omega_2^*(\cdot), v^*(\cdot)) &= \max_{u \in U} l_1(v^*(\omega_1^*(u_1), \omega_2^*(u_2))) \\ &\geq l_1(v^*(\omega_1^*(u_1), \omega_2^*(u_2))) \\ &= K_1(u, \omega_1^*(\cdot), \omega_2^*(\cdot), v^*(\cdot)) \end{aligned}$$

for all $u \in U$. Since $\omega_1^*(u_1^*), \omega_2^*(u_2^*)$ form a Nash equilibrium in the auxiliary game $\Gamma(u_1^*, u_2^*)$ the relationships

$$\begin{aligned} K_2(u^*, \omega_1^*(\cdot), \omega_2^*(\cdot), v^*(\cdot)) &= l_2(v^*(\omega_1^*(u_1^*), \omega_2^*(u_2^*))) \\ &\geq l_2(v^*(\omega_1(u_1^*), \omega_2^*(u_2^*))) \\ &= K_2(u^*, \omega_1(\cdot), \omega_2^*(\cdot), v^*(\cdot)) \end{aligned}$$

hold for any function $\omega_1(\cdot) \in B_1$, $\omega_1(u^*) \in B_1(u_1^*)$.

A similar inequality holds for Player B_2.

By the definition of the function v^*, from (5.7.6) we have

$$\begin{aligned} K_4(u^*, \omega_1^*(\cdot), \omega_2^*(\cdot), v^*(\cdot)) &= l_4(v^*(\omega_1^*(u_1^*), \omega_2^*(u_2^*))) \\ &= \max_{v \in C(\omega_1^*(u_1^*), \omega_2^*(u_2^*))} l_4(v) \geq l_4(v) \\ &= K_4(u^*, \omega_1^*(\cdot), \omega_2^*(\cdot), v(\cdot)) \end{aligned}$$

for any function $v(\cdot) \in C$, $v(\omega_1^*(u_1^*), \omega_2^*(u_2^*)) = v \in C(\omega_1^*(u_1^*), \omega_2^*(u_2^*))$.

This completes the proof of lemma.

5.7.6. Applying the maximin approach, for every coalition $S \subset \{A_0, B_1, B_2, C\}$ we define $v'(S)$ to be the guaranteed maximum payoff to S in a zero-sum game between coalition S acting as the maximizer, and coalition $S' = \{A_0, B_1, B_2, C\} \setminus S$. Suppose there exists $v_0 \in C(\omega_1, \omega_2)$ such that $l_i(v_0) = 0$, $i = 1, 2, 3, 4$ for all ω_1, ω_2.

We shall distinguish two forms of coalitions: 1) $S : C \notin S$; 2) $S : C \in S$.

In the first case $S \subset \{A_0, B_1, B_2\}$ and Player C, the member of coalition $N \setminus S$, may choose strategy $v_0 : l_i(v_0) = 0$, $i = 1, 2, 3, 4$, therefore $v'(S) = 0$.

In the second case the characteristic function $v'(S)$ is defined by the following equalities:

a) $S = \{C\}$

$$v'(S) = \min_{u \in U} \min_{\omega_1 \in B_1(u_1)} \min_{\omega_2 \in B_2(u_2)} \max_{v \in C(\omega_1, \omega_2)} l_4(v),$$

(in what follows we assume that all max and min are achieved);

b) $S = \{A_0, C\}$

$$v'(S) = \max_{u \in U} \min_{\omega_1 \in B_1(u_1)} \min_{\omega_2 \in B_2(u_2)} \max_{v \in C(\omega_1, \omega_2)} (l_1(v) + l_4(v));$$

c) $S = \{B_1, C\}$

$$v'(S) = \min_{u \in U} \max_{\omega_1 \in B_1(u_1)} \min_{\omega_2 \in B_2(u_2)} \max_{v \in C(\omega_1, \omega_2)} (l_2(v) + l_4(v));$$

d) $S = \{B_2, C\}$

$$v'(S) = \min_{u \in U} \max_{\omega_2 \in B_2(u_2)} \min_{\omega_1 \in B_1(u_1)} \max_{v \in C(\omega_1, \omega_2)} (l_3(v) + l_4(v));$$

e) $S = \{B_1, B_2, C\}$

$$v'(S) = \min_{u \in U} \max_{\omega_1 \in B_1(u_1)} \max_{\omega_2 \in B_2(u_2)} \max_{v \in C(\omega_1, \omega_2)} \sum_{i=2,3,4} l_i(v);$$

f) $S = \{A_0, B_1, C\}$

$$v'(S) = \max_{u \in U} \max_{\omega_1 \in B_1(u_1)} \min_{\omega_2 \in B_2(u_2)} \max_{v \in C(\omega_1, \omega_2)} \sum_{i=1,2,4} l_i(v);$$

g) $S = \{A_0, B_2, C\}$

$$v'(S) = \max_{u \in U} \max_{\omega_2 \in B_2(u_2)} \min_{\omega_1 \in B_1(u_1)} \max_{v \in C(\omega_1, \omega_2)} \sum_{i=1,3,4} l_i(v);$$

h) $S = \{A_0, B_1, B_2, C\}$

$$v'(S) = \max_{u \in U} \max_{\omega_1 \in B_1(u_1)} \max_{\omega_2 \in B_2(u_2)} \max_{v \in C(\omega_1, \omega_2)} \sum_{i=1}^{4} l_i(v).$$

By this definition, the characteristic function is super-additive, i.e. the inequality $v(S \cup R) \geq v(S) + v(R)$ holds for any $S, R \subset \{A_0, B_1, B_2, C\}$ for which $S \cap R = \oslash$.

5.8 Multistage Games with Incomplete Information

5.8.1. In Secs. 5.1–5.4, we considered multistage games with perfect information defined in terms of a finite tree graph $G = (X, F)$ in which each of the players exactly knows at his move the position or the tree node where he stays. That is why we were able to introduce the notion of player i's strategy as a single-valued function $u_i(x)$ defined on the set of personal positions X_i with its values in the set F_x. If, however, we wish to study a multistage game in which the players making their choices have no exact knowledge of positions in which they make their moves or may merely speculate that this position belongs to some subset A of personal positions X_i, then the realization of player's strategy as a function of position $x \in X_i$ turns out to be impossible. In this manner, the wish to complicate the information structure of a game inevitably involves changes in the notion of a strategy. In order to provide exact formulations, we should first formalize the notion of information in the game. Here

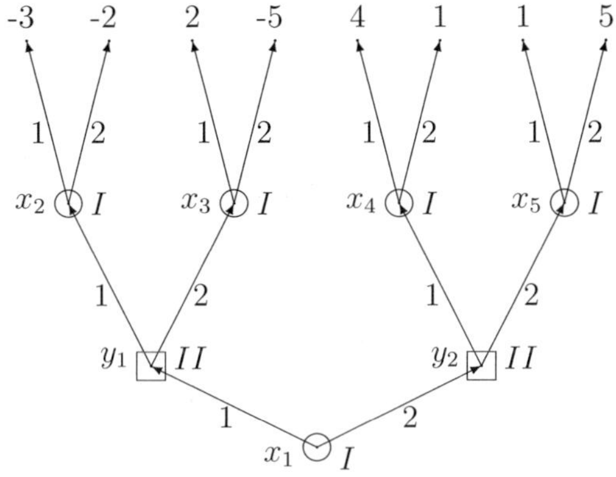

Figure 5.10

the notion of an information set plays an important role. This will be illustrated with some simple, already classical examples from texts on game theory [McKinsey (1952)].

Example 7. Zero-sum game. Player 1 selects at the first move a number from the set $\{1, 2\}$. The second move is made by Player 2. He is informed about Player 1's choice and selects a number from the set $\{1, 2\}$. The third move is again to be made by Player 1. He knows Player 2's choice, remembers his own choice and selects a number from the set $\{1, 2\}$. At this point the game terminates and Player 1 receives a payoff H (Player 2 receives a payoff $(-H)$, i.e. the game is zero-sum), where the function H is defined as follows:

$$
\begin{aligned}
&H(1,1,1) = -3,\ H(2,1,1) = 4,\\
&H(1,1,2) = -2,\ H(2,1,2) = 1,\\
&H(1,2,1) = 2,\ H(2,2,1) = 1,\\
&H(1,2,2) = -5,\ H(2,2,2) = 5,
\end{aligned}
\tag{5.8.1}
$$

The graph $G = (X, F)$ of the game is depicted in Fig. 5.10. The circles in the graph represent positions in which Player 1 makes a move, whereas the blocks represent positions in which Player 2 makes a move.

If the set X_1 is denoted by X, the set X_2 by Y and the elements of these sets by $x \in X$, $y \in Y$, respectively, then Player 1's strategy $u_1(\cdot)$ is given by the five-dimensional vector $u_1(\cdot) = \{u_1(x_1), u_1(x_2), u_1(x_3), u_1(x_4), u_1(x_5)\}$ prescribing the choice of one of the two numbers $\{1, 2\}$ in each position of the set X. Similarly, Player 2's strategy $u_2(\cdot)$ is a two-dimensional vector $u_2(\cdot) = \{u_2(y_1), u_2(y_2)\}$ prescribing the choice of one of the two numbers $\{1, 2\}$ in each of the positions of the set Y. Now, in this game Player 1 has 32 strategies and Player 2 has 4 strategies. The corresponding normal form of the game has a 32×4 matrix which (this follows from the Theorem in 5.2.1) has an equilibrium in pure strategies. It can be seen that the value of this game is 4. Player 1 has four optimal pure strategies: (2,1,1,1,2), (2,1,2,1,2), (2,2,1,1,2), (2,2,2,1,2). Player 2 has two optimal strategies: (1,1), (2,1).

Example 8. We shall slightly modify the information conditions of Example 7. The game is zero-sum. The first move is made by Player 1. He selects a number from the set $\{1, 2\}$. The second move is made by Player 2. He is informed about Player 1's choice and selects a number from the set $\{1, 2\}$. The third move is made by Player 1. Without knowledge of Player 2's choice and with no memory of his own choice he chooses a number of the set $\{1, 2\}$. At this point the game terminates and the payoff is determined by formula (5.7.1) in the same way as Example 7.

The graph of the game, $G = (X, F)$, remains unaffected. In the nodes x_2, x_3, x_4, x_5 (at the third move in the game) Player 1 cannot identify exactly the node in which he actually stays. With the knowledge of the priority of his move (the third move), he can be sure that he is not in the node x_1. In the graph G the nodes x_2, x_3, x_4, x_5 are traced by dashed line (Fig. 5.11).

The node x_1 is enclosed in a circle, which may be interpreted for Player 1 as an exact knowledge of this node when he stayed in it. The nodes y_1, y_2 are enclosed in blocks, which also means that

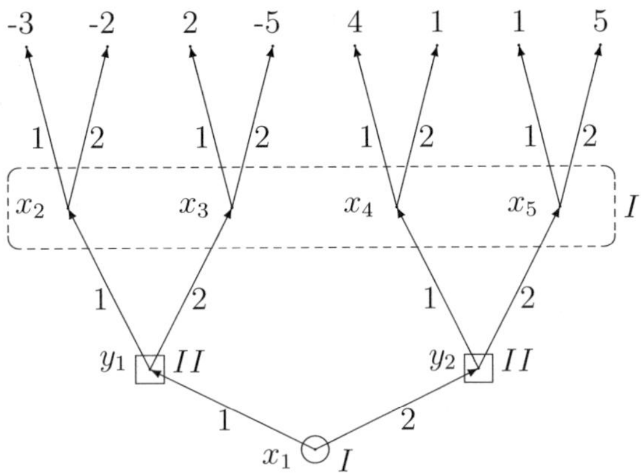

Figure 5.11

Player 2 staying in one of them at his move can distinguish this node from the other. Combining the nodes x_2, x_3, x_4, x_5 into one set, we shall illustrate the fact that they are indistinguishable for Player 1.

The sets into which the nodes are collected in this way are called *information sets.*

We shall now describe strategies. The information state of Player 2 remains unchanged; therefore the set of his strategies is the same as in Example 7, i.e. it consists of four vectors (1,1), (1,2), (2,1), (2,2). The information state of Player 1 changed. At the third step of the game he knows only the number of this step, but does not know the position in which he stays. Therefore, he cannot realize the choice of the next node (or the choice of a number from the set $\{1, 2\}$) depending on the position in which he stays at the third step. For this reason irrespective of the actually realized position he has to choose at the third step one of the two numbers $\{1, 2\}$. Thus the strategy for him is a pair of numbers (i, j), $i, j \in \{1, 2\}$, where the number i is chosen in position x_1 while the number j at the third step is the same in all positions x_2, x_3, x_4, x_5. Now the choice of a number j turns out to be a function of the set and can be written as $u\{x_2, x_3, x_4, x_5\} = j$. In this game both players have four strategies and the matrix of the game is

$$\begin{array}{c} \\ (1.1) \\ (1.2) \\ (2.1) \\ (2.2) \end{array} \begin{array}{c} \begin{array}{cccc} (1.1) & (1.2) & (2.1) & (2.2) \end{array} \\ \begin{bmatrix} -3 & -3 & 2 & 2 \\ -2 & -2 & -5 & -5 \\ 4 & 1 & 4 & 1 \\ 1 & 5 & 1 & 5 \end{bmatrix} \end{array}.$$

This game has no equilibium in pure stategies. The value of the game is 19/7, an optimal mixed strategy for Player 1 is the vector $(0, 0, 4/7, 3/7)$, and an optimal mixed stategy for Player 2 is $(4/7, 3/7, 0, 0)$. The guaranteed payoff to Player 1 is reduced as compared to the one in Example 7. This is due to the degradation of his information state.

It is interesting to note that the game in Example 8 has a 4×4 matrix, whereas the game in Example 7 has a 32×4 matrix. The deterioration of available information thus reduces the size of the payoff matrix and hence facilitates the solution of the game itself. But this contradicts the wide belief that the deterioration of information results in complication of decision making.

Modifying information conditions we may obtain other variants of the game described in Example 7.

Example 9. Player 1 chooses at the first move a number from the set $\{1, 2\}$. The second move is made by Player 2, who, without knowing Player 1's choice, chooses a number from the set $\{1, 2\}$. Further, the third move is made by Player 1. Being informed about Player 2's choice and with the memory of his own choice on the first step he chooses a number from the set $\{1, 2\}$. The payoff is determined in the same way as in Example 7 (Fig. 5.12).

Since on the third move the player knows the position in which he stays, the positions of the third level are enclosed in circles and the two nodes, in which Player 2 makes his move, are traced by the dashed line and are included in one information set.

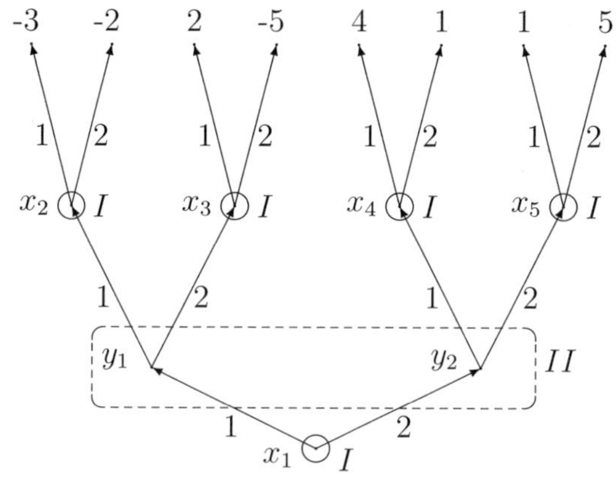

Figure 5.12

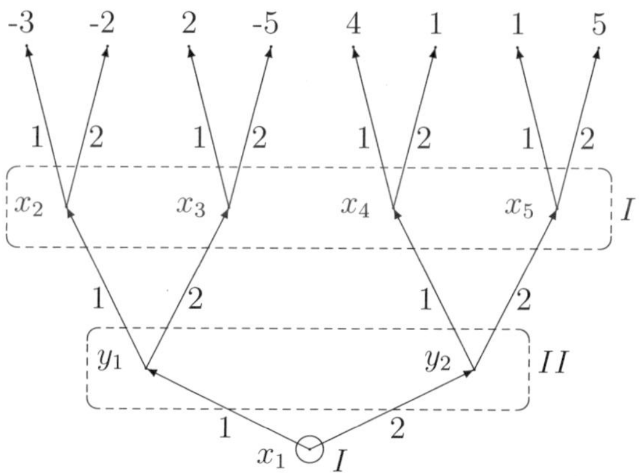

Figure 5.13

Example 10. Player 1 chooses a number from the set $\{1, 2\}$ on the first move. The second move is made by Player 2 without being informed about Player 1's choice. Further, on the third move Player 1 chooses a number from the set $\{1, 2\}$ without knowing Player 2's choice and with no memory of his own choice at the first step. The payoff is determined in the same way as in Example 7 (Fig. 5.13).

Here the strategy of Player 1 consists of a pair of numbers (i, j), the i-th choice is at the first step, and jth choice is at the third step; the strategy of Player 2 is a choice of number j at the second step of the game. Now, Player 1 has four strategies and Player 2 has two strategies. The game in normal form has a 4×2 matrix:

$$\begin{array}{c} \\ (1.1) \\ (1.2) \\ (2.1) \\ (2.2) \end{array} \begin{array}{c} \begin{array}{cc} 1 & \quad 2 \end{array} \\ \begin{bmatrix} -3 & 2 \\ -2 & -5 \\ 4 & 1 \\ 1 & 5 \end{bmatrix} \end{array}.$$

The value of the game is 19/7, an optimal mixed strategy for Player 1 is $(0, 0, 4/7, 3/7)$, whereas an optimal strategy for Player 2 is $(4/7, 3/7)$.

In this game the value is found to be the same as in Example 8, i.e. it turns out that the deterioration of information conditions for Player 2 did not improve the state of Player 1. This condition is random in nature and is accountable to special features of the payoff function.

Example 11. In the previous example the players fail to distinguish among positions placed at the same level of the game tree, but they do know the move to be made. It is possible to construct the game in which the players may reveal their ignorance to a greater extent.

Let us consider a zero-sum two-person game in which Player 1 is one person, whereas Player 2 is the team of two persons, A and B. All three persons are placed in different rooms and cannot communicate with each other. At the start of the game a mediator comes to Player 1 and suggests that he should choose a number from the set $\{1, 2\}$. If Player 1 chooses 1, the mediator suggests that A should be the first to make his choice. However, if Player 1 chooses 2, the mediator suggests that B should be the first to make his choice. Once these three numbers have been chosen, Player 1 wins an amount $K(x, y, z)$, where x, y, z are the choices made by Player 1 and members of Team 2,

A and B, respectively. The payoff function $K(x, y, z)$ is defined as follows:

$$K(1,1,1) = 1,\ K(1,1,2) = 3,$$
$$K(1,2,1) = 7,\ K(1,2,2) = 9,$$
$$K(2,1,1) = 5,\ K(2,1,2) = 1,$$
$$K(2,2,1) = 6,\ K(2,2,2) = 7.$$

From the rules of the game it follows that when a member of the team, A or B, is suggested that he should make his choice he does not know whether he makes his choice at the second or at the third step of the game. The structure of the game is shown in Fig. 5.14.

Now, the information sets of Player 2 contain the nodes belonging to different levels, this means that Player 2 does not know on which step (second or third) he makes a mave. Here Player 1 has two strategies, whereas Player 2 has four strategies composed of all possible choices by members of the team, A and B, i.e. strategies for him are the pairs $(1,1), (1,2), (2,1), (2,2)$.

In order to understand how the elements of the payoff matrix are determined, we consider a situation $(2, (2, 1))$. Since Player 1 has chosen 2, the mediator goes to B who, in accordance with strategy

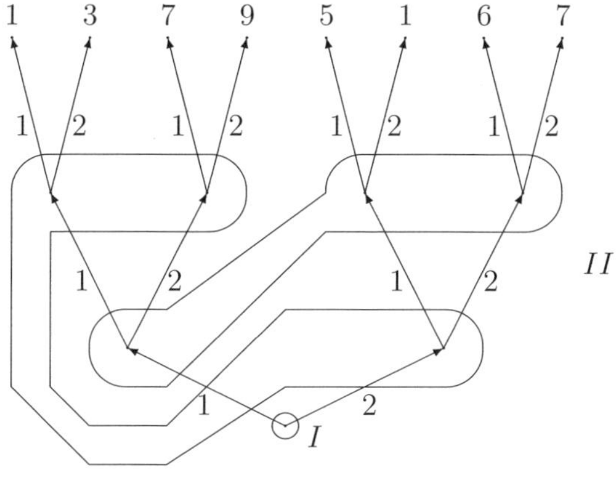

Figure 5.14

(2, 1), chooses 1. Then the mediator goes to A who chooses 2. Thus the payoff in situation $(2,(2,1))$ is $K(2,1,2)=1$. The payoff matrix for the game in normal form is

$$\begin{array}{c} \\ 1 \\ 2 \end{array}\begin{array}{c} \begin{array}{cccc} (1.1) & (1.2) & (2.1) & (2.2) \end{array} \\ \begin{bmatrix} 1 & 3 & 7 & 9 \\ 5 & 6 & 1 & 7 \end{bmatrix} \end{array}.$$

The value of the game is 17/5 and optimal mixed strategies for Players 1 and 2 respectively are $(2/5, 3/5)$, $(3/5, 0, 2/5, 0)$.

Note that in multistage games with perfect information (see Theorem in 5.2.1) there exists a Nash equilibrium in the class of pure strategies, while in multistage zero-sum games there exists an equilibrium in pure strategies. Yet all the games with incomplete information discussed in Examples 8–11 have no equilibrium in pure strategies.

5.8.2. We shall now give a formal definition of a multistage *game in extensive form.*

Definition [Kuhn (1953)]. The n-person game in extensive form is defined by

(1) Specifying the tree graph $G=(X,F)$ with the initial vertex x_0 referred to as the initial position of the game.
(2) Partition the sets of all vertices X into $n+1$ sets $X_1, X_2, \dots, X_n, X_{n+1}$, where the set X_i is called the set of personal positions of the i-th player, $i=1,\dots,n$, and the set $X_{n+1}=\{x : F_x=\oslash\}$ is called the set of final positions.
(3) Specifying the vector function $K(x)=(K_1(x),\dots,K_n(x))$ on the set of final positions $x \in X_{n+1}$; the function $K_i(x)$ is called the payoff to the ith player.
(4) Subpartition of the set X_i, $i=1,\dots,n$ into nonoverlapping subsets X_i^j referred to as information sets of the ith player. In this case, for any position of one and the same information set the set of its subsequent vertices should contain one and the same number of vertices, i.e. for any $x, y \in X_i^j |F_x| = |F_y|$ ($|F_x|$ is the

number of elements of the set F_x), and no vertex of the information set should follow another vertex of this set, i.e. if $x \in X_i^j$, then there is no other vertex $y \in X_i^j$ such that $y \in \hat{F}_x$ (see 5.1.2).

The definition of a multistage game with perfect information (see 5.1.4) is distinguished from the one given here only by condition 4, where additional partitions of players' personal positions X_i into information sets are introduced. As may be seen from the above examples, the conceptual meaning of such a partition is that when player i makes his move in position $x \in X_i$ in terms of incomplete information he does not know the position x itself, but knows that this position is in a certain set $X_i^j \subset X_i$ $(x \in X_i^j)$. Some restrictions are imposed by condition 4 on the players' information sets. The requirement $|F_x| = |F_y|$ for any two vertices of the same information set are introduced to make vertices $x, y \in X_i^j$ indistinguishable. In fact, with $|F_x| \neq |F_y|$ Player i could distinguish among the vertices $x, y \in X_i^j$ by the number of arcs emanating therefrom. If one information set could have two vertices x, y such that $y \in \hat{F}_x$ this would mean that a play of the game can intersect twice an information set, but this in turn is equivalent to the fact that player j has no memory of the index of his move in this play which can hardly be conceived in the actual play of the game.

5.9 Behavior Strategy

We shall continue examination of the game in extensive form and show that in the complete memory case for all players it has an equilibrium in behavior strategies.

5.9.1. For the purposes of further discussion we need to introduce some additional notions.

Definition. The arcs incidental with x, i.e. $\{(x, y) : y \in F_x\}$, are called alternatives at the vertex $x \in X$.

If $|F_x| = k$, at the vertex x there are k alternatives. We assume that if at the vertex x there are k alternatives, then they

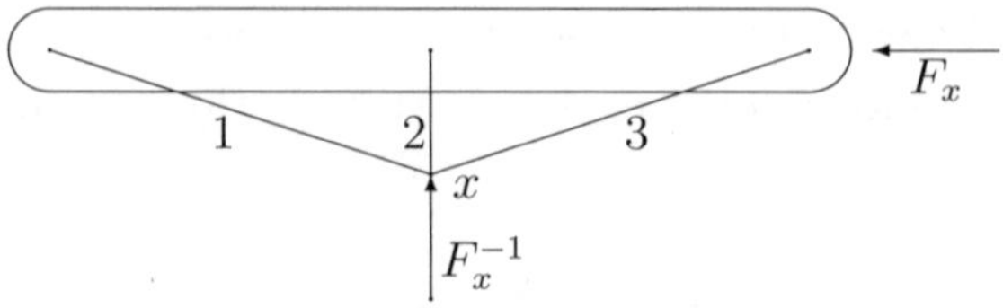

Figure 5.15

are designated by integers $1, \ldots, k$ with the vertex x bypassed in a clockwise sense. The first alternative at the vertex x_0 is indicated in an arbitrary way. If some vertex $x \neq x_0$ is bypassed in a clockwise sense, then an alternative which follows a single arc (F_x^{-1}, x) entering into x (Fig. 5.15) is called the first alternative at x.

Suppose that in the game Γ all alternatives are enumerated as above. Let A_k be the set of all vertices $x \in X$ having exactly k alternatives, i.e. $A_k = \{x : |F_x| = k\}$. Let $I_i = \{X_i^j : X_i^j \subset X_i\}$ be the set of all information sets for player i. By definition the *pure strategy* of player i means the function u_i mapping I_i into the set of positive numbers so that $u_i(X_i^j) \leq k$ if $X_i^j \subset A_k$. We say that the strategy u_i chooses alternative l in position $x \in X_i^j$ if $u_i(X_i^j) = l$, where l is the number of the alternative.

As in 5.1.4, we may show that to each situation $u(\cdot) = (u_1(\cdot), \ldots, u_n(\cdot))$ uniquely corresponds a play (path) ω, and hence the payoff in the final position of this play (path).

Let $x \in X_{n+1}$ be a final position and ω is the only path (F is the tree) leading from x_0 to x. The condition that the position y belongs to the path ω will be written as $y \in \omega$ or $y < x$.

Definition. Position $x \in X$ is called possible for $u_i(\cdot)$ if there exists a situation $u(\cdot)$ containing $u_i(\cdot)$ such that the path ω containing position x is realized in situation $u(\cdot)$, i.e. $x \in \omega$. The information set X_i^j is called relevant for $u_i(\cdot)$ if some position $x \in X_i^j$ is possible for $u_i(\cdot)$.

The set of positions, possible for $u_i(\cdot)$, is denoted by $Possu_i(\cdot)$, while the collection of information sets that are relevant for $u_i(\cdot)$ are denoted by $Relu_i(\cdot)$.

Lemma. *Position $x \in X$ is possible for $u_i(\cdot)$ if and only if $u_i(\cdot)$ chooses alternatives lying on the segment of the path ω_x from x_0 to x in all its information sets intersecting ω_x.*

Proof. Let $x \in Possu_i(\cdot)$. Then there exists a situation $u(\cdot)$ containing $u_i(\cdot)$ such that the path ω realized in this situation passes through x, which exactly means that in all its information sets intersecting the segment of the path ω_x the strategy $u_i(\cdot)$ chooses alternatives (arcs) belonging to ω_x.

Now let $u_i(\cdot)$ choose all alternatives for player i in ω_x. In order to prove the possibility of x for $u_i(\cdot)$ we need to construct a situation $u(\cdot)$ containing $u_i(\cdot)$ in which the path would pass through x. For player $k \neq i$, we construct a strategy $u_k(\cdot)$ which, in the information sets X_k^j intersecting the segment of the path ω_x, chooses alternatives (arcs) lying on this path and is arbitrary otherwise. Since each information set only intersects once the path ω, this can always be done. In the resulting situation $u(\cdot)$ the path ω necessarily passes through x; hence we have shown that $x \in Possu_i(\cdot)$.

5.9.2. Mixed strategies in the games in extensive form Γ are defined in the same way as in 1.4.2, for finite games.

Definition. The probability distribution over the set of pure strategies of player i which prescribes to every pure strategy $u_i(\cdot)$ the probability $q_{u_i}(\cdot)$ (for simplicity we write q_{u_i}) is called a mixed strategy μ_i for player i.

The situation $\mu = (\mu_1, \dots, \mu_n)$ in mixed strategies determines the probability distribution over all plays (paths) ω (hence, in final positions X_{n+1} as well) by the formula

$$P_\mu(\omega) = \sum_u q_{u_1} \dots q_{u_n} P_u(\omega),$$

where $P_u(\omega) = 1$ if the play (path) ω is realized in situation $u(\cdot)$ and $P_u(\omega) = 0$ otherwise.

Lemma. *Denote by $P_\mu(x)$ the probability that the position x is realized in situation μ. Then we have*

$$P_\mu(x) = \sum_{\{u(\cdot):x\in Possu_i(\cdot),i=1,\dots,n\}} q_{u_1}\cdots q_{u_n} = \prod_{i=1}^{n} \sum_{\{u_i:x\in Possu_i\}} q_{u_i}. \tag{5.9.1}$$

The proof of this statement immediately follows from Lemma in (5.9.1). The mathematical expectation of the payoff $E_i(\mu)$ for player i in situation μ is

$$E_i(\mu) = \sum_{x\in X_{n+1}} K_i(x)P_\mu(x), \tag{5.9.2}$$

where $P_\mu(x)$ is computed by formula (5.9.1).

Definition. Position $x \in X$ is possible for μ_i if there exists a mixed strategy situation μ containing μ_i such that $P_\mu(x) > 0$. The information set X_i^j for player i is called relevant for μ_i if some $x \in X_i^j$ is possible for μ_i.

The set of positions, possible for μ_i, is denoted by $Poss\mu_i$ and the collection information sets essential for μ_i is denoted by $Rel\mu_i$

5.9.3. Examination of multistage games with perfect information (see 5.3.3) shows that strategies can be chosen at each step in a suitable position of the game, while in the solution of specific problems it is not necessary (and it is not feasible) previously determine a strategy, i.e. a complete set of the recommended behavior in all positions (informations sets), since such a rule (see Example in 5.2.2) "suffers from strong redudancy". The question now arises of whether a similar simplification is feasible in the games with incomplete information. In other words, is it possible to form a strategy as it arise at a suitable information set rather than to construct the strategy as a certain previously fixed rule for selection in all information sets? It turns out that in the general case it is not feasible. However, there exists a class of games in extensive form where such a simplification is feasible. Let us introduce the notion of a behavior strategy.

Definition. By definition the behavior strategy β_i for player i means the rule which places each information set $X_i^j \subset A_k$ for player i in correspondence with a system of k numbers $b(X_i^j, \nu) \geq 0$, $\nu = 1, \dots, k$ such that

$$\sum_{\nu=1}^{k} b\left(X_i^j, \nu\right) = 1,$$

where $A_k = \{x : |F_x| = k\}$.

The numbers $b(X_i^j, \nu)$ can be interpreted as the probabilities of choosing alternative ν in the information set $X_i^j \subset A_k$ each position of which contains exactly k alternatives.

Any behavior strategy set $\beta = (\beta_1, \dots, \beta_n)$ for n players determines the probability distribution over the plays (paths) of the game and in final positions as follows:

$$P_\beta(\omega) = \prod_{X_i^j \cap \omega \neq \oslash, \nu \in \omega} b\left(X_i^j, \nu\right). \tag{5.9.3}$$

Here the product is taken over all X_i^j, ν such that $X_i^j \cup \omega \neq \oslash$ and the choice in the point $X_i^j \cap \omega$ of an alternative numbered as ν leads to a position belonging to the path ω.

In what follows it is convenient to interpret the notion of a "path" not only as a set of its component positions, but also as a set of suitable alternatives (arcs).

The expected payoff $E_i(\beta)$ in the behavior strategy situation $\beta = (\beta_1, \dots, \beta_n)$ is defined to be the expectation

$$E_i(\beta) = \sum_{x \in X_{n+1}} K_i(x) P_\beta(\omega_x), \; i = 1, \dots, n,$$

where ω_x is the play (path) terminating in position $x \in X_{n+1}$.

5.9.4. For every mixed strategy μ_i there can be a particular behavior strategy β_i.

Definition. The behavior strategy β_i corresponding to player i's mixed strategy $\mu_i = \{q_{u_i}\}$ is the behavior strategy defined as follows.

If $X_i^j \in Rel\mu_i$, then

$$b(X_i^j, \nu) = \frac{\sum_{\{u_i : X_i^j \in Relu_i,\ u_i(X_i^j)=\nu\}} q_{u_i}}{\sum_{\{u_i : X_i^j \in Relu_i\}} q_{u_i}}. \tag{5.9.4}$$

If $X_i^j \notin Rel\mu_i$, then on the set X_i^j the strategy β_i can be defined as distinct from (5.9.4) in an arbitrary way. (In the case $X_i^j \notin Rel\mu_i$ the denominator in (5.9.4) goes to zero.) For definiteness, let

$$b(X_i^j, \nu) = \sum_{\{u_i : u_i(X_i^j)=\nu\}} q_{u_i}. \tag{5.9.5}$$

We shall present the following result without proof.

Lemma. *Let β_i be a behavior strategy for player i and $\mu_i = \{q_{u_i}\}$ be a mixed strategy determined by*

$$q_{u_i} = \prod_{X_i^j} b\left(X_i^j, u_i\left(X_i^j\right)\right).$$

Then β_i is the behavior strategy corresponding to μ_i.

5.9.5. Definition [Kuhn (1953)]. The game Γ is called a game with perfect recall for the ith player if for any $u_i(\cdot), X_i^j, x$ from the conditions $X_i^j \in Relu_i$ and $x \in X_i^j$ it follows that $x \in Possu_i$.

From the definition it follows that in the perfect recall game for the ith player any position from the information set relevant for $u_i(\cdot)$ is also possible for $u_i(\cdot)$. The term "perfect recall" underlines the fact that, appearing in any one of his information sets the ith player can exactly reconstruct which of the alternatives (i.e. numbers) he has chosen on all his previous moves (by one-to-one correspondence) and remembers everything he has known about his opponents. The perfect recall game for all players becomes the game with perfect information if all its information sets contain one vertex each.

5.9.6. Lemma. *Let Γ be a perfect recall game for all players with ω as a play in Γ. Suppose $x \in X_i^j$ is the final position of the path ω in which player i makes his move, and suppose he chooses in x an arc ν. Let*

$$T_i(\omega) = \left\{ u_i : X_i^j \in Relu_i,\ u_i\left(X_i^j\right) = \nu \right\}.$$

If ω has no positions from X_i, then we denote by $T_i(\omega)$ the set of all pure strategies for player i. Then the play ω is realized only in those situations $u(\cdot) = (u_1(\cdot), \dots, u_n(\cdot))$ for which $u_i \in T_i(\omega)$.

Proof. Sufficiency. It suffices to show that if $u_i \in T_i(\omega)$, then the strategy u_i chooses all the arcs (alternatives) for player i appearing in the play ω (if player i has a move in ω). However, if $u_i \in T_i(\omega)$ then $X_i^j \in Relu_i$, and since the game Γ has perfect recall, $x \in Possu_i$ ($x \in \omega$). Thus, by Lemma in 5.8.1, the strategy u_i chooses all the alternatives for player i appearing in the play ω.

Necessity. Suppose the play ω is realized in situation $u(\cdot)$, where $u_i \notin T_i(\omega)$ for some i. Since $X_i^j \in Relu_i$, this means that $u_i(X_i^j) \neq \nu$. But then the path ω is not realized. This contradiction completes the proof of the lemma.

5.9.7. Lemma. *Let Γ be a perfect recall game for all players. Suppose ν is an alternative (arc) in a play ω that is incidental to $x \in X_i^j$, where $x \in \omega$, and the next position for player i (if any) on the path ω is $y \in X_i^k$. Consider the sets S and T, where*

$$S = \left\{ u_i : X_i^j \in Relu_i,\ u_i\left(X_i^j\right) = \nu \right\},$$

$$T = \left\{ u_i : X_i^k \in Relu_i \right\}.$$

Then $S = T$.

Proof. Let $u_i \in S$. Then $X_i^j \in Relu_i$, and since Γ has perfect recall $x \in Possu_i$. By Lemma 5.9.1, it follows that the strategy u_i chooses all the arcs incidental to Player i's positions on the path from x_0 to x, though $u_i(X_i^j) = \nu$. Thus, u_i chooses all the arcs incidental to Player i's positions on the path from x_0 to y, i.e. $y \in Possu_i$, $X_i^k \in Relu_i$ and $u_i \in T$.

Let $u_i \in T$. Then $X_i^k \in Relu_i$, and since Γ has perfect recall $y \in Possu_i$. But this means that $x \in Possu_i$ and $u_i(X_i^j) = \nu$, i.e. $u_i \in S$. This completes the proof of the lemma.

5.9.8. Theorem. *Let β be a situation in behavior strategies corresponding to a situation in mixed strategies μ in the game Γ (in which all positions have at least two alternatives). Then for*

$$E_i(\beta) = E_i(\mu), \ i = 1, \dots, n,$$

it is necessary and sufficient that Γ be a perfect recall game for all players.

Proof. Sufficiency. Let Γ be a perfect recall game for all players. Fix an arbitrary μ. It suffices to show that $P_\beta(\omega) = P_\mu(\omega)$ for all plays ω. If in ω there exists a position for player i belonging to the information set that is irrelevant for μ_i, then there is $X_i^j \in Rel\mu_i$, $X_i^j \cap \omega \neq \oslash$ such that the equality $b(X_i^j, \nu) = 0$ where $\nu \in \omega$ holds for the behavior strategy β_i corresponding to μ_i. Hence we have $P_\beta(\omega) = 0$. The validity of relationship $P_\mu(\omega) = 0$ in this case is obvious.

We now assume that all the information sets for the ith player through which the play ω passes, are relevant for μ_i, $i = 1, 2, \dots, n$. Suppose Player i in the play ω makes his succeeding moves in the positions belonging to the sets $X_i^1, \dots, X_i^s$ and chooses in the set X_i^j an alternative ν_j, $i = 1, \dots, s$. Then, by formula (5.9.4) and Lemma 5.9.7, we have

$$\prod_{j=1}^{s} b\left(X_i^j, \nu_j\right) = \sum_{u_i \in T_i(\omega)} q_{u_i}.$$

Indeed, since in the play ω player i makes his first move from the set X_i^1, it is relevant for all $u_i(\cdot)$, therefore the denominator in formula (5.9.4) for $b(X_i^1, \nu_1)$ is equal to 1. Further, by Lemma 5.9.7, the numerator $b(X_i^j, \nu_j)$ in formulas (5.9.4) is equal to the denominator $b(X_i^{j+1}, \nu_{j+1})$, $i = 1, \ldots, s$. By formula (5.9.3), we finally get

$$P_\beta(\omega) = \prod_{i=1}^{n} \sum_{u_i \in T_i(\omega)} q_{u_i},$$

where $T_i(\omega)$ is determined in Lemma 5.8.6.

By Lemma 5.9.6

$$P_\mu(\omega) = \sum_{u(\cdot)} q_{u_1} \ldots q_{u_n} P_u(\omega) = \sum_{u: u_i \in T_i(\omega),\ i=1,\ldots,n} q_{u_1} \ldots q_{u_n},$$

i.e. $P_\mu(\omega) = P_\beta(\omega)$. This proves the sufficiency part of the theorem.

Necessity. Suppose Γ is not a perfect recall game for all players. Then there exist Player i, a strategy u_i, an information set $X_i^j \in Relu_i$ and two positions $x, y \in X_i^j$ such that $x \in Possu_i$, $y \notin Possu_i$. Let u_i' be a strategy for player i for which $y \in Possu_i'$ and ω is the corresponding play passing through y in situation u'. Denote by μ_i a mixed strategy for player i which prescribes with probability 1/2 the choice of strategy u_i or u_i'. Then $P_{u'\|\mu_i}(y) = P_{u'\|\mu_i}(\omega) = 1/2$ (here $u'\|\mu_i$ is a situation in which the pure strategy u_i' is replaced by the mixed strategy μ_i). From the condition $y \notin Possu_i$ it follows that the path $\overline{\omega}$ realized in situation $u'\|u_i$ does not pass through y. This means that there exists X_i^k such that $X_i^k \cap \omega = X_i^k \cap \overline{\omega} \neq \oslash$ and $u_i(X_i^k) \neq u_i'(X_i^k)$. Hence, in particular, it follows that $X_i^k \in Relu_i$, $X_i^k \in Relu_i'$. Let β_i be the behavior strategy corresponding to μ_i. Then $b(X_i^k, u_i'(X_i^k)) = 1/2$. We may assume without loss of generality that $u_i(X_i^j) \neq u_i'(X_i^j)$. Then $b(X_i^j, u_i'(X_i^j)) = 1/2$. Denote by β a situation in behavior strategies corresponding to a mixed strategy situation $u'\|\mu_i$. Then $P_\beta(\omega) \leq 1/4$, whereas $P_{u'\|\mu_i}(\omega) = 1/2$. This completes the proof of the theorem.

From Theorem 5.9.8, in particular, it follows that in order to find an equilibrium in the games with perfect recall it is sufficient to restrict ourselves to the class of behavior strategies.

5.9.9. *Battle of the Sexes with Incomplete information* [Peters (2008)]. This example is a version of Battle of the Sexes game considered in Sec. 3.1.4. In this setting the second player (the woman) may go with him (man) even when their choices do not coincide. But the man (Player 1) does not exactly know whether this will happens or not. Thus he did not know what game he is playing, game

$$A = \begin{array}{c} \\ a_1 \\ a_2 \end{array} \begin{array}{c} \begin{array}{cc} b_1 & b_2 \end{array} \\ \begin{bmatrix} (4,2) & (0,0) \\ (0,0) & (2,4) \end{bmatrix} \end{array},$$

or game

$$B = \begin{array}{c} \\ a_1 \\ a_2 \end{array} \begin{array}{c} \begin{array}{cc} b_1 & b_2 \end{array} \\ \begin{bmatrix} (4,2) & (3,1) \\ (2,3) & (2,4) \end{bmatrix} \end{array}.$$

Payoff in solution (a_1, b_2) in game B can be interpreted for Player 1 in the following way: Player 1 gets 3 since his partner is coming with less pleasure, similar payoff 1 shows this loss of pleasure for woman visiting Soccer game against her general tastes. Similarly one can explain the payoffs in game B in situation (a_2, b_1).

Suppose that games A and B can occur with probabilities $\frac{1}{2}$, $\frac{1}{2}$. And this probabilities are known to both players and also player 2 knows with probability 1 which game A or B she is playing.

Then this game can be presented as game in extensive form with incomplete information on the Fig. 5.16.

The normal form of this game is

$$\begin{array}{c} \\ a_1 \\ a_2 \end{array} \begin{array}{c} \begin{array}{cccc} (b_1, b_1) & (b_1, b_2) & (b_2, b_1) & (b_2, b_2) \end{array} \\ \begin{bmatrix} (4;2) & (3,5;1,5) & (2;1) & (1,5;0,5) \\ (1;1,5) & (1;1,5) & (2,5;3,5) & (2,4) \end{bmatrix} \end{array}.$$

In this game we get also two Nash equilibrium $[a_1; (b_1, b_1)]$ and $[a_2; (b_2, b_2)]$ with payoffs (4, 2) and (2, 4).

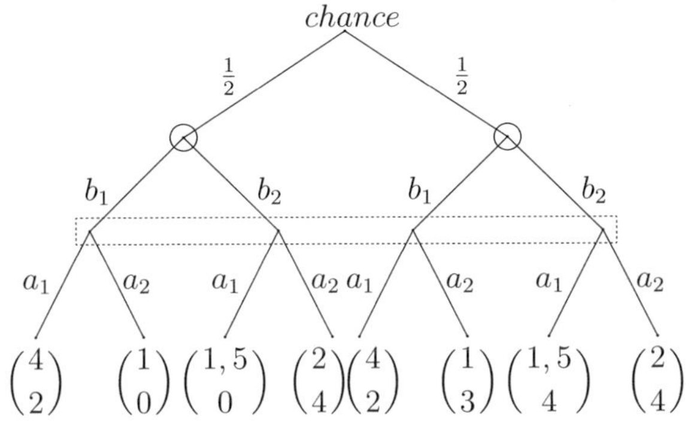

Figure 5.16

5.10 Functional Equations for Simultaneous Multistage Games

The behavior strategy theorem proved in the preceding section fails, in the general case, to provide a means of immediately solving games with perfect recall. However, when information sets have a simple structure this theorem provides a basis for derivation of functional equations for the value of the game and the methods of finding optimal strategies based on these equations. The simplest games with perfect recall, excluding games with perfect information, are the so-called *multistage zero-sum games.* We shall derive a functional equation for the value of such games and consider some of the popular examples [Diubin and Suzdal (1981), Owen (1968)] where these equations are solvable.

5.10.1. Conceptually, a repeated game (see Sec. 5.4) is a multistage zero-sum game, where at each step of the game the Players 1 and 2 choose their actions simultaneously, i.e. without being informed about the opponent's choice at this moment. After the choices have been made they become known to both players, and the players again make their choices simultaneously, and so on.

Such a game can be represented with the help of a graph which may have one of the two representations *a*) or *b*) in Fig. 5.17.

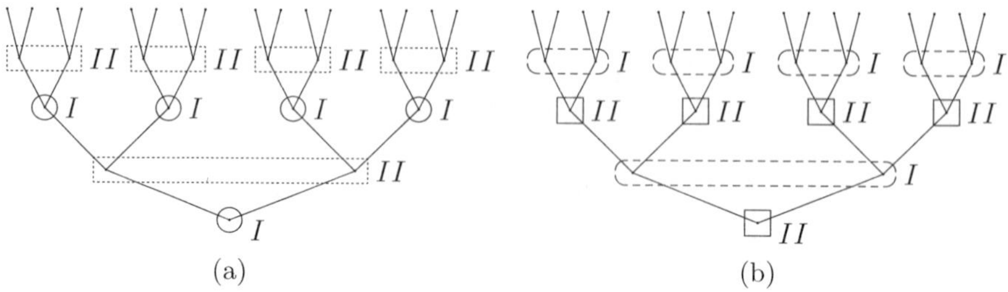

Figure 5.17

The graph represents an alternating game with an even number of moves, where the information sets for a player who makes the first move are single-element, while the information sets for the other player are two-element. In such a game Γ, the two players have perfect recall. Therefore, in this game, by Theorem 5.9.8, the search for an equilibrium may be restricted to the class of behavior strategies.

For definiteness, we assume that the first move in Γ is made by Player 1 and for every $x \in X_1$ there is a subgame Γ_x which has the same informational structure as the game Γ. The normal form of any multistage zero-sum game with incomplete information is a matrix game, i.e. a zero-sum game with a finite number of strategies; therefore in all subgames Γ_x, $x \in X_1$ (including the game $\Gamma = \Gamma_{x_0}$) there exists an equilibrium in the class of mixed strategies. By Theorem 5.9.8, such an equilibrium also exists in the class of behavior strategies and the values of the game (i.e. the values of the payoff function in a mixed strategy equilibrium or in a behavior strategy equilibrium) are equal.

Denote the value of the game Γ_x by $v(x)$, $x \in X_1$ and set up functional equations for $v(x)$.

Each information set of Player 1 consists of one position $x \in X_1$. For each $x \in X_1$ the next position x' (if any), in which Player 1 makes his move, belongs to the set F_x^2. Position x' is realized as a result of two consecutive choices: first by Player 1 of an arc incidental to vertex x, and then by Player 2 of an arc in positions $y \in F_x$ forming information sets for Player 2 (each information set of Player 2 coincides with F_x, for some $x \in X_1$). Hence we may say that the position x' results from the mapping of T_x depending on the choices of α, β by the Players 1 and 2, i.e.

$$x' = T_x(\alpha, \beta).$$

Since the number of various alternatives α and β is finite, for every $x \in X_1$, we may consider a matrix game with the payoff matrix $A_x = \{v[T_x(\alpha, \beta)]\}$. Let $\beta_I^*(x) = \{b_I^*(x, \alpha)\}$, $\beta_{II}^*(x) = \{b_{II}^*(x, \beta)\}$ be optimal mixed strategies in the game with the matrix A_x. Then we have the following theorem for the structure of optimal strategies in the game Γ_x.

Theorem. *In the game Γ an optimal behavior strategy for Player 1 at the point x (each information set of Player 1 in the game Γ consists of one position $x \in X_1$) assigns probability to each alternative α in accordance with an optimal mixed strategy of Player 1 in the matrix game $A_x = \{v(T_x(\alpha, \beta))\}$ that is*

$$b_1(x, \alpha) = b_I^*(x, \alpha).$$

An optimal behavior strategy $\{b_2(X_2^j, \beta)\}$ of Player 2 in the game Γ assigns probability to each alternative β in accordance with an optimal mixed strategy of Player 2 in the game with the matrix A_x, i.e.

$$b_2(X_2^j, \beta) = b_{II}^*(x, \beta),$$

where $x = F_y^{-1}$ if $y \in X_2^j = F_x$.

The value of the game satisfies the following functional equation:

$$v(x) = Val\{v[T_x(\alpha, \beta)]\}, \ x \in X_1, \tag{5.10.1}$$

with the initial condition

$$v(x)|_{x \in X_3} = H(x). \tag{5.10.2}$$

(Here $Val A$ is the value of the game with matrix A.)

The proof is carried out by induction and is completely analogous to the proof of Theorem 5.2.1.

5.10.2. *Example 12. Game of inspection* [Diubin and Suzdal (1981)]. Player E (Violator) wishes to take a wrongful action. There are N periods of time during which this action can be performed. Player P (Inspector) wishes to prevent this action, but can perform only one inspection during any one of these periods. The payoff to Player E is 1 if the wrongful action remains undetected after it has been performed, and is (-1) if the violator has been detained (this is possible when he chooses for his action the same period of time as the inspector for his inspection); the payoff is zero if the violator takes no action. Denote this N-step game by Γ_N.

Each player has two alternatives during the first period (at the 1st step). Player E may or may not take an action; Player P may or may not perform inspection. If Player E acts and Player P inspects, then the game terminates and the payoff is -1. If Player E acts, while Player P fails to inspect, the game terminates and the payoff is 1. If Player E does not act, while Player P inspects, then Player E may take action during the next period of time (assuming that $N > 1$) and the payoff will also be 1. If Player E does not act and Player P does not inspect, they pass to the next step which differs from the previous one only in that there are less periods left before the end of the game, i.e. they pass to a subgame Γ_{N-1}. Therefore, the game matrix for the 1st step is as follows:

$$\begin{bmatrix} -1 & 1 \\ 1 & v_{N-1} \end{bmatrix}. \tag{5.10.3}$$

Equation (5.9.1) then becomes

$$v_N = Val \begin{bmatrix} -1 & 1 \\ 1 & v_{N-1} \end{bmatrix}. \tag{5.10.4}$$

Here $v(x)$ is the same for all game positions of the same level and hence depends only on the number of periods until the end of the game. For this reason we write v_N in place of $v(x)$. In what follows it will be shown that $v_{N-1} < 1$; hence the matrix in (5.10.4) does not have a saddle point, i.e. the game with matrix (5.10.4) is completely mixed. From this (see 1.9.1) we obtain the recursive equation

$$v_N = \frac{v_{N-1} + 1}{-v_{N-1} + 3}, \tag{5.10.5}$$

which together with the initial condition

$$v_1 = Val \begin{pmatrix} -1 & 1 \\ 0 & 0 \end{pmatrix} = 0 \tag{5.10.6}$$

determines v_N. Let us transform equation (5.10.5) by substituting $t_N = \frac{1}{v_{N-1}}$. We obtain a new recursive equation $t_N = t_{N-1} - 1/2$, $t_1 = -1$. This equation has an obvious solution $t_N = -(N+1)/2$, hence we have

$$v_N = \frac{N-1}{N+1}. \tag{5.10.7}$$

We may now compute optimal behavior strategies at each step of the game. In fact, the game matrix (5.10.4) becomes

$$\begin{bmatrix} -1 & 1 \\ 1 & [N-2]/N \end{bmatrix}$$

and the optimal behavior strategies are

$$b_1^N = \left(\frac{1}{N+1}, \frac{N}{N+1}\right), \quad b_0^N = \left(\frac{1}{N+1}, \frac{N}{N+1}\right).$$

Example 13. Game-theoretic features of optimal use of resource. Suppose that initially the Players 1 and 2 have respectively r and

$R - r$ units of some resource and two pure strategies each. We also assume that if the players choose the same pure strategies, then Player 2's resource is reduced by a unit. If, however, the players choose different pure strategies, then Player 1's resource is reduced by unit. The game terminates after the resource of one of the players has become zero. In this case, the payoff to Player 1 is 1 if the resource of Player 2 is zero. The payoff to him is -1 if his resource is zero.

Denote by $\Gamma_{k,l}$ a multistage game in which Player 1 has k $(k = 1, 2, \dots, r)$ units, and Player 2 has l $(l = 1, \dots, R - r)$ units of resource. Then

$$Val\Gamma_{k,l} = Val \begin{bmatrix} Val\Gamma_{k,l-1} & Val\Gamma_{k-1,l} \\ Val\Gamma_{k-1,l} & Val\Gamma_{k,l-1} \end{bmatrix},$$

where $Val\Gamma_{k,0} = 1,\ Val\Gamma_{0,l} = -1$.

Consider the 1st step from the end, i.e. when both players are left with 1 unit of resource each. Evidently, at this step the following matrix game is played:

$$\Gamma_{1,1} = \begin{bmatrix} 1 & -1 \\ -1 & 1 \end{bmatrix}.$$

The game $\Gamma_{1,1}$ is symmetric, its value denoted by $v_{1,1}$ is zero, optimal strategies for players coincide and are equal to $(1/2, 1/2)$.

At the 2nd step from the end, i.e. when the players are left with 3 units of resource, one of the two matrix games is played: $\Gamma_{1,2}$ or $\Gamma_{2,1}$. In this case

$$v_{1,2} = Val\Gamma_{1,2} = Val \begin{bmatrix} v_{1,1} & -1 \\ -1 & v_{1,1} \end{bmatrix} = \frac{v_{1,1} - 1}{2} = -\frac{1}{2},$$

$$v_{2,1} = Val\Gamma_{2,1} = Val \begin{bmatrix} 1 & v_{1,1} \\ v_{1,1} & 1 \end{bmatrix} = \frac{v_{1,1} + 1}{2} = \frac{1}{2}.$$

At the 3rd step from the end (i.e. the players have a total of 4 units of resource) one of the following three games is played: $\Gamma_{1,3}$,

$\Gamma_{2,2}$, $\Gamma_{3,1}$. In this case

$$v_{1,3} = Val\Gamma_{1,3} = Val \begin{bmatrix} v_{1,2} & -1 \\ -1 & v_{1,2} \end{bmatrix} = \frac{v_{1,2} - 1}{2} = -\frac{3}{4},$$

$$v_{2,2} = Val\Gamma_{2,2} = Val \begin{bmatrix} v_{2,1} & v_{1,2} \\ v_{1,2} & v_{2,1} \end{bmatrix} = \frac{v_{2,1} + v_{1,2}}{2} = 0,$$

$$v_{3,1} = Val\Gamma_{3,1} = Val \begin{bmatrix} 1 & v_{2,1} \\ v_{2,1} & 1 \end{bmatrix} = \frac{v_{2,1} + 1}{2} = \frac{3}{4}.$$

Continuing analogous computations up to the Nth step from the end we obtain the following expression for the value of the original game:

$$v_{r,R-r} = Val\Gamma_{r,R-r} = Val \begin{bmatrix} v_{r,R-r-1} & v_{r-1,R-r} \\ v_{r-1,R-r} & v_{r,R-r-1} \end{bmatrix}.$$

By the symmetry of the payoff matrix of the game $\Gamma_{r,R-r}$ we have

$$v_{r,R-r} = \frac{1}{2}(v_{r,R-r-1} + v_{r-1,R-r}),$$

optimal behavior strategies for the players at each step coincide and are equal to $(1/2, 1/2)$.

Example 14 [Blackwell and Girshick (1954)] (Owen, G., 1982). This jocky game is played by two teams — Player 1 (m_1 women and m_2 cats) and Player 2 (n_1 mouses and n_2 men). Each team chooses at each step his representative. One of the two chosen representatives is "removed" by the following rule: woman "removes" man; man "removes" cat; mouse "removes" woman; cat "removes" mouse. The game is continued until only players of one type remain in one of the groups. When a group has nothing to choose the other group evidently wins.

Denote the value of the original game as $v(m_1, m_2, n_1, n_2)$. Let

$$v(m_1, m_2, n_1, 0) = v(m_1, m_2, 0, n_2) = 1, \text{ if } m_1, m_2 > 0, \quad (5.10.8)$$

$$v(m_1, 0, n_1, n_2) = v(0, m_2, n_1, n_2) = -1, \text{ if } n_1, n_2 > 0.$$

Let us introduce the following notation: $v(m_1 - 1) = v(m_1 - 1, m_2, n_1, n_2)$, $v(m_2 - 1) = v(m_1, m_2 - 1, n_1, n_2)$, $v(n_1 - 1) = v(m_1, m_2, n_1 - 1, n_2)$, $v(n_2 - 1) = v(m_1, m_2, n_1, n_2 - 1)$. By Theorem 5.10.1, the following relationship holds:

$$v(m_1, m_2, n_1, n_2) = Val \begin{bmatrix} v(m_1 - 1) & v(n_2 - 1) \\ v(n_1 - 1) & v(m_2 - 1) \end{bmatrix}.$$

It can be shown that this game is completely mixed. By Theorem 5.10.1, we have

$$v(m_1, m_2, n_1, n_2) = \frac{v(m_1 - 1)v(m_2 - 1) - v(n_1 - 1)v(n_2 - 1)}{v(m_1 - 1) + v(m_2 - 1) - v(n_1 - 1) - v(n_2 - 1)}.$$

In terms of the boundary condition (5.10.8) we obtain

$$v(m_1, 1, 1, 1) = \frac{v(m_1 - 1) + 1}{v(m_1 - 1) + 3}$$

and $v(1, 1, 1, 1) = 0$. But these equations coincide with equations (5.10.5), (5.10.6), hence $v(m, 1, 1, 1) = (m - 1)/(m + 1)$ and optimal strategies in this case also coincide with those in Example 12.

5.10.3. *Repeated Evolutionary Games.* To define evolutionary games in extensive form it is necessary to define the symmetry of the extensive form game. We shall restrict our attention to two-person games with perfect recall and without chance moves. Following Selten (1983), a symmetry of game Γ is defined as a one to one correspondence (mapping $(\cdot)^T$) from alternatives to alternatives with the following properties. Let M_i denote the set of alternatives (choices) of player i in Γ.

If $m \in M_i$, then $m^T \in M_j$ $(i \neq j \in \{1, 2\})$,

$$(m^T)^T = m \text{ for all } m.$$

For every information set u there exists an information set u^T such that every alternative at u is mapped onto a choice at u^T, for every endpoint $x \in X_{n+1}$ there exists an endpoint $x^T \in X_{n+1}$ such that if x is reached by the sequence $m_1, m_2, \ldots, m_n$, then x^T is reached by (a permutation of) $m_1^T, m_2^T, \ldots, m_k^T$, and the payoffs $H_1(x) = H_2(x^T)$ for every endpoint $x \in X_{n+1}$, $x^T \in X_{n+1}$.

A symmetric game in extensive form is a pair (Γ, T) where Γ is a game in extensive form and where T is a symmetry of Γ. If b is a behavior strategy of Player 1 in (Γ, T), then the symmetric image of b is the behavior strategy b^T of Player 2 defined by

$$b_u^T(m) = b_{u_T}(m^T) \qquad (u \in U_1, m \in M_u).$$

If b_1, b_2 are behavior strategies of Player 1, then the probability that the endpoint x is reached when (b_1, b_2^T) is played is equal to the probability that x^T is reached when (b_2, b_1^T) is played. Therefore, the expected payoff to Player 1 when (b_1, b_2^T) is played is equal to Player 2's expected payoff when (b_2, b_1^T) is played

$$E_1(b_1, b_2^T) = E_2(b_2, b_1^T).$$

This equation defines the symmetric normal form of (Γ, T) if restricted to the pure strategies. Following van Damme (1991) define the direct ESS of (Γ, T), as a behavior strategy $\bar{b}$ of Player 1 that satisfies

$$E_1(\bar{b}, \bar{b}^T) = \max_{b \in B_1} E_1(b, \bar{b}^T)$$

and if $b \in B_2 \neq \bar{b}$ and $E_1(b, \bar{b}^T) = E_2(\bar{b}, \bar{b}^T)$, then

$$E_1(b, b^T) < E_1(\bar{b}, b^T)$$

(here B_i is the set of all behavior strategies of Player i). Van Damme (1991) notes that in many games intuitively acceptable solutions will fail to satisfy the condition of direct ESS.

We give here a "refinement" of this definition which will not reject intuitively acceptable solution. First consider an example.

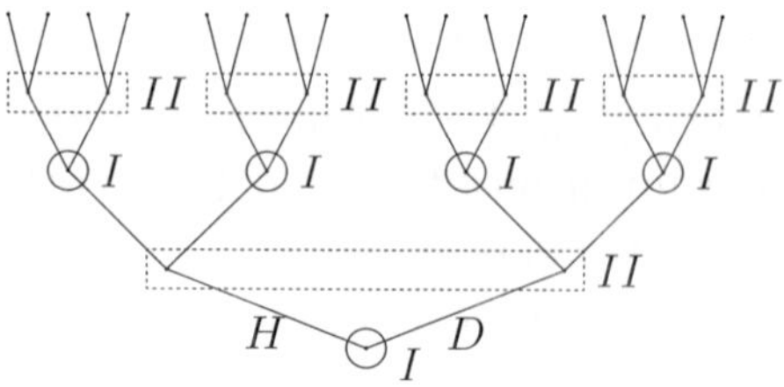

Figure 5.18

Example 15. The repeated Hawk and Dove game. The matrices of this bimatrix game have the form

$$I = \begin{matrix} & H & D \\ H & \multicolumn{2}{c}{\left[\begin{matrix} 1/2(V-C) & V \\ 0 & 1/2V \end{matrix}\right]} \\ D & & \end{matrix}, \quad II = \begin{matrix} & H & D \\ H & \multicolumn{2}{c}{\left[\begin{matrix} 1/2(V-C) & 0 \\ V & 1/2V \end{matrix}\right]} \\ D & & \end{matrix}.$$

If $v > c$, then (H, H) is ESS in this bimatrix game Γ. The game tree is represented on Fig. 5.18.

In the two-stage game the strategy of the Player 1(2) is the rule which selects H or D in each of his information sets. Player 1(2) has five information sets. Thus Player 1(2) has 32 strategies which consists from the sequences of the form

$$(H, H, D, H, D)$$

and so on. Denote the strategy of player by $u(\cdot)$.

Consider a strategy $u(\cdot) = (H, H, H, H, H)$ which is formed from the ESS strategies $(v > c)$ in each of the subgames (one stage games). It would be preferable if this strategy will be ESS in the two-stage game Γ. Unfortunately by the definition of the direct ESS for games in extensive form it is not.

It is easily seen that the first condition of direct ESS is satisfied since $u(\cdot)$ is a Nash equilibrium in Γ, but we can find a strategy

$$v(\cdot) = (H, H, D, D, D)$$

for which the payoff (in pure strategies the expected payoff E coincides with the payoff function K) in the situation $(v(\cdot), u(\cdot))$ – $K(v(\cdot), u(\cdot))$ is equal to the payoff in the situation $(u(\cdot), u(\cdot))$ – $K(u(\cdot), u(\cdot))$.

$$K(v(\cdot), u(\cdot)) = K(u(\cdot), u(\cdot)) = v - c,$$

but $K(v(\cdot), v(\cdot))$ is also equal to $K(u(\cdot), v(\cdot))$

$$K(v(\cdot), v(\cdot)) = K(u(\cdot), v(\cdot)) = v - c$$

and the second condition of direct ESS is not satisfied.

This unnatural arrangement happens because of bad definition of *direct ESS* for positional games.

We purpose a new definition.

Definition. The pair $(u(\cdot), u(\cdot))$ is a direct ESS if

1. $K(u(\cdot), u(\cdot)) \geq K(v(\cdot), u(\cdot))$ for all $v(\cdot)$.
2. If $v(\cdot)$ is such that in the situation $(u(\cdot), u(\cdot)), (v(\cdot), u(\cdot))$ the realized paths in Γ are different (the terminal positions in the game Γ are different), then if

$$K(u(\cdot), u(\cdot)) = K(v(\cdot), u(\cdot)),$$

then

$$K(v(\cdot), v(\cdot)) < K(u(\cdot), v(\cdot)).$$

By this definition the strategy $u(\cdot) = (H, H, H, H, H)$ is direct ESS, since the strategy $v(\cdot) = (H, H, D, D, D)$ giving the same payoff against $u(\cdot)$ as $u(\cdot)$ itself is excluded by the point 2 of this definition from the consideration.

The slight modification of this definition for the games with random moves (as example in van Damme (1991)) shows that the situation natural to be ESS and excluded there from the direct ESS is direct ESS in our sense.

5.11 Cooperative Multistage Games with Perfect Information

5.11.1. In what follows as basic model we shall consider the game in extensive form with perfect information.

Suppose that finite oriented treelike graph G with the root x_0 is given.

For simplicity, we shall use the following notations. Let x be some vertex (position). We denote by G(x) a subtree of G with root in x. We denote by $Z(x)$ immediate successors of x. As before the vertices y, directly following after x, are called alternatives in x ($y \in Z(x)$). The player who makes a decision in x (who selects the next alterative position in x), will be denoted by $i(x)$. The choice of player $i(x)$ in position x will be denoted by $\bar{x} \in Z(x)$.

Following 5.1.4 we may give the definition.

Definition. A game in extensive form with perfect information (see []) $\Gamma(x_0)$ is a graph tree $G(x_0)$, with the following additional properties:

1. The set of vertices (positions) is split up into $n+1$ subsets $X_1, X_2, \ldots, X_{n+1}$, which form a partition of the set of all vertices of the graph tree $G(x_0)$. The vertices (positions) $x \in X_i$ are called players i personal positions, $i = 1, \cdots, n$; vertices (positions) $x \in X_{n+1}$ are called terminal positions.
2. For each vertex $x \notin X_{n+1}$ and $y \in Z(x)$ define an arc (x, y) on the graph $G(x_0)$. On each arc (x, y) n real numbers (payoffs of players on this arc) $h_i(x, y)$, $i = 1, \ldots, n$, $h_i \geq 0$ are defined, and also terminal payoffs $g_i(x) \geq 0$, for $x \in X_{n+1}$, $i = 1, \ldots, n$.

Definition. A strategy of player i is a mapping $U_i(\cdot)$, which associate to each position $x \in X_i$ a unique alternative $y \in Z(x)$. Denote by $H_i(x; u_1(\cdot), \ldots, u_n(\cdot))$ the payoff function of player $i \in N$ in the subgame $\Gamma(x)$ (see 5.1.5) starting from the position x.

$$H_i(x;u_1(\cdot),\ldots,u_n(\cdot)) = \sum_{k=0}^{l-1} h_i(x_k,x_{k+1}) + g_i(x_l), \quad h_i \geq 0, \quad g_i \geq 0, \tag{5.11.1}$$

where $x_l \in X_{n+1}$ is the last vertex (position) in the path $\tilde{x} = (x_0, x_1, \ldots, x_l)$ realized in subgame $\Gamma(x)$, and $x_0 = x$, when n-tuple of strategies $(u_1(\cdot),\ldots,u_n(\cdot))$ is played.

Denote by $\bar{u}(\cdot) = (\bar{u}_1(\cdot),\ldots,\bar{u}_n(\cdot))$ the n-tuple of strategies and the trajectory (path) $\bar{x} = (\bar{x}_0, \bar{x}_1, \ldots, \bar{x}_l)$, $\bar{x}_l \in P_{n+1}$ such that

$$\begin{aligned}
&\max_{u_1(\cdot),\ldots,u_n(\cdot)} \sum_{i=1}^{n} H_i(x_0;u_1(\cdot),\ldots,u_n(\cdot)) \\
&= \sum_{i=1}^{n} H_i(x_0;\bar{u}_1(\cdot),\ldots,\bar{u}_n(\cdot)) \\
&= \sum_{i=1}^{n} \left(\sum_{k=0}^{l-1} h_i(\bar{x}_k,\bar{x}_{k+1}) + g_i(\bar{x}_l) \right).
\end{aligned} \tag{5.11.2}$$

The path $\bar{x} = (\bar{x}_0, \bar{x}_1, \ldots, \bar{x}_l)$ satisfying (5.11.2) we shall call "optimal cooperative trajectory".

Define in $\Gamma(x_0)$ characteristic function in a classical way

$$\begin{aligned}
V(x_0;N) &= \sum_{i=1}^{n} \left(\sum_{k=0}^{l-1} h_i(\bar{x}_k,\bar{x}_{k+1}) + g_i(\bar{x}_l) \right), \\
V(x_0;\emptyset) &= 0, \\
V(x_0;S) &= Val\Gamma_{S,N\setminus S}(x_0),
\end{aligned}$$

where $Val\Gamma_{S,N\setminus S}(x_0)$ is a value of zero-sum game played between coalition S acting as first player and coalition $N\setminus S$ acting as Player 2, with payoff of player S equal to

$$\sum_{i\in S} H_i(x_0;u_1(\cdot),\ldots,u_n(\cdot)).$$

If the characteristic function is defined then we can define the set of imputations in the game $\Gamma(x_0)$

$$C(x_0) = \left\{\xi = (\xi_1, \dots, \xi_n): \quad \xi_i \geq V(x_0; \{i\}), \quad \sum_{i \in N} \xi_i = V(x_0; N)\right\},$$

the core $M(x_0) \subset C(x_0)$

$$M(x_0) = \left\{\xi = (\xi_1, \dots, \xi_n): \quad \sum_{i \in S} \xi_i \geq V(x_0; S), \quad S \subset N\right\} \subset C(x_0),$$

NM solution, Shapley value and other optimality principles of classical game theory. In what follows we shall denote by $M(x_0) \subset C(x_0)$ anyone of this optimality principles.

Suppose at the beginning of the game players agree to use the optimality principle $M(x_0) \subset C(x_0)$ as the basis for the selection of the "optimal" imputation $\bar{\xi} \in M(x_0)$.

This means that playing cooperatively by choosing the strategy maximizing the common payoff each one of them is waiting to get the payoff $\bar{\xi}_i$ from the optimal imputation $\bar{\xi} \in M(x_0)$ after the end of the game (after the maximal common payoff $V(x_0; N)$ is really earned by the players).

But when the game Γ actually develops along the "optimal" trajectory $\bar{x} = (\bar{x}_0, \bar{x}_1, \dots, \bar{x}_l)$ at each vertex $\bar{x}_k$ the players find themselves in the new multistage game with perfect information $\Gamma_{\bar{x}_k}$, $k = 0, \dots, l$, which is the subgame of the original game Γ starting from $\bar{x}_k$ with payoffs

$$H_i(\bar{x}_k; u_1(\cdot), \dots, u_n(\cdot)) = \sum_{j=k}^{l-1} h_i(x_j, x_{j+1}) + g_i(x_l), \quad i = 1, \dots, n.$$

It is important to mention that for the problem (5.11.2) the Bellman optimality principle holds and the part $\bar{x}^k = (\bar{x}_k, \dots, \bar{x}_j, \dots, \bar{x}_l)$ of the trajectory $\bar{x}$, starting from $\bar{x}_k$ maximizes the sum of the payoffs in the subgame $\Gamma_{\bar{x}_k}$, i.e.

$$\max_{x_k,\dots,x_j,\dots,x_l} \sum_{i=1}^{n} \left[\sum_{j=k}^{l-1} h_i(x_j, x_{j+1}) + g_i(x_l) \right]$$

$$= \sum_{i=1}^{n} \left[\sum_{j=k}^{l-1} h_i(\bar{x}_j, \bar{x}_{j+1}) + g_i(\bar{x}_l) \right],$$

which means that the trajectory $\bar{x}^k = (\bar{x}_k, \dots, \bar{x}_j, \dots, \bar{x}_l)$ is also "optimal" in the subgame $\Gamma_{\bar{x}_k}$.

Before entering the subgame $\Gamma_{\bar{x}_k}$ each of the players i have already earned the amount

$$H_i^{\bar{x}_k} = \sum_{j=0}^{k-1} h_i(\bar{x}_j, \bar{x}_{j+1}).$$

At the same time at the beginning of the game $\Gamma = \Gamma(x_0)$ the player i was oriented to get the payoff $\bar{\xi}_i$ — the ith component of the "optimal" imputation $\bar{\xi} \in M(x_0) \subset C(x_0)$. From this it follows that in the subgame $\Gamma_{\bar{x}_k}$ he is expected to get the payoff equal to

$$\bar{\xi}_i - H_i^{\bar{x}_k} = \bar{\xi}_i^{\bar{x}_k}, \quad i = 1, \dots, n$$

and then the question arises whether the new vector $\bar{\xi}^{\bar{x}_k} = (\bar{\xi}_1^{\bar{x}_k}, \dots, \bar{\xi}_i^{\bar{x}_k}, \dots, \bar{\xi}_n^{\bar{x}_k})$ remains to be optimal in the same sense in the subgame $\Gamma_{\bar{x}_k}$ as the vector $\bar{\xi}$ was in the game $\Gamma(\bar{x}_0)$. If this will not be the case, it will mean that the players in the subgame $\Gamma_{\bar{x}_k}$ will not orient themselves on the same optimality principle as in the game $\Gamma(\bar{x}_0)$ which may enforce them to go out from the cooperation by changing the chosen cooperative strategies $\bar{u}_i(\cdot)$, $i = 1, \dots, n$ and thus changing the optimal trajectory $\bar{x}$ in the subgame $\Gamma(\bar{x}_k)$. Try now formalize this reasoning.

Introduce in the subgame $\Gamma(\bar{x}_k)$, $k = 1, \dots, l$, the characteristic function $V(\bar{x}_k; S)$, $S \subset N$ in the same manner as it was done in the game $\Gamma = \Gamma(x_0)$. Based on the characteristic function $V(\bar{x}_k; S)$ we can introduce the set of imputations

$$C(\bar{x}_k) = \left\{ \xi = (\xi_1, \dots, \xi_n) : \ \xi_i \geq V(\bar{x}_k; \{i\}), \ \sum_{i \in N} \xi_i = V(\bar{x}_k; N) \right\},$$

the core $M(\bar{x}_k) \subset C(\bar{x}_k)$

$$M(\bar{x}_k) = \left\{ \xi = (\xi_1, \dots, \xi_n) : \ \sum_{i \in S} \xi_i \geq V(\bar{x}_k; S), \ \ S \subset N \right\} \subset C(\bar{x}_k),$$

NM solution, Shapley value and other optimality principles of classical game theory. Denote by $M(\bar{x}_k) \subset C(\bar{x}_k)$ the optimality principle $M \subset C$ (which was selected by players in the game $\Gamma(x_0)$) considered in the subgame $\Gamma(\bar{x}_k)$.

If we suppose that the players in the game $\Gamma(x_0)$ when moving along the optimal trajectory $(\bar{x}_0, \bar{x}_1, \dots, \bar{x}_l)$ follow the same ideology of optimal behavior then the vector $\bar{\xi}^{\bar{x}_k} = \bar{\xi} - H^{\bar{x}_k}$ must belong to the set $M(\bar{x}_k)$ — the corresponding optimality principle in the cooperative game $\Gamma(\bar{x}_k)$, $k = 0, \dots, l$.

It is clearly seen that it is very difficult to find games and corresponding optimality principles for which this condition is satisfied. Try to illustrate this on the following example.

Suppose that in the game $\Gamma h_i(x_k, x_{k+1}) = 0, k = 0, \dots, l-1$, $g_i(x_l) \neq 0$ (the game Γ is the game with terminal payoff). Then the last condition would mean that

$$\bar{\xi} = \bar{\xi}^{\bar{x}_k} \in M(\bar{x}_k), \quad k = 0, \dots, l,$$

which gives us

$$\bar{\xi} \in \bigcap_{k=0}^{l} M(\bar{x}_k). \tag{5.11.3}$$

For $k = l$ we shall have that

$$\bar{\xi} \in M(\bar{x}_l).$$

But $M(\bar{x}_l) = C(\bar{x}_l) = \{g_i(\bar{x}_l)\}$. And this condition have to be valid for all imputations of the set $M(\bar{x}_0)$ and for all optimality principles $M(x_0) \subset C(x_0)$, which means that in the cooperative game with terminal payoffs the only reasonable optimality principle will be

$$\bar{\xi} = \{g_i(\bar{x}_l)\},$$

the payoff vector obtained at the end point of the cooperative trajectory in the game $\Gamma(x_0)$. At the same time the simplest examples show that the intersection (5.11.3) except the "dummy" cases, is void for the games with terminal payoffs.

How to overcome this difficulty. The plausible way of finding the outcome is to introduce a special rule of payments (stage salary) on each stage of the game in such a way that the payments on each stage will not exceed the common amount earned by the players on this stage and the payments received by the players starting from the stage k(in the subgame $\Gamma(\bar{x}_k)$) will belong to the same optimality principle as the imputation $\bar{\xi}$ on which players agree in the game $\Gamma(x_0)$ at the beginning of the game. Whether it is possible or not we shall consider now.

Introduce the notion of the imputation distribution procedure (IDP).

Definition. Suppose that $\xi = \{\xi_1, \dots, \xi_i, \dots, \xi_n\} \in M(x_0)$. Any matrix $\beta = \{\beta_{ik}\}$, $i = 1, \dots, n$, $k = 0, \dots, n$ such that

$$\xi_i = \sum_{k=0}^{l} \beta_{ik}, \tag{5.11.4}$$

is called IDP.

Denote $\beta_k = (\beta_{1k}, \dots, \beta_{nk})$, $\beta(k) = \sum_{m=0}^{k-1} \beta_m$. The interpretation of IDP β is: β_{ik} is the payment to player i on the stage k of the game Γ_{x_0}, i.e. on the first stage of the subgame $\Gamma(\bar{x}_k)$. From the definition (5.11.4) it follows that in the game $\Gamma(x_0)$ each player i gets the amount ξ_i, $i = 1, \dots, n$, which he expects to get as the ith component of the optimal imputation $\xi_i \in M(x_0)$ in the game $\Gamma(x_0)$.

The interpretation of $\beta_i(k)$ is: $\beta_i(k)$ is the amount received by player i on the first k stages of the game Γ_{x_0}.

Definition. The optimality principle $M(x_0)$ is called time-consistent if for every $\xi \in M(x_0)$ there exists IDP β such that

$$\xi^k = \xi - \beta(k) \in M(\bar{x}_k), \quad k = 0, 1, \dots, l. \tag{5.11.5}$$

Definition. The optimality principle $M(x_0)$ is called strongly time-consistent if for every $\xi \in M(x_0)$ there exists IDP β such that

$$\beta(k) \oplus M(\bar{x}_k) \subset M(x_0), \quad k = 0, 1, \dots, l.$$

Here $a \oplus A = \{a + a' : \ a' \in A, a \in R^n, A \subset R^n\}$.

The time-consistency of the optimality principle $M(x_0)$ implies that for each imputation $\xi \in M$ there exits such IDP β that if the payments on each arc $(\bar{x}_k, \bar{x}_{k+1})$ on the optimal trajectory $\bar{x}$ will be made to the players according to IDP β, in every subgame $\Gamma(\bar{x}_k)$ the players may expect to receive the payments $\bar{\xi}^k$ which are optimal in the subgame $\Gamma(\bar{x}_k)$ in the same sense as it was in the game $\Gamma(x_0)$.

The strongly time-consistency means that if the payments are made according to IDP β then after earning on the stage k amount $\beta(k)$ the players (if they are oriented in the subgame $\Gamma(\bar{x}_k)$ on the same optimality principle as in $\Gamma(x_0)$) start with reconsidering of the imputation in this subgame (but optimal) they will get as a result in the game $\Gamma(x_0)$ the payments according to some imputation, optimal in the previous sense, i.e. the imputation belonging to the set $M(x_0)$.

For any optimality principle $M(x_0) \subset C(x_0)$ and for every $\bar{\xi} \in M(x_0)$ we can define β_{ik} by the following formulas

$$\begin{aligned} \bar{\xi}_i^{\bar{x}_k} - \bar{\xi}_i^{\bar{x}_{k+1}} &= \beta_{ik}, \quad i = 1, \dots, n, \quad k = 0, \dots, l-1, \\ \bar{\xi}_i^{\bar{x}_l} &= \beta_{il}. \end{aligned} \tag{5.11.6}$$

From the definition it follows that

$$\sum_{k=0}^{l}\beta_{ik}=\sum_{k=0}^{l-1}(\bar{\xi}_i^{\bar{x}_k}-\bar{\xi}_i^{\bar{x}_{k+1}})+\bar{\xi}_i^{\bar{x}_l}=\bar{\xi}_i^{\bar{x}_0}=\bar{\xi}_i.$$

And at the same time

$$\bar{\xi}-\beta(k)=\bar{\xi}^{\bar{x}_k}\in M(\bar{x}_k),k=0,\ldots,l.$$

The last inclusion would mean the time consistency $M(x_0)$.

Unfortunately the elements β_{ik} may take in many cases negative values, which may stimulate questions about the use of this payment mechanism in real life situations. Because this means that players in some cases have to pay to support time-consistency. We understand that this argument can be waved since the total amount the player gets in the game is equal to the component ξ_i of the optimal imputation, and he can borrow the money to cover the side payment β_{ik} on stage k.

But we have another approach which enables us to use only nonnegative IDP's, and get us result not only time-consistent, but strongly time-consistent solution. For this reason some integral transformation of characteristic function is needed.

Example 16. Time inconsistency of the Shapley value. Consider the cooperative version of the game on the Fig. 5.19, 5.2.5 in the case when there are only three players. The following coalitions are possible $\{1,2,3\}$, $\{1,2\}$, $\{1,3\}$, $\{2,3\}$, $\{1\}$, $\{2\}$, $\{3\}$. The characteristic

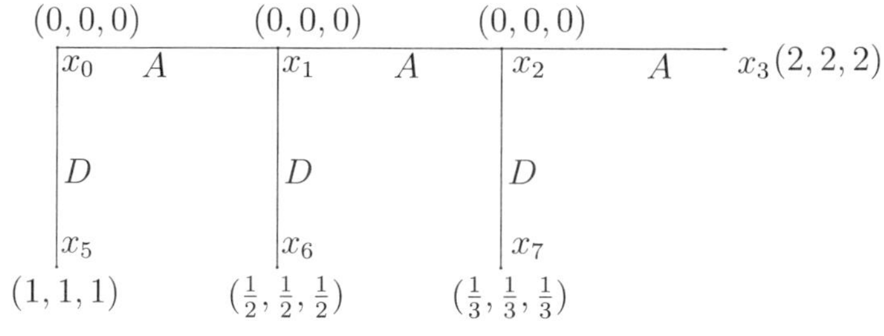

Figure 5.19

function has the form $v(\{1,2,3\}) = 6$, $v(\{1,2\}) = 2$, $v(\{1,3\}) = 2$, $v(\{2,3\}) = 2$, $v(\{1\}) = 1$, $v(\{2\}) = \frac{1}{2}$, $v(\{3\}) = \frac{1}{2}$. Computing the Shapley value we get

$$Sh(x_0) : Sh_1 = \frac{26}{12},\ Sh_2 = \frac{23}{12},\ Sh_3 = \frac{23}{12}.$$

Suppose the game develops along the optimal cooperative trajectory, which corresponds to the choices (A, A, A), and coincides with the path $\bar{x} = (x_0, x_1, x_2, x_3)$. As we have seen $v(x_0; \{1,2,3\}) = 6$, $v(x_0; \{1,2\}) = v(x_0; \{1,3\}) = v(x_0; \{2,3\}) = 2$, $v(x_0; \{1\}) = 1$, $v(x_0; \{2\}) = v(x_0; \{3\}) = \frac{1}{2}$. Consider now the subgame starting on cooperative trajectory from vertex $\bar{x}_1$. It can be easily seen that $v(\bar{x}_1; \{1,2,3\}) = 6$, $v(\bar{x}_1; \{1,2\}) = 1$, $v(\bar{x}_1; \{1,3\}) = 1$, $v(\bar{x}_1; \{2,3\}) = 4$, $v(\bar{x}_1; \{1\}) = \frac{1}{3}$, $v(\bar{x}_1; \{2\}) = \frac{1}{2}$, $v(\bar{x}_1; \{3\}) = \frac{1}{2}$. And the Shapley value in the subgame $\Gamma(\bar{x}_1)$ is equal to

$$Sh(\bar{x}_1) = \left(\frac{34}{36}, \frac{91}{36}, \frac{91}{36}\right),$$

and we see that $Sh(x_0) \neq Sh(\bar{x}_1)$.

Consider now the subgame starting on cooperative trajectory from vertex $\bar{x}_2$. It can be easily seen that $v(\bar{x}_2; \{1,2,3\}) = 6$, $v(\bar{x}_2; \{1,2\}) = \frac{2}{3}$, $v(\bar{x}_2; \{1,3\}) = 4$, $v(\bar{x}_2; \{2,3\}) = 4$, $v(\bar{x}_2; \{1\}) = \frac{1}{3}$, $v(\bar{x}_2; \{2\}) = \frac{1}{3}$, $v(\bar{x}_2; \{3\}) = 2$. And the Shapley value in the subgame $\Gamma(\bar{x}_2)$ is equal to

$$Sh(\bar{x}_2) = \left(\frac{21}{18}, \frac{21}{18}, \frac{66}{18}\right),$$

and we see that $Sh(\bar{x}_0) \neq Sh(\bar{x}_1) \neq Sh(\bar{x}_2)$. It is obvious that $Sh(\bar{x}_3) = (2, 2, 2)$.

IPD for Shapley value in this game can be easily calculated

$$Sh(\bar{x}_0) = \left(\frac{44}{36}, -\frac{22}{36}, -\frac{22}{36}\right) + Sh(\bar{x}_1),$$

$$Sh(\bar{x}_1) = \left(-\frac{8}{36}, \frac{49}{36}, -\frac{41}{36}\right) + Sh(\bar{x}_2),$$

$$Sh(\bar{x}_2) = \left(-\frac{15}{18},\ -\frac{15}{18},\ \frac{30}{18}\right) + Sh(\bar{x}_3),$$

$$Sh(\bar{x}_3) = (2,\ 2,\ 2).$$

The strongly time consistency condition is more obligatory. We cannot even derive the formula like (5.11.6).

5.11.2. Now introduce the following functions

$$\beta_i^0 = \frac{\xi_i^0 \sum_{i=1}^n h_i(\overline{x}_0, \overline{x}_1)}{V(N; x_0)},$$

where $\xi^0 \in C(x_0)$

$$\beta_i^1 = \frac{\xi_i^1 \sum_{i=1}^n h_i(\overline{x}_1, \overline{x}_2)}{V(N, \overline{x}_1)}, \quad \xi^1 \in C(\overline{x}_1);$$

$$\dots$$

$$\beta_i^k = \frac{\xi_i^k \sum_{i=1}^n h_i(\overline{x}_k, \overline{x}_{k+1})}{V(N, \overline{x}_k)}, \quad \xi^k \in C(\overline{x}_k);$$

$$\dots$$

$$\beta_i^{l-1} = \frac{\xi_i^l \sum_{i=1}^n h_i(\overline{x}_{l-1}, \overline{x}_l)}{V(N, \overline{x}_l)}, \quad \xi^l \in C(\overline{x}_l). \tag{5.11.7}$$

$$\beta_i^l = q_i(\overline{x}_l).$$

Define the IDP $\overline{\beta}^k = \{\beta_i^k,\ i = 1, \dots, n\}$, $k = 0, \dots, l$. It is easily seen that $\overline{\beta}^k \geq 0$. Consider the formula (5.11.7). For different imputations $\xi^k \in C(\overline{z}_{k-1})$ we get different values of β_i^k and, hence, different values of $\overline{\beta}$. Let B^k be the set of all possible $\overline{\beta}^k$ for all $\xi^k \in C(\overline{z}_k)$, $k = 1, \dots, l$.

Consider the set:

$$\tilde{C}(z_0) = \left\{ \overline{\xi} : \overline{\xi} = \sum_{k=0}^{l} \overline{\beta}^k,\ \overline{\beta}^k \in B^k \right\}$$

and the sets $\tilde{C}(\overline{z}_k) = \{\overline{\xi}^k : \overline{\xi}^k = \sum_{m=k}^{l} \overline{\beta}^m, \overline{\beta}^m \in B^m\}$.

The set $\tilde{C}(z_0)$ is called the regularized OP $\overline{C}(z_0)$ and, correspondingly, $\tilde{C}(\overline{z}_k)$ is a regularized OP $\overline{C}(\overline{z}_k)$.

We consider $\tilde{C}(z_0)$ as a new optimality principle in the game $\Gamma(z_0)$.

Theorem. *If the IDP* β *is defined as a* $\overline{\beta}$, $k = 1, \dots, l$, *then always*:

$$\beta(k) \oplus \tilde{C}(\overline{z}_k) \subset \tilde{C}(\overline{z}_0),$$

i.e. the OP $\tilde{C}(z_0)$ *is strongly time consistent.*

Proof. Suppose

$$\overline{\xi} \in \beta(k) \oplus \tilde{C}(\overline{z}_k),$$

then $\overline{\xi} = \beta(k) + \sum_{m=k}^{l} \beta^m$, for some $\beta^m \in B^m$, $m = k, \dots, l$.

But $\beta(k) = \sum_{m=0}^{k-1} \beta'^m$ for some $\beta'^m \in B^m$, $m = 0, \dots, k-1$.

Consider

$$(\beta'')^m = \begin{cases} \beta'^m, & m = 0, \dots, k-1, \\ \beta^m, & m = k, \dots, l, \end{cases}$$

then $(\beta'')^m \in B^m$ and $\overline{\xi} = \sum_{m=0}^{l} (\beta'')^m$ and thus $\overline{\xi} \in \tilde{C}(z_0)$. The theorem is proved.

The defined IDP has the advantage (compared with β defined by (5.11.6)):

$$\beta_i^k \geq 0, \quad \sum_{i=1}^{n} \beta_i^k = 1, \quad k = 0, \dots, l,$$

and thus

$$\sum_{i=1}^{n} \beta_i(\Theta) = \sum_{i=1}^{n} \sum_{k=0}^{\Theta-1} h_i(\overline{z}_{k+1}), \tag{5.11.8}$$

which is the actual amount to be divided between the players on the first $\Theta + 1$ stages and which is as it is seen by the formula (5.11.4) exactly equal to the amount earned by them on this stages.

5.11.3. Regularized game Γ_α. For every $\alpha \in M(x_0)$ define the noncooperative game $\Gamma_\alpha(x_0)$, which differs from the game $\Gamma(x_0)$ only by payoffs defined along optimal cooperative path $\bar{x} = (\bar{x}_0, \dots, \bar{x}_l)$.

Let $\alpha \in M(x_0)$. Define IDP as function $\beta_k = (\beta_1(k), \ldots, \beta_n(k))$, $k = 0, 1, \ldots, m$ such that

$$\alpha_i = \sum_{k=0}^{l} \beta_i(k). \tag{5.11.9}$$

Consider current subgame $\Gamma(\bar{x}_k)$ along the optimal path $\bar{x}$ and current imputation sets $C(\bar{x}_k)$. Let $\alpha^k \in C(\bar{x}_k)$.

Define by $H_i^\alpha(x_0; u_1(\cdot), \ldots, u_n(\cdot))$ the payoff function in the game $\Gamma_\alpha(x_0)$ and by $\bar{x} = \{\bar{x}_0, \ldots, \bar{x}_l\}$ the cooperative path.

Suppose $x = (x_1, x_2, \ldots, x_l)$ is the path resulting from the initial state x_0, when the situation $(u_1(\cdot), \ldots, u_n(\cdot))$ is used, and suppose that m is the maximal index for which $x_k = \bar{x}_k$ (the maximal number of stages in which the path coincides with cooperative path $\bar{x}$). Then

$$H_i^\alpha(x_0; u_1(\cdot), \ldots, u_n(\cdot)) = \sum_{k=0}^{m-1} \beta_{ik} + \sum_{k=m}^{l-1} h_i(x_k, x_k + 1) + g_i(x_l)$$

and

$$H_i^\alpha(x_0; \bar{u}_1(\cdot), \ldots, \bar{u}_n(\cdot)) = \alpha_i.$$

By the definition of the payoff function in the game $\Gamma_\alpha(x_0)$ we get that the payoffs along the optimal cooperative trajectory are equal to the components of the imputation $\alpha = (\alpha_1, \ldots, \alpha_n)$.

Definition. The game $\Gamma_\alpha(x_0)$ is called regularization of the game $\Gamma(x_0)$ (α-regularization) if the IDP β is defined in such a way that

$$\alpha_i^k = \sum_{j=k}^{l} \beta_i(j)$$

or $\beta_i(j) = \alpha_i^j - \alpha_i^{j+1}$, $i \in N$, $k = 0, 1, \ldots, l-1$, $\beta_i(l) = \alpha_i^l$, $\alpha_i^0 = \alpha_i$.

Theorem. *In the regularization of the game $\Gamma_\alpha(x_0)$ there exist a Nash equilibrium with payoffs $\alpha = (\alpha_1, \ldots, \alpha_n)$.*

Proof. The proof is similar to the proof of corresponding theorem from (5.4.4), by using the theorem from (5.4.3).

Construct the analogue of penalty strategies defined in (5.4.1), (5.4.2). Let $y \in X_i$, $y \notin \bar{x}$, and let $\bar{x}_r \in \bar{x}$ is such vertex on cooperative path $\bar{x}$, that $(\Gamma_y^{-1})^m = \bar{x}_r$ and m is minimal. This means that y belongs to the trajectory $x' = (\bar{x}_0, \ldots, \bar{x}_r, x_{r+1}, \ldots, x_{l'})$ which has exactly r vertexes on cooperative trajectory $(y = x_{r+m})$. And suppose that $y \in X_j$. Consider now the subgame Γ_y, and on basis of this subgame construct a zero-sum game $\Gamma_y(k; N \setminus \{k\})$ with player $k \in N$ as first player and the coalition $N \setminus \{k\}$ as second player since the game is zero-sum, the aim of coalition $N \setminus \{k\}$ is to minimize the payoff of player k in the game $\Gamma_y(k; N \setminus \{k\})$.

Denote the optimal pure strategy of player $N \setminus \{k\}$ (the game is with perfect information and there exists a saddle point in pure strategies) by $\{\tilde{u}_l^{N\setminus\{k\}}(\cdot), l \in N \setminus \{k\}\}$.

Definition. The strategy $\tilde{u}_i(\cdot)$ is called penalty strategy of player i if

$$
\begin{aligned}
\tilde{u}_i(\bar{x}_s) = \bar{x}_{s+1} \quad & \text{for } \bar{x}_x \in \bar{x}, \\
\tilde{u}_i(y) = \tilde{u}_i^{N\setminus\{k\}}(y) \quad & \text{for } i \neq k,\ y \notin \bar{x}, \\
\tilde{u}_k(y) \ \text{arbitrary} \quad & \text{for } i = k,\ y \notin \bar{x}.
\end{aligned}
\tag{5.11.10}
$$

We have to prove that

$$
\begin{aligned}
H_i^\alpha(x_0, \tilde{u}_i(\cdot)) &= H_i^\alpha(x_0; \tilde{u}_1(\cdot), \ldots, \tilde{u}_{i-1}(\cdot), \tilde{u}_i(\cdot), \tilde{u}_{i+1}(\cdot), \ldots, \tilde{u}_n(\cdot)) \\
&\geq H_i^\alpha(x_0; \tilde{u}_1(\cdot), \ldots, \tilde{u}_{i-1}(\cdot), u_i(\cdot), \tilde{u}_{i+1}(\cdot), \ldots, \tilde{u}_n(\cdot)) \\
&= H_i^\alpha(x_0, \tilde{u}_i(\cdot)||u_i(\cdot)),
\end{aligned}
\tag{5.11.11}
$$

for all $u_i(\cdot)$, $i \in N$.

Here two cases are possible.

Case 1. $u_i(\bar{x}_k) = \tilde{u}_i(\bar{x}_k)$, for all $x_k \in \bar{x}$, then in this case $H_i^\alpha(x_0, \tilde{u}_i(\cdot))= H_i^\alpha(x_0, \tilde{u}_i(\cdot)||u_i(\cdot))$, $i = 1, \ldots, n$, and (5.11.11) holds.

Case 2. There exists $\bar{x}_k \in \bar{x} \cap X_i$, such that $u_i(\bar{x}_k) \neq \tilde{u}_i(\bar{x}_k) = \bar{x}_{k+1}$, suppose that k is the minimal integer for which $u_i(\bar{x}_k) \neq \tilde{u}_i(\bar{x}_k)$, then we shall have

$$H_i^{\alpha}(x_0; \tilde{u}_1(\cdot), \ldots, \tilde{u}_{i-1}(\cdot), u_i(\cdot), \tilde{u}_{i+1}(\cdot), \ldots, \tilde{u}_n(\cdot)) \tag{5.11.12}$$
$$= H_i^{\alpha}(x_0, \tilde{u}_i(\cdot)||u_i(\cdot)) = \sum_{j=0}^{k-1} \beta_i(j) + \sum_{j=k}^{l-1} h_i(x_j, x_{j+1}) + g_i(x_l),$$

where $x_k = \bar{x}_k$.

But since the strategies $\tilde{u}_j(\cdot)$ (see 5.11.10) are constructed in such a way that in any subgame Γ_x, where $x \in Z(\bar{x}_k)$ players $j \neq i$ use behavior prescribed by strategies $\tilde{u}_j^{N\setminus\{i\}}(\cdot)$ optimal in zero-sum game played by coalition $N \setminus \{i\}$ against player i, player i cannot get in this subgame more than the value of this subgame which is equal to $v(\bar{x}_k; \{i\})$. And we have

$$\sum_{j=k}^{l-1} h_i(x_j, x_{j+1}) + g_i(x_l) \leq v(\bar{x}_k; \{i\}),$$

and finally

$$= H_i^{\alpha}(x_0, \tilde{u}_i(\cdot)||u_i(\cdot)) = \sum_{j=0}^{k-1} \beta_i(j) + \sum_{j=k}^{l-1} h_i(x_j, x_{j+1}) + g_i(x_l)$$
$$\leq \sum_{j=0}^{k-1} \beta_i(j) + v(\bar{x}_k; \{i\}). \tag{5.11.13}$$

But

$$= H_i^{\alpha}(x_0, \tilde{u}_i(\cdot)) = \alpha_i = \sum_{j=0}^{l} \beta_i(j) = \sum_{j=0}^{k-1} \beta_i(j) + \sum_{j=k}^{l} \beta_i(j)$$
$$\geq \sum_{j=0}^{k-1} \beta_i(j) + v(\bar{x}_k; \{i\}), \; i = 1, \ldots, n, \; k = 0, \ldots, l.$$

This follows from the condition that along the cooperative path we have

$$\alpha_i^k \geq V(\bar{x}_k; \{i\}), \quad i \in N, \quad k = 0, 1, \ldots, l,$$

since $\alpha^k = (\alpha_1^k, \ldots, \alpha_n^k) \in C(\bar{x}_k)$ is an imputation in $\Gamma(\bar{x}_k)$ (note that here $V(\bar{x}_k; \{i\})$ is computed in the subgame $\Gamma(\bar{x}_k)$ but not $\Gamma_\alpha(\bar{x}_k)$). In the same time

$$\alpha_i^k = \sum_{j=k}^{l} \beta_i(j)$$

and we get

$$\sum_{j=k}^{l} \beta_i(j) \geq V(\bar{x}_k; \{i\}), \quad i \in N, \quad k = 0, 1, \ldots, l. \tag{5.11.14}$$

And we have from (5.11.13), (5.11.14)

$$\begin{aligned} \alpha_i = H_i^\alpha(x_0, \tilde{u}_i(\cdot)) &\geq \sum_{j=0}^{k-1} \beta_i(j) + v(\bar{x}_k; \{i\}) \\ &\geq H_i^\alpha(x_0, \tilde{u}_i(\cdot)||u_i(\cdot)), \ i \in N. \end{aligned}$$

The theorem is proved.

Thus we constructed the Nash equilibrium in penalty strategies with payoffs $\alpha = (\alpha_1, \ldots, \alpha_n)$ and resulting cooperative path $\bar{x} = (\bar{x}_0, \ldots, \bar{x}_l)$.

Example 17. In this example as an imputation we shall consider the Shapley value. Using the proposed regularization of the game we shall see that there exists a Nash equilibrium with payoffs equal to the components of Shapley value.

In the game Γ_1 (see Fig. 5.20), $N = \{1, 2, 3, 4\}$, $P_1 = \{x_1\}$, $P_2 = \{x_2\}$, $P_3 = \{x_3\}$, $P_4 = \{x_4\}$, $P_5 = \{x_5, x_6, x_7, x_8, x_9\}$. $h(x_5) =$

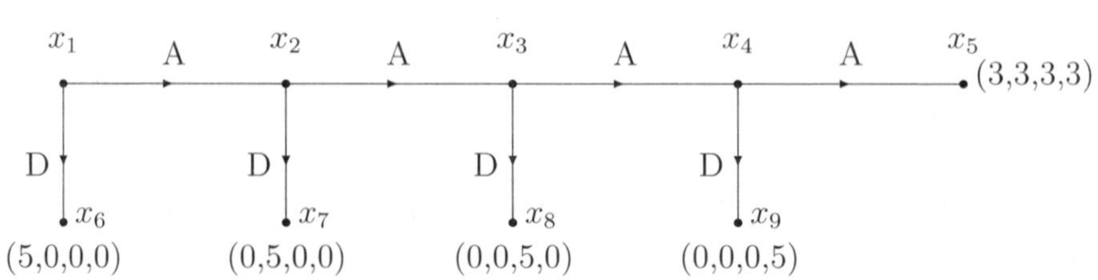

Figure 5.20 Game Γ_1

$(3, 3, 3, 3)$, $h(x_6) = (5, 0, 0, 0)$, $h(x_7) = (0, 5, 0, 0)$, $h(x_8) = (0, 0, 5, 0)$, $h(x_9) = (0, 0, 0, 5)$. The cooperative path is

$$\bar{x} = \{x_1, x_2, x_3, x_4, x_5\}.$$

It can be easily seen that (D, D, D, D) is Nash equilibrium, but (A, A, A, A) is not Nash equilibrium.

Characteristic function of the game Γ_1 (C.f. of Γ_1)
$V_1(1, 2, 3, 4) = 12$, $V_1(1, 2, 3) = 5$, $V_1(1, 3, 4) = 5$, $V_1(2, 3, 4) = 0$, $V_1(1, 2, 4) = 5$, $V_1(1, 2) = 5$, $V_1(1, 3) = 5$, $V_1(1, 4) = 5$, $V_1(2, 3) = 0$, $V_1(2, 4) = 0$, $V_1(3, 4) = 0$, $V_1(1) = 5$, $V_1(2) = 0$, $V_1(3) = 0$, $V_1(4) = 0$.

Shapley value for Γ_1 is

$$Sh^1 = \left(\frac{27}{4}, \frac{7}{4}, \frac{7}{4}, \frac{7}{4}\right).$$

Consider now the subgame $\Gamma_2 = \Gamma(x_2)$ (see Fig. 5.21).

C.f. of Γ_2
$V_2(1, 2, 3, 4) = 12$, $V_2(1, 2, 3) = 5$, $V_2(1, 3, 4) = 0$, $V_2(2, 3, 4) = 9$, $V_2(1, 2) = 5$, $V_2(1, 3) = 0$, $V_2(1, 4) = 0$, $V_2(2, 3) = 5$, $V_2(2, 4) = 5$, $V_2(3, 4) = 0$, $V_2(1) = 0$, $V_2(2) = 5$, $V_2(3) = 0$, $V_3(4) = 0$.

Shapley value for Γ_2 is

$$Sh^2 = \left(\frac{3}{4}, \frac{85}{12}, \frac{25}{12}, \frac{25}{12}\right),$$

and $Sh^1 \neq Sh^2$.

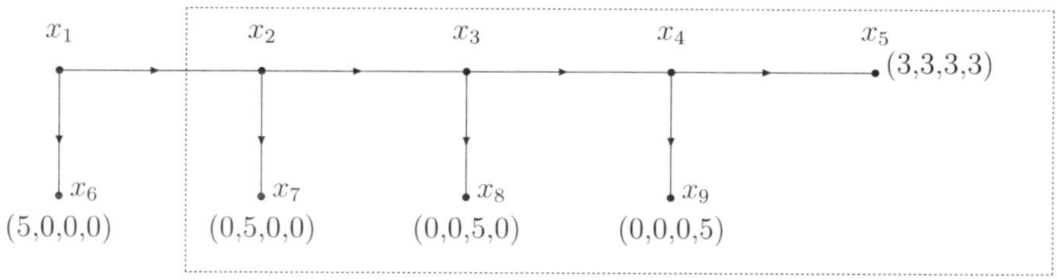

Figure 5.21 Subgame Γ_2

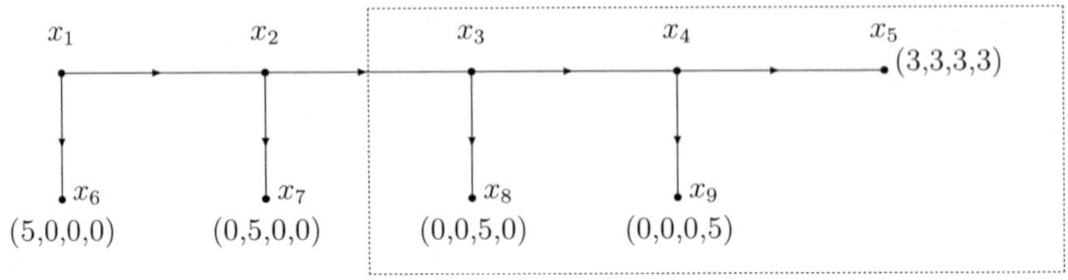

Figure 5.22 Subgame $\Gamma_3 = \Gamma(x_3)$

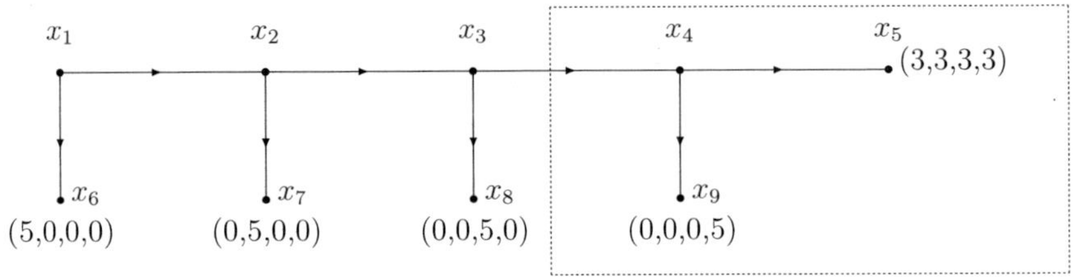

Figure 5.23 Subgame $\Gamma_4 = \Gamma(x_4)$

C.f. of Γ_3 reference to Fig. 5.22.
$V_3(1,2,3,4) = 12$, $V_3(1,2,3) = 5$, $V_3(1,3,4) = 9$, $V_3(2,3,4) = 9$, $V_3(1,2,4) = 0$, $V_3(1,2) = 0$, $V_3(1,3) = 5$, $V_3(1,4) = 0$, $V_3(2,3) = 5$, $V_3(2,4) = 0$, $V_3(3,4) = 6$, $V_3(1) = 0$, $V_3(2) = 0$, $V_3(3) = 5$, $V_2(4) = 0$.

Shapley value for Γ_3 is

$$Sh^3 = \left(1, 1, \frac{90}{12}, \frac{30}{12}\right),$$

and $Sh^2 \neq Sh^3$.

C.f. of Γ_4 reference to Fig. 5.23.
$V_4(1,2,3,4) = 12$, $V_4(1,2,3) = 0$, $V_4(1,3,4) = 9$, $V_4(2,3,4) = 9$, $V_4(1,2,4) = 9$, $V_4(1,2) = 0$, $V_4(1,3) = 0$, $V_4(1,4) = 6$, $V_4(2,3) = 0$, $V_4(2,4) = 6$, $V_4(3,4) = 6$, $V_4(1) = 0$, $V_4(2) = 0$, $V_4(3) = 0$, $V_4(4) = 5$.

Shapley value for Γ_4 is

$$Sh^4 = \left(\frac{16}{12}, \frac{16}{12}, \frac{16}{12}, \frac{96}{12}\right),$$

and $Sh^4 \neq Sh^3$.

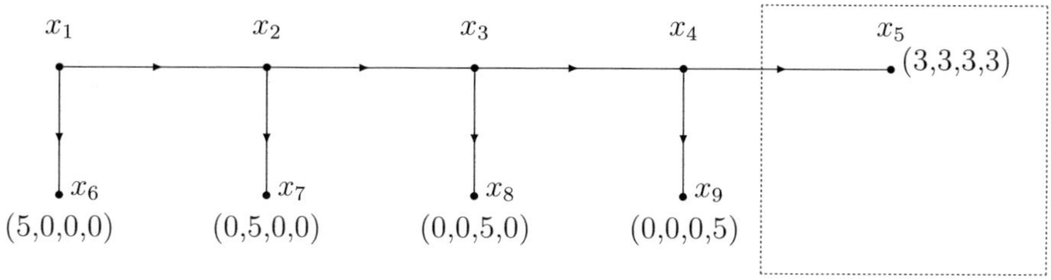

Figure 5.24 Subgame Γ_5

C.f. of Γ_5 reference to Fig. 5.24.
$V_5(1,2,3,4) = 12$, $V_5(1,2,3) = V_5(1,3,4) = V_5(2,3,4) = V_5(1,2,4) = 9$, $V_5(1,2) = V_5(1,3) = V_5(1,4) = V_5(2,3) = V_5(2,4) = V_5(3,4) = 6$, $V_5(1) = V_5(2) = V_5(3) = V_5(4) = 3$.

Shapley value for Γ_4 is

$$Sh^5 = (3,3,3,3),$$

and $Sh^5 \neq Sh^4$.

Compute now the IDP (imputation distribution procedure)

$$Sh^1 = \beta_1 + Sh^2, \quad Sh^2 = \beta_2 + Sh^3, \ldots, Sh^4 = \beta_4 + Sh^5,$$

$$\beta_1 = (Sh^1 - Sh^2),\ \beta_2 = (Sh^2 - Sh^3),$$

$$\beta_3 = (Sh^3 - Sh^4),\ \beta_4 = (Sh^4 - Sh^5),\ \beta_5 = Sh^5.$$

$$\sum_{k=1}^{5} \beta_k = Sh^1, \sum_{k=2}^{5} \beta_k = Sh^2, \sum_{k=3}^{5} \beta_k = Sh^3,$$

$$\sum_{k=4}^{5} \beta_k = Sh^4, \sum_{k=5}^{5} \beta_k = Sh^5,$$

$$\beta_1 = \left(\frac{72}{12}, -\frac{64}{12}, -\frac{4}{12}, -\frac{4}{12}\right),$$

$$\beta_2 = \left(-\frac{3}{12}, \frac{73}{12}, -\frac{65}{12}, -\frac{5}{12}\right),$$

$$\beta_3 = \left(-\frac{4}{12}, -\frac{4}{12}, \frac{74}{12}, -\frac{66}{12}\right),$$

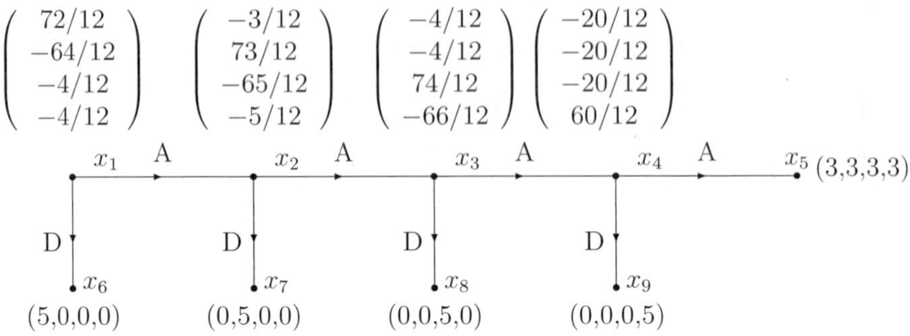

Figure 5.25 Game Γ_α

$$\beta_4 = \left(-\frac{20}{12}, -\frac{20}{12}, -\frac{20}{12}, \frac{60}{12}\right),$$
$$\beta_5 = (3, 3, 3, 3).$$

In regularization Γ_α of the game Γ_1 (wheere α is the Shapley value in Γ_1) the payoffs β_1, β_2, β_3, β_4 are defined on arcs $(1, 2)$, $(2, 3)$, $(3, 4)$, $(4, 5)$ correspondingly.

We can see that the inequalities (5.11.14) hold in this game Γ_α reference to Fig. 5.25.

$$\sum_{j=1}^{4} \beta_1(j) = \frac{72}{12} - \frac{3}{12} - \frac{4}{12} - \frac{20}{12} + 3 > 5 = V(\bar{x}_1; \{1\}),$$

$$\sum_{j=2}^{4} \beta_2(j) = \frac{73}{2} - \frac{4}{12} - \frac{20}{12} + 3 > 5 = V(\bar{x}_2; \{2\}),$$

$$\sum_{j=2}^{4} \beta_3(j) = \frac{74}{12} - \frac{20}{12} + 3 > 5 = V(\bar{x}_3; \{3\}),$$

$$\sum_{j=3}^{4} \beta_4(j) = \frac{60}{12} + 3 > 5 = V(\bar{x}_4; \{4\}).$$

This means that in Nash equilibrium (A, A, A, A) the payoffs in G_α are $(\frac{27}{4}, \frac{7}{4}, \frac{7}{4}, \frac{7}{4})$ exactly equal to Shapley value. Thus the computed

Nash equilibrium supports the cooperative outcome (payoffs) in original game which are redistributed according to IDP guaranteeing the time-consistency of the Shapley value.

5.12 One-Way Flow Two-Stage Network Games

5.12.1. As we have seen in the theory of dynamic cooperative games, time-consistency of a solution is the key problem. Namely, having agreed on the particular solution before the game starts, players have to get the payoff prescribed by this solution at the end of the game. Such a problem is quite common for dynamic cooperative games, since during the game in the case of time-inconsistency, players may break initial agreement by their actions. Time-consistency of the cooperative solution based on a special payment scheme stimulates players to follow agreed upon cooperative behavior.

The two stage network games [Jackson and Watts (2002)] are considered. Players form a network at the first stage, and then at the second stage they choose admissible controls. In particular, it was proved that in the cooperative setting, the cooperative solution — the Shapley value Petrosyan and Sedakov (2014) — is time-inconsistent Petrosyan, Sedakov and Bochkarev (2013).

5.12.2. *The model.* Let $N = \{1, \ldots, n\}$ be a finite set of players, and g be a given network — the set of pairs $(i, j) \in N \times N$ where $(i, j) \in g$ means that there is a direct link connecting players i and j and such a link generates communication of player i with player j. The network under consideration is one-way flow network, that is any link (i, j) is direct link, i.e. $(i, j) \neq (j, i)$.

Consider a two-stage model. At the first stage, players choose their partners — players with whom they want to form links. Once the partners are chosen, a communication structure, i.e. a network is formed. At the second stage, players choose admissible control variables, which together with the formed network, affect their payoffs. Consider the model in details.

5.12.3. *First stage: network formation.* A network is formed as a result of simultaneous choices of all players. Let $M_i \subseteq N \setminus \{i\}$ be the set of players to whom player $i \in N$ can offer a link, and $a_i \in \{0, \ldots, n-1\}$ be the maximal number of links which player i can offer. At the first stage, behavior of player $i \in N$ is a profile $g_i = (g_{i1}, \ldots, g_{in})$ where

$$g_{ij} = \begin{cases} 1, & \text{if player } i \text{ offers a link to player } j \in M_i, \\ 0, & \text{otherwise,} \end{cases} \tag{5.12.1}$$

subject to the constraint:

$$\sum_{j \in N} g_{ij} \leq a_i. \tag{5.12.2}$$

From (5.12.1) we get $g_{ii} = 0$, $i \in N$, which excludes loops from the network, whereas the condition (5.12.2) shows that the number of "offers" is limited. Note that if $M_i = N \setminus \{i\}$, player i can offer a link to any player, whereas if $a_i = n - 1$, he can offer any number of links.

Denote the set of all possible behaviors of player $i \in N$ at the first stage satisfying (5.12.1), (5.12.2) by G_i. The set $\prod_{i \in N} G_i$ is the set of behavior profiles at the first stage. Supposing that players choose their behaviors $g_i \in G_i$, $i \in N$, simultaneously and independently from each other, the behavior profile $(g_1, \ldots, g_n)$ is formed. A resulting network g consists of directed links (i, j) s.t. $g_{ij} = 1$.

Define the closure of network g as an undirected network $\bar{g}$ where $\bar{g}_{ij} = \max\{g_{ij}, g_{ji}\}$. Denote neighbors of Player i in the network g by $N_i(g) = \{j \in N \setminus \{i\} : (i, j) \in g\}$, whereas neighbors of Player i in the closure $\bar{g}$ are denoted by $N_i(\bar{g}) = \{j \in N \setminus \{i\} : (i, j) \in \bar{g}\}$.

Example 18. Consider a four player case. Let $N = \{1, 2, 3, 4\}$ and players choose the following behaviors: $g_1 = (0, 0, 0, 1)$, $g_2 = (0, 0, 1, 0)$, $g_3 = (0, 1, 0, 1)$, $g_4 = (1, 0, 0, 0)$. The network g consists of five links $g = \{(1, 4), (2, 3), (3, 2), (3, 4), (4, 1)\}$, whereas its closure $\bar{g}$ consists of three undirected links $\bar{g} = \{(1, 4), (2, 3), (3, 4)\}$ (see figure). Note that, for instance, $N_4(g) = \{1\}$ while $N_4(\bar{g}) = \{1, 3\}$.

5.12.4. *Second stage: controls.* To allow players to break formed links at the first stage (we introduce this possibility to punish neighbors from the complementary coalition in the case of zero-sum game which can appear at coalition formation stage), we define an n-dimensional profile $d_i(g)$ as follows:

$$d_{ij}(g) = \begin{cases} 1, & \text{if player } i \text{ keeps the link formed at the first stage} \\ & \quad \text{with player } j \in N_i(g) \text{ in network } g, \\ 0, & \text{otherwise.} \end{cases} \tag{5.12.3}$$

Denote all profiles $d_i(g)$ satisfying (5.12.3) by $D_i(g)$, $i \in N$. At the second stage players simultaneously and independently choose $d_i(g)$, $i \in N$, thus the profile $(d_1(g), \ldots, d_n(g))$ changes network g forming a new network which is denoted by g^d.

Example 19. Suppose that players choose their profiles g_i, $i \in N$ as in Example 1 forming the network $g = \{(1,4),(2,3),(3,2),(3,4),(4,1)\}$. Let $d_1(g) = (0,0,0,1)$, $d_2(g) = (0,0,0,0)$, $d_3(g) = (0,0,0,1)$, $d_4(g) = (1,0,0,0)$, i.e. Player 1 keeps the link with Player 4, Player 2 breaks the link with Player 3, Player 3 breaks the link with Player 2, and Player 4 keeps the link with Player 1. Then we have a new network $g^d = \{(1,4),(3,4),(4,1)\}$. The closure $\bar{g}^d$ consists of two undirected links $\bar{g}^d = \{(1,4),(3,4)\}$ (see Fig. 5.26c and d).

Also at the second stage Player $i \in N$ chooses control u_i from a given set U_i. Then behavior of Player $i \in N$ at the second stage is a pair $(d_i(g), u_i)$.

Payoff function K_i of Player i depends on network g^d, his control u_i and controls u_j, $j \in N_i(\bar{g}^d)$ of his neighbors in the closure $\bar{g}^d$. More formally,

$$K_i(u_i, u_{N_i(\bar{g}^d)}) : U_i \times \prod_{j \in N_i(\bar{g}^d)} U_j \mapsto R^+, \quad i \in N$$

is a real-valued function where notation $u_{N_i(\bar{g}^d)}$ means the profile of chosen controls u_j of all player $j \in N_i(\bar{g}^d)$ in network $\bar{g}^d$.

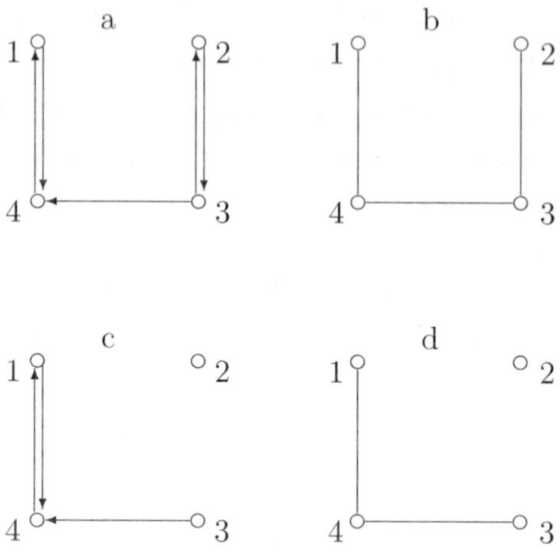

Figure 5.26

5.12.5. *Cooperation in one-way flow two-stage network games.* In this section, we study the cooperative case: we allow players to coordinate their actions and choose behaviors jointly. Players being rational, choose their behaviors $g_i \in G_i$, $(d_i(g^d), u_i) \in D_i(g) \times U_i$, $i \in N$, to maximize the joint payoff, the value:

$$\sum_{i \in N} K_i(u_i, u_{N_i(\bar{g}^d)}). \tag{5.12.4}$$

It can be easily seen that to maximize the total sum (5.12.4) of players' payoffs (supposing that maximum in (5.12.4) exists), it is sufficient to form the network at the first stage without changing it at the second stage, i.e. $d_i(g) \equiv g_i$, for all $i \in N$ and

$$\max_{\substack{(g_i, d_i(g), u_i) \in G_i \times D_i(g) \times U_i, \\ i \in N}} \sum_{i \in N} K_i(u_i, u_{N_i(\bar{g}^d)})$$

$$= \max_{\substack{(g_i, u_i) \in G_i \times U_i, \\ i \in N}} \sum_{i \in N} K_i(u_i, u_{N_i(\bar{g})}).$$

The profile (g_i^*, u_i^*), $i \in N$, maximizing (5.11.4) we call the cooperative profile. Behavior profile $(g_1^*, \ldots, g_n^*)$ forms the network g^* and

$$\sum_{i \in N} K_i(u_i^*, u^*_{N_i(\bar{g}^*)}) = \max_{\substack{(g_i, u_i) \in G_i \times U_i, \\ i \in N}} \sum_{i \in N} K_i(u_i, u_{N_i(\bar{g})}).$$

To allocate the maximal sum of players' payoffs according to some solution concept, one needs to construct a cooperative TU-game (N, V). The characteristic function V is defined in the sense of von Neumann and Morgenstern as:

$$V(N) = \sum_{i \in N} K_i(u_i^*, u^*_{N_i(\bar{g}^*)}),$$

$$V(S) = \max_{\substack{(g_i, d_i(g), u_i) \in G_i \times D_i(g) \times U_i, \\ i \in S}} \min_{\substack{(g_j, d_j(g), u_j) \in G_j \times D_j(g) \times U_j, \\ j \in N \setminus S}} \times \sum_{i \in S} K_i(u_i, u_{N_i(\bar{g}^d)}),$$

$$V(\emptyset) = 0,$$

where the network g is formed by profile $(g_1, \ldots, g_n)$ and the network g^d is formed by profile $(d_1(g), \ldots, d_n(g))$.

Consider a non-empty coalition $S \subset N$. Denote a network, formed by profiles g_i, $i \in N$, s.t. $g_j = (0, \ldots, 0)$ for all $j \in N \setminus S$, by g_S. Let $\bar{g}_S$ be the closure of g_S. For any controls u_i, $i \in S$ let controls $\tilde{u}_j(u_S)$, $j \in N \setminus S$, where $u_S = \{u_i\}$, $i \in S$, solve the following optimization problem

$$\sum_{i \in S} K_i\left(u_i, u_{N_i(\bar{g}_S) \cap S}, \tilde{u}_{(N \setminus S) \cap N_i(\bar{g}_S)}(u_S)\right) = \min_{u_j, j \in (N \setminus S) \cap N_i(\bar{g}_S)} \sum_{i \in S} K_i\left(u_i, u_{N_i(\bar{g}_S) \cap S}, u_{(N \setminus S) \cap N_i(\bar{g}_S)}\right).$$

Here $u_{N_i(\bar{g}_S) \cap S}$ is the profile of controls chosen by all neighbors of Player i from coalition S in the network $\bar{g}_S$, and $\tilde{u}_{(N \setminus S) \cap N_i(\bar{g}_S)}(u_S)$ is a profile of controls chosen by all players from coalition $N \setminus S$ who are neighbors of player i in the network $\bar{g}_S$.

Theorem. *Suppose that functions K_i, $i \in N$, are non-negative and satisfy the following property: for any two networks g and g' s.t. $g' \subseteq g$, controls $(u_i, u_{N_i(\bar{g})}) \in U_i \times \prod_{j \in N_i(\bar{g})} U_j$ and Player i, the inequality $K_i(u_i, u_{N_i(\bar{g})}) \geq K_i(u_i, u_{N_i(\bar{g}')})$ holds. Then for all $S \subset N$ we have*

$$V(S) = \max_{\substack{(g_i,u_i)\in G_i\times U_i,\\ i\in S}} \sum_{i\in S} K_i\left(u_i, u_{N_i(\bar{g}_S)\cap S}, \tilde{u}_{(N\setminus S)\cap N_i(\bar{g}_S)}(u_S)\right).$$

Proof. Consider the maxmin value for coalition $S \subset N$:

$$V(S) = \max_{\substack{(g_i,d_i(g),u_i)\in G_i\times D_i(g)\times U_i,\\ i\in S}} \min_{\substack{(g_j,d_j(g),u_j)\in G_j\times D_j(g)\times U_j,\\ j\in N\setminus S}} \times \sum_{i\in S} K_i(u_i, u_{N_i(\bar{g}^d)}).$$

Since the presence of a link $(j, i) \in g$, $i \in S$, $j \in N\setminus S$, increases payoff of coalition S according to the property formulated in the statement of the Proposition 1, therefore, Player $j \in N \setminus S$, as a neighbor of i, changes his component $d_{ji}(g)$ from 1 to 0 in the profile $d_j(g)$, i.e. removes link (j, i) to minimize the payoff of coalition S. Thus, to minimize the value $\sum_{i\in S} K_i(u_i, u_{N_i(\bar{g}^d)})$ players from $N \setminus S$ remove all links with players from S and use controls $\tilde{u}_j(u_S)$, $j \in N \setminus S$. Note that it is not important for coalition S how players from its complement $N \setminus S$ connect with each other. Therefore, without loss of generality assuming that $d_j(g) = (0, \ldots, 0)$ for all $j \in N \setminus S$, we obtain

$$V(S) = \max_{\substack{(g_i,d_i(g),u_i)\in G_i\times D_i(g)\times U_i,\\ i\in S}} \times \sum_{i\in S} K_i\left(u_i, u_{N_i(\bar{g}_S)\cap S}, \tilde{u}_{(N\setminus S)\cap N_i(\bar{g}_S)}(u_S)\right).$$

To maximize the sum $\sum_{i\in S} K_i(u_i, u_{N_i(\bar{g}_S)\cap S}, \tilde{u}_{(N\setminus S)\cap N_i(\bar{g}_S)(u_S)})$, it is sufficient for players from coalition S to form the network at the first stage without changing it at the second stage, i.e. $d_i(g) \equiv g_i$, for all $i \in S$. Then we get

$$V(S) = \max_{\substack{(g_i, u_i) \in G_i \times U_i, \\ i \in S}} \sum_{i \in S} K_i(u_i, u_{N_i(\bar{g}_S) \cap S}, \tilde{u}_{(N \setminus S) \cap N_i(\bar{g}_S)}(u_S)),$$

which proves the statement.

An imputation in the cooperative two-stage network game is a profile $\xi(V) = (\xi_1(V), \ldots, \xi_n(V))$ s.t. $\sum_{i \in N} \xi_i(V) = V(N)$ and $\xi_i(V) \geq V(\{i\})$ for all $i \in N$. We denote the set of all imputations in the game (N, V) by $I(V)$. A solution concept (or simply solution) of TU-game (N, V) is a rule that uniquely assigns a subset of $I(V)$ to the game (N, V). For example, if the solution concept is the Shapley value $\phi(V) = (\phi_1(V), \ldots, \phi_n(V))$, its components can be calculated as

$$\phi_i(V) = \sum_{S \subseteq N, i \in S} \frac{(|N| - |S|)!(|S| - 1)!}{|N|!} \times [V(S) - V(S \setminus \{i\})] \text{ for all } i \in N.$$

5.12.6. *Time-consistency problem.*

In this section we study time-consistency of the Shapley value $\phi(V)$. We already found behavior profiles (g_i^*, u_i^*), $i \in N$, of players which maximize the sum (5.12.4) allowing players to get the value $V(N)$. Allocating $V(N)$ according to the Shapley value, we obtain the solution $\phi(V) = (\phi_1(V), \ldots, \phi_n(V))$. In other words, in the cooperative two-stage network game player $i \in N$ should receive the amount of $\phi_i(V)$ as his payoff.

After the first stage (after forming network g^*) players may recalculate the solution according to the same solution concept. To find the new, recalculated solution, one needs to consider the subgame (one-stage game) starting from the second stage, provided that players chose behavior profile $(g_1^*, \ldots, g_n^*)$ at the first stage, and therefore formed network g^*. Consider this subgame. The characteristic function for the subgame will depend on parameter — the network g^* — formed at the first stage, and we denote this function as $v(g^*, S)$ to stress the dependence from the network. The characteristic function $v(g^*, S)$ in the sense of von Neumann and Morgenstern is defined as follows:

$$v(g^*, N) = \sum_{i \in N} K_i(u_i^*, u^*_{N_i(\bar{g}^*)}) = V(N),$$

$$v(g^*, S) = \max_{\substack{(d_i(g^*), u_i) \in D_i(g^*) \times U_i, \\ i \in S}} \min_{\substack{(d_j(g^*), u_j) \in D_j(g^*) \times U_j, \\ j \in N \setminus S}} \sum_{i \in S} K_i(u_i, u_{N_i(\bar{g}^{d*})}),$$

$$v(g^*, \emptyset) = 0.$$

The following proposition can be proved similarly to Proposition 1.

Theorem. *If functions K_i, $i \in N$, are non-negative and satisfy the property stated in Proposition 1, the value $v(g^*, S)$ can be calculated by formula*

$$v(g^*, S) = \max_{\substack{u_i \in U_i, \\ i \in S}} \sum_{i \in S} K_i(u_i, u_{N_i(\bar{g}^*_S) \cap S}, \tilde{\tilde{u}}_{(N \setminus S) \cap N_i(\bar{g}^*_S)}(u_S)),$$

where $\tilde{\tilde{u}}_j(u_S)$, $j \in N \setminus S$, solve the following optimization problem:

$$\sum_{i \in S} K_i \left(u_i, u_{N_i(\bar{g}^*_S) \cap S}, \tilde{\tilde{u}}_{(N \setminus S) \cap N_i(\bar{g}^*_S)}(u_S) \right)$$

$$= \min_{u_j, j \in (N \setminus S) \cap N_i(\bar{g}^*_S)} \sum_{i \in S} K_i \left(u_i, u_{N_i(\bar{g}^*_S) \cap S}, u_{(N \setminus S) \cap N_i(\bar{g}^*_S)} \right)$$

and $\bar{g}^*_S$ is the closure of network g^*_S, formed by profiles g^*_i, $i \in N$, s.t. $g^*_j = (0, \ldots, 0)$ for all $j \in N \setminus S$.

In the subgame, an imputation $\xi(g^*, v) = (\xi_1(g^*, v), \ldots, \xi_n(g^*, v))$ satisfies two conditions: $\sum_{i \in N} \xi_i(g^*, v) = v(g^*, N)$ and $\xi_i(g^*, v) \geq v(g^*, \{i\})$, $i \in N$. Recalculate players' payoffs in the subgame using the same solution concept — the Shapley value $\phi(g^*, v) = (\phi_1(g^*, v), \ldots, \phi_n(g^*, v))$, where its components can be computed as

$$\phi_i(g^*, v) = \sum_{S \subseteq N, i \in S} \frac{(|N| - |S|)!(|S| - 1)!}{|N|!} \left[v(g^*, S) - v(g^*, S \setminus \{i\}) \right],$$

for all $i \in N$.

Definition. The Shapley value $\phi(V)$ is time-consistent if:

$$\phi_i(V) = \phi_i(g^*, v), \quad i \in N. \tag{5.12.5}$$

The equality (5.12.5) means that if we use the imputation $\xi(V) = \phi(V)$ at the first stage, and then at the second stage recalculate players' payoffs according to the same solution concept $\xi(g^*, v) = \phi(g^*, v)$, i.e. calculate a new imputation $\xi(g^*, v) = \phi(g^*, v)$, subject to formed network g^*, players' payoffs prescribed by this imputation will not change. Since in most games the condition (5.11.5) is not satisfied (in our setting characteristic functions $V(S)$ and $v(g^*, S)$ are different), the time-consistency problem arises: player $i \in N$, who initially expected his payoff to be equal to $\phi_i(V)$, can receive different payoff $\phi_i(g^*, v)$. To avoid such situation in the game, we propose a stage payments mechanism — imputation distribution procedure for the Shapley value $\phi(V)$.

Definition. Imputation distribution procedure for $\phi(V)$ in the cooperative two-stage network game is a matrix

$$\beta = \begin{pmatrix} \beta_{11} & \beta_{12} \\ \vdots & \vdots \\ \beta_{n1} & \beta_{n2} \end{pmatrix},$$

s.t. $\phi_i(V) = \beta_{i1} + \beta_{i2}$ for all $i \in N$.

The value β_{ik} is a payment to player i at stage $k = 1, 2$. Therefore, the following payment scheme is applied: player $i \in N$ at the first stage of the game receives the payment β_{i1}, at the second stage of the game he receives the payment β_{i2} in order to his total payment received on both stages $\beta_{i1} + \beta_{i2}$ would be equal to the component of the Shapley value $\phi_i(V)$, which he initially wanted to get in the game as the payoff.

Definition. Imputation distribution procedure β for the Shapley value $\phi(V)$ is time-consistent if for all $i \in N$

$$\phi_i(V) - \beta_{i1} = \phi_i(g^*, v).$$

It is obvious that time-consistent imputation distribution procedure for $\phi(V)$ in the cooperative two-stage network game can be defined as follows:

$$\beta_{i1} = \phi_i(V) - \phi_i(g^*, v),$$
$$\beta_{i2} = \phi_i(g^*, v), \quad i \in N. \tag{5.12.6}$$

5.12.7. *Numerical Example.* To illustrate the theoretical results obtained in the previous sections, consider a three-person game as an example. Let $N = \{1, 2, 3\}$ be the set of players. We suppose that Player 1 can establish a link with Player 3, Player 2 can establish links with Players 1 and 3, and, finally, Player 3 can establish links with Players 1 and 2. Therefore, we have: $M_1 = \{3\}$, $M_2 = \{1, 3\}$, $M_3 = \{1, 2\}$. Moreover, we suppose that each player can offer a limited number of links: Players 1 and 3 can offer only one link, while Player 2 can offer two links that is $a_1 = a_3 = 1$, $a_2 = 2$. Thus, at the first stage the sets of behaviors of players are: $G_1 = \{(0,0,0); (0,0,1)\}$, $G_2 = \{(0,0,0); (1,0,0); (0,0,1); (1,0,1)\}$; $G_3 = \{(0,0,0); (1,0,0); (0,1,0)\}$.

Let u_i be behavior of player $i \in N$ at the second stage, and $U_i = [0, \infty)$ for all $i \in N$. The payoff function of Player i depends on players connected with Player i as well as on players with whom Player i established links and have the following form (the expression of the payoff function is justified in above mentioned papers and is used in network models of public goods):

$$K_i(g, u) = \ln\left(1 + u_i + \sum_{j \in N_i(\bar{g})} u_j\right) - c_i u_i - k|N_i(g)|,$$

where parameters $c_1 = 0.2$, $c_2 = 0.25$, $c_3 = 0.4$, $k = 0.75$, and network g is formed by the profile (g_1, g_2, g_3), and $u = (u_1, u_2, u_3)$. To find cooperative behavior, one needs to maximize the total payoff,

Table 5.1

Network g	$\sum_{i=1}^{3} K_i$	Network g	$\sum_{i=1}^{3} K_i$
$\emptyset$	1.7620	$\{(1,3),(3,2)\}$	2.2870
$\{(1,3)\}$	2.6915	$\{(2,1)\}$	2.3715
$\{(1,3),(2,1)\}$	3.8242	$\{(2,1),(2,3)\}$	3.2047
$\{(1,3),(2,1),(2,3)\}$	3.0742	$\{(2,1),(2,3),(3,1)\}$	3.0742
$\{(1,3),(2,1),(2,3),(3,1)\}$	2.3242	$\{(2,1),(2,3),(3,2)\}$	2.4547
$\{(1,3),(2,1),(2,3),(3,2)\}$	2.3242	$\{(2,1),(3,1)\}$	3.8242
$\{(1,3),(2,1),(3,1)\}$	3.0742	$\{(2,1),(3,2)\}$	3.2047
$\{(1,3),(2,1),(3,2)\}$	3.0742	$\{(2,3)\}$	2.4683
$\{(1,3),(2,3)\}$	2.2870	$\{(2,3),(3,1)\}$	2.2870
$\{(1,3),(2,3),(3,1)\}$	1.5370	$\{(2,3),(3,2)\}$	1.7183
$\{(1,3),(2,3),(3,2)\}$	1.5370	$\{(3,1)\}$	2.6915
$\{(1,3),(3,1)\}$	1.9415	$\{(3,2)\}$	2.4683

i.e. to solve the optimization problem:

$$\max_{\substack{(g_i,u_i)\in G_i\times U_i,\\ i=1,2,3}} \sum_{i=1}^{3} K_i(g,u)$$

$$= \max_{\substack{(g_i,u_i)\in G_i\times U_i,\\ i=1,2,3}} \sum_{i=1}^{3} \left[\ln\left(1 + u_i + \sum_{j\in N_i(\bar{g})} u_j\right) - c_i u_i - k|N_i(\bar{g})| \right].$$

From Table 5.1, we conclude that $V(N) = 3.8242$ which is attained at two different profiles:

$$\begin{aligned} g_1^* &= (0,0,1),\ u_1^* = 14, \\ g_2^* &= (1,0,0),\ u_2^* = 0, \\ g_3^* &= (0,0,0),\ u_3^* = 0, \end{aligned} \quad \text{and} \quad \begin{aligned} g_1^* &= (0,0,0),\ u_1^* = 14, \\ g_2^* &= (1,0,0),\ u_2^* = 0, \\ g_3^* &= (1,0,0),\ u_3^* = 0, \end{aligned}$$

which form networks $\{(1,3),(2,1)\}$ and $\{(2,1),(3,1)\}$ at the first stage respectively.

By the definition of characteristic function $V(S)$, find its values for all $S \subset N$. For $i \in N$ we have

$$V(\{i\}) = \max_{(g_i,d_i(g),u_i)\in G_i\times D_i(g)\times U_i} \min_{\substack{(g_j,d_j(g),u_j)\in G_j\times D_j(g)\times U_j,\\ j\neq i}} K_i(g,u)$$

$$= \max_{(g_i,u_i)\in G_i\times U_i} K_i(g,u)|_{g_j=0,u_j=0,\ j\neq i}$$

$$= \max_{u_i\in U_i} [\ln(1+u_i) - c_i u_i].$$

For all $i,\ j \in \{1,2,3\}$, such that either $i \in M_j$ or $j \in M_i$, $m = N \setminus \{i,j\}$,

$$V(\{i,j\}) = \max_{\substack{(g_i,d_i(g),u_i)\in G_i\times D_i(g)\times U_i\\ (g_j,d_j(g),u_j)\in G_j\times D_j(g)\times U_j}} \min_{(g_m,d_m(g),u_m)\in G_m\times D_m(g)\times U_m}$$

$$\times [K_i(g,u) + K_j(g,u)]$$

$$= \max_{\substack{(g_i,d_i(g),u_i)\in G_i\times D_i(g)\times U_i\\ (g_j,d_j(g),u_j)\in G_j\times D_j(g)\times U_j}} [K_i(g,u) + K_j(g,u)]_{g_m=0,u_m=0}$$

$$= \max\{ \max_{\substack{u_i\in U_i\\ u_j\in U_j}} [\ln(1+u_i) - c_i u_i + \ln(1+u_j) - c_j u_j];$$

$$\times \max_{\substack{u_i\in U_i\\ u_j\in U_j}} [2\ln(1+u_i+u_j) - c_i u_i - c_j u_j - k]\}$$

$$= \max\left\{ V(\{i\}) + V(\{j\});\ \max_{\substack{u_i\in U_i\\ u_j\in U_j}} [2\ln(1+u_i+u_j)\right.$$

$$\left. -c_i u_i - c_j u_j - k] \right\}.$$

Thus after solving the corresponding maximization problems, we obtain values of characteristic function $V(S)$:

S	$\{1,2,3\}$	$\{1,2\}$	$\{1,3\}$	$\{2,3\}$	$\{1\}$	$\{2\}$	$\{3\}$
$V(S)$	3.8242	2.0552	2.0552	1.6589	0.8094	0.6363	0.3163

The Shapley value $\phi(V) = (\phi_1(V), \phi_2(V), \phi_3(V))$, calculated for characteristic function $V(S)$, is

$$\phi(V) = (1.5179,\ 1.2331,\ 1.0731), \qquad (5.12.7)$$

i.e. choosing behaviors jointly at both stages, players get the total payoff of 3.8242 and after allocation of the amount according to the Shapley value at the end of the game, each player gets $\phi_i(V)$, $i = 1, 2, 3$, as his payoff in the game.

To show that the Shapley value $\phi(V)$ is time-inconsistent, consider the subgame of the two-stage game, starting from the second stage, provided that players chose the cooperative behaviors at the first stage. Select the cooperative profile $g_1^* = (0, 0, 1)$, $g_2^* = (1, 0, 0)$, $g_3^* = (0, 0, 0)$, and $u^* = (u_1^*, u_2^*, u_3^*) = (14, 0, 0)$. The cooperative behaviors at the first stage (g_1^*, g_2^*, g_3^*) form the network $g^* = \{(1, 3), (2, 1)\}$. To prove that the Shapley value $\phi(V)$ is time-inconsistent it is sufficient to compute the Shapley value $\phi(g^*, v)$ and show that $\phi(V) \neq \phi(g^*, v)$. For this purpose calculate characteristic function $v(g^*, S)$ in the subgame for all $S \subset N$. Note that $v(g^*, S) = V(S) = 3.8242$. For $i \in \{1, 2, 3\}$, we have

$$\begin{aligned} v(g^*, \{i\}) &= \max_{(d_i(g^*), u_i) \in D_i(g^*) \times U_i} \ \min_{\substack{(d_j(g^*), u_j) \in D_j(g^*) \times U_j, \\ j \neq i}} K_i(g^*, u) \\ &= \max_{(d_i(g^*), u_i) \in D_i(g^*) \times U_i} K_i(g^*, u)|_{d_j(g^*) = 0, u_j = 0,\ j \neq i} \\ &= \max_{u_i \in U_i} [\ln(1 + u_i) - c_i u_i] = V(\{i\}). \end{aligned}$$

For all i, $j \in \{1, 2, 3\}$, $m = N \setminus \{i, j\}$ we get:

$$\begin{aligned} &v(g^*, \{i, j\}) \\ &\quad = \max_{\substack{(d_i(g^*), u_i) \in D_i(g^*) \times U_i \\ (d_j(g^*), u_j) \in D_j(g^*) \times U_j}} \ \min_{(d_m(g^*), u_m) \in D_m(g^*) \times U_m} [K_i(g^*, u) + K_j(g^*, u)] \end{aligned}$$

$$= \max_{\substack{(d_i(g^*),u_i)\in D_i(g^*)\times U_i \\ (d_j(g^*),u_j}} [K_i(g^*,u) + K_j(g^*,u)]_{d_m(g^*)=0,u_m=0}$$

$$= \begin{cases} \max\limits_{\substack{u_i\in U_i \\ u_j\in U_j}} [\ln(1+u_i) - c_i u_i + \ln(1+u_j) - c_j u_j)], & \{i,j\} = \{2,3\}, \\ \max\{ \max\limits_{\substack{u_i\in U_i \\ u_j\in U_j}} [\ln(1+u_i) - c_i u_i + \ln(1+u_j) - c_j u_j]; & \\ \qquad \max\limits_{\substack{u_i\in U_i \\ u_j\in U_j}} [2\ln(1+u_i+u_j) - c_i u_i - c_j u_j - k]\}, & \text{otherwise.} \end{cases}$$

Thus after solving the corresponding maximization problems, we obtain values of characteristic function $v(g^*, S)$, $g^* = \{(1,3),(2,1)\}$:

S	$\{1,2,3\}$	$\{1,2\}$	$\{1,3\}$	$\{2,3\}$	$\{1\}$	$\{2\}$	$\{3\}$
$v(g^*,S)$	3.8242	2.0552	2.0552	0.9526	0.8094	0.6363	0.3163

Using the values $v(g^*, S)$, the Shapley value $\phi(g^*, v) = (\phi_1(g^*, v), \phi_2(g^*, v), \phi_3(g^*, v))$ is computed:

$$\phi(g^*, v) = (1.7533,\ 1.1154,\ 0.9554),$$

and from (5.12.7) we conclude that $\phi(V) \neq \phi(g^*, v)$. This shows time-inconsistency of the Shapley value $\phi(V)$. Time-consistent imputation distribution procedure β of the Shapley value $\phi(V)$ can be computed by formulas (5.11.6):

$$\beta = \begin{pmatrix} \beta_{11} & \beta_{12} \\ \beta_{21} & \beta_{22} \\ \beta_{31} & \beta_{32} \end{pmatrix} = \begin{pmatrix} -0.2354 & 1.7533 \\ 0.1177 & 1.1154 \\ 0.1177 & 0.9554 \end{pmatrix}. \tag{5.12.8}$$

Similarly, it can be seen that the Shapley value $\phi(V)$ is time-inconsistent also for the second cooperative behavior profile: $g_1^* = (0,0,0)$, $g_2^* = (1,0,0)$, $g_3^* = (1,0,0)$, and $u^* = (u_1^*, u_2^*, u_3^*) = (14,0,0)$. The cooperative behaviors at the first stage (g_1^*, g_2^*, g_3^*) form the network $g^* = \{(2,1),(3,1)\}$. One can show that the characteristic function $v(g^*, S)$ in the subgame, calculated for the given network $g^* = \{(2,1),(3,1)\}$, coincides with the characteristic function $v(g, S)$

calculated for network $g = \{(1,3),(2,1)\}$. Therefore, given the network $g^* = \{(2,1),(3,1)\}$, we have time-inconsistency of the Shapley value $\phi(V)$, and time-consistent imputation procedure β for the Shapley value $\phi(V)$ will be the same as in (5.12.8).

5.12.8. Two-stage games with network formation at the first stage are considered. One of assumptions is that the payoff of any player depends only on his behavior and behavior of his neighbors in the network. The model deals with directed network that influenced the construction of characteristic function of the game. It is shown that the Shapley value — proposed cooperative solution of the two-stage game — is time-inconsistent, but with the use of a newly introduced payment mechanism — imputation distribution procedure — one can guarantee the realization of such solution in the game.

5.13 Exercises and Problems

1. Find all absolute Nash equilibria in Example 4, 5.2.2.

2. Prove that in a finite-stage two-person zero-sum game with perfect information the payoffs are equal in all "favorable" ("unfavorable") Nash equilibria.

3. Let $v_1(x), v_2(x), \ldots, v_n(x)$ be the values of the payoff functions for players $1, 2, \ldots, n$ in an absolute equilibrium in the subgame Γ_x, which is unique in each subgame.

(a) Show that the functions $v_i(x)$, $i = 1, 2, \ldots, n$, satisfy the following system of functional equations:

$$v_i(x) = \max_{x' \in \Gamma_x} v_j(x'), \; x \in X_j, \; i = 1, 2, \ldots, n, \qquad (5.13.1)$$

with the boundary condition

$$v_i(x)|_{x \in X_{n+1}} = H_i(x). \qquad (5.13.2)$$

(b) Give an example of the game in which the payoffs to players in a penalty strategy equilibrium do not satisfy the system of functional equations (5.12.1) with the boundary condition (5.12.2).

Consider the version of Battle of the Sexes game in which Player I (woman) will go with Player II man independ of what strategy he chooses with probability $\frac{1}{2}$ and will behave as in the game from Sec. 3.1.4 in either case.

4. Construct an example of a multistage two-person nonzero-sum game where, in a penalty strategy equilibrium, the penalizing player penalizes his opponent for deviation from the chosen path and thus penalizes himself to a greater extent.

5. Construct pareto-optimal sets in the game from Example 4, 5.2.2.

6. Construct an example of multistage nonzero-sum game where none of the Nash equilibria leads to a pareto-optimal solution.

7. Construct the map T which sets up a correspondence between each subgame Γ_z of the game Γ some subset of situations U_z in this subgame. Let $T(\Gamma) = U_{x_0}$.

We say that the map T is dynamically stable (time-consistent) if from $u(\cdot) \in U_{x_0}$ it follows that $u^{z_k}(\cdot) \in U_{z_k}$ where $u^{z_k}(\cdot) = (u_1^{z_k}(\cdot), \ldots, u_n^{z_k}(\cdot))$ is the truncation of situation $u(\cdot)$ to the subgame Γ_{z_k}, and $\omega_0 = \{x_0, z_1, \ldots, z_k, \ldots\}$ is the play realized in situation $u(\cdot) \in U_{x_0}$.

Show that if the map T places each subgame Γ_{z_k} in correspondence with the set of pareto-optimal situations U_z^P, then it is dynamically stable (time-consistent).

8. The map T defined in Example 7 is called strongly dynamic stable (strongly time-consistent) if for any situation $u(\cdot) \in U_{x_0}$, any $z_k \in \{z_i\} = \omega$, where $\{z_i\} = \omega$ is a play in situation $u(\cdot)$, situation $\hat{u}^{z_k}(\cdot) \in U_{z_k}$ there exists a situation $\hat{u}(\cdot) \in U_{x_0}$ for which the situation $\hat{u}^{z_k}(\cdot)$ is its truncation on positions of the subgame Γ_{z_k}.

Show that if the map T places each subgame Γ_{z_k} in correspondence with the set of Nash equilibria, then it is strongly dynamic stable.

9. Construct an example where the map T placing each subgame Γ_z in correspondence with the set of pareto-optimal equilibria is not strongly dynamic stable.

10. For each subgame Γ_z we introduce the quantities $v(\{i\})$, $i = 1, \ldots, n$ representing a maximal guaranteed payoff to the i-th player in the subgame Γ_z, i.e. $v(\{i\}, z)$ is the value of the game constructed in terms of the graph of the subgame Γ_z between Player i and Players $N \setminus i$ acting as one player. In this case, a strategy set for the coalition of players $N \setminus i$ is the Cartesian product of strategy sets for each of the players $k \in \{N \setminus i\}$, $u_{N \setminus i} \in \prod_{k \in N} u_i$, the payoff function for player i in situation $(u_i, u_{N\setminus i})$ is defined to be $H_i^z(u_i, u_{N\setminus i})$, and the payoff function for coalition $N \setminus i$ is taken to be $[-H_i^z(u_i, u_{N\setminus i})]$.

Construct the functions $v(\{i\}, z)$ for all subgames Γ_z of the game from Example 4, 5.2.2.

11. Show that if in a multistage nonzero-sum game Γ with non-negative payoffs ($H_i \geq 0$, $i = 1, \ldots, n$), $v(\{i\}, z) = 0$ for all $i = 1, \ldots, n$ and $z \in \cup_{i=1}^n X_i$, then any play can be realized in some penalty strategy equilibrium.

12. Formalize the k-level control tree-like system as a hierarchical game in which a control center at the i-th level ($i = 1, \ldots, k-1$) allocate resources among subordinate control centers at the next level with $i < k - 1$ and among its subordinate production divisions with $i = k - 1$. The payoff to each production division depends only on its output, while the payoff to the control centers depends on their subordinate production divisions.

13. Find a Nash equilibrium in the tree-like hierarchical k-level game constructed in Exercise 12.

14. Show that the payoff vector $\alpha = \{\overline{v}(N), 0, \ldots, 0\}$ belongs to the core of a tree-like hierarchical two-level game with the characteristic function $\overline{v}(S)$. Show that the equilibrium constructed in the tree-like hierarchical two-level game is also a strong equilibrium.

15. In a diamond-shaped hierarchical game construct a characteristic function by using a Nash equilibrium.

16. Describe the set of all Nash equilibria in a tree-like hierarchical two-level game. Take into account the possibility that the players $B_1, \ldots, B_n$ can "penalize" center A_0 (e.g. by stopping production when the allocation of resources runs counter to the interests of player i).

17. Construct the payoff matrix for players in the game of Example 7, 5.7.1. Find optimal pure strategies and the value of the matrix game obtained.

18. Convert the game from Example 9, 5.7.1, to the matrix form and solve it.

19. Consider the following multistage zero-sum game with delayed information about the positions of one of the players. The game is played by two players: target E and shooter P. The target can move only by the point of the Ox axis with coordinates $0, 1, 2, \ldots$. In this case, if player E is at the point i, then at the next moment he can move only to the points $i+1, i-1$ or stay where he is. Shooter P has j bullets, $(j = 0, 1, \ldots)$ and can fire no more than one bullet at each time instant. It is assumed that the shooter hits the point at which he is aiming.

At each time instant player P knows exactly the position of player E at the previous step, i.e. if player E has been in the point i at the previous step then player P has to aim at the points $i+1$, i and $i-1$. Player E is informed about the number of bullets that player P has at each time instant, but he does not know where Player P is aiming at. The payoff to shooter P is determined by his accuracies, and so the objective of shooter P is to maximize the number of his accurate hits before target E can reach a "bunker". The objective of the target is the opposite one. Here "bunker" means the point 0 where the target is inaccessible for player P.

Denote this game by $\Gamma_{(i,j)}$ with the proviso that at the initial time instant target E is at the point with the coordinate i, while shooter P has j bullets. Denote by $v(i,j)$ the value of the game (if any). It can be readily seen that $v(i,0) = 0$, $i = 1, 2, \ldots$,

$v(1,j) = 0$, $j = 1, 2, \ldots$. At each step of the game $\Gamma_{i,j}$, $i = 2, 3, \ldots$, $j = 1, 2, \ldots$ the shooter has four strategies (actually he has more strategies, but they are not rational), whereas player E has three strategies. The strategies for shooter P are: shooting at the point $i-1$, shooting at the point i, shooting at the point $i+1$, no shooting at this step. The strategies for the target are: move to the point $i-1$, stay at the point i, move to the point $i+1$. Thus at each step of the game we have the matrix game with the payoff matrix

$$A = \begin{bmatrix} 1+v(i-1,j-1) & v(i,j-1) & v(i+1,j-1) \\ v(i-1,j-1) & 1+v(i,j-1) & v(i+1,j-1) \\ v(i-1,j-1) & v(i,j-1) & 1+v(i+1,j-1) \\ v(i-1,j) & v(i,j) & v(i+1,j) \end{bmatrix}.$$

Denote by $x_1(i,j), x_2(i,j), x_3(i,j), x_4(i,j)$ the probabilities that shooter P will use his 1st, 2nd, 3rd and 4th strategies. Also, denote by $y_1(i,j), y_2(i,j), y_3(i,j)$ the probabilities that target E will use its 1st, 2nd and 3rd strategies (behavior strategies for Players P and E respectively are functions of the pairs $\{i,j\}$).

(a) Show that the value of the game $v(i,j)$ and the optimal behavior strategies for shooter $P(x_1(i,j), x_2(i,j), x_3(i,j), x_4(i,j))$ and target $E(y_1(i,j), y_2(i,j), y_3(i,j))$ are connected by the following inequalities:

$$\begin{aligned}
&(1+v(i-1,j-1))x_1 + v(i-1,j-1)x_2 + v(i-1,j-1)x_3 \\
&\qquad +v(i-1,j)x_4 \geq v(i,j), \\
&v(i,j-1)x_1 + (1+v(i,j-1))x_2 + v(i,j-1)x_3 \\
&\qquad +v(i,j)x_4 \geq v(i,j), \\
&v(i+1,j-1)x_1 + v(i+1,j-1)x_2 + (1+v(i+1,j-1))x_3 \\
&\qquad +v(i+1,j)x_4 \geq v(i,j),
\end{aligned}$$

$$x_1 + x_2 + x_3 + x_4 = 1,\ x_1 \geq 0,\ x_2 \geq 0,\ x_3 \geq 0,\ x_4 \geq 0;$$

$$(1 + v(i-1, j-1))y_1 + v(i, j-1)y_2$$
$$+v(i+1, j-1)y_3 \leq v(i, j),$$
$$v(i-1, j-1)y_1 + v(i, j-1)y_2$$
$$+(1 + v(i+1, j-1))y_3 \leq v(i, j),$$
$$v(i-1, j-1)y_1 + (1 + v(i, j-1))y_2$$
$$+v(i+1, j-1)y_3 \leq v(i, j),$$
$$v(i-1, j)y_1 + v(i, j)y_2 + v(i+1, j)y_3 \leq v(i, j),$$
$$y_1 + y_2 + y_3 = 1,\ y_1 \geq 0,\ y_2 \geq 0,\ \ y_3 \geq 0.$$

Hint. The difficulty associated with this game is that in order to determine $v(i, j)$ we need to know $v(i+1, j)$, in order to determine $v(i+1, j)$ we need to know $v(i+2, j)$ and so on. The exercises below provide a solution to the game $\Gamma_{i,j}$ and some of its properties.

(b) Let $\varphi(i, j)$, $i = 1, 2, \ldots$, $j = 0, 1, \ldots$ be the double sequence defined by the relationships

$$\varphi(i, 0) = 0,\ i = 1, 2, \ldots;\ \varphi(1, j) = 0,\ j = 1, 2, \ldots,$$
$$\varphi(i, j) = \min\{(1 + \varphi(i-1, j-1) + \varphi(i, j-1)$$
$$+\varphi(i+1, j-1))/3,$$
$$\times(1 + \varphi(i-1, j-1) + \varphi(i, j-1))/2\}.$$

1) Prove that $v(i, j) = \varphi(i, j)$, and if $v(i, j) = (1 + v(i-1, j-1) + v(i, j-1) + v(i+1, j-1))/3$, then

$$x_1(i, j) = v(i, j) - v(i-1, j-1),$$
$$x_2(i, j) = v(i, j) - v(i, j-1),$$
$$x_3(i, j) = v(i, j) - v(i+1, j-1),$$
$$x_4(i, j) = 0,$$
$$y_1(i, j) = y_2(i, j) = y_3(i, j) = 1/3;$$

2) Prove that $v(i,j) = \varphi(i,j)$, and if $v(i,j) = (1 + v(i-1, j-1) + v(i, j-1))/2$, then

$$x_1(i,j) = v(i,j) - v(i-1, j-1),$$

$$x_2(i,j) = v(i,j) - v(i, j-1),$$

$$x_3(i,j) = x_4(i,j) = 0,$$

$$y_1(i,j) = y_2(i,j) = 1/2,$$

$$y_3(i,j) = 0;$$

(c) Prove that the following relationships hold for any $j = 0, 1, 2, \ldots$:

1) $v(i,j) = j/3,\ i = j+1, j+2, \ldots;$
2) $v(i,j) \leq v(i+1, j),\ i = 1, 2, \ldots;$
3) $v(i,j) \leq v(i, j+1),\ i = 2, 3, \ldots;$
4) $v(i,j) + v(i+2, j) \leq 2v(i+1, j),\ i = 1, 2, \ldots$

(d) Prove that:

1) $\lim_{i \to +\infty} v(i,j) = j/3$ for any fixed $j = 0, 1, 2, \ldots;$
2) $\lim_{j \to -\infty} v(i,j) = i - 1$ for any fixed $i = 1, 2, \ldots.$

20. Consider an extension of the game of shooter and target, where target E is in position i, from where it can move at most k units to the right or the left, i.e. it can move to each of the points: $i-k, i-k+1, \ldots, i, \ldots, i+k$. The other objectives and possibilitilies for shooter P and target E remain unaffected in terms of the new definition of a strategy for Player E.

Denote by $G(i,j)$ the game with the proviso that at the initial time instant the target is at the ith point and the shooter has j bullets. Further, denote by $v(i,j)$ the value of the game $G(i,j)$. From the definition of $G(i,j)$ we have

$$v(i,0) = 0,\ i = 1, 2, \ldots,$$

$$v(i,j) = 0,\ i = 1, 2, \ldots, k,\ j = 1, 2, \ldots.$$

At each step of the game $G(i,j)$, $i = k+1, \ldots$, $j = 1, \ldots$ shooter P has $(2k+2)$ pure strategies, whereas target E has $(2k+1)$

pure strategies. The pure strategies for player P are: shooting at the point $i-k$, shooting at the point $i-k+1,\ldots$, shooting at the point $i+k$, no shooting at this step. The strategies for player E are: move to the point $i-k$, move to the point $i-k+1,\ldots$, move to the point $i+k$.

Thus, at each step of the game we have the game with the $(2k+2)\times(2k+1)$ matrix $\{\alpha_{mn}(i,j)\}$, where

$$\alpha_{mn}(i,j)=\begin{cases} 1+v(i+n-k-1,j-1), & \text{if } m=n \\ & \quad =1,\ldots,2k+1, \\ v(i+n-k-1,j-1), & \text{if } m\neq n;\ m,n \\ & \quad =1,\ldots,2k+1, \\ v(i+n-k-1,j), & \text{if } m=2k+2, \\ & \quad n=1,\ldots,2k+1. \end{cases}$$

(a) Show that the game $G(i,j)$ has the value equal to $v(i,j)$ if and only if there exist $(x_1,x_2,\ldots,x_{2k+2})$, $(y_1,y_2,\ldots,y_{2k+1})$ such that

$$\sum_{m=1}^{2k+2}\alpha_{mn}(i,j)x_m\geq v(i,j),\ n=1,\ldots,2k+1,$$

$$\sum_{m=1}^{2k+2}x_m=1,\ x_m\geq 0,\ m=1,\ldots,2k+2,$$

$$\sum_{n=1}^{2k+1}\alpha_{mn}(i,j)y_n\leq v(i,j),\ m=1,\ldots,2k+1,$$

$$\sum_{n=1}^{2k+1}y_n=1,\ y_n\geq 0,\ n=1,\ldots,2k+1.$$

Hint. Denote by $x_1(i,j),x_2(i,j),\ldots,x_{2k+2}(i,j)$ the optimal behavior strategies for shooter P and by $y_1(i,j)$, $y_2(i,j),\ldots,y_{2k+1}(i,j)$ the optimal behavior strategies for target E. The exercises below provide a solution of the game $G(i,j)$ and its properties.

(b) Denote by $\varphi(i,j)$, $j = 0,1,\ldots$, $i = 1,2,\ldots$, the following double sequence:

$$\varphi(i,0) = 0, \; i = 1,2,\ldots;$$

$$\varphi(i,j) = 0, \; i = 1,2,\ldots,k; \; j = 1,2,\ldots;$$

$$\varphi(i,j) = \min_{r=1,\ldots,k+1}\left(\left(1+\sum_{t=1}^{k+r} v(i+t-k-1,j-1)\right)\Big/(k+2)\right),$$

$$i = k+1,k+2,\ldots, \; j = 1,2,\ldots. \tag{5.13.3}$$

Prove that:

1) $v(i,j) = \varphi(i,j)$;
2) for $i = k+1,\ldots$; $j = 1,2,\ldots$, we have $x_m(i,j) = v(i,j) - v(i+m-k-1,j-1)$ with $m = 1,\ldots,k+r^*$, otherwise $y_n = 0$. Here $r = r^*$ is the point at which the minimum is attained in (5.12.3).

(c) Prove that for $j = 0,1,\ldots$:

1) $v(i,j) \geq 0, \; i = 1,2,\ldots;$
2) $v(i,j) = j/(2k+1), \; i = kj+1, kj+2,\ldots;$
3) $v(i,j) \leq v(i+1,j), \; i = 1,2,\ldots;$
4) $v(i,j) \leq v(i,j+1), \; i = k+1,k+2,\ldots;$
5) $v(i,j+1) \leq v(i,j) + 1/(2k+1), \; i = 1,2,\ldots;$

(d) The game $G(i,\infty)$. Prove that $\lim_{j\to\infty} v(i,j) = w(i)$ for each $i = 1,2,\ldots$, where $w(i)$ is a solution to the linear difference equation

$$kw(i) - \sum_{p=1}^{k} w(i-p) = 1, \; i = k+1,k+2,\ldots$$

with initial conditions

$$w(1) = w(2) = \ldots = w(k) = 0.$$

21. Consider a repeated l stage game, where on each stage a two-level hierarchical game is played (see 5.6). Describe the set of all possible Nash equilibria in this game, taking in account the penalty strategies of center A_0 (he can prescribe to Players B_i some fixed behavior and on the next step penalize those B_i, who do not follow his recommendations by giving them 0 recourses) and threat of players B_i to A_0 to stop production if he will not satisfy their requirements in giving the resources to each of B_i.

Chapter 6

N-Person Differential Games

6.1 Optimal Control Problem

6.1.1 Consider the optimal control problem in which the single decision-maker:

$$\max_u \left\{ \int_{t_0}^{T} g[s, x(s), u(s)]ds + q(x(T)) \right\}, \tag{6.1.1}$$

subject to the vector-valued differential equation:

$$\dot{x}(s) = f[s, x(s), u(s)]ds, \quad x(t_0) = x_0, \tag{6.1.2}$$

where $x(s) \in X \subset R^m$ denotes the state variables of game, and $u \in U$ is the control.

The functions $f[s, x, u]$, $g[s, x, u]$ and $q(x)$ are differentiable functions.

Dynamic programming and optimal control are used to identify optimal solutions for the problem (6.1.1)–(6.1.2) .

6.1.2. Dynamic Programming. A frequently adopted approach to dynamic optimization problems is the technique of dynamic programming. The technique was developed by Bellman (1957). The technique is given in Theorem below.

Theorem. *A set of controls* $u^*(t) = \phi^*(t, x)$ *constitutes an optimal solution to the control problem* (6.1.1)–(6.1.2) *if there exist continuously differentiable functions* $V(t, x)$ *defined on* $[t_0, T] \times R^m \to R$ *and satisfying the following Bellman equation:*

$$-V_t(t, x) = \max_u \{g[t, x, u] + V_x(t, x) f[t, x, u]\}$$
$$= \{g[t, x, \phi^*(t, x)] + V_x(t, x) f[t, x, \phi^*(t, x)]\},$$
$$V(T, x) = q(x).$$

Proof. Define the maximized payoff at time t with current state x as a *value Bellman function* in the form:

$$V(t, x) = \max_u \left[\int_t^T g(s, x(s), u(s)) ds + q(x(T)) \right]$$
$$= \int_t^T g[s, x^*(s), \phi^*(s, x^*(s))] ds + q(x^*(T))$$

satisfying the boundary condition

$$V(T, x^*(T)) = q(x^*(T)),$$

and

$$\dot{x}^*(s) = f[s, x^*(s), \phi^*(s, x^*(s))], \quad x^*(t_0) = x_0.$$

If in addition to $u^*(s) \equiv \phi^*(s, x)$, we are given another set of strategies, $u(s) \in U$, with the corresponding terminating trajectory $x(s)$, then from conditions of the theorem we have

$$g(t, x, u) + V_x(t, x) f(t, x, u) + V_t(t, x) \leq 0,$$

and

$$g(t, x^*, u^*) + V_{x^*}(t, x^*) f(t, x^*, u^*) + V_t(t, x^*) = 0.$$

Integrating the above expressions from t_0 to T, we obtain

$$\int_{t_0}^{T} g(s, x(s), u(s))ds + V(T, x(T)) - V(t_0, x_0) \leq 0,$$

and

$$\int_{t_0}^{T} g(s, x^*(s), u^*(s))ds + V(T, x^*(T)) - V(t_0, x_0) = 0.$$

Elimination of $V(t_0, x_0)$ yields

$$\int_{t_0}^{T} g(s, x(s), u(s))ds + q(x(T)) \leq \int_{t_0}^{T} g(s, x^*(s), u^*(s))ds + q(x^*(T)),$$

from which it readily follows that u^* is the optimal strategy.

Upon substituting the optimal strategy $\phi^*(t, x)$ into (6.1.2) yields the dynamics of optimal state trajectory as

$$\dot{x}(s) = f[s, x(s), \phi^*(s, x(s))]ds, \quad x(t_0) = x_0. \tag{6.1.3}$$

Let $x^*(t)$ denote the solution to (6.1.3). The optimal trajectory $\{x^*(t)\}_{t=t_0}^{T}$ can be expressed as:

$$x^*(t) = x_0 + \int_{t_0}^{t} f\left[s, x^*(s), \psi^*(s, x^*(s))\right] ds. \tag{6.1.4}$$

For notational convenience, we use the terms $x^*(t)$ and x_t^* interchangeably.

The value (Bellman) function $V(t, x)$ where $x = x_t^*$ can be expressed as

$$V(t, x_t^*) = \int_{t}^{T} g[s, x^*(s), \phi^*(s)]ds + q(x^*(T)).$$

Example 1. Consider the optimization problem:

$$\max_u \left\{ \int_0^T \exp[-rs] \left[-x(s) - cu(s)^2\right] ds + \exp[-rT]qx(T) \right\} \tag{6.1.5}$$

subject to

$$\dot{x}(s) = a - u(s)(x(s))^{1/2}, \quad x(0) = x_0, \quad u(s) \geq 0, \tag{6.1.6}$$

where a, c, and x_0 are positive parameters.

Invoking the above theorem we have

$$-V_t(t,x) = \max_u \left\{ \left[-x - cu^2\right] \exp[-rt] + V_x(t,x) \left[a - ux^{1/2}\right] \right\}, \tag{6.1.7}$$

and

$$V(T,x) = \exp\left[-rT\right] qx.$$

Performing the indicated maximization in (6.1.7) yields

$$\phi(t,x) = \frac{-V_x(t,x)x^{1/2}}{2c} \exp[rt].$$

Substituting $\phi(t,x)$ into (6.1.7) and upon solving (6.1.7), one obtains

$$V(t,x) = \exp[-rt][A(t)x + B(t)],$$

where $A(t)$ and $B(t)$ satisfy:

$$\dot{A}(t) = rA(t) - \frac{A(t)^2}{4c} + 1,$$
$$\dot{B}(t) = rB(t) - aA(t),$$
$$A(T) = q \quad \text{and} \quad B(T) = 0.$$

The optimal control can be solved explicitly as

$$\phi(t,x) = \frac{-A(t)x^{1/2}}{2c} \exp[rt].$$

Now, consider the infinite-horizon dynamic optimization problem with a constant discount rate

$$\max_u \left\{ \int_{t_0}^{\infty} g[x(s), u(s)] \exp[-r(s - t_0)] ds \right\}, \tag{6.1.8}$$

subject to the vector-valued differential equation

$$\dot{x}(s) = f[x(s), u(s)]ds, \quad x(t_0) = x_0. \tag{6.1.9}$$

Since s does not appear in $g[x(s), u(s)]$ and the state dynamics explicitly, the set of problem (6.1.8)–(6.1.9) is an autonomous problem.

Consider the alternative problem:

$$\max_u \int_t^{\infty} g[x(s), u(s)] \exp[-r(s - t)] ds, \tag{6.1.10}$$

subject to

$$\dot{x}(s) = f[x(s), u(s)], \quad x(t) = x. \tag{6.1.11}$$

The infinite-horizon autonomous problem (6.1.10)–(6.1.11) is independent of the choice of t and dependent only upon the state at the starting time, that is x.

Define the value function to problems (6.1.8)–(6.1.9) by

$$V(t, x) = \max_u \left\{ \int_t^{\infty} g[x(s), u(s)] \exp[-r(s - t_0)] ds | x(t) = x = x_t^* \right\},$$

where x_t^* is the state at time t along the optimal trajectory. Moreover, we can write

$$V(t, x) = \exp[-r(t - t_0)] \max_u \times \left\{ \int_t^{\infty} g[x(s), u(s)] \exp[-r(s - t)] ds | x(t) = x = x_t^* \right\}.$$

Since the problem

$$\max_u \left\{ \int_t^{\infty} g[x(s), u(s)] \exp[-r(s-t)] ds | x(t) = x = x_t^* \right\}$$

depends on the current state x only, we can write

$$W(x) = \max_u \left\{ \int_t^{\infty} g[x(s), u(s)] \exp[-r(s-t)] ds | x(t) = x = x_t^* \right\}.$$

It follows that

$$\begin{aligned} V(t, x) &= \exp[-r(t-t_0)] W(x), \\ V_t(t, x) &= -r \exp[-r(t-t_0)] W(x), \quad \text{and} \\ V_x(t, x) &= -r \exp[-r(t-t_0)] W_x(x). \end{aligned} \tag{6.1.12}$$

Substituting the results from (6.1.12) into Bellman equation from the theorem yields

$$rW(x) = \max_u \{g[x, u] + W_x(x) f[x, u]\}. \tag{6.1.13}$$

Since time is not explicitly involved (6.1.13), the derived control u will be a function of x only. Hence one can obtain the theorem.

Theorem. *A set of controls $u = \phi^*(x)$ constitutes an optimal solution to the infinite-horizon control problem* (6.1.10)–(6.1.11) *if there exists continuously differentiable function $W(x)$ defined on $R^m \to R$ which satisfies the following equation:*

$$\begin{aligned} rW(x) &= \max_u \{g[x, u] + W_x(x) f[x, u]\} \\ &= \{g[x, \phi^*(x)] + W_x(x) f[x, \phi^*(x)]\}. \end{aligned}$$

Substituting the optimal control into (6.1.9) yields the dynamics of the optimal state path as

$$\dot{x}(s) = f[x(s), \phi^*(x(s))] ds, \quad x(t_0) = x_0.$$

Solving the above dynamics yields the optimal state trajectory $\{x^*(t)\}_{t \geq t_o}$ as

$$x^*(t) = x_0 + \int_{t_0}^{t} f[x^*(s), \psi^*(x^*(s))]ds, \quad \text{for } t \geq t_0.$$

We denote term $x^*(t)$ by x_t^*. The optimal control to the infinite-horizon problem (6.1.8)–(6.1.9) can be expressed as $\psi^*(x_t^*)$ in the time interval $[t_0, \infty)$.

Example 2. Consider the infinite-horizon dynamic optimization problem:

$$\max_u \int_0^\infty \exp[-rs] \left[-x(s) - cu(s)^2\right] ds \tag{6.1.14}$$

subject to dynamics (6.1.6).

Invoking previous theorem we have

$$rW(x) = \max_u \left\{ \left[-x - cu^2\right] + W_x(x) \left[a - ux^{1/2}\right] \right\}. \tag{6.1.15}$$

Performing the indicated maximization in (6.1.15) yields

$$\phi^*(x) = \frac{-V_x(x)x^{1/2}}{2c}.$$

Substituting $\phi(x)$ into (6.1.15) and upon solving (6.1.15), one obtains

$$V(t, x) = \exp[-rt][Ax + B],$$

where A and B satisfy

$$0 = rA - \frac{A^2}{4c} + 1 \quad \text{and} \quad B = \frac{-a}{r}A.$$

Solving A to be $2c[r \pm (r^2 + c^{-1})^{1/2}]$. For a maximum, the negative root of A holds. The optimal control can be obtained as

$$\phi^*(x) = \frac{-Ax^{1/2}}{2c}.$$

Substituting $\phi^*(x) = -Ax^{1/2}/(2c)$ into (6.1.6) yields the dynamics of the optimal trajectory as

$$\dot{x}(s) = a + \frac{A}{2c}(x(s)), \quad x(0) = x_0.$$

Upon the above dynamical equation yields the optimal trajectory $\{x^*(t)\}_{t \geq t_0}$ as

$$x^*(t) = \left[x_0 + \frac{2ac}{A}\right] \exp\left(\frac{A}{2c}t\right) - \frac{2ac}{A} = x_t^*, \quad \text{for } t \geq t_0.$$

The optimal control of problem (6.1.14)–(6.1.15) is then

$$\phi^*(x_t^*) = \frac{-A(x_t^*)^{1/2}}{2c}.$$

6.1.3. The Optimal Control. maximum principle of optimal control was developed by Pontryagin (details in Pontryagin *et al.* (1962)). Consider again the dynamic optimization problem (6.1.1)–(6.1.2).

Theorem. Pontryagin's Maximum Principle. *A set of controls $u^*(s) = \zeta^*(s, x_0)$ provides an optimal solution to control problem (6.1.1)–(6.1.2), and $\{x^*(s), t_0 \leq s \leq T\}$ is the corresponding state trajectory, if there exist costate functions $\Lambda(s) : [t_0, T] \to R^m$ such that the following relations are satisfied:*

$$\zeta^*(s, x_0) \equiv u^*(s) = \arg\max_u \{g[s, x^*(s), u(s)] + \Lambda(s) f[s, x^*(s), u(s)]\},$$

$$\dot{x}^*(s) = f[s, x^*(s), u^*(s)], \quad x^*(t_0) = x_0,$$

$$\dot{\Lambda}(s) = -\frac{\partial}{\partial x}\{g[s, x^*(s), u^*(s)] + \Lambda(s) f[s, x^*(s), u^*(s)]\},$$

$$\Lambda(T) = \frac{\partial}{\partial x^*} q(x^*(T)).$$

Proof. First define the function (Hamiltonian)

$$H(t, x, u) = g(t, x, u) + V_x(t, x) f(t, x, u).$$

The theorem from 6.1.2 gives us

$$-V_t(t,x) = \max_u H(t,x,u).$$

This yields the first condition of the above theorem. Using u^* to denote the payoff maximizing control, we obtain

$$H(t,x,u^*) + V_t(t,x) \equiv 0,$$

which is an identity in x. Differentiating this identity partially with respect to x yields

$$V_{tx}(t,x) + g_x(t,x,u^*) + V_x(t,x)f_x(t,x,u^*) + V_{xx}(t,x)f(t,x,u^*)$$
$$+ [g_u(t,x,u^*) + V_x(t,x)f_u(t,x,u^*)]\frac{\partial u^*}{\partial x} = 0.$$

If u^* is an interior point, then $[g_u(t,x,u^*) + V_x(t,x)f_u$ $(t,x,u^*)] = 0$ according to the condition $-V_t(t,x) = \max_u H(t,x,u)$. If u^* is not an interior point, then it can be shown that

$$[g_u(t,x,u^*) + V_x(t,x)f_u(t,x,u^*)]\frac{\partial u^*}{\partial x} = 0$$

(because of optimality, $[g_u(t,x,u^*)..+V_x(t,x)f_u(t,x,u^*)]$ and $\partial u^*/\partial x$ are orthogonal; and for specific problems we may have $\partial u^*/\partial x = 0$). Moreover, the expression $V_{tx}(t,x) + V_{xx}(t,x)f(t,x,u^*) \equiv V_{tx}(t,x) + V_{xx}(t,x)\dot{x}$ can be written as $[dV_x(t,x)](dt)^{-1}$. Hence, we obtain:

$$\frac{dV_x(t,x)}{dt} + g_x(t,x,u^*) + V_x(t,x)f_x(t,x,u^*) = 0.$$

By introducing the *costate vector*, $\Lambda(t) = V_{x^*}(t,x^*)$, where x^* denotes the state trajectory corresponding to u^*, we arrive at

$$\frac{dV_x(t,x^*)}{dt} = \dot{\Lambda}(s) = -\frac{\partial}{\partial x}\{g[s,x^*(s),u^*(s)] + \Lambda(s)f[s,x^*(s),u^*(s)]\}.$$

Finally, the boundary condition for $\Lambda(t)$ is determined from the terminal condition of optimal control in Theorem from 6.1.2 as

$$\Lambda(T) = \frac{\partial V(T, x^*)}{\partial x} = \frac{\partial q(x^*)}{\partial x}.$$

Hence theorem follows.

Example 3. Consider the problem in Example 1. Invoking theorem, we first solve the control $u(s)$ that satisfies

$$\arg\max_u \left\{ \left[-x^*(s) - cu(s)^2\right] \exp[-rs] + \Lambda(s) \left[a - u(s)x^*(s)^{1/2}\right] \right\}.$$

Performing the indicated maximization:

$$u^*(s) = \frac{-\Lambda(s)x^*(s)^{1/2}}{2c} \exp[rs]. \tag{6.1.16}$$

We also obtain

$$\dot{\Lambda}(s) = \exp[-rs] + \frac{1}{2}\Lambda(s)u^*(s)x^*(s)^{-1/2}. \tag{6.1.17}$$

Substituting $u^*(s)$ from (6.1.16) into (6.1.6) and (6.1.17) yields a pair of differential equations:

$$\begin{aligned} \dot{x}^*(s) &= a + \frac{1}{2c}\Lambda(s)(x^*(s)) \exp[rs], \\ \dot{\Lambda}(s) &= \exp[-rs] + \frac{1}{4c}\Lambda(s)^2 \exp[rs], \end{aligned} \tag{6.1.18}$$

with boundary conditions:

$$x^*(0) = x_0 \quad \text{and} \quad \Lambda(T) = \exp[-rT]q.$$

Solving (6.1.18) yields

$$\begin{aligned} \Lambda(s) = 2c &\left(\theta_1 - \theta_2 \frac{q - 2c\theta_1}{q - 2c\theta_2} \exp\left[\frac{\theta_1 - \theta_2}{2}(T - s)\right]\right) \exp(-rs) \\ &\div \left(1 - \frac{q - 2c\theta_1}{q - 2c\theta_2} \exp\left[\frac{\theta_1 - \theta_2}{2}(T - s)\right]\right), \end{aligned}$$

and

$$x^*(s) = \varpi(0,s)\left[x_0 + \int_0^s \varpi^{-1}(0,t)a\,dt\right], \quad \text{for } s \in [0,T],$$

where

$$\theta_1 = r - \sqrt{r^2 + \frac{1}{c}} \quad \text{and} \quad \theta_2 = r + \sqrt{r^2 + \frac{1}{c}},$$

$$\varpi(0,s) = \exp\left[\int_0^s H(\tau)d\tau\right],$$

and

$$H(\tau) = \left(\theta_1 - \theta_2 \frac{q - 2c\theta_1}{q - 2c\theta_2}\exp\left[\frac{\theta_1 - \theta_2}{2}(T - \tau)\right]\right)$$
$$\div\left(1 - \frac{q - 2c\theta_1}{q - 2c\theta_2}\exp\left[\frac{\theta_1 - \theta_2}{2}(T - \tau)\right]\right).$$

Upon substituting $\Lambda(s)$ and $x^*(s)$ into (6.1.16) yields $u^*(s) = \zeta^*(s, x_0)$ which is a function of s and x_0.

Consider the infinite-horizon dynamic optimization problem (6.1.8)–(6.1.9). The Hamiltonian function can be expressed as

$$H(t,x,u) = g(x,u)\exp[-r(t - t_0)] + \Lambda(t)f(x,u).$$

Define $\lambda(t) = \Lambda(t)\exp[r(t - t_0)]$ and the current value Hamiltonian

$$\begin{aligned}\hat{H}(t,x,u) &= H(t,x,u)\exp[r(t - t_0)] \\ &= g(x,u) + \lambda(t)f(x,u). \qquad (6.1.19)\end{aligned}$$

Using the previous theorem and (6.1.19) we get the maximum principle for the game (6.1.10)–(6.1.11).

Theorem. *A set of controls* $u^*(s) = \zeta^*(s, x_t)$ *provides an optimal solution to the infinite-horizon control problem* (6.1.10)–(6.1.11), *and* $\{x^*(s),\ s \geq t\}$ *is the corresponding state trajectory, if there exist costate functions* $\lambda(s) : [t, \infty) \to R^m$ *such that the following relations* *are satisfied*

$$\zeta^*(s, x_t) \equiv u^*(s) = \arg\max_u \{g[x^*(s), u(s)] + \lambda(s) f[x^*(s), u(s)]\},$$
$$\dot{x}^*(s) = f[x^*(s), u^*(s)], \quad x^*(t) = x_t,$$
$$\dot{\lambda}(s) = r\lambda(s) - \frac{\partial}{\partial x}\{g[x^*(s), u^*(s)] + \lambda(s) f[x^*(s), u^*(s)]\}.$$

Example 4. Consider the infinite-horizon problem in Example 2. Invoking the previous theorem we have

$$\begin{aligned} \zeta^*(s, x_t) &\equiv u^*(s) \\ &= \arg\max_u \{[-x^*(s) - cu(s)^2] + \lambda(s)[a - u(s)x^*(s)^{1/2}]\}, \\ \dot{x}^*(s) &= a - u^*(s)(x^*(s))^{1/2}, \quad x^*(t) = x_t, \\ \dot{\lambda}(s) &= r\lambda(s) + \left[1 + \frac{1}{2}\lambda(s)u^*(s)x^*(s)^{-1/2}\right]. \end{aligned} \tag{6.1.20}$$

Performing the indicated maximization yields

$$u^*(s) = \frac{-\lambda(s)x^*(s)^{1/2}}{2c}.$$

Substituting $u^*(s)$ into (6.1.20), one obtains

$$\begin{aligned} \dot{x}^*(s) &= a + \frac{\lambda(s)}{2c}u^*(s)x^*(s), \quad x^*(t) = x_t, \\ \dot{\lambda}(s) &= r\lambda(s) + \left[1 - \frac{1}{4c}\lambda(s)^2\right]. \end{aligned} \tag{6.1.21}$$

Solving (6.1.21) in a manner similar to that in Example 3 yields the solutions of $x^*(s)$ and $\lambda(s)$. Upon substituting them into $u^*(s)$ gives the optimal control of the problem.

6.2 Differential Games and Their Solution Concepts

6.2.1. One particularly complex but fruitful branch of game theory is dynamic or differential games, which investigates interactive decision-making over time under different assumptions regarding pre-commitment (of actions), information, and uncertainty. The origin of differential games traces back to the late 1940s. Rufus Isaacs (whose work was published in 1965) formulated missile versus enemy aircraft pursuit schemes in terms of descriptive and navigation variables (state and control), and established a fundamental principle: the tenet of transition. The seminal contributions of Isaacs together with the classic research of Bellman on dynamic programming and Pontryagin *et al.* on optimal control laid the foundations of deterministic differential zero-sum games.

Differential games or continuous-time infinite dynamic games study a class of decision problems, under which the evolution of the state is described by a differential equation and the players act throughout a time interval. In this section, we follow Basar and Olsder (1984), Yeung and Petrosyan (2006), see also Kleimenov (1993), Vaisbord and Zhukovsky (1980).

In particular, in the general n-person differential game, Player i seeks to:

$$\max_{u_i} \int_{t_0}^{T} g^i[s, x(s), u_1(s), u_2(s), \ldots, u_n(s)]ds + q^i(x(T)),$$
$$\text{for } i \in N = \{1, 2, \ldots, n\}, \tag{6.2.1}$$

subject to the deterministic dynamics

$$\dot{x}(s) = f[s, x(s), u_1(s), u_2(s), \ldots, u_n(s)], \quad x(t_0) = x_0, \tag{6.2.2}$$

where $x(s) \in X \subset R^m$ denotes the state variables of game, and $u_i \in U^i$ is the control of Player i, for $i \in N$.

The functions $f[s, x, u_1, u_2, \ldots, u_n]$, $g^i[s, \cdot, u_1, u_2, \ldots, u_n]$ and $q^i(\cdot)$, for $i \in N$, and $s \in [t_0, T]$ are differentiable functions.

A set-valued function $\eta^i(\cdot)$ defined for each $i \in N$ as

$$\eta^i(s) = \left\{x(t), \quad t_0 \le t \le \epsilon_s^i\right\}, \quad t_0 \le \epsilon_s^i \le s,$$

where ϵ_s^i is nondecreasing in s, and $\eta^i(s)$ determines the state information gained and recalled by Player i at time $s \in [t_0, T]$. Specification of $\eta^i(\cdot)$(in fact, ϵ_s^i in this formulation) characterizes the *information structure* of Player i and the collection (over $i \in N$) of these information structures is the *information structure* of the game.

A sigma-field N_s^i in S_0 generated for each $i \in N$ by the cylinder sets $\{x \in S_0,\ x(t) \in B\}$ where B is a Borel set in S^0 and $0 \leq t \leq \epsilon_s$. N_s^i, $s \geq t_0$, is called the *information field* of Player i.

A pre-specified class Γ^i of mappings $\upsilon_i : [t_0, T] \times S_0 \to S^i$, with the property that $u_i(s) = \upsilon_i(s, x)$ is n_s^i-measurable (i.e. it is adapted to the information field N_s^i). U^i is the strategy space of Player i and each of its elements υ_i is a permissible strategy for Player i.

Definition. A set of strategies $\{\upsilon_1^*(s), \upsilon_2^*(s), \ldots, \upsilon_n^*(s)\}$ is said to constitute a noncooperative Nash equilibria solution for the n-person differential game (6.2.1)–(6.2.2), if the following inequalities are satisfied for all $\upsilon_i(s) \in U^i$, $i \in N$:

$$\int_{t_0}^{T} g^1[s, x^*(s), \upsilon_1^*(s), \upsilon_2^*(s), \ldots, \upsilon_n^*(s)]ds + q^1(x^*(T))$$

$$\geq \int_{t_0}^{T} g^1[s, x^{[1]}(s), \upsilon_1(s), \upsilon_2^*(s), \ldots, \upsilon_n^*(s)]ds + q^1(x^{[1]}(T)),$$

$$\int_{t_0}^{T} g^2[s, x^*(s), \upsilon_1^*(s), \upsilon_2^*(s), \ldots, \upsilon_n^*(s)]ds + q^2(x^*(T))$$

$$\geq \int_{t_0}^{T} g^2[s, x^{[2]}(s), \upsilon_1^*(s), \upsilon_2(s), \upsilon_3^*(s), \ldots, \upsilon_n^*(s)]ds + q^2(x^{[2]}(T)),$$

$$\vdots \qquad \vdots$$

$$\vdots \qquad \vdots$$

$$\int_{t_0}^{T} g^n[s, x^*(s), \upsilon_1^*(s), \upsilon_2^*(s), \ldots, \upsilon_n^*(s)]ds + q^n(x^*(T))$$

$$\geq \int_{t_0}^{T} g^n[s, x^{[n]}(s), \upsilon_1^*(s), \upsilon_2^*(s), \ldots, \upsilon_{n-1}^*(s), \upsilon_n(s)]ds$$

$$+ q^n(x^{[n]}(T));$$

where on the time interval $[t_0, T]$:

$$\begin{aligned}
\dot{x}^*(s) &= f[s, x^*(s), \upsilon_1^*(s), \upsilon_2^*(s), \ldots, \upsilon_n^*(s)], \quad x^*(t_0) = x_0, \\
\dot{x}^{[1]}(s) &= f[s, x^{[1]}(s), \upsilon_1(s), \upsilon_2^*(s), \ldots, \upsilon_n^*(s)], \quad x^{[1]}(t_0) = x_0, \\
\dot{x}^{[2]}(s) &= f[s, x^{[2]}(s), \upsilon_1^*(s), \upsilon_2(s), \upsilon_3^*(s), \ldots, \upsilon_n^*(s)], \quad x^{[2]}(t_0) = x_0, \\
&\vdots \qquad\qquad \vdots \\
&\vdots \qquad\qquad \vdots \\
\dot{x}^{[n]}(s) &= f[s, x^{[n]}(s), \upsilon_1^*(s), \upsilon_2^*(s), \ldots, \upsilon_{n-1}^*(s), \upsilon_n(s)], \quad x^{[n]}(t_0) = x_0.
\end{aligned}$$

The set of strategies $\{\upsilon_1^*(s), \upsilon_2^*(s), \ldots, \upsilon_n^*(s)\}$ is known as a *Nash equilibria* of the game.

6.2.2. Open-loop Nash Equilibria. If the players choose to commit their strategies from the outset, the players' information structure can be seen as an *open-loop* pattern in which $\eta^i(s) = \{x_0\}$, $s \in [t_0, T]$. Their strategies become functions of the initial state x_0 and time s, and can be expressed as $\{u_i(s) = \vartheta_i(s, x_0)$, for $i \in N\}$. In particular, an open-loop Nash equilibria for the game (6.2.1) and (6.2.2) is characterized as:

Theorem. *A set of strategies* $\{u_i^*(s) = \zeta_i^*(s, x_0)$, *for* $i \in N\}$ *provides an open-loop Nash equilibria solution to the game* (6.2.1)–(6.2.2), *and* $\{x^*(s),\ t_0 \leq s \leq T\}$ *is the corresponding state trajectory, if there exist* m *costate functions* $\Lambda^i(s) : [t_0, T] \to R^m$, *for* $i \in N$, *such that the following relations are satisfied*:

$$\begin{aligned}
\zeta_i^*(s, x_0) &\equiv u_i^*(s) \\
&= \arg\max_{u_i \in U^i} \Big\{ g^i\big[s, x^*(s), u_1^*(s), u_2^*(s), \ldots, u_{i-1}^*(s), u_i(s), \\
&\qquad\qquad u_{i+1}^*(s), \ldots, u_n^*(s)\big] + \Lambda^i(s) f[s, x^*(s), u_1^*(s), \\
&\qquad\qquad u_2^*(s), \ldots, u_{i-1}^*(s), u_i(s), u_{i+1}^*(s), \ldots, u_n^*(s)] \Big\},
\end{aligned}$$

$$\dot{x}^*(s) = f[s, x^*(s), u_1^*(s), u_2^*(s), \ldots, u_n^*(s)], \quad x^*(t_0) = x_0,$$

$$\dot{\Lambda}^i(s) = -\frac{\partial}{\partial x^*}\{g^i[s, x^*(s), u_1^*(s), u_2^*(s), \ldots, u_n^*(s)].$$
$$+\Lambda^i(s) f[s, x^*(s), u_1^*(s), u_2^*(s), \ldots, u_n^*(s)]\},$$
$$\Lambda^i(T) = \frac{\partial}{\partial x^*} q^i(x^*(T)), \quad \textit{for } i \in N.$$

Proof. Consider the ith equality in conditions of the theorem, which states that $\upsilon_i^*(s) = u_i^*(s) = \zeta_i^*(s, x_0)$ maximizes

$$\int_{t_0}^{T} g^i[s, x(s), u_1^*(s), u_2^*(s), \ldots, u_{i-1}^*(s), u_i(s), u_{i+1}^*(s), \ldots, u_n^*(s)]ds$$
$$+ q^i(x(T)),$$

over the choice of $\upsilon_i(s) \in U^i$ subject to the state dynamics

$$\dot{x}(s) = f[s, x(s), u_1^*(s), u_2^*(s), \ldots, u_{i-1}^*(s), u_i(s), u_{i+1}^*(s), \ldots, u_n^*(s)],$$
$$x(t_0) = x_0, \quad \text{for } i \in N.$$

This is standard optimal control problem for Player i, since $u_j^*(s)$, for $j \in N$ and $j \neq i$, are open-loop controls and hence do not depend on $u_i^*(s)$. These results then follow directly from the maximum principle of Pontryagin as stated in 6.1.3.

6.2.3. Closed-loop Nash Equilibria. Under the memoryless perfect state information, the players' information structures follow the pattern $\eta^i(s) = \{x_0, x(s)\}$, $s \in [t_0, T]$. The players' strategies become functions of the initial state x_0, current state $x(s)$ and current time s, and can be expressed as $\{u_i(s) = \vartheta_i(s, x, x_0), \text{for } i \in N\}$. The following theorem provides a set of necessary conditions for any closed-loop no-memory Nash equilibria solution to satisfy.

Theorem. *A set of strategies* $\{u_i(s) = \vartheta_i(s, x, x_0), \textit{ fori } \in N\}$ *provides a closed-loop no memory Nash equilibria solution to the game* (6.2.1)–(6.2.2), *and* $\{x^*(s), t_0 \leq s \leq T\}$ *is the corresponding state trajectory, if there exist* N *costate functions* $\Lambda^i(s) : [t_0, T] \to R^m$, *for* $i \in N$, *such that the following relations* *are satisfied:*

$$\begin{aligned}
\vartheta_i^*(s, x^*, x_0) &\equiv u_i^*(s) \\
&= \arg\max_{u_i \in U^i} \{g^i[s, x^*(s), u_1^*(s), u_2^*(s), \ldots, u_{i-1}^*(s), u_i(s), \\
&\qquad u_{i+1}^*(s), \ldots, u_n^*(s)] + \Lambda^i(s) f[s, x^*(s), \\
&\qquad u_1^*(s), u_2^*(s), \ldots, u_{i-1}^*(s), u_i(s), \\
&\qquad u_{i+1}^*(s), \ldots, u_n^*(s)]\}, \\
\dot{x}^*(s) &= f[s, x^*(s), u_1^*(s), u_2^*(s), \ldots, u_n^*(s)], \quad x^*(t_0) = x_0, \\
\dot{\Lambda}^i(s) &= -\frac{\partial}{\partial x^*}\{g^i[s, x^*(s), \vartheta_1^*(s, x^*, x_0), \\
&\qquad \vartheta_2^*(s, x^*, x_0), \ldots, \vartheta_{i-1}^*(s, x^*, x_0), u_i^*(s), \\
&\qquad \vartheta_{i+1}^*(s, x^*, x_0), \ldots, \vartheta_n^*(s, x^*, x_0)] \\
&\qquad + \Lambda^i(s) f[s, x^*(s), \vartheta_1^*(s, x^*, x_0), \\
&\qquad \vartheta_2^*(s, x^*, x_0), \ldots, \vartheta_{i-1}^*(s, x^*, x_0), \\
&\qquad u_i^*(s), \vartheta_{i+1}^*(s, x^*, x_0), \ldots, \vartheta_n^*(s, x^*, x_0)], \\
\Lambda^i(T) &= \frac{\partial}{\partial x^*} q^i(x^*(T)); \quad \textit{for } i \in N.
\end{aligned}$$

Proof. Consider the ith equality in conditions of the theorem, which fixed all players' strategies (except those of the ith player) at $u_j^*(s) = \vartheta_j^*(s, x^*, x_0)$, for $j \neq i$ and $j \in N$, and constitutes an optimal control problem for Player i. Therefore, the above conditions follow from the maximum principle of Pontryagin, and Player i maximizes

$$\int_{t_0}^{T} g^i[s, x(s), u_1^*(s), u_2^*(s), \ldots, u_{i-1}^*(s), u_i(s), u_{i+1}^*(s), \ldots, u_n^*(s)] ds \\ + q^i(x(T)),$$

over the choice of $v_i(s) \in U^i$ subject to the state dynamics:

$$\dot{x}(s) = f[s, x(s), u_1^*(s), u_2^*(s), \ldots, u_{i-1}^*(s), u_i(s), u_{i+1}^*(s), \ldots, u_n^*(s)], \\ x(t_0) = x_0, \quad \text{for } i \in N.$$

Note that the partial derivative with respect to x in the costate equations receives contributions from dependence of the other $n-1$ players' strategies on the current value of x. This is a feature absent from the costate equations in 6.2.1. The set of equations in Theorem (see 6.2.2) in general admits of an uncountable number of solutions, which correspond to "*informationally nonunique*" Nash equilibria solutions of differential games under memoryless perfect state information pattern. Derivation of nonunique closed-loop Nash equilibria can be found in Basar (1977) and Mehlmann and Willing (1984).

6.2.4. Feedback Nash Equilibria. To eliminate information nonuniqueness in the derivation of Nash equilibria, one can constrain the Nash solution further by requiring it to satisfy the feedback Nash equilibrium property. In particular, the players' information structures follow either a *closed-loop perfect state* (CLPS) pattern in which $\eta^i(s) = \{x(s), t_0 \leq t \leq s\}$ or a *memoryless perfect state* (MPS) pattern in which $\eta^i(s) = \{x_0, x(s)\}$. Moreover, we require the following feedback Nash equilibrium condition to be satisfied.

Definition. For the n-person differential game (6.1.1)–(6.1.2) with MPS or CLPS information, an n-tuple of strategies $\{u_i^*(s) = \phi_i^*(s,x) \in U^i$, for $i \in N\}$ constitutes a *feedback Nash equilibrium solution* if there exist functionals $V^i(t,x)$ defined on $[t_0, T] \times R^m$ and satisfying the following relations for each $i \in N$:

$$
\begin{aligned}
V^i(T,x) &= q^i(x), \\
V^i(t,x) &= \int_t^T g^i\left[s, x^*(s), \phi_1^*(s,\eta_s), \phi_2^*(s,\eta_s),\right. \\
&\qquad \left. \ldots, \phi_n^*(s,\eta_s)\right] ds + q^i(x^*(T)) \\
&\geq \int_t^T g^i[s, x^{[i]}(s), \phi_1^*(s,\eta_s), \phi_2^*(s,\eta_s), \\
&\qquad \ldots, \phi_{i-1}^*(s,\eta_s), \phi_i(s,\eta_s), \phi_{i+1}^*(s,\eta_s), \\
&\qquad \ldots, \phi_n^*(s,\eta_s)] ds + q^i(x^{[i]}(T)), \quad \forall \phi_i(\cdot,\cdot) \in \Gamma^i, x \in R^n
\end{aligned}
$$

where on the interval $[t_0, T]$,

$$\dot{x}^{[i]}(s) = f\left[s, x^{[i]}(s), \phi_1^*(s, \eta_s), \phi_2^*(s, \eta_s), \ldots, \phi_{i-1}^*(s, \eta_s),\right.$$
$$\left.\phi_i(s, \eta_s), \phi_{i+1}^*(s, \eta_s), \ldots, \phi_n^*(s, \eta_s)\right], \quad x^{[1]}(t) = x;$$
$$\dot{x}^*(s) = f[s, x^*(s), \phi_1^*(s, \eta_s), \phi_2^*(s, \eta_s), \ldots, \phi_n^*(s, \eta_s)], \quad x(s) = x$$

and η_s stands for either the data set $\{x(s), x_0\}$ or $\{x(\tau), \tau \leq s\}$, depending on whether the information pattern is MPS or CLPS.

One salient feature of the concept introduced above is that if an n-tuple $\{\phi_i^*; i \in N\}$ provides a feedback Nash equilibrium solution (FNES) to an N-person differential game with duration $[t_0, T]$, its restriction to the time interval $[t, T]$ provides an FNES to the same differential game defined on the shorter time interval $[t, T]$, with the initial state taken as $x(t)$, and this being so for all $t_0 \leq t \leq T$. An immediate consequence of this observation is that, under either MPS or CLPS information pattern, feedback Nash equilibrium strategies will depend only on the time variable and the current value of the state, but not on memory (including the initial state x_0). Therefore, the players' strategies can be expressed as $\{u_i(s) = \phi_i(s, x), \text{ for } i \in N\}$. The following theorem provides a set of necessary conditions characterizing a feedback Nash equilibrium solution for the game (6.2.1) and (6.2.2) is characterized as follows:

Theorem. *An n-tuple of strategies $\{u_i^*(s) = \phi_i^*(t, x) \in U^i$, for $i \in N\}$ provides a feedback Nash equilibrium solution to the game (6.2.1)–(6.2.2) if there exist continuously differentiable functions $V^i(t, x) : [t_0, T] \times R^m \to R$, $i \in N$, satisfying the following set of partial differential equations:*

$$-V_t^i(t, x) = \max_{u_i} \{g^i[t, x, \phi_1^*(t, x), \phi_2^*(t, x), \ldots$$
$$\ldots, \phi_{i-1}^*(t, x), u_i(t, x), \phi_{i+1}^*(t, x), \ldots, \phi_n^*(t, x)]$$

$$
\begin{aligned}
&+ V_x^i(t,x) f[t, x, \phi_1^*(t,x), \phi_2^*(t,x), \ldots \\
&\ldots, \phi_{i-1}^*(t,x), u_i(t,x), \phi_{i+1}^*(t,x), \ldots, \phi_n^*(t,x)]\} \\
= &\{g^i[t, x, \phi_1^*(t,x), \phi_2^*(t,x), \ldots, \phi_n^*(t,x)] \\
&+ V_x^i(t,x) f[t, x, \phi_1^*(t,x), \phi_2^*(t,x), \ldots, \phi_n^*(t,x)]\},
\end{aligned}
$$

$$
V^i(T,x) = q^i(x), \quad i \in N
$$

Proof. By (6.1.2), $V^i(t,x)$ is the value function associated with the optimal control problem of Player i, $i \in N$. Together with the ith expression in Definition, the conditions in the Theorem imply a Nash equilibrium.

Consider the two-person zero-sum version of the game (6.2.1)–(6.2.2) in which the payoff of Player 1 is the negative of that of Player 2. Under either MPS or CLPS information pattern, a feedback saddle-point is characterized as follows.

Theorem. *A pair of strategies* $\{\phi_i^*(t,x); i = 1,2\}$ *provides a feedback saddle-point solution to the zero-sum version of the game* (6.2.1)–(6.2.2) *if there exists a function* $V : [t_0, T] \times R^m \to R$ *satisfying the partial differential equation*

$$
\begin{aligned}
-V_t(t,x) &= \min_{u_1 \in S^1} \max_{u_2 \in S^2} \{g[t, x, u_1(t), u_2(t)] + V_x f[t, x, u_1(t), u_2(t)]\} \\
&= \max_{u_2 \in S^2} \min_{u_1 \in S^1} \{g[t, x, u_1(t), u_2(t)] + V_x f[t, x, u_1(t), u_2(t)]\} \\
&= \{g[t, x, \phi_1^*(t,x), \phi_2^*(t,x)] + V_x f[t, x, \phi_1^*(t,x), \phi_2^*(t,x)]\},
\end{aligned}
$$

$$
V(T,x) = q(x).
$$

Proof. This result follows as a special case of the previous Theorem by taking $n = 2$, $g^1(\cdot) = -g^2(\cdot) \equiv g(\cdot)$, and $q^1(\cdot) = -q^2(\cdot) \equiv q(\cdot)$, in which case $V^1 = -V^2 \equiv V$ and existence of a saddle point is equivalent to interchangeability of the min max operations.

The partial differential equation in this Theorem was first obtained by Isaacs [see, Isaacs (1965)], and is therefore called the Isaacs equation.

6.3 Application of Differential Games in Economics

In this section, we consider application of differential games in competitive advertising.

6.3.1. Open-loop Solution in Competitive Advertising Consider the competitive dynamic advertising Jorgensen and Sorger (1990) game in Sorger (1989). There are two firms in a market and the profit of firm 1 and that of 2 are respectively

$$\int_0^T \left[q_1 x(s) - \frac{c_1}{2} u_1(s)^2\right] \exp(-rs)ds + \exp(-rT)S_1 x(T) \quad \text{and}$$
$$\int_0^T \left[q_2(1 - x(s)) - \frac{c_2}{2} u_2(s)^2\right] \exp(-rs)ds + \exp(-rT)S_2[1 - x(T)], \tag{6.3.1}$$

where r, q_i, c_i, S_i, for $i \in \{1, 2\}$, are positive constants, $x(s)$ is the market share of firm 1 at time s, $[1 - x(s)]$ is that of firm 2's, $u_i(s)$ is advertising rate for firm $i \in \{1, 2\}$.

It is assumed that market potential is constant over time. The only marketing instrument used by the firms is advertising. Advertising has diminishing returns, since there are increasing marginal costs of advertising as reflected through the quadratic cost function. The dynamics of firm 1's market share is governed by

$$\dot{x}(s) = u_1(s)[1 - x(s)]^{1/2} - u_2(s)x(s)^{1/2}, \quad x(0) = x_0. \tag{6.3.2}$$

There are saturation effects, since u_i operates only on the buyer market of the competing firm j.

Consider that the firms would like to seek an open-loop solution. Using open-loop strategies requires the firms to determine their action paths at the outset. This is realistic only if there are restrictive commitments concerning advertising. Invoking 6.2.2, an open-loop solution to the game (6.3.1)–(6.3.2) has to satisfy the following conditions

$$u_1^*(s) = \arg\max_{u_1} \left\{ \left[q_1 x^*(s) - \frac{c_1}{2} u_1(s)^2 \right] \exp(-rs) \right.$$
$$\left. + \Lambda^1(s) \left(u_1(s)[1 - x^*(s)]^{1/2} - u_2(s) x^*(s)^{1/2} \right) \right\},$$
$$u_2^*(s) = \arg\max_{u_2} \left\{ \left[q_2(1 - x^*(s)) - \frac{c_2}{2} u_2(s)^2 \right] \exp(-rs) \right.$$
$$\left. + \Lambda^2(s) \left(u_1(s)[1 - x^*(s)]^{1/2} - u_2(s) x^*(s)^{1/2} \right) \right\},$$
$$\dot{x}^*(s) = u_1^*(s)[1 - x^*(s)]^{1/2} - u_2^*(s) x^*(s)^{1/2}, \quad x^*(0) = x_0,$$
$$\dot{\Lambda}^1(s) = \left\{ -q_1 \exp(-rs) + \Lambda^1(s) \right.$$
$$\left. \times \left(\frac{1}{2} u_1^*(s)[1 - x^*(s)]^{-1/2} + \frac{1}{2} u_2^*(s) x^*(s)^{-1/2} \right) \right\},$$
$$\dot{\Lambda}^2(s) = \left\{ q_2 \exp(-rs) + \Lambda^2(s) \right.$$
$$\left. \times \left(\frac{1}{2} u_1^*(s)[1 - x^*(s)]^{-1/2} + \frac{1}{2} u_2^*(s) x^*(s)^{-1/2} \right) \right\},$$
$$\begin{aligned} \Lambda^1(T) &= \exp(-rT) S_1, \\ \Lambda^2(T) &= -\exp(-rT) S_2. \end{aligned} \tag{6.3.3}$$

Using (6.3.3), we obtain

$$u_1^*(s) = \frac{\Lambda^1(s)}{c_1} [1 - x^*(s)]^{1/2} \exp(rs),$$

and

$$u_2^*(s) = \frac{\Lambda^2(s)}{c_2} [x^*(s)]^{1/2} \exp(rs).$$

Upon substituting $u_1^*(s)$ and $u_2^*(s)$ into (6.3.3) yields:

$$\begin{aligned} \dot{\Lambda}^1(s) &= \left\{ -q_1 \exp(-rs) + \left(\frac{[\Lambda^1(s)]^2}{2c_1} + \frac{\Lambda^1(s)\Lambda^2(s)}{2c_2} \right) \right\}, \\ \dot{\Lambda}^2(s) &= \left\{ q_2 \exp(-rs) + \left(\frac{[\Lambda^2(s)]^2}{2c_2} + \frac{\Lambda^1(s)\Lambda^2(s)}{2c_1} \right) \right\}, \end{aligned} \tag{6.3.4}$$

with boundary conditions

$$\Lambda^1(T) = \exp(-rT)S_1 \quad \text{and} \quad \Lambda^2(T) = -\exp(-rT)S_2.$$

The game equilibrium state dynamics becomes

$$\begin{aligned}
\dot{x}^*(s) &= \frac{\Lambda^1(s)\exp(rs)}{c_1}[1 - x^*(s)] - \frac{\Lambda^2(s)\exp(rs)}{c_2}x^*(s), \\
x^*(0) &= x_0. && (6.3.5)
\end{aligned}$$

Solving the block recursive system of differential equations (6.3.4)–(6.3.5) gives the solutions to $x^*(s)$, $\Lambda^1(s)$ and $\Lambda^2(s)$. Upon substituting them into $u_1^*(s)$ and $u_2^*(s)$ yields the open-loop game equilibrium strategies.

6.3.2. Feedback Solution in Competitive Advertising

A feedback solution which allows the firm to choose their advertising rates contingent upon the state of the game is a realistic approach to the problem (6.3.1)–(6.3.2). Invoking 6.2.4, a feedback Nash equilibrium solution to the game (6.3.1)–(6.3.2) has to satisfy the following conditions:

$$\begin{aligned}
-V_t^1(t,x) &= \max_{u_1}\left\{\left[q_1 x - \frac{c_1}{2}u_1^2\right]\exp(-rt)\right. \\
&\qquad\qquad \left. + V_x^1(t,x)\left(u_1[1-x]^{1/2} - \phi_2^*(t,x)x^{1/2}\right)\right\}, \\
-V_t^2(t,x) &= \max_{u_2}\left\{\left[q_2(1-x) - \frac{c_2}{2}u_2^2\right]\exp(-rt)\right. \\
&\qquad \left. + V_x^2(t,x)\left(\phi_1^*(t,x)[1-x]^{1/2} - u_2 x^{1/2}\right)\right\}, \\
V^1(T,x) &= \exp(-rT)S_1 x, \\
V^2(T,x) &= \exp(-rT)S_2(1-x). && (6.3.6)
\end{aligned}$$

Performing the indicated maximization in (6.3.6) yields:

$$\phi_1^*(t,x) = \frac{V_x^1(t,x)}{c_1}[1-x]^{1/2}\exp(rt) \quad \text{and}$$

$$\phi_2^*(t,x) = \frac{V_x^2(t,x)}{c_2}[x]^{1/2}\exp(rt).$$

Upon substituting $\phi_1^*(t,x)$ and $\phi_2^*(t,x)$ into (6.3.6) and solving (6.3.6) we obtain the value functions:

$$V^1(t,x) = \exp[-r(t)][A_1(t)x + B_1(t)] \quad \text{and}$$
$$V^2(t,x) = \exp[-r(t)][A_2(t)(1-x) + B_2(t)], \tag{6.3.7}$$

where $A_1(t)$, $B_1(t)$, $A_2(t)$ and $B_2(t)$ satisfy:

$$\dot{A}_1(t) = rA_1(t) - q_1 + \frac{A_1(t)^2}{2c_1} + \frac{A_1(t)A_2(t)}{2c_2},$$
$$\dot{A}_2(t) = rA_2(t) - q_2 + \frac{A_2(t)^2}{2c_2} + \frac{A_1(t)A_2(t)}{2c_1},$$
$$A_1(T) = S_1, \quad B_1(T) = 0, \quad A_2(T) = S_2 \quad \text{and} \quad B_2(T) = 0.$$

Upon substituting the relevant partial derivatives of $V^1(t,x)$ and $V^2(t,x)$ from (6.3.7) into (6.3.6) yields the feedback Nash equilibrium strategies

$$\phi_1^*(t,x) = \frac{A_1(t)}{c_1}[1-x]^{1/2} \quad \text{and} \quad \phi_2^*(t,x) = \frac{A_2(t)}{c_2}[x]^{1/2}. \tag{6.3.8}$$

6.4 Infinite-Horizon Differential Games

6.4.1. Consider the infinite-horizon autonomous game problem with constant discounting, in which T approaches infinity and where the objective functions and state dynamics are both autonomous. In particular, the game becomes:

$$\max_{u_i} \int_{t_0}^{\infty} g^i[x(s), u_1(s), u_2(s), \ldots, u_n(s)] \exp[-r(s-t_0)]ds,$$
$$\text{for } i \in N, \tag{6.4.1}$$

subject to the deterministic dynamics

$$\dot{x}(s) = f[x(s), u_1(s), u_2(s), \ldots, u_n(s)], \quad x(t_0) = x_0, \tag{6.4.2}$$

where r is a constant discount rate.

6.4.2. Game Equilibrium Solutions. Now consider the alternative game to (6.4.1)–(6.4.2)

$$\max_{u_i} \int_t^\infty g^i[x(s), u_1(s), u_2(s), \ldots, u_n(s)] \exp[-r(s-t)]ds,$$
$$\text{for } i \in N, \tag{6.4.3}$$

subject to the deterministic dynamics

$$\dot{x}(s) = f[x(s), u_1(s), u_2(s), \ldots, u_n(s)], \quad x(t) = x. \tag{6.4.4}$$

The infinite-horizon autonomous game (6.4.3)–(6.4.4) is independent of the choice of t and dependent only upon the state at the starting time, that is x.

In the infinite-horizon optimization problem in 6.1.2, the control is shown to be a function of the state variable x only. With the validity of the game equilibrium $\{u_i^*(s) = \phi_i^*(x) \in U^i, \text{ for } i \in N\}$ to be verified later, we define.

Definition. For the n-person differential game (6.4.1)–(6.4.2) with MPS or CLPS information, an n-tuple of strategies

$$\{u_i^*(s) = \phi_i^*(x) \in U^i, \quad \text{for } i \in N\}$$

constitutes a *feedback Nash equilibrium solution* if there exist functionals $V^i(t,x)$ defined on $[t_0, \infty) \times R^m$ and satisfying the following relations for each $i \in N$:

$$\begin{aligned} V^i(t,x) &= \int_t^\infty g^i[x^*(s), \phi_1^*(\eta_s), \phi_2^*(\eta_s), \ldots, \phi_n^*(\eta_s)] \exp[-r(s-t_0)]ds \\ &\geq \int_t^\infty g^i[x^{[i]}(s), \phi_1^*(\eta_s), \phi_2^*(\eta_s), \ldots, \phi_{i-1}^*(\eta_s)\phi_i(\eta_s)\phi_{i+1}^*(\eta_s), \\ &\qquad \ldots, \phi_n^*(\eta_s)] \exp[-r(s-t_0)]ds \quad \forall \phi_i(\cdot,\cdot) \in \Gamma^i, \quad x \in R^n, \end{aligned}$$

where on the interval $[t_0, \infty)$,

$$\begin{aligned} \dot{x}^{[i]}(s) = f\Big[&s^{[i]}(s), \phi_1^*(\eta_s), \phi_2^*(\eta_s), \\ &\ldots, \phi_{i-1}^*(\eta_s)\phi_i(\eta_s)\phi_{i+1}^*(\eta_s), \ldots, \phi_n^*(\eta_s)\Big], \end{aligned}$$

$$x^{[1]}(t) = x;$$

$$\dot{x}^*(s) = f[x^*(s), \phi_1^*(\eta_s), \phi_2^*(\eta_s), \ldots, \phi_n^*(\eta_s)], \quad x^*(s) = x;$$

and η_s stands for either the data set $\{x(s), x_0\}$ or $\{x(\tau), \ \tau \leq s\}$, depending on whether the information pattern in MPS or CLPS.

We can write

$$V^i(t,x) = \exp[-r(t-t_0)] \int_t^\infty g^i[x^*(s), \phi_1^*(\eta_s), \phi_2^*(\eta_s), \ldots, \phi_n^*(\eta_s)] \times \exp[-r(s-t)]ds, \quad \text{for } x(t) = x = x_t^* = x^*(t).$$

Since

$$\int_t^\infty g^i[x^*(s), \phi_1^*(\eta_s), \phi_2^*(\eta_s), \ldots, \phi_n^*(\eta_s)] \exp[-r(s-t)]ds$$

depends on the current state x only, we can write:

$$W^i(x) = \int_t^\infty g^i[x^*(s), \phi_1^*(\eta_s), \phi_2^*(\eta_s), \ldots, \phi_n^*(\eta_s)] \exp[-r(s-t)]ds.$$

It follows that:

$$\begin{aligned} V^i(t,x) &= \exp[-r(t-t_0)]W^i(x), \\ V_t^i(t,x) &= -r\exp[-r(t-t_0)]W^i(x), \quad \text{and} \\ V_x^i(t,x) &= \exp[-r(t-t_0)]W_x^i(x), \quad \text{for } i \notin N. \end{aligned} \tag{6.4.5}$$

A feedback Nash equilibrium solution for the infinite-horizon autonomous game (6.4.3) and (6.4.4) can be characterized as follows:

Theorem. *An n-tuple of strategies* $\{u_i^*(s) = \phi_i^*(\cdot) \in U^i; \ \text{for } i \in N\}$ *provides a feedback Nash equilibrium solution to the infinite-horizon game (6.4.3) and (6.4.4)if there exist continuously differentiable functions* $W^i(x) : R^m \to R, \ i \in N$, *satisfying the following set of partial differential equations:*

$$rW^i(x) = \max_{u_i} \{g^i[x, \phi_1^*(x), \phi_2^*(x), \ldots, \phi_{i-1}^*(x), u_i, \phi_{i+1}^*(x), \ldots, \phi_n^*(x)] + W_x^i(x) f[x, \phi_1^*(x), \phi_2^*(x), \ldots, \phi_{i-1}^*(x), u_i(x), \phi_{i+1}^*(x), \ldots, \phi_n^*(x)]\}$$

$$= \{g^i[x, \phi_1^*(x), \phi_2^*(x), \ldots, \phi_n^*(x)] + W_x^i(x) f[x, \phi_1^*(x), \phi_2^*(x), \ldots, \phi_n^*(x)]\}, \qquad \textit{for } i \in N.$$

Proof. By 6.1.2, $W^i(x)$ is the value function associated with the optimal control problem of Player i, $i \in N$. Together with the ith expression in Definition, the conditions in Theorem imply a Nash equilibrium.

Since time s is not explicitly involved the partial differential equations in Theorem, the validity of the feedback Nash equilibrium $\{u_i^* = \phi_i^*(x), \text{for } \ i \in N\}$ are functions independent of time is obtained.

Substituting the game equilibrium strategies in Theorem into (6.4.2) yields the game equilibrium dynamics of the state path as:

$$\dot{x}(s) = f[x(s), \phi_1^*(x(s)), \phi_2^*(x(s)), \ldots, \phi_n^*(x(s))], \quad x(t_0) = x_0.$$

Solving the above dynamics yields the optimal state trajectory $\{x^*(t)\}_{t \geq t_0}$ as

$$x^*(t) = x_0 + \int_{t_0}^{t} f[x^*(s), \phi_1^*(x^*(s)), \phi_2^*(x^*(s)), \ldots, \phi_n^*(x^*(s))] ds, \quad \text{for } \ t \geq t_0.$$

We denote term $x^*(t)$ by x_t^*. The feedback Nash equilibrium strategies for the infinite-horizon game (6.4.1)–(6.4.2) can be obtained as

$$[\phi_1^*(x_t^*), \phi_2^*(x_t^*), \ldots, \phi_n^*(x_t^*)], \quad \text{for } \ t \geq t_0.$$

Following the above analysis and using Theorems 6.4 and 6.5, we can characterize an open loop equilibrium solution to the infinite-horizon game (6.4.3) and (6.4.4) as:

Theorem. *A set of strategies* $\{u_i^*(s) = \zeta_i^*(s, x_t),$ *for* $i \in N\}$ *provides an open-loop Nash equilibrium solution to the infinite-horizon game* (6.4.3) *and* (6.4.4), *and* $\{x^*(s),\ t \leq s \leq T\}$ *is the corresponding state trajectory, if there exist* m *costate functions* $\Lambda^i(s) : [t, T] \rightarrow R^m$, *for $i \in N$, such that the following relations are satisfied:*

$$\zeta_i^*(s,x) \equiv u_i^*(s)$$
$$= \arg\max_{u_i \in U^i} \{g^i[x^*(s), u_1^*(s), u_2^*(s), \ldots, u_{i-1}^*(s), u_i(s), u_{i+1}^*(s),$$
$$\ldots, u_n^*(s)] + \lambda^i(s) f[x^*(s), u_1^*(s), u_2^*(s), \ldots, u_{i-1}^*(s),$$
$$u_i(s), u_{i+1}^*(s), \ldots, u_n^*(s)]\},$$
$$\dot{x}^*(s) = f[x^*(s), u_1^*(s), u_2^*(s), \ldots, u_n^*(s)], \quad x^*(t) = x_t,$$
$$\dot{\lambda}^i(s) = r\lambda(s) - \frac{\partial}{\partial x^*}\{g^i[x^*(s), u_1^*(s), u_2^*(s), \ldots, u_n^*(s)]$$
$$+ \lambda^i(s) f[x^*(s), u_1^*(s), u_2^*(s), \ldots, u_n^*(s)]\}, \quad \textit{for } i \in N.$$

Proof. Consider the ith equality in the above Theorem, which states that $v_i^*(s) = u_i^*(s) = \zeta_i^*(s, x_t)$ maximizes

$$\int_{t_0}^{\infty} g^i[x(s), u_1^*(s), u_2^*(s), \ldots, u_{i-1}^*(s), u_i(s), u_{i+1}^*(s), \ldots, u_n^*(s)]ds,$$

over the choice of $v_i(s) \in U^i$ subject to the state dynamics:

$$\dot{x}(s) = f[x(s), u_1^*(s), u_2^*(s), \ldots, u_{i-1}^*(s), u_i(s), u_{i+1}^*(s), \ldots, u_n^*(s)],$$
$$x(t) = x_t, \quad \text{for } i \in N.$$

This is the infinite-horizon optimal control problem for Player i, since $u_j^*(s)$, for $j \in N$ and $j \neq i$, are open-loop controls and hence do not depend on $u_i^*(s)$. These results are stated in 6.1.3.

6.4.3. Infinite-horizon Duopolistic Competition. Consider a dynamic duopoly in which there are two publicly listed firms selling a homogeneous good. Since the value of a publicly listed firm is the present value of its discounted expected future earnings, the terminal time of the game, T, may be very far in the future and nobody knows when the firms will be out of business. Therefore, setting $T = \infty$ may very well be the best approximation for the true game horizon. Even if the firm's management restricts itself to considering profit maximization over the next year, it should value its asset positions at the end of the year by the earning potential of these assets in the years to come. There is a lag in price adjustment so the evolution of market price over time is assumed to be a function of the current market price and the price specified by the current demand condition. In particular, we follow Tsytsui and Mino (1990) and assume that

$$\dot{P}(s) = k[a - u_1(s) - u_2(s) - P(s)], \quad P(t_0) = P_0, \tag{6.4.6}$$

where $P(s)$ is the market price at time s, $u_i(s)$ is output supplied firm $i \in \{1, 2\}$, current demand condition is specified by the instantaneous inverse demand function $P(s) = [a - u_1(s) - u_2(s)]$, and $k > 0$ represents the price adjustment velocity.

The payoff of firm i is given as the present value of the stream of discounted profits

$$\int_{t_0}^{\infty} \{P(s)u_i(s) - cu_i(s) - (1/2)[u_i(s)]^2\} \exp[-r(s - t_0)]ds,$$
$$\text{for } \; i \in \{1, 2\}, \tag{6.4.7}$$

where $cu_i(s) + (1/2)[u_i(s)]^2$ is the cost of producing output $u_i(s)$ and r is the interest rate.

Once again, we consider the alternative game

$$\max_{u_i} \int_{t_0}^{\infty} \{P(s)u_i(s) - cu_i(s) - (1/2)[u_i(s)]^2\} \exp[-r(s - t)]ds,$$
$$\text{for } \; i \in \{1, 2\}, \tag{6.4.8}$$

subject to

$$\dot{P}(s) = k[a - u_1(s) - u_2(s) - P(s)], \quad P(t) = P. \tag{6.4.9}$$

The infinite-horizon game (6.4.8)–(6.4.9) has autonomous structures and a constant rate. Therefore, we can apply the second theorem from 6.4.2 to characterize a feedback Nash equilibrium solution as

$$rW^i(P) = \max_{u_i}\{[Pu_i - cu_i - (1/2)(u_i)^2] \\ + W^i_P[k(a - u_i - \phi^*_j(P) - P)]\}, \quad \text{for } i \in \{1, 2\}. \tag{6.4.10}$$

Performing the indicated maximization in (6.4.10), we obtain:

$$\phi^*_i(P) = P - c - kW^i_P(P), \quad \text{for } i \in \{1, 2\}. \tag{6.4.11}$$

Substituting the results from (6.4.11) into (6.4.10), and upon solving (6.4.10) yields:

$$W^i(P) = \frac{1}{2}AP^2 - BP + C, \tag{6.4.12}$$

where

$$A = \frac{r + 6k - \sqrt{(r + 6k)^2 - 12k^2}}{6k^2},$$

$$B = \frac{-akA + c - 2kcA}{r - 3k^2A + 3k}, \quad \text{and}$$

$$C = \frac{c^2 + 3k^2B^2 - 2kB(2c + a)}{2r}.$$

Again, one can readily verify that $W^i(P)$ in (6.4.12) indeed solves (6.4.10) by substituting $W^i(P)$ and its derivative into (6.4.10) and (6.4.11).

The game equilibrium strategy can then be expressed as:

$$\phi^*_i(P) = P - c - k(AP - B), \quad \text{for } i \in \{1, 2\}.$$

Substituting the game equilibrium strategies above into (6.4.6) yields the game equilibrium state dynamics of the game (6.4.6)–(6.4.7) as:

$$\dot{P}(s) = k[a - 2(c + kB) - (3 - kA)P(s)], \quad P(t_0) = P_0.$$

Solving the above dynamics yields the optimal state trajectory as

$$P^*(t) = \left[P_0 - \frac{k[a - 2(c + kB)]}{k(3 - kA)}\right]$$
$$\times \exp[-k(3 - kA)t] + \frac{k[a - 2(c + kB)]}{k(3 - kA)}.$$

We denote term $P^*(t)$ by P_t^*. The feedback Nash equilibrium strategies for the infinite-horizon game (6.4.6)–(6.4.7) can be obtained as

$$\phi_i^*(P_t^*) = P_t^* - c - k(AP_t^* - B), \quad \text{for } \{1, 2\}.$$

6.5 Cooperative Differential Games in Characteristic Function Form

We begin with the basic formulation of cooperative differential games in characteristic function form and the solution imputations. For another approach to cooperative differential games see Leitmann (1974), Leitmann and Schmitendorf (1978).

6.5.1. Game Formulation. Consider a general N-person differential game in which the state dynamics has the form

$$\dot{x}(s) = f[s, x(s), u_1(s), u_2(s), \ldots, u_n(s)], \quad x(t_0) = x_0. \tag{6.5.1}$$

The payoff of Player i is:

$$\int_{t_0}^{T} g^i[s, x(s), u_1(s), u_2(s), \ldots, u_n(s)]ds + q^i(x(T)),$$
$$\text{for } i \in N = \{1, 2, \ldots, n\}, \tag{6.5.2}$$

where $x(s) \in X \subset R^m$ denotes the state variables of game, and $u_i \in U^i$ is the control of Player i, for $i \in N$. In particular, the players' payoffs are transferable. Invoking 6.2.4, a feedback Nash equilibrium solution can be characterized if the players play noncooperatively.

Now consider the case when the players agree to cooperate. Let $\Gamma_c(x_0, T - t_0)$ denote a cooperative game with the game structure of $\Gamma(x_0, T - t_0)$ in which the players agree to act according to an agreed upon optimality principle. The agreement on how to act cooperatively and allocate cooperative payoff constitutes the solution optimality principle of a cooperative scheme. In particular, the solution optimality principle for a cooperative game $\Gamma_c(x_0, T - t_0)$ includes

(i) an agreement on a set of cooperative strategies/controls, and
(ii) a mechanism to distribute total payoff among players.

The solution optimality principle will remain in effect along the cooperative state trajectory path $\{x_s^*\}_{s=t_0}^T$. Moreover, group rationality requires the players to seek a set of cooperative strategies/controls that yields a pareto-optimal solution. In addition, the allocation principle has to satisfy individual rationality in the sense that neither player would be no worse off than before under cooperation.

To fulfill group rationality in the case of transferable payoffs, the players have to maximize the sum of their payoffs

$$\sum_{j=1}^{N} \left\{ \int_{t_0}^{T} g^j[s, x(s), u_1(s), u_2(s), \ldots, u_n(s)]ds + q^j(x(T)) \right\}, \tag{6.5.3}$$

subject to (6.5.1).

Invoking Pontryagin's Maximum Principle, a set of optimal controls $u^*(s) = [u_1^*(s), u_2^*(s), \ldots, u_n^*(s)]$ can be characterized as in 6.1.3. Substituting this set of optimal controls into (6.5.1) yields the optimal trajectory $\{x^*(t)\}_{t=t_0}^T$, where

$$x^*(t) = x_0 + \int_{t_0}^{t} f[s, x^*(s)u^*(s)]ds, \quad \text{for } t \in [t_0, T]. \tag{6.5.4}$$

For notational convenience in subsequent exposition, we use $x^*(t)$ and x_t^* interchangeably.

We denote

$$\sum_{j=1}^{n} \left\{ \int_{t_0}^{T} g^j[s, x^*(s), u^*(s)]ds + q^j(x^*(T)) \right\}$$

by $v(N; x_0, T - t_0)$. Let $S \subseteq N$ and $v(S; x_0, T - t_0)$ stands for a characteristic function reflecting the payoff of coalition S. The quantity $v(S; x_0, T - t_0)$ yields the maximized payoff to coalition S as the rest of the players form a coalition $N \backslash S$ to play against S. Calling on the super-additivity property of characteristic functions, $v(S; x_0, T - t_0) \geq v(S'; x_0, T - t_0)$ for $S' \subset S \subseteq N$. Hence, it is advantageous for the players to form a maximal coalition N and obtain a maximal total payoff $v(N; x_0, T - t_0)$ that is possible in the game.

6.5.2. Solution Imputation. One of the integral parts of cooperative game is to explore the possibility of forming coalitions and offer an "agreeable" distribution of the total cooperative payoff among players. In fact, the characteristic function framework displays the possibilities of coalitions in an effective manner and establishes a basis for formulating distribution schemes of the total payoffs that are acceptable to participating players.

We can use $\Gamma_v(x_0, T - t_0)$ to denote a *cooperative differential game in characteristic function form.* The optimality principle for a cooperative game in characteristic function form includes

(i) an agreement on a set of cooperative strategies/controls

$$u^*(s) = [u_1^*(s), u_2^*(s), \ldots, u_n^*(s)], \quad \text{for } s \in [t_0, T], \quad \text{and}$$

(ii) a mechanism to distribute total payoff among players.

A set of payoff distributions satisfying the optimality principle is called a solution imputation to the cooperative game. We will now examine the solutions to $\Gamma_v(x_0, T - t_0)$. Denote by $\xi_i(x_0, T - t_0)$ the share of the player $i \in N$ from the total payoff $v(N; x_0, T - t_0)$.

Definition. A vector

$$\xi(x_0, T - t_0) = [\xi_1(x_0, T - t_0), \xi_2(x_0, T - t_0), \ldots, \xi_n(x_0, T - t_0)]$$

that satisfies the conditions:

(i)

$$\xi_i(x_0, T - t_0) \geq v(\{i\}; x_0, T - t_0), \quad \text{for } i \in N, \text{ and}$$

(ii)

$$\sum_{j \in N} \xi_j(x_0, T - t_0) = v(N; x_0, T - t_0)$$

is called an imputation of the game $\Gamma_v(x_0, T - t_0)$.

Part (i) of Definition guarantees individual rationality in the sense that each player receives at least the payoff he or she will get if play against the rest of the players. Part (ii) ensures Pareto optimality and hence group rationality.

Theorem. *Suppose the function* $w : 2^n \times R^m \times R^1 \to R^1$ *is additive in* $S \in 2^n$*, that is for any* S*,* $A \in 2^n$*,* $S \cap A = \emptyset$ *we have* $w(S \cup A;\ x_0, T - t_0) = w(S; x_0, T - t_0) + w(A; x_0, T - t_0)$*. Then in the game* $\Gamma_w(x_0, T - t_0)$ *there is a unique imputation* $\xi_i(x_0, T - t_0) = w(\{i\}; x_0, T - t_0)$*, for* $i \in N$*.*

Proof. From the additivity of w *we immediately obtain*

$$w(N; x_0, T - t_0) = w(\{1\}; x_0, T - t_0) + w(\{2\}; x_0, T - t_0) + \cdots + w(\{n\}; x_0, T - t_0),$$

Hence the Theorem follows.

Games with additive characteristic functions are called inessential and games with superadditive characteristic functions are called essential. In an essential game $\Gamma_v(x_0, T - t_0)$ there is an infinite set of imputations. Indeed, any vector of the form

$$[v(\{1\}; x_0, T - t_0) + \alpha_1, v(\{2\}; x_0, T - t_0) + \alpha_2, \ldots \\ \ldots, v(\{n\}; x_0, T - t_0) + \alpha_n], \quad \text{for } \alpha_i \geq 0, \quad i \in N \text{ and}$$

$$\sum_{i \in N} \alpha_i = v(N; x_0, T - t_0) - \sum_{i \in N} v(\{i\}; x_0, T - t_0), \tag{6.5.5}$$

is an imputation of the game $\Gamma_v(x_0, T - t_0)$. We denote the imputation set of $\Gamma_v(x_0, T - t_0)$ by $E_v(x_0, T - t_0)$.

Definition. The imputation $\xi(x_0, T-t_0)$ dominates the imputation $\eta(x_0, T-t_0)$ in the coalition S, or $\xi(x_0, T-t_0) \overset{S}{\succ} \eta(x_0, T-t_0)$, if

(i)

$$\xi_i(x_0, T-t_0) \geq \eta_i(x_0, T-t_0), \quad i \in S; \text{ and}$$

(ii)

$$\sum_{i \in S} \xi_i(x_0, T-t_0) \leq v(S; x_0, T-t_0).$$

The imputation $\xi(x_0, T-t_0)$ is said to dominate the imputation $\eta(x_0, T-t_0)$, or $\xi(x_0, T-t_0) \succ \eta(x_0, T-t_0)$, if there does not exist any $S \subset N$ such that $\eta(x_0, T-t_0) \overset{S}{\succ} \xi(x_0, T-t_0)$ but there exists coalition $S \subset N$ such that $\xi(x_0, T-t_0) \overset{S}{\succ} \eta(x_0, T-t_0)$. It follows from the definition of imputation that domination in single-element coalition and coalition N, is not possible.

Definition. The set of undominated imputations is called the core of the game $\Gamma_v(x_0, T-t_0)$ and is denoted by $C_v(x_0, T-t_0)$.

Definition. The set $L_v(x_0, T-t_0) \subset E_v(x_0, T-t_0)$ is called the Neumann–Morgenstern solution (the NM-solution) of the game $\Gamma_v(x_0, T-t_0)$ if

(i)

$$\xi(x_0, T-t_0), \eta(x_0, T-t_0) \in L_v(x_0, T-t_0), \text{ implies}$$
$$\xi(x_0, T-t_0) \not\succ \eta(x_0, T-t_0),$$

(ii)

$$\text{for } \eta(x_0, T-t_0) \notin L_v(x_0, T-t_0) \text{ there exists}$$
$$\xi(x_0, T-t_0) \in L_v(x_0, T-t_0) \text{ such that}$$
$$\xi(x_0, T-t_0) \succ \eta(x_0, T-t_0).$$

Note that the NM-solution always contains the core.

Definition. The vector

$$\Phi^v(x_0, T - t_0) = \{\Phi_i^v(x_0, T - t_0),\ i = 1, \ldots, n\}$$

is called the Shapley value if it satisfies the following conditions:

$$\begin{aligned}&\Phi_i^v(x_0, T - t_0)\\&\quad = \sum_{S \subset N (S \ni i)} \frac{(n-s)!(s-1)!}{n!}[v(S; x_0, T - t_0) - v(S \backslash i; x_0, T - t_0)];\\&\quad i = 1, \ldots, n.\end{aligned}$$

The components of the Shapley value are the players' shares of the cooperative payoff. The Shapley value is unique and is an imputation [see Shapley (1953)]. Unlike the core and NM-solution, the Shapley value represents an optimal distribution principle of the total gain $v(N; x_0, T - t_0)$ and is defined without using the concept of domination.

6.6 Imputation in a Dynamic Context

Section 6.5.2 characterizes the solution imputation at the outset of the game. In dynamic games, the solution imputation along the cooperative trajectory $\{x^*(t)\}_{t=t_0}^T$ would be of concern to the players. In this section, we focus our attention on the dynamic imputation brought about by the solution optimality principle.

Let an optimality principle be chosen in the game $\Gamma_v(x_0, T - t_0)$. The solution of this game constructed in the initial state $x(t_0) = x_0$ based on the chosen principle of optimality contains the solution imputation set $W_v(x_0, T - t_0) \subseteq E_v(x_0, T - t_0)$ and the conditionally optimal trajectory $\{x^*(t)\}_{t=t_0}^T$ which maximizes

$$\sum_{j=1}^n \left\{ \int_{t_0}^T g^j[s, x^*(s), u^*(s)]ds + q^j(x^*(T)) \right\}.$$

Assume that $W_v(x_0, T - t_0) \neq \emptyset$.

Definition. Any trajectory $\{x^*(t)\}_{t=t_0}^{T}$ of the system (6.5.1) such that

$$\sum_{j=1}^{n}\left\{\int_{t_0}^{T} g^j[s,x^*(s),u^*(s)]ds + q^j(x^*(T))\right\} = v(N;x_0,T-t_0)$$

is called a conditionally optimal trajectory in the game $\Gamma_v(x_0,T-t_0)$.

Definition suggests that along the conditionally optimal trajectory the players obtain the largest total payoff. For exposition sake, we assume that such a trajectory exits. Now we consider the behavior of the set $W_v(x_0,T-t_0)$ along the conditionally optimal trajectory $\{x^*(t)\}_{t=t_0}^{T}$. Towards this end, in each current state $x^*(t) \equiv x_t^*$ the current subgame $\Gamma_v(x_t^*,T-t)$ is defined as follows. At time t with state $x^*(t)$, we define the characteristic function

$$v(S;x_t^*,T-t) = \begin{cases} 0, & S=\emptyset \\ Val\ \Gamma_S(x_t^*,T-t), & \text{if} \quad S\subset N \\ K_N(x^*(t),u^*(\cdot),T-t) & S=N \end{cases} \tag{6.6.1}$$

where

$$K_N(x_t^*,u^*(\cdot),T-t) = \sum_{j=1}^{n}\left\{\int_{t}^{T} g^j[s,x^*(s),u^*(s)]ds + q^j(x^*(T))\right\}$$

is the total payoff of the players over the time interval $[t,T]$ along the conditionally optimal trajectory $\{x^*(s)\}_{s=t}^{T}$; and $Val\ \Gamma_S(x_t^*,T-t)$ is the value of the zero-sum differential game $\Gamma_S(x_t^*,T-t)$ between coalitions S and $N\backslash S$ with initial state $x^*(t) \equiv x_t^*$, duration $T-t$ and the S coalition being the maximizer.

The imputation set in the game $\Gamma_v(x_t^*,T-t)$ is of the form:

$$E_v(x_t^*,T-t) = \left\{\xi \in R^n | \xi_i \geq v(\{i\};x_t^*,T-t),\ i=1,2,\ldots,n; \right.$$

$$\left. \sum_{i\in N}\xi_i = v(N;x_t^*,T-t)\right\}, \tag{6.6.2}$$

where

$$v(N; x_t^*, T-t)$$
$$= v(N; x_0, T-t_0) - \sum_{j=1}^{n} \left\{ \int_{t_0}^{t} g^j[s, x^*(s), u^*(s)]ds + q^j(x^*(T)) \right\}.$$

The quantity

$$\sum_{j=1}^{n} \left\{ \int_{t_0}^{t} g^j[s, x^*(s), u^*(s)]ds + q^j(x^*(T)) \right\}$$

denoted the cooperative payoff of the players over the time interval $[t_0, t]$ along the trajectory $\{x^*(s)\}_{s=t_0}^{T}$.

Consider the family of current games

$$\{\Gamma_v(x_t^*, T-t),\ t_0 \leq t \leq T\},$$

and their solutions $W_v(x_t^*, T-t) \subset E_v(x_t^*, T-t)$ generated by the same principle of optimality that yields the initially solution $W_v(x_0, T-t_0)$.

Lemma. The set $W_v(x_T^*, 0)$ is a solution of the current game $\Gamma_v(x_T^*, 0)$ at time T and is composed of the only imputation

$$q(x^*(T)) = \{q^1(x^*(T)), q^2(x^*(T)), \ldots, q^n(x^*(T))\}$$
$$= \{q^1(x_T^*), q^2(x_T^*), \ldots, q^n(x_T^*)\}.$$

Proof. Since the game $\Gamma_v(x_T^*, 0)$ is of zero-duration, then for all $i \in N$, $v(\{i\}; x_T^*, 0) = q^i(x_T^*)$. Hence

$$\sum_{i \in N} v(\{i\}; x_T^*, 0) = \sum_{i \in N} q^i(x_T^*) = v(N; x_T^*, 0),$$

and the characteristic function of the game $\Gamma_v(x_T^*, 0)$ is additive for S and, by Theorem,

$$E_v(x_T^*, 0) = q(x_T^*) = W_v(x_T^*, 0).$$

This completes the proof of Lemma.

6.7 Principle of Dynamic Stability

Formulation of optimal behaviors for players is a fundamental element in the theory of cooperative games. The players' behaviors satisfying some specific optimality principles constitute a solution of the game. In other words, the solution of a cooperative game is generated by a set of optimality principles (for instance, the Shapley value (1953), the von Neumann–Morgenstern solution (1944) and the Nash bargaining solution (1953)). For dynamic games, an additional stringent condition on their solutions is required: the specific optimality principle must remain optimal at any instant of time throughout the game along the optimal state trajectory chosen at the outset. This condition is known as *dynamic stability or time consistency.* Assume that at the start of the game the players adopt an optimality principle (which includes the consent to maximize the joint payoff and an agreed upon payoff distribution principle). When the game proceeds along the "optimal" trajectory, the state of the game changes and the optimality principle may not be feasible or remain optimal to all players. Then, some of the players will have an incentive to deviate from the initially chosen trajectory. If this happens, instability arises. In particular, the dynamic stability of a solution of a cooperative differential game is the property that, when the game proceeds along an "optimal" trajectory, at each instant of time the players are guided by the same optimality principles, and yet do not have any ground for deviation from the previously adopted "optimal" behavior throughout the game.

The question of dynamic stability in differential games has been explored rigorously in the past three decades. Haurie (1976) discussed the problem of instability in extending the Nash bargaining solution to differential games. Petrosyan (1977) formalized mathematically the notion of dynamic stability in solutions of differential games. Petrosyan and Danilov (1979 and 1982) introduced the notion of "imputation distribution procedure" for cooperative solution. In particular, the method of regularization was introduced to construct time consistent solutions. Yeung and Petrosyan (2001) designed a time consistent solution in differential games and characterized the conditions that the allocation distribution procedure must satisfy. (see also Yeung D.W.K (1989) and Zaccour (2008).

6.8 Dynamic Stable Solutions

Let there exist solutions $W_v(x_t^*, T-t) \neq \emptyset$, $t_0 \leq t \leq T$ along the conditionally optimal trajectory $\{x^*(t)\}_{t=t_0}^T$. If this condition is not satisfied, it is impossible for the players to adhere to the chosen principle of optimality, since at the very first instant t, when $W_v(x_t^*, T-t) \neq \emptyset$, the players have no possibility to follow this principle. Assume that at time t_0 when the initial state x_0 is the players agree on the imputation

$$\begin{aligned}\xi(x_0, T-t_0) &= [\xi_1(x_0, T-t_0), \xi_2(x_0, T-t_0), \ldots, \xi_n(x_0, T-t_0)] \\ &\in W_v(x_0, T-t_0).\end{aligned}$$

This means that the players agree on an imputation of the gain in such a way that the share of the ith player over the time interval $[t_0, T]$ is equal to $\xi_i(x_0, T-t_0)$. If according to $\xi(x_0, T-t_0)$ Player i is supposed to receive a payoff equaling $\varpi_i[\xi(x_0, T-t_0); x^*(\cdot), t-t_0]$ over the time interval $[t_0, t]$, then over the remaining time interval $[t, T]$ according to the $\xi(x_0, T-t_0)$ Player i is supposed to receive

$$\begin{aligned}&\eta_i[\xi(x_0, T-t_0); x^*(t), T-t] \\ &\quad = \xi_i(x_0, T-t_0) - \varpi_i[\xi(x_0, T-t_0); x^*(\cdot), t-t_0].\end{aligned} \qquad (6.8.1)$$

Theorem. Let $\eta[\xi(x_0, T-t_0); x^*(t), T-t]$ be the vector containing

$$\eta_i[\xi(x_0, T-t_0); x^*(t), T-t], \quad \text{for } i \in \{1, 2, \ldots, n\}.$$

For the original imputation agreement (that is the imputation $\xi(x_0, T-t_0)$) to remain in force at the instant t, it is essential that the vector

$$\eta[\xi(x_0, T - t_0); x^*(t), T - t] \in W_v(x_t^*, T - t), \tag{6.8.2}$$

and $\eta[\xi(x_0, T - t_0); x^*(t), T - t]$ is indeed a solution of the current game $\Gamma_v(x_t^*, T - t)$. If such a condition is satisfied at each instant of time $t \in [t_0, T]$ along the trajectory $\{x^*(t)\}_{t=t_0}^{T}$, then the imputation $\xi(x_0, T - t_0)$ is dynamical stable.

Along the trajectory $x^*(t)$ over the time interval $[t, T]$, $t_0 \leq t \leq T$, the coalition N obtains the payoffs

$$v(N; x^*(t), T - t) = \sum_{j=1}^{n} \left\{ \int_t^T g^j[s, x^*(s), u^*(s)]ds + q^j(x^*(T)) \right\}. \tag{6.8.3}$$

Then the difference

$$v(N; x_0, T - t_0) - v(N; x^*(t), T - t) = \sum_{j=1}^{n} \left\{ \int_{t_0}^{t} g^j[s, x^*(s), u^*(s)]ds \right\}$$

is the payoff the coalition N obtains on the time interval $[t_0, t]$.

Dynamic stability or time consistency of the solution imputation $\xi(x_0, T - t_0)$ guarantees that the extension of the solution policy to a situation with a later starting time and along the optimal trajectory remains optimal. Moreover, group and individual rationalities are satisfied throughout the entire game interval. A payment mechanism leading to the realization of this imputation scheme must be formulated. This will be done in the next section.

6.9 Payoff Distribution Procedure

A PDP proposed by Petrosyan (1997) will be formulated so that the agreed upon dynamically stable imputations can be realized. Let the payoff Player i receive over the time interval $[t_0, t]$ be expressed as

$$\varpi_i[\xi(x_0(\cdot), T - t_0); x^*(\cdot), t - t_0] = \int_{t_0}^{t} B_i(s)ds, \tag{6.9.1}$$

where

$$\sum_{j\in N} B_j(s) = \sum_{j\in N} g^j[s, x^*(s), u^*(s)], \quad \text{for } t_0 \le s \le t \le T.$$

From (6.9.1) we get

$$\frac{d\varpi_i}{dt} = B_i(t). \tag{6.9.2}$$

This quantity may be interpreted as the instantaneous payoff of the Player i at the moment t. Hence it is clear the vector $B(t) = [B_1(t), B_2(t), \ldots, B_n(t)]$ prescribes distribution of the total gain among the members of the coalition N. By properly choosing $B(t)$, the players can ensure the desirable outcome that at each instant $t \in [t_0, T]$ there will be no objection against realization of the original agreement (the imputation $\xi(x_0, T - t_0)$).

Definition. The imputation $\xi(x_0, T - t_0) \in W_v(x_0, T - t_0)$ is dynamically stable in the game $\Gamma_v(x_0, T - t_0)$ if the following conditions are satisfied:

1. there exists a conditionally optimal trajectory $\{x^*(t)\}_{t=t_0}^{T}$ along which $W_v(x^*(t), T - t) \neq \emptyset, \;\; t_0 \le t \le T$,
2. there exists function $B(t) = [B_1(t), B_2(t), \ldots, B_n(t)]$ integrable along $[t_0, T]$ such that

$$\sum_{j\in N} B_j(t) = \sum_{j\in N} g^j[t, x^*(t), u^*(t)] \quad \text{for } t_0 \le s \le t \le T, \quad \text{and}$$

$$\xi(x_0, T - t_0) \in \bigcap_{t_0 \le t \le T} (\varpi[\xi(x_0(\cdot), T - t_0); x^*(\cdot), t - t_0] \oplus W_v(x^*(t), T - t))$$

where $\varpi[\xi(x_0(\cdot), T - t_0); x^*(\cdot), t - t_0]$ is the vector of

$$\varpi_i[\xi(x_0(\cdot), T - t_0); x^*(\cdot), t - t_0], \;\; \text{for } i \in N;$$

and $W_v(x^*(t), T - t)$ is a solution of the current game $\Gamma_v(x^*(t), T - t)$; and the operator $\oplus$ defines the operation: for $\eta \in R^n$ and $A \subset R^n$, $\eta \oplus A = \{\pi + a | a \in A\}$.

The cooperative differential game $\Gamma_v(x_0, T - t_0)$ has a dynamic stable solution $W_v(x_0, T - t_0)$ if all of the imputations $\xi(x_0, T - t_0) \in W_v(x_0, T - t_0)$ are dynamically stable. The conditionally optimal trajectory along which there exists a dynamically stable solution of the game $\Gamma_v(x_0, T - t_0)$ is called an *optimal trajectory*.

From Definition we have

$$\xi(x_0, T - t_0) \in (\varpi[\xi(x_0(\cdot), T - t_0); x^*(\cdot), T - t_0] \oplus W_v(x^*(T), 0))$$

where $W_v(x^*(T), 0) = q(x^*(T))$ is a solution of the game $\Gamma_v(x^*(T), 0)$. Therefore, we can write

$$\xi(x_0, T - t_0) = \int_{t_0}^{T} B(s)ds + q(x^*(T)).$$

The dynamically stable imputation $\xi(x_0, T - t_0) \in W_v(x_0, T - t_0)$ may be realized as follows. From Definition at any instant $t_0 \leq t \leq T$ we have

$$\xi(x_0, T - t_0) \in (\varpi[\xi(x_0(\cdot), T - t_0); x^*(\cdot), t - t_0] \oplus W_v(x^*(t), T - t)), \tag{6.9.3}$$

where $\varpi[\xi(x_0(\cdot), T - t_0); x^*(\cdot), t - t_0] = \int_{t_0}^{t} B(s)ds$ is the payoff vector on the time interval $[t_0, t]$.

Player i's payoff over the same interval can be expressed as:

$$\varpi_i[\xi(x_0(\cdot), T - t_0); x^*(\cdot), t - t_0] = \int_{t_0}^{t} B(s)ds.$$

When the game proceeds along the optimal trajectory, over the time interval $[t_0, t]$ the players share the total gain $\int_{t_0}^{t} \sum_{j \in N} g^j[s, x^*(s), u^*(s)]ds$ so that the inclusion

$$\xi(x_0, T - t_0) - \varpi[\xi(x_0(\cdot), T - t_0);\ x^*(\cdot), t - t_0] \in W_v(x^*(t), T - t) \tag{6.9.4}$$

is satisfied. Condition (6.9.4) implies the existence of a vector $\xi(x_t^*, T-t) \in W_v(x^*(t), T-t)$ satisfying

$$\xi(x_0, T-t_0) = \varpi[\xi(x_0(\cdot), T-t_0); x^*(\cdot), t-t_0] + \xi(x_t^*, T-t).$$

Thus in the process of choosing $B(s)$, the vector of the gains to be obtained by the players at the remaining game interval $[t, T]$ has to satisfy:

$$\begin{aligned}\xi(x_t^*, T-t) &= \xi(x_0, T-t_0) - \varpi[\xi(x_0(\cdot), T-t_0); x^*(\cdot), t-t_0] \\ &= \int_t^T B(s)ds + q(x^*(T)),\end{aligned}$$

where

$$\sum_{j\in N} B_j(s) = \sum_{j\in N} g^j[s, x^*(s), u^*(s)]ds \quad \text{for } t \le s \le T, \text{ and}$$

$$\xi(x_t^*, T-t) \in W_v(x^*(t), T-t).$$

By varying the vector $\varpi[\xi(x_0(\cdot), T-t_0); x^*(\cdot), t-t_0]$ restricted by the condition

$$\sum_{j\in N} \varpi_j[\xi(x_0(\cdot), T-t_0); x^*(\cdot), t-t_0] = \int_{t_0}^t \sum_{j\in N} g^j[s, x^*(s), u^*(s)]ds$$

the players ensure displacement of the set

$$(\varpi[\xi(x_0(\cdot), T-t_0); x^*(\cdot), t-t_0] \oplus W_v(x^*(t), T-t))$$

in such a way that (6.9.3) is satisfied.

For any vector $B(\tau)$ satisfying condition (6.9.3) and (6.9.4) at each time instant $t_0 \le t \le T$ the players are guided by the same optimality principle that leads to the imputation $\xi(x_t^*, T-t) \in W_v(x^*(t), T-t)$ throughout the game, and hence the players have no reason to dispute the previous agreement. In general, it is fairly easy to see that the vectors $B(\tau)$ satisfying conditions (6.9.3) and (6.9.4) may not be unique. Thus, there exist multiple sharing methods satisfying the condition of dynamic stability.

Dynamic instability of the solution of the cooperative differential game leads to abandonment of the original optimality principle generating this solution, because none of the imputations from the solution set $W_v(x_0, T-t_0)$ remain optimal until the game terminates. Therefore, the set $W_v(x_0, T-t_0)$ may be called a solution to the game $\Gamma_v(x_0, T-t_0)$ only if it is dynamically stable. Otherwise the game $\Gamma_v(x_0, T-t_0)$ is assumed to have no solution.

6.10 An Analysis in Pollution Control

6.10.1. Consider the pollution model in Petrosyan and Zaccour (2003). Let N denote the set of countries involved in the game of emission reduction. Emission of player $i \in \{1, 2, \ldots, n\} = N$ at time $t(t \in [0, \infty))$ is denoted by $m_i(t)$. Let $x(t)$ denote the stock of accumulated pollution by time t. The evolution of this stock is governed by the following differential equation:

$$\frac{dx(t)}{dt} = \dot{x}(t) = \sum_{i \in I} m_i(t) - \delta x(t), \quad \text{given } x(0) = x_0, \tag{6.10.1}$$

where δ denotes the natural rate of pollution absorption.

Each player seeks to minimize a stream of discounted sum of emission reduction cost and damage cost. The latter depends on the stock of pollution. We omit from now on the time argument when no ambiguity may arise, $C_i(m_i)$ denotes the emission reduction cost incurred by country i when limiting its emission to level m_i, and $D_i(x)$ its damage cost. We assume that both functions are continuously differentiable and convex, with $C'(m_i) < 0$ and $D'(x) > 0$. The optimization problem of country i is

$$\min_{m_i} J^i(m, x) = \int_0^\infty \exp(-rs)\{C_i(m_i(s)) + D_i(x(s))\} ds \tag{6.10.2}$$

subject to (6.10.1), where $m = (m_1, m_2, \ldots, m_n)$ and r is the common social discount rate.

This formulation is chosen due to the following motivations. First, a simple formulation of the ecological economics problem is chosen to put the emphasis on the cost sharing issue and the mechanism for allocating the total cost over time in a desirable manner. Second, obviously this model still captures one of the main ingredients of the problem, that is, each player's cost depends on total emissions and on inherited stock of pollution. Third, the convexity assumption and the sign assumption for the first derivatives of the two cost functions seem natural. Indeed, the convexity of $C_i(e_i)$ means that the marginal cost of reduction emissions is higher for lower levels of emissions (see Germain *et al.* (1998) for a full discussion). Fourth, for the sake of mathematical tractability it is assumed that all countries discount their costs at the same rate. Finally, again for tractability, it is worth noticing that this formulation implies that reduction of pollution can only be achieved by reducing emissions.

6.10.2. Decomposition Over Time of the Shapley Value. A cooperative game methodology to deal with the problem of sharing the cost of emissions reduction is adopted. The steps are as follows:

(i) Computation of the characteristic function values of the cooperative game.
(ii) Allocation among the players of the total cooperative cost based on the Shapley value.
(iii) Allocation over time of each player's Shapley value having the property of being time-consistent.

The Shapley value is adopted as a solution concept for its fairness and uniqueness merits. The first two steps are classical and provide the individual total cost for each player as a lump sum. The third step aims to allocate over time this total cost in a time-consistent way. The definition and the computation of a time-consistent distribution scheme are dealt with below after introducing some notation.

Let the state of the game be defined by the pair (t, x) and denote by $\Gamma_v(x, t)$ the subgame starting at date t with stock of pollution x. Denote by $x^N(t)$ the trajectory of accumulated pollution under full cooperation (grand coalition N). In the sequel, we use $x^N(t)$ and x_t^N interchangeably. Let $\Gamma_v(x_t^N, t)$ denote a subgame that starts along the cooperative trajectory of the state. The characteristic function value for a coalition $K \subseteq N$ in subgame $\Gamma_v(x,t)$ is defined to be its minimal cost and is denoted $v(K; x, t)$. With this notation, the total cooperative cost to be allocated among the players is then $v(N; x, 0)$ which is the minimal cost for the grand coalition N and its characteristic function value in the game $\Gamma_v(x, 0)$. Let $\Phi^v(x,t) = [\Phi_1^v(x,t), \Phi_2^v(x,t), \ldots, \Phi_n^v(x,t)]$ denote the Shapley value in subgame $\Gamma_v(x,t)$. Finally, denote by $\beta_i(t)$ the cost to be allocated to Player i at instant of time t and $\beta(t) = (\beta_1(t), \ldots, \beta_n(t))$.

Let the vector $B(t) = [B_1(t), B_2(t), \ldots, B_n(t)]$ denote an imputation distribution procedure (IDP) so that

$$\Phi_1^v(x,0) = \int_0^\infty \exp(-rt)\beta_i(t)dt, \quad i = 1, \ldots, n. \tag{6.10.3}$$

The interpretation of this definition is obvious: a time function $B_i(t)$ qualifies as an IDP if it decomposes over time the total cost of Player i as given by the Shapley value component for the whole game $\Gamma_v(x,0)$, i.e. the sum of discounted instantaneous costs is equal to $\Phi_i^v(x,0)$.

The vector $B(t)$ is a time-consistent IDP if at (x_t^N, t), $\forall t \in [0, \infty)$ the following condition holds

$$\Phi_i^v(x_0, 0) = \int_0^t \exp(-r\tau)\beta_i(\tau)d\tau + \exp(-rt)\Phi_i^v(x_t^N, t). \tag{6.10.4}$$

To interpret condition (6.10.4), assume that the players wish to renegotiate the initial agreement reached in the game $\Gamma_v(x,0)$ at (any) intermediate instant of time t. At this moment, the state of the system is $x^N(t)$, meaning that cooperation has prevailed from initial time until t, and that each Player i would have been allocated a sum of discounted stream of monetary amounts given by the first right-hand side term. Now, if the subgame $\Gamma_v(x_t^N, t)$, starting with initial condition $x(t) = x^N(t)$, is played cooperatively, then Player i will get his Shapley value component in this game given by the second right-hand side term of (6.10.4). If what he has been allocated until t and what he will be allocated from this date onward sum up to his cost in the original agreement, i.e. his Shapley value $\Phi_i^v(x_0, 0)$, then this renegotiation would leave the original agreement unaltered. If one can find an IDP $B(t) = [B_1(t), B_2(t), \ldots, B_n(t)]$ such that (6.10.4) holds true, then this IDP is time-consistent. An "algorithm" to build a time-consistent IDP in the case where the Shapley value is differentiable over time is suggested below. One can provide in such a case a simple expression for $B(t) = [B_1(t), B_2(t), \ldots, B_n(t)]$ having an economically appealing interpretation.

6.10.3. A Solution Algorithm. The first three steps compute the necessary elements to define the characteristic function given in the fourth step. In the next two, the Shapley value and the functions $B_i(t)$, $i = 1, 2, \ldots, n$, are computed.

Step 1: Minimize the total cost of the grand coalition.

The grand coalition solves a standard dynamic programming problem consisting of minimizing the sum of all players' costs subject to pollution accumulation dynamics, that is:

$$\min_{m_1, m_2, \ldots, m_n} \sum_{i \in N} \int_t^{\infty} \exp[-r(\tau - t)]\{C_i(m_i(\tau)) + D_i(x(\tau))\} d\tau$$

$$\text{s.t. } \dot{x}(s) = \sum_{i \in N} m_i(s) - \delta x(s), \quad x(t) = x^N(t).$$

Denote by $W(N, x, t)$ the (Bellman) value function of this problem, where the first entry refers to the coalition for which the optimization has been performed, here is the grand coalition N. The outcome of this optimization is a vector of emission strategy $m^N(x^N(\tau)) = [m_1^N(x^N(\tau)), \ldots, m_n^N(x^N(\tau))]$, where $x^N(\tau)$ refers to the accumulated pollution under the scenario of full cooperation (grand coalition).

Step 2. Compute a feedback Nash equilibrium.

Since the game is played over an infinite time horizon, stationary strategies will be sought. To obtain a feedback Nash equilibrium, assuming differentiability of the value function, the following Isaacs–Bellman equations (see 6.4.2) must be satisfied

$$r\overline{V}^i(x) = \min_{m_i} \left\{ C_i(m_i) + D_i(x) + \overline{V}^i_x(x) \left[\sum_{i\in I} m_i - \delta x \right] \right\}, \quad i \in N.$$

Denote by $m^*(x) = [m_1^*(x), m_2^*(x), \ldots, m_n^*(x)]$ any feedback Nash equilibrium of this noncooperative game. This vector can be interpreted as a business-as-usual emission scenario in the absence of an international agreement. It will be fully determined later on with special functional forms. For the moment, we need to stress that from this computation we can get the player's Nash outcome (cost-to-go) in game $\Gamma_v(x_0, 0)$, that we denote $V^i(0, x_0) = \overline{V}^i(x_0)$, and his outcome in subgame $\Gamma_v(x_t^N, t)$, that we denote $V^i(t, x_t^N) = \overline{V}^i(x_t^N)$.

Step 3: Compute outcomes for all remaining possible coalitions.

The optimal total cost for any possible subset of players containing more than one player and excluding the grand coalition (there will be $2^n - n - 2$ subsets) is obtained in the following way. The objective function is the sum of objectives of players in the subset (coalition) considered. In the objective and in the constraints of this optimization problem, we insert for the left-out players the values of their decision variables (strategies) obtained at Step 2, that is their Nash values. Denote by $W(K, x, t)$ the value function for coalition K. This value is formally obtained as follows

$$\begin{aligned} W(K, x, t) = {} & \min_{m_i, i\in K} \sum_{i\in K} \\ & \times \left\{ \int_t^{\infty} \exp[-r(\tau - t)]\{C_i(m_i(\tau)) + D_i(x(\tau))\} d\tau \right\} \end{aligned}$$

$$\text{s.t. } \dot{x}(s) = \sum_{i \in N} m_i(s) - \delta x(s), \quad x(t) = x^N(t).$$

$$m_j = m_j^N \quad \text{for } j \in I \backslash K.$$

Step 4: Define the characteristic function.

The characteristic function $v(K; x, t)$ of the cooperative game is defined as follows:

$$v(\{i\}; x, t) = V^i(x, t) = \overline{V}^i(x), \quad i = 1, \ldots, n;$$

$$v(K; x, t) = W(K; x, t), \quad K \subseteq I.$$

Step 5: Compute the Shapley value.

Denote by $\Phi^v(x, t) = [\Phi_1^v(x, t), \Phi_2^v(x, t), \ldots, \Phi_n^v(x, t)]$ the Shapley value in game $\Gamma_v(x, t)$. Component i is given by

$$\Phi_i^v(x, t) = \sum_{K \ni i} \frac{(n-k)!(k-1)!}{n!} [W(K; x, t) - W(K \backslash \{i\}; x, t)],$$

where k denotes the number of players in coalition K. In particular, if cooperation is in force for the whole duration of the game, then the total cost of Player i would be given by his Shapley value in the game $\Gamma_v(x_0, 0)$, that is

$$\Phi_i^v(x_0, 0) = \sum_{K \ni i} \frac{(n-k)!(k-1)!}{n!} [W(K; x_0, 0) - W(K \backslash \{i\}; x_0, 0)].$$

Justification for the use of this nonstandard definition of the characteristic function is provided in 6.10.4.

Step 6: Define a time consistent IDP.

Allocate to Player i, $i = 1, \ldots, n$, at instant of time $t \in [0, \infty)$, the following amount:

$$B_i(t) = \Phi_i^v(x_t^N, t) - \frac{d}{dt} \Phi_i^v(x_t^N, t). \tag{6.10.5}$$

The formula (6.10.5) allocates at instant of time t to Player i a cost corresponding to the interest payment (interest rate times his cost-to-go under cooperation given by his Shapley value) minus the variation over time of this cost-to-go.

The following proposition shows that $\beta(t) = (\beta_1(t), \ldots, \beta_n(t))$, as given by (6.10.5), is indeed a time-consistent IDP.

Proposition. *The vector* $\beta(t) = (\beta_1(t), \ldots, \beta_n(t))$ *where* $\beta_i(t)$ *is given by* (6.10.5) *is a time-consistent IDP.*

Proof. We first show that it is an IDP, that is

$$\int_0^\infty \exp(-rt)\beta_i(t)dt = \Phi_i^v(x_0, 0).$$

Multiply (6.10.5) by the discount factor $\exp(-rt)$ and integrate

$$\int_0^\infty \exp(-rt)\beta_i(t)dt = \int_0^\infty \exp(-rt)[r\Phi_i^v(x^N, t) - \frac{d}{dt}\Phi_i^v(x^N, t)]dt$$
$$= -exp(-rt)\Phi_i^v(x^N, t)|_0^\infty = \Phi_i^v(x_0, 0).$$

Repeating the above integration for $\Phi_i^v(x_t^N, t)$, one can readily obtain

$$\Phi_i^v(x_0, 0) = \int_0^t \exp(-r\tau)\beta_i(\tau)d\tau + \exp(-rt)\Phi_i^v(x_t^N, t),$$

which satisfies the time-consistent property.

6.10.4. Rationale for the Algorithm and the Special Characteristic Function.

This section discusses the rationale for the algorithm proposed in 6.10.3 and the nonstandard definition of the characteristic function adopted.

As Petorsian and Zaccour pointed out while formulating the solution algorithm, a central element in formal negotiation theories is the status quo, which gives what a player would obtain if negotiation fails. It is a measure of the strategic force of a player when acting alone. The same idea can be extended to subsets of players. To measure the strategic force of a subset of players (coalition), one may call upon the concept of characteristic function, a mathematical tool that is precisely intended to provide such measure. All classical cooperative game solutions (core, the Shapley value, etc.) use the characteristic function to select a subset of imputations that satisfy the requirements embodied in the solution concept adopted. For instance, the core selects the imputations that cannot be blocked by any coalition whereas Shapley value selects one imputation satisfying some axioms, among them fairness. If the set of imputations is not a singleton, the players may negotiate to select one of them. In a dynamic game, the computed imputations usually correspond to the payoffs (here the sum of discounted costs) for the whole duration of the game. In this case, an interesting problem emerges which is how to allocate these amounts over time. One basic requirement is that the distribution over time is feasible, that is, the amounts allocated to each player sum up to his entitled total share (see the definition of an IDP). Obviously, one may construct an infinite number of intertemporal allocations that satisfy this requirement, but not all these streams are conceptually and intuitively appealing. The approach pursued here is to decompose the total individual cost over time so that if the players renegotiate the agreement at any intermediate instant of time along the cooperative state trajectory, then they would obtain the same outcomes.

It is also noted that the computation of the characteristic function values is not standard. The assumption that left-out players $(I\backslash K)$ stick to their feedback Nash strategies when the characteristic function value is computed for coalition K is made. Basically, there are few current options in game theory literature regarding this issue.

One option is the one offered by Von Neumann and Morgenstern (1944) where they assumed that the left-out players maximize the cost of the considered coalition. This approach, which gives the minimum guaranteed cost, does not seem to be the best one in our context. Indeed, it is unlikely that if a subset of countries form a coalition to tackle an environmental problem, then the remaining countries would form an anti-coalition to harm their efforts. For instance, the Kyoto Protocol permits that countries fulfill jointly their abatement obligation (this is called "joint implementation" in the Kyoto Protocol). The question is then why, e.g. the European Union would wish to maximize abatement cost of say a collation formed of Canada and USA if they wish to take the joint implementation option? We believe that the Von Neumann–Morgenstern determination of the characteristic function has a great historic value and the advantage of being easy to compute but would not suit well the setting we are dealing with.

Next option is to assume that the characteristic function value for a coalition is its Nash equilibrium total cost in the noncooperative game between this coalition and the other players acting individually or forming an anti-coalition.

One "problem" with this approach is computation. Indeed, this approach requires solving $2^n - 2$ dynamic equilibrium problems (that is, as many as the number of nonvoid coalitions and excluding the grand one). Here we solve only one equilibrium problem, all others being standard dynamic optimization problems. Therefore, the computational burden is not at all of the same magnitude since solving a dynamic feedback equilibrium problem is much harder than dealing with a dynamic optimization one.

Now, assume that Nash equilirbia exist for all partitions of the set of players (clearly this is far from being automatically guaranteed). First, recall that we aim to compute the Shapley value for Player i. This latter involves his marginal contributions, which are differences between values of the characteristic function of the form $v(K, S, t) - v(K\backslash\{i\}, S, t)$. In the equilibrium approach, these values correspond to Nash outcomes of a game between players in coalition K and the remaining players in $I\backslash K$ (acting individually or collectively is not an issue at this level). If in any of the $2^n - 2$ equilibrium problems that must be solved the equilibrium is not unique, then we face an extremely hard selection problem. In our approach, the coalition computes its value with the assumption that left-out players will continue to adopt a noncooperative emission strategies. In the event that our Step 2 gives multiple equilibria, we could still compute the Shapley value for each of them without having to face a selection problem.

Finally global environmental problems involve by definition all countries around the globe. Although few of them are heavy weights in the sense that their environmental policies can make a difference on the level of pollution accumulation, many countries can be seen as nonatomistic players. It is intuitive to assume that probably these countries will follow their business-as-usual strategy, i.e. by sticking to their Nash emissions, even when some (possibly far away) countries are joining effort.

6.11 Illustration with Specific Functional Forms

Consider the following specification of (6.10.2) in which

$$C_i(m_i) = \frac{\gamma}{2}[m_i - \overline{m}_i]^2, \quad 0 \le m_i \le \overline{m}_i, \quad \gamma > 0 \quad \text{and} \quad i \in \{1, 2, 3\};$$
$$D_i(x) = \pi x, \quad \pi > 0.$$

Computation of optimal cost of grand coalition (Step 1).
The value function $W(N, x, t)$ must satisfy the Bellman equation

$$rW(N,x,t) = \min_{m_1,m_2,m_3} \left\{ \sum_{i=1}^{3} \left(\frac{\gamma}{2}[m_i - \overline{m}_i]^2 + \pi x \right) + W_x(N,x,t) \left[\sum_{i=1}^{3} m_i - \delta x \right] \right\}. \tag{6.11.1}$$

Performing the indicated minimization in (6.11.1) yields

$$m_i^N = \overline{m}_i - \frac{1}{\gamma} W_x(N, x, t), \quad \text{for } i \in \{1, 2, 3\}.$$

Substituting m_i^N in (6.11.1) and upon solving yields

$$\begin{aligned} W(N, x, t) &= \overline{W}(N, x) \\ &= \frac{3\pi}{r(r+\delta)} \left\{ \left[\sum_{i=1}^{3} \overline{m}_i - \frac{3^2 \pi}{2\gamma(r+\delta)} \right] + rx \right\}, \quad \text{and} \end{aligned} \tag{6.11.2}$$

$$m_i^N = \overline{m}_i - \frac{3\pi}{\gamma(r+\delta)}, \quad \text{for } i \in \{1, 2, 3\}. \tag{6.11.3}$$

The optimal trajectory of the stock of pollution can be obtained as

$$x^N(t) = \exp(-\delta t) x(0) + \frac{1}{\delta} \left\{ \left[\sum_{i=1}^{3} m_i^N \right] [1 - \exp(-\delta t)] \right\}. \tag{6.11.4}$$

Computation of feedback Nash equilibrium (*Step 2*).

To solve a feedback Nash equilibrium for the noncooperative game (6.10.1)–(6.10.2), we follow 6.4.2 and obtain the Bellman equation

$$r\overline{V}^i(x) = \min_{m_i} \left\{ \frac{\gamma}{2} [m_i - \overline{m}_i]^2 + \pi x + \overline{V}_x^i(x) \left[\sum_{\substack{j \in [1,2,3] \\ i \neq j}} m_j^* + m_i - \delta x \right] \right\},$$
$$\text{for } i \in \{1, 2, 3\}. \tag{6.11.5}$$

Performing the indicated minimization yields

$$m_i^* = \overline{m}_i - \frac{1}{\gamma} \overline{V}_x^i(x), \quad \text{for } i \in \{1, 2, 3\}. \tag{6.11.6}$$

Substituting (6.11.6) into (6.11.5) and upon solving yield

$$\overline{V}^i(x) = \frac{\pi}{r(r+\delta)}\left\{\frac{\pi}{2\gamma(r+\delta)} + \sum_{i=1}^{3}\overline{m}_i - \frac{3\pi}{\gamma(r+\delta)} + rx\right\},$$

$$\text{for } i \in \{1,2,3\}. \tag{6.11.7}$$

The Nash equilibrium feedback level of emission can then be obtained as:

$$m_i^* = \overline{m}_i^* - \frac{\pi}{\gamma(r+\delta)}, \quad \text{for } i \in \{1,2,3\}. \tag{6.11.8}$$

The difference between Nash equilibrium emissions and those obtained for the grand coalition is that player takes into account the sum of marginal damage costs of all coalition members and not only his own one.

Computation of optimal cost for intermediate coalitions (*Step 3*).

The value function $W(K,x,t)$ for any coalition K of two players must satisfy the following Bellman equation

$$rW(K,x,t) = \min_{m_1, i\in K}\left\{\sum_{i\in K}\left(\frac{\gamma}{2}[m_i - \overline{m}_i]^2 + \pi x\right)\right.$$

$$\left. + W_x(K,x,t)\left[\sum_{i\in K} m_i + m_j^* - \delta x\right]\right\}, \tag{6.11.9}$$

where $j \notin K$.

Following similar procedure adopted for solving for the grand coalition, one can obtain:

$$W(K,x,t) = \overline{W}(K,x)$$

$$= \frac{2\pi}{r(r+\delta)}\left\{\sum_{i\in K}\overline{m}_i - \frac{4\pi}{2\gamma(r+\delta)} - \frac{\pi}{\gamma(r+\delta)} + rx\right\}. \tag{6.11.10}$$

The corresponding emission of coalition K is:

$$m_i^K = \overline{m}_i - \frac{2\pi}{\gamma(r+\delta)}, \quad i \in K. \tag{6.11.11}$$

Definition of the characteristic function (Step 4).

The characteristic function values are given by

$$\begin{aligned} v(\{i\};x,t) &= V^i(x,t) = \overline{V}^i(x) = \frac{\pi}{r(r+\delta)} \\ &\quad \times \left\{ \frac{\pi}{2\gamma(r+\delta)} + \sum_{i=1}^{3} \overline{m}_i - \frac{3\pi}{\gamma(r+\delta)} + rx \right\}, \quad i = 1,2,3; \\ v(K;x,t) &= W(K,x,t) = \overline{W}(K,x) \\ &= \frac{2\pi}{r(r+\delta)} \left\{ \sum_{i\in K} \overline{m}_i - \frac{4\pi}{2\gamma(r+\delta)} - \frac{\pi}{\gamma(r+\delta)} + rx \right\}, \\ &\quad K \subseteq \{1,2,3\}. \end{aligned}$$

Computation of the Shapley value (Step 5).

Assuming symmetric $\overline{m}_i$, the Shapley value of the game can be expressed as

$$\begin{aligned} \Phi_i^v(x,t) &= \sum_{K \ni i} \frac{(n-k)!(k-1)!}{n!} [v(K;x,t) - v(K\backslash\{i\};x,t)] \\ &= \frac{1}{2r(r+\delta)} \left\{ 2\pi \left(\sum_{i=1}^{3} \overline{m}_i + \rho S \right) - \frac{9\pi^2}{\gamma(r+\delta)} \right\}, \\ &\quad i = 1,2,3. \end{aligned} \tag{6.11.12}$$

Computation of IDP functions (Step 6).

To provide a allocation that sustains the Shapley value $\Phi_i^v(x,t)$ over time along the optimal trajectory $x^N(t)$ in (6.11.4), we recall from (6.10.5) that the IDP functions are given by

$$B_i(t) = \Phi_i^v\left(x_t^N, t\right) - \frac{d}{dt}\Phi_i^v\left(x_t^N, t\right).$$

Straightforward calculations lead to

$$B_i(t) = \pi x^N(t) + \frac{9\pi^2}{2\gamma(r+\delta)^2}, \quad i = 1,2,3. \tag{6.11.13}$$

To verify that $B_i(t)$ indeed brings about the Shapley value of Player i we note that

$$\Phi_i^v(x^N, 0) = \frac{1}{2r(r+\delta)}\left\{2\pi\left(\sum_{i=1}^{3}\overline{m}_i + rx(0)\right) - \frac{9\pi^2}{\gamma(r+\delta)}\right\},$$
$$i = 1,2,3. \tag{6.11.14}$$

Multiply both sides of (6.11.13) by the discount factor and integrate

$$\int_0^\infty \exp(-rt)B_i(t)dt = \int_0^\infty \exp(-rt)\left[\pi x^N(t) + \frac{9\pi^2}{2\gamma(r+\delta)^2}\right]dt,$$
$$i = 1,2,3, \tag{6.11.15}$$

where from (6.11.3)–(6.11.4)

$$x^N(t) = \exp(-\delta t)x(0) + \frac{1}{\delta}\left\{\left[\sum_{j=1}^{3}\left(\overline{m}_j - \frac{3\pi}{2\gamma(r+\delta)}\right)\right][1 - \exp(-\delta t)]\right\}.$$

Substituting $x^N(t)$ into (6.11.15) yields:

$$\int_0^\infty \exp(-rt)\beta_i(t)dt$$
$$= \int_0^\infty \exp[-(r+\delta)t]\pi x_0 dt + \int_0^\infty \exp(-rt)\frac{\pi}{\delta}$$
$$\times\left(\sum_{i=1}^{3}\overline{m}_i - \frac{9\pi}{\gamma(r+\delta)}\right)dt + \int_0^\infty \exp[-(r+\delta)t]\frac{\pi}{\delta}$$
$$\times\left(\sum_{i=1}^{3}\overline{m}_i - \frac{9\pi}{\gamma(r+\delta)}\right)dt + \int_0^\infty \exp(-rt)\frac{9\pi^2}{2\gamma(r+\delta)^2}dt.$$

Upon integrating

$$\int_0^\infty \exp(-rt)\beta_i(t)dt$$

$$= \frac{1}{2r(r+\delta)}\left\{2\pi\left(\sum_{i=1}^{3}\overline{m}_i + rx(0)\right) - \frac{9\pi^2}{\gamma(r+\delta)}\right\} = \Phi_i^v(x_0,0),$$

for $i = 1,2,3$.

6.12 Exercises and Problems

1. Consider the dynamic optimization problem

$$\int_{t_0}^{T}\left[u(s)^{1/2} - \frac{c}{x(s)^{1/2}}u(s)\right]\exp[-r(s-t_0)]ds$$
$$+\exp[-r(T-t_0)]qx(T)^{1/2},$$

subject to

$$\dot{x}(s) = \left[ax(s)^{1/2} - bx(s) - u(s)\right], \quad x(t_0) = x_0 \in X.$$

Use Bellman's techniques of dynamic programming to solve the problem.

2. Consider again the dynamic optimization problem

$$\int_{t_0}^{T}\left[u(s)^{1/2} - \frac{c}{x(s)^{1/2}}u(s)\right]\exp[-r(s-t_0)]ds$$
$$+\exp[-r(T-t_0)]qx(T)^{1/2},$$

subject to

$$\dot{x}(s) = \left[ax(s)^{1/2} - bx(s) - u(s)\right], \quad x(t_0) = x_0 \in X.$$

(a) If $c = 1$, $q = 2$, $r = 0.01$, $t_0 = 0$, $T = 5$, $a = 0.5$, $b = 1$ and $x_0 = 20$, use optimal control theory to solve the optimal controls, the optimal state trajectory and the costate trajectory $\{\Lambda(s)\}_{t=t_0}^{T}$.

(b) If $c = 1$, $q = 2$, $r = 0.01$, $t_0 = 0$, $T = 5$, $a = 0.5$, $b = 1$ and $x_0 = 30$, use optimal control theory to solve the optimal controls, the optimal state trajectory and the costate trajectory $\{\Lambda(s)\}_{t=t_0}^{T}$.

3. Consider the game

$$\max_{u_i} \left\{ \int_0^{10} \left[10u_i(s) - \frac{u_i(s)^2}{x(s)} \right] \exp[-0.05s]ds + \exp(-0.5)2x(T) \right\},$$

for $i \in \{1, 2, \ldots, 6\}$

subject to

$$\dot{x}(s) = 15 - \frac{1}{2}x(s) - \sum_{j=1}^{6} u_j(s), \quad x(0) = 25.$$

(a) Obtain an open-loop solution for the game.
(b) Obtain a feedback Nash equilibrium for the game.

4. Consider a time-continuous version of the ecological economics model involving three countries. Emission of country $i \in \{1, 2, 3\}$ at time $t (t \in [0, \infty))$ is denoted by $m_i(t)$. Let $x(t)$ denote the stock of accumulated pollution by time t. The evolution of this stock is governed by the following differential equation:

$$\frac{dx(t)}{dt} = \dot{x}(t) = \sum_{i=1}^{3} m_i(t) - \delta x(t), \quad \text{given } x(0) = 100,$$

where $\delta = 0.05$ denotes the natural rate of pollution absorption.

Each country seeks to minimize a stream of discounted sum of emission reduction cost and damage cost.

$$\min_{m_i} J^i(m, x) = \int_0^{\infty} \exp(-rs)\{C_i(m_i(s)) + D_i(x(s))\}ds,$$

where

$$C_i(m_i) = \frac{1}{2}\left[m_i - \overline{m}_i\right]^2, \quad 0 \le m_i \le \overline{m}_i, \quad \gamma > 0,$$

$$D_i(x) = x.$$

The countries agree to cooperate and share the cost of emissions reduction based on the Shapley value. Following Petrosyan and Zaccour's way of defining characteristic functions, derive a cooperative solution and its IDP.

5. Consider the game of pollution control from 6.10. Suppose that players form coalitions and we have a coalitional partition of the set $N = S_1 \cup S_2 \cup \cdots \cup S_n$. The costs of each coalition is equal to the sum of costs of its members. Find the Nash equilibrium for this case when players are coalitions.

6. In Example 5, find the optimal allocation of the payoff inside coalitions according to the Shapley value.

7. Introduce a two step optimality principle (two step Shapley value) in the game from Example 5 for coalitions on the first step and players inside the coalitions on the second.

8. Find the Nash bargaining solution for the game of pollution control. Investigate the dynamic stability (time-consistency) of the solution.

Chapter 7

Zero-Sum Differential Games

7.1 Differential Zero-Sum Games with Prescribed Duration

Differential games are a generalization of multistage games to the case where the number of steps in a game is infinite (continuum) and the Players 1 and 2 (denoted as E and P, respectively) have the possibility of taking decisions continuously in time. In this setting the trajectories of players' motion are a solution of the system of differential equations whose right-hand sides depend on parameters that are under control of the players.

7.1.1. Let $x \in R^n$, $y \in R^n$, $u \in U \subset R^k$, $v \in V \subset R^l$, $f(x,u), g(y,v)$ be the vector functions of dimension n given on $R^n \times U$ and $R^n \times V$, respectively. Consider two systems of ordinary differential equations

$$\dot{x} = f(x,u), \tag{7.1.1}$$

$$\dot{y} = g(y,v) \tag{7.1.2}$$

with initial conditions x_0, y_0. Player P (E) starts his motion from the phase state x_0 (y_0) and moves in the phase space R^n in accordance with (7.1.1) and (7.1.2), choosing at each instant of time the value of parameter $u \in U (v \in V)$ to suit his objectives and in terms of information available in each current state.

The simplest to describe is the case of perfect information. In the differential game this means that at each time instant t the players choosing parameters $u \in U$, $v \in V$ know the time t and their own and the opponent's phase states x, y. Sometimes one of the players, say, Player P, is required to know at each current instant t the value of the parameter $v \in V$ chosen by Player E at the same instant of time. In this case Player E is said to be discriminated and the game is called the *game with discrimination* against Player E.

The parameters $u \in U$, $v \in V$ are called controls for the players P and E, respectively. The functions $x(t), y(t)$ which satisfy equations (7.1.1), (7.1.2) and initial conditions that are called *trajectories* for the players P, E.

7.1.2. Objectives in the differential game are determined by the payoff which may defined by the realized trajectories $x(t), y(t)$ in a variety of ways. For example, suppose it is assumed that the game is played during a prescribed time T. Let $x(T), y(T)$ be phase coordinates of players P and E at the time instant T the game terminates. Then the payoff to Player E is taken to be $H(x(T), y(T))$, where $H(x, y)$ is some function given on $R^n \times R^n$. In the specific case, when

$$H(x(T), y(T)) = \rho(x(T), y(T)), \tag{7.1.3}$$

where $\rho(x(T), y(T)) = \sqrt{\sum_{i=1}^{n}(x_i(T) - y_i(T))^2}$ is the Euclidean distance between the points $x(T), y(T)$, the game describes the process of pursuit during which the objective of Player E is to avoid Player P by moving a maximum distance from him by the time the game ends. In all cases the game is assumed to be a differential zero-sum game. Under condition (7.1.3), this means that the objective of Player P is to come within the shortest distance of Player E by the time T the game ends.

With such a definition, the payoff depends only on final states and the results obtained by each player during the game until the time T are not scored. It is of interest to state the problem in which the payoff to Player E is defined as a minimum distance between the players during the game:

$$\min_{0 \le t \le T} \rho(x(t), y(t)).$$

There exist games in which the constraint on the game duration is not essential and the game continues until the players obtain a particular result. Let an m-dimensional surface F be given in R^{2n}. This surface will be called terminal. Let

$$t_n = \{\min t : (x(t), y(t)) \in F\}, \tag{7.1.4}$$

i.e. t_n is the first time instant when the point $(x(t), y(t))$ falls on F. If for all $t \ge 0$ the point $(x(t), y(t)) \notin F$, then t_n is $+\infty$. For the realized paths $x(t), y(t)$ the payoff to Player E is t_n (the payoff to Player P is $-t_n$). In particular, if F is a sphere of radius $l \ge 0$ given by the equation

$$\sqrt{\sum_{i=1}^{n} (x_i - y_i)^2} = l,$$

then we have the game of pursuit in which Player P seeks to come within a distance $l \ge 0$ to Player E as soon as possible. If $l = 0$, then the capture is taken to mean the coincidence of phase coordinates for the players P and E, in which case Player E seeks to postpone the capture time. Such games of pursuit are called the *time-optimal games of pursuit.*

The theory of differential games also deals with the problems of determining the set of initial states for the players from which Player P can ensure the capture of Player E within a distance l. And a definition is provided for the set of initial states of the players from which Player E can avoid in a finite time the encounter with Player P within a distance l. One set is called a *capture zone* (C, Z) and the other an *escape zone* (E, Z). It is apparent that these zones do not meet. However, a critical question arises of whether the closure of the union of the capture and the escape zones spans the entire phase space. Also the answer to this question is provided below, we now note that in order to adequately describe this process, it suffices to define the payoff as follows. If there exists $t_n < \infty$ (see (7.1.4)), then the payoff to Player E is -1. If, however, $t_n = \infty$, then the payoff is $+1$ (the payoff to Player P is equal to the payoff to Player E but opposite in sign, since the game is zero-sum). The games of pursuit with such a payoff are called the *pursuit games of kind.*

7.1.3. *Phase constraints.* If we further require that the phase point (x, y) would not leave some set $F \subset R^{2n}$ during the game, then we obtain a differential game with *phase constraints.* A special case of such a game is the "Life-line" game. The "Life-line" game is a zero-sum game of kind in which the payoff to Player E is $+1$ if he reaches the boundary of the set F ("Life-line") before Player P captures him. Thus, the objective of Player E is to reach the boundary of the set F before being captured by Player P (coming within a distance l, $l \geq 0$ with Player P). The objective of Player P, however, is to come within a distance l with Player E while the latter is still in the set F. It is assumed that Player P cannot abandon the set F.

7.1.4. *Example 1.* (*Simple motion.*) The game is played on a plane. Motions of the players P and E are described by the system of differential equations

$$\begin{gathered}
\dot{x}_1 = u_1,\ \dot{x}_2 = u_2,\ u_1^2 + u_2^2 \leq \alpha^2, \\
\dot{y}_1 = v_1,\ \dot{y}_2 = v_2,\ v_1^2 + v_2^2 \leq \beta^2, \\
x_1(0) = x_1^0,\ x_2(0) = x_2^0,\ y_1(0) = y_1^0,\ y_2(0) = y_2^0, \alpha \geq \beta. \qquad (7.1.5)
\end{gathered}$$

The physical implication of equation (7.1.5) is that the players P and E are moving in a plane at limited velocities, the maximum velocities α and β are constant in value and the velocity of Player E does not exceed the velocity of Player P.

Player P can change the direction of his motion (the velocity vector) by choosing at each time instant the control $u = (u_1, u_2)$ constrained by $u_1^2 + u_2^2 \leq \alpha^2$ (the set U). Similarly, Player E can also change the direction of his motion by choosing at each time instant the control $v = (v_1, v_2)$ constrained by $v_1^2 + v_2^2 \leq \beta^2$ (the set V). It is obvious that if $\alpha > \beta$, then the capture zone (C, Z) coincides with the entire space, i.e. Player P can always ensure the capture of Player E within any distance l in a finite time. To this end, it suffices to choose the motion with the maximum velocity α and to direct the velocity vector at each time instant t towards the pursued point $y(t)$, i.e. to carry out pursuit along the pursuit line. If $\alpha \leq \beta$, the escape set (E, Z) coincides with the entire space of the game except the points (x, y) for which $\rho(x, y) \leq l$. Indeed, if at the initial instant $\rho(x_0, y_0) > l$, then Player E can always avoid capture by moving away from Player P along the straight line joining the initial points x_0, y_0, the maximum velocity being β.

The special property manifested here will also be encountered in what follows. In order to form the control which ensures the avoidance of capture for Player E, we need only to know the initial states x_0, y_0 while, to form the control which ensures the capture of Player E in the case $\alpha > \beta$, Player P needs information on his own and the opponent's state at each current instant of time.

Example 2. The players P and E are the material points with unit masses moving on the plane under the control of the modulus-constrained and frictional forces. The equations of motion for the players are

$$\begin{aligned}
&\dot{x}_1 = x_3, \ \dot{x}_2 = x_4, \ \dot{x}_3 = \alpha u_1 - k_P x_3, \\
&\dot{x}_4 = \alpha u_2 - k_P x_4, \ u_1^2 + u_2^2 \leq \alpha^2, \\
&\dot{y}_1 = y_3, \ \dot{y}_2 = y_4, \ \dot{y}_3 = \beta v_1 - k_E y_3, \\
&\dot{y}_4 = \beta v_2 - k_E y_4, \ v_1^2 + v_2^2 \leq \beta^2,
\end{aligned} \tag{7.1.6}$$

where $(x_1, x_2), (y_1, y_2)$ are geometric coordinates, $(x_3, x_4), (y_3, y_4)$ are respectively momenta of the points P and E, k_P and k_E are friction coefficients, α and β are maximum forces which can be applied to the material points P and E. The motion starts from the states $x_i(0) = x_i^0$, $y_i(0) = y_i^0$, $i = 1, 2, 3, 4$. Here, by the state is meant not the locus of the players P and E, but their phase state in the space of coordinates and momenta. The sets U, V are the circles $U = \{u = (u_1, u_2) : u_1^2 + u_2^2 \leq \alpha^2\}$, $V = \{v = (v_1, v_2) : v_1^2 + v_2^2 \leq \beta^2\}$. This means that at each instant the players P and E may choose the direction of applied forces. However, the maximum values of these forces are restricted by the constants α and β. In this formulation as shown below, the condition $\alpha > \beta$ (power superiority) is not adequate for Player P to accomplish pursuit from any initial state.

7.1.5. We did not define yet the ways of selecting controls $u \in U$, $v \in V$ by the players P and E in terms of the incoming information. In other words, the notion of a strategy in the differential game remains to be defined.

Although there exist several approaches to this notion, we shall focus on those intuitively obvious game-theoretic properties which the notion must possess. As noted in Chapter 5, the strategy must describe the behavior of a player in all information states in which he may find himself during the game. In what follows the information state of each player will be determined by the phase vectors $x(t), y(t)$ at the current time instant t. Then it would be natural to regard the strategy for Player P (E) as a function $u(x, y, t)$ $(v(x, y, t))$ with values in the set of controls U (V). That is how the strategy is defined in Isaacs (1965). Strategies of this type are called *synthesizing*. However, this method of defining a strategy suffers from some grave disadvantages. Indeed, suppose the players P and E have chosen strategies $u(x, y, t), v(x, y, t)$, respectively. Then, to determine the paths for the players, and hence the payoff (which is dependent of the paths), we substitute the functions $u(x, y, t), v(x, y, t)$ into equations (7.1.1),(7.1.2) in place of the control parameters u, v and integrate them with initial conditions x_0, y_0 on the time interval $[0, T]$.

We obtain the following system of ordinary differential equations:

$$\dot{x} = f(x, u(x, y, t)), \quad \dot{y} = g(y, v(x, y, t)). \tag{7.1.7}$$

For the existence and uniqueness of the solution to system (7.1.7) it is essential that some conditions be imposed on the functions $f(x, u), g(y, v)$ and the strategies $u(x, y, t), v(x, y, t)$. The first group of conditions places no limitations on the players' capabilities, refers to the statement of the problem and is justified by the physical nature of the process involved. The case is different from the constraints on the class of functions (strategies) $u(x, y, t), v(x, y, t)$. Such constraints on the players' capabilities contradict the notions adopted in game theory that the players are at liberty to choose a behavior. In some cases this leads to substantial impoverishment of the sets of strategies. For example, if we restrict ourselves to continuous functions $u(x, y, t), v(x, y, t)$, the problems arise where there are no solutions in the class of continuous functions. The assumption of a more general class of strategies makes impossible the unique solution of system (7.1.7) on the interval $[0, T]$. At times, to overcome this difficulty, one considers the sets of strategies $u(x, y, t), v(x, y, t)$ under which the system (7.1.7) has a unique solution extendable to the interval $[0, T]$. However, such an approach (aside from the nonconstructivity of the definition of the strategy sets) is not adequately justified, since the set of all pairs of strategies $u(x, y, t), v(x, y, t)$ under which the system (7.1.7) has a unique solution is found to be nonrectangular.

7.1.6. We shall consider the strategies in the differential game to be *piecewise open-loop strategies.*

The piecewise open-loop strategy $u(\cdot)$ for Player P consists of a pair $\{\sigma, a\}$, where σ is some partition $0 = t'_0 < t'_1 < \ldots < t'_n < \ldots$ of the time interval $[0, \infty)$ by the points t'_k which have no finite accumulation points; a is the map which places each point t'_k and phase coordinates $x(t'_k), y(t'_k)$ in correspondence with some measureable open-loop control $u(t) \in U$ for $t \in [t'_k, t'_{k+1})$ (the measurable function $u(t)$ taking values from the set U). Similarly, the piecewise open-loop strategy $v(\cdot)$ for Player E consists of a pair $\{\tau, b\}$ where τ is some partition $0 = t''_0 < t''_1 < \ldots < t''_n < \ldots$ of the time interval $[0, \infty)$ by the points t''_k which have no accumulation points; b is the map which places each point t''_k and positions $x(t''_k), y(t''_k)$ in correspondence with some measurable open-loop control $v(t) \in V$ on the interval $[t''_k, t''_{k+1})$ (the measurable function $v(t)$ taking values from the set V). Using a piecewise open-loop strategy the player responds to changes in information not continuously in time, but at the time instants $t_k \in \tau$ which are determined by the player himself.

Denote the set of all piecewise open-loop strategies for Player P by P, and the set of all possible piecewise open-loop strategies for Player E by E.

Let $u(t), v(t)$ be a pair of measurable open-loop controls for the players P and E (measurable functions with values in the control sets U, V). Consider a system of ordinary differential equations

$$\dot{x} = f(x, u(t)),\ \dot{y} = g(y, v(t)),\ t \geq 0. \tag{7.1.8}$$

Impose the following constraints on the right-hand sides of system (7.1.8). The vector functions $f(x, u), g(y, v)$ are continuous in all their independent variables and are uniformly bounded, i.e. $f(x, u)$ is continuous on the set $R^n \times U$, while $g(y, v)$ is continuous on the set $R^n \times V$ and $\|f(x, u)\| \leq \alpha$, $\|g(y, v)\| \leq \beta$ (here $\|z\|$ is the vector norm in R^n). Furthermore, the vector functions $f(x, u)$ and $g(y, v)$ satisfy the Lipschitz condition in x and y uniformly with respect to u and v, respectively, that is

$$\|f(x_1, u) - f(x_2, u)\| \leq \alpha_1 \|x_1 - x_2\|,\ u \in U,$$
$$\|g(y_1, v) - g(y_2, v)\| \leq \beta_1 \|y_1 - y_2\|,\ v \in V.$$

From the Karatheodory existence and uniqueness theorem it follows that, under the above conditions, for any initial states x_0, y_0 any measurable open-loop controls $u(t), v(t)$ given on the interval $[t_1, t_2]$, $0 \leq t_1 < t_2$, there exist unique vector functions $x(t)$,$y(t)$ which satisfy the following system of differential equations almost everywhere (i.e.

everywhere except the set of measure zero). On the interval $[t_1, t_2]$

$$\dot{x}(t) = f(x(t), u(t)), \ \dot{y}(t) = g(y(t), v(t)) \qquad (7.1.9)$$

and the initial conditions are $x(t_1) = x_0$, $y(t_1) = y_0$ (see Kolmogorov and Fomin (1981), Sansone (1954)).

7.1.7. Let (x_0, y_0) be a pair of initial conditions for equations (7.1.8). The system $S = \{x_0, y_0; u(\cdot), v(\cdot)\}$, where $u(\cdot) \in P$, $v(\cdot) \in E$, is called a *situation* in the differential game. For each situation S there is a unique pair of paths $x(t), y(t)$ such that $x(0) = x_0$, $y(0) = y_0$ and relationships (7.1.9) hold for almost all $t \in [0, T]$, $T > 0$.

Indeed, let $u(\cdot) = \{\delta, a\}$, $v(\cdot) = \{\tau, b\}$. Furthermore, let $0 = t_0 < t_1 < \ldots < t_k < \ldots$ be a partition of the interval $[0, \infty)$ that is the union of partitions δ, τ. The solution to system (7.1.9) is constructed as follows. On each interval $[t_k, t_{k+1})$, $k = 0, 1, \ldots$, the images of the maps a, b are the measurable open-loop controls $u(t), v(t)$; hence on the interval $[t_0, t_1)$ the system of equations (7.1.9) with $x(0) = x_0$, $y(0) = y_0$ has a unique solution. On the interval $[t_1, t_2)$ with $x(t_1) = \lim_{t \to t_1 - 0} x(t)$, $y(t_1) = \lim_{t \to t_1 - 0} y(t)$ as initial conditions, we construct the solution to (7.1.9) by reusing the measurability of controls $u(t), v(t)$ as images of the maps a and b on the intervals $[t_k, t_{k+1})$, $k = 1, 2, \ldots$ Setting $x(t_2) = \lim_{t \to t_2 - 0} x(t)$, $y(t_2) = \lim_{t \to t_2 - 0} y(t)$ we continue this process to find a unique solution $x(t), y(t)$ such that $x(0) = x_0$, $y(0) = y_0$. Any path $x(t)$ ($y(t)$) corresponding to some situation $\{x_0, y_0, u(\cdot), v(\cdot)\}$ is called the path of the Player P (Player E).

7.1.8. *Payoff function.* As shown above, each situation $S = \{x_0, y_0; u(\cdot), v(\cdot)\}$ in piecewise open-loop strategies uniquely determines the paths $x(t), y(t)$ for Players P and E, respectively. The priority degree of these paths will be estimated by the payoff function K which places each situation in correspondence with some real number — a payoff to Player E. The payoff to Player P is $-K$ (this means that the game is zero-sum, since the sum of payoffs to players P and E in each situation is zero). We shall consider the games with payoff functions of four types.

Terminal payoff. Let there be given some number $T > 0$ and a function $H(x, y)$ that is continuous in (x, y). The payoff in each situation $S = \{x_0, y_0; u(\cdot), v(\cdot)\}$ is determined as follows:

$$K(x_0, y_0; u(\cdot), v(\cdot)) = H(x(T), y(T)),$$

where $x(T) = x(t)|_{t=T}$, $y(T) = y(t)|_{t=T}$ (here $x(t), y(t)$ are the paths of players P and E in a situation S). We have the game of pursuit when the function $H(x, y)$ is a Euclidean distance between the points x and y.

Minimum result. Let $H(x, y)$ be a real-valued continuous function. In the situation $S = \{x_0, y_0; u(\cdot), v(\cdot)\}$ the payoff to Player E is taken to be $\min_{0 \le t \le T} H(x(t), y(t))$, where $T > 0$ is a given number. If $H(x, y) = \rho(x, y)$ then the game describes the process of pursuit.

Integral payoff. Some manifold F of dimension m and a continuous function $H(x, y)$ are given in $R^n \times R^n$. Suppose in the situation $S = \{x_0, y_0; u(\cdot), v(\cdot)\}$, t_n is the first instant at which the path $(x(t), y(t))$ falls on F. Then

$$K(x_0, y_0; u(\cdot), v(\cdot)) = \int_0^{t_n} H(x(t), y(t))dt$$

(if $t_n = \infty$, then $K \triangleq \infty$), where $x(t), y(t)$ are the paths of Players P and E corresponding to the situation S. In the case $H \equiv 1, K = t_n$, we have the time optimal game of pursuit.

Qualitative payoff. The payoff function K can take only one of the three values $+1, 0, -1$ depending on a position of $(x(t), y(t))$ in $R^n \times R^n$. Two manifolds F and L of dimensions m_1 and m_2 respectively are given in $R^n \times R^n$. Suppose that in the situation $S = \{x_0, y_0; u(\cdot), v(\cdot)\}$, t_n is the first instant at which the path $(x(t), y(t))$ falls on F. Then

$$K(x_0, y_0; u(\cdot), v(\cdot)) = \begin{cases} -1, & \text{if } (x(t_n), y(t_n)) \in L, \\ 0, & \text{if } t_n = \infty, \\ +1, & \text{if } (x(t_n), y(t_n)) \notin L. \end{cases}$$

7.1.9. Having defined the strategy sets for Players P and E and the payoff function, we may define the differential game as the game in normal form. In 1.1.1, we interpreted the normal form Γ as the triple $\Gamma = < X, Y, K >$, where $X \times Y$ is the space of pairs of all possible strategies in the game Γ, and K is the payoff function defined on $X \times Y$. In the case involved, the payoff function is defined not only on the set of pairs of all possible strategies in the game, but also on the set of all pairs of initial positions x_0, y_0. Therefore, for each pair $(x_0, y_0) \in R^n \times R^n$ there is the corresponding game in normal form, i.e. in fact some family of games in normal form that are dependent on parameters $(x_0, y_0) \in R^n \times R^n$ are defined.

Definition. The normal form of the differential game $\Gamma(x_0, y_0)$ given on the space of strategy pairs $P \times E$ means the system

$$\Gamma(x_0, y_0) = \langle x_0, y_0; P, E, K(x_0, y_0; u(\cdot), v(\cdot)) \rangle ,$$

where $K(x_0, y_0; u(\cdot), v(\cdot))$ is the payoff function defined by any one of the above methods.

Note that in this chapter maximizing player (1 Player) is Player E, and K is the payoff of Player E. If the payoff function K (payoff of E, payoff of P being $(-K)$) in the game Γ is terminal, then the corresponding game Γ is called the game with terminal payoff. If the function K is defined by the second method, then we have the game for achievement of a minimum result. If the function K in the game Γ is integral, then the corresponding game Γ is called the game with integral payoff. When the payoff function in the game Γ is qualitative, then the corresponding game Γ is called the *game of kind.*

7.1.10. It appears natural that optimal strategies cannot exist in the class of piecewise open-loop strategies (in view of the open structure of the class). However, we can show that in a sufficiently large number of cases, for any $\epsilon > 0$ there is an ϵ-equilibrium point.

Recall the definition of the ϵ-equilibrium point (see 2.2.3).

Definition. Let $\epsilon > 0$ be given. The situation $s_\epsilon = \{x_0, y_0; u_\epsilon(\cdot), v_\epsilon(\cdot)\}$ is called an ϵ-equilibrium in the game $\Gamma(x_0, y_0)$ if for all $u(\cdot) \in P$ and $v(\cdot) \in E$ there is

$$K(x_0, y_0; u(\cdot), v_\epsilon(\cdot)) + \epsilon \geq K(x_0, y_0; u_\epsilon(\cdot), v_\epsilon(\cdot)) \qquad (7.1.10)$$
$$\geq K(x_0, y_0; u_\epsilon(\cdot), v(\cdot)) - \epsilon.$$

The strategies $u_\epsilon(\cdot), v_\epsilon(\cdot)$ determined in (7.1.10) are called *ϵ-optimal strategies* for players P and E respectively.

The following Lemma is rephrasing of Theorem 2.2.5 for differential games.

Lemma. *Suppose that in the game $\Gamma(x_0, y_0)$ for every $\epsilon > 0$ there is an ϵ-equilibrium. Then there exists a limit*

$$\lim_{\epsilon \to 0} K(x_0, y_0; u_\epsilon(\cdot), v_\epsilon(\cdot)).$$

Definition. The function $V(x, y)$ defined at each point (x, y) of some set $D \subset R^n \times R^n$ by the rule

$$\lim_{\epsilon \to 0} K(x, y; u_\epsilon(\cdot), v_\epsilon(\cdot)) = V(x, y), \qquad (7.1.11)$$

is called the value of the game $\Gamma(x, y)$ with initial conditions $(x, y) \in D$.

The existence of an ϵ-equilibrium in the game $\Gamma(x_0, y_0)$ for any $\epsilon > 0$ is equivalent to the fulfilment of the equality

$$\sup_{v(\cdot) \in E} \inf_{u(\cdot) \in P} K(x_0, y_0; u(\cdot), v(\cdot)) = \inf_{u(\cdot) \in P} \sup_{v(\cdot) \in E} K(x_0, y_0; u(\cdot), v(\cdot)).$$

If in the game $\Gamma(x_0, y_0)$ for any $\epsilon > 0$ there are ϵ-optimal strategies for Players P and E, then the game $\Gamma(x_0, y_0)$ is said to *have a solution.*

Definition. Let $u^*(\cdot), v^*(\cdot)$ be the pair of strategies such that

$$K(x_0, y_0; u(\cdot), v^*(\cdot)) \geq K(x_0, y_0; u^*(\cdot), v^*(\cdot)) \geq K(x_0, y_0; u^*(\cdot), v(\cdot)) \qquad (7.1.12)$$

for all $u(\cdot) \in P$ and $v(\cdot) \in E$. The situation $s^* = (x_0, y_0; u^*(\cdot), v^*(\cdot))$ is then called an equilibrium in the game $\Gamma(x_0, y_0)$. The strategies

$u^*(\cdot) \in P$ and $v^*(\cdot) \in E$ from (7.1.12) are called optimal strategies for players P and E, respectively.

The existence of an equilibrium in the game $\Gamma(x_0, y_0)$ is equivalent (see 1.3.4) to the fulfilment of the equality

$$\max_{v(\cdot)\in E} \inf_{u(\cdot)\in P} K(x_0, y_0; u(\cdot), v(\cdot)) = \min_{u(\cdot)\in E} \sup_{v(\cdot)\in P} K(x_0, y_0; u(\cdot), v(\cdot)).$$

Clearly, if there exists an equilibrium, then for any $\epsilon > 0$ it is also an ϵ-equilibrium, i.e. here the function $V(x, y)$ merely coincides with $K(x, y; u^*(\cdot), v^*(\cdot))$ (see 2.2.3).

7.1.11. We shall now consider a synthesizing strategies.

Definition. The pair $(u^*(x, y, t), v^*(x, y, t))$ is called a synthesizing strategy equilibrium in the differential game, if the inequality

$$\begin{aligned} K(x_0, y_0; u(x, y, t), v^*(x, y, t)) &\geq K(x_0, y_0; u^*(x, y, t), v^*(x, y, t) \\ &\geq K(x_0, y_0; u^*(x, y, t), v(x, y, t)) \end{aligned} \tag{7.1.13}$$

holds for all situations $(u(x, y, t), v^*(x, y, t))$ and $(u^*(x, y, t), v(x, y, t))$ for which there exists a unique solution to system (7.1.7) that can be prolonged on $[0, \infty)$ from the initial states x_0, y_0. The strategies $u^*(x, y, t), v^*(x, y, t)$ are called optimal strategies for Players P and E.

A distinction must be made between the notions of an equilibrium in piecewise open-loop and synthesizing strategies. Note that in the ordinary sense the equilibrium in the class of functions $u(x, y, t), v(x, y, t)$ cannot be defined because of the nonrectangularity of the space of situations, i.e. in the synthesizing strategies it is impossible to require that the inequality (7.1.13) holds for all strategies $u(x, y, t), v(x, y, t)$, since some pairs $(u^*(x, y, t), v(x, y, t))$, $(u(x, y, t), v^*(x, y, t))$ cannot be admissible (in the corresponding situation the system of equations (7.1.7) may have no solution or may have no unique solution).

In what follows we shall consider the classes of piecewise open-loop strategies, unless otherwise indicated. Preparatory to proving the existence of an ϵ-equilibrium in the differential game we will first consider one auxiliary class of multistage games with perfect information.

7.2 Multistage Perfect-Information Games with an Infinite Number of Alternatives

7.2.1. We shall consider the class of multistage games with perfect information that is a generalization of the games with perfect information from Sec. 5.1. The game proceeds in the n-dimensional Euclidean space R^n. Denote by $x \in R^n$ the position of Player 1, and by $y \in R^n$ the position of Player 2. Suppose that the sets U_x, V_y are defined for each $x \in R^n, y \in R^n$, respectively. These are taken to be the compact sets in the Euclidean space R^n. The game starts from a position x_0, y_0. At the 1st step the players 1 and 2 choose the points $x_1 \in U_{x_0}$ and $y_1 \in V_{y_0}$. In this case the choice by Player 2 is made known to Player 1 before he chooses the point $x_1 \in U_{x_0}$. At the points x_1, y_1 the players 1 and 2 choose $x_2 \in U_{x_1}$ and $y_2 \in V_{y_1}$ and Player 2's choice is made known to Player 1 before he chooses the point $x_2 \in U_{x_1}$ and so on. In positions x_{k-1}, y_{k-1} at the kth step the players choose $x_k \in U_{x_{k-1}}$, $y_k \in V_{y_{k-1}}$ and Player 2's choice is made known to Player 1 before he chooses the point $x_k \in U_{x_{k-1}}$. This process terminates at the Nth step by choosing $x_N \in U_{x_{N-1}}$, $y_N \in V_{y_{N-1}}$ and passing to the state x_N, y_N.

The family of sets U_x, V_y, $x \in R^n$, $y \in R^n$ is taken to be continuous in x, y in Hausdorff metric. This means that for any $\epsilon > 0$ there is $\delta > 0$ such that for $|x - x_0| < \delta$ ($|y - y_0| < \delta$)

$$(U_{x_0})_\epsilon \supset U_x, \qquad (U_x)_\epsilon \supset U_{x_0};$$

$$(V_{y_0})_\epsilon \supset V_y, \qquad (V_y)_\epsilon \supset V_{y_0}.$$

Here U_ϵ (V_ϵ) is an ϵ-neighborhood of the set U (V).

The following result is well known in analysis (see Petrosyan (1993)).

Lemma. *Let $f(x', y')$ be a continuous function on the Cartesian product $U_x \times V_y$. If the families $\{U_x\}, \{V_y\}$ are Hausdorff-continuous in x, y, then the functionals*

$$F_1(x, y) = \max_{y' \in V_y} \min_{x' \in U_x} f(x', y'),$$

$$F_2(x, y) = \min_{x' \in U_x} \max_{y' \in V_y} f(x', y')$$

are continuous in x, y.

Let $\overline{x} = (x_0, \ldots, x_N)$ and $\overline{y} = (y_0, \ldots, y_N)$ be the respective paths of players 1 and 2 realized during the game. The payoff to Player 2 is

$$\max_{0 \le k \le N} f(x_k, y_k) = F(\overline{x}, \overline{y}), \tag{7.2.1}$$

where $f(x, y)$ is a continuous function of x, y. The payoff to Player 1 is $-F$ (the game is zero-sum).

We assume that this game is a perfect-information game, i.e. at each moment (at each step) the players know the positions x_k, y_k and the time instant $k+1$, moreover, Player 1 is informed about the choice y_{k+1} by Player 2.

Strategies for Player 1 are all possible functions $u(x, y, t)$ such that $u(x_{k-1}, y_k, k) \in U_{x_{k-1}}$. Strategies for Player 2 are all possible functions $v(x, y, t)$ such that $v(x_{k-1}, y_{k-1}, k) \in V_{y_{k-1}}$. These strategies are called pure strategies (as distinct from mixed strategies).

Suppose that players 1 and 2 use pure strategies $u(x, y, t)$, $v(x, y, t)$. In situation $(u(\cdot), v(\cdot))$ the game proceeds as follows. At the first step Player 2 passes from the state y_0 to the state $y_1 = v(x_0, y_0, 1)$, while Player 1 passes from the state x_0 to the state $x_1 = u(x_0, y_1, 1) = u(x_0, v(x_0, y_0, 1), 1)$ (because Player 1 is informed about the choice by Player 2). At the 2nd step the players pass to the states $y_2 = v(x_1, y_1, 2)$, $x_2 = u(x_1, y_2, 2) = u(x_1, v(x_1, y_1, 2), 2)$ and so on. At the kth step players 1 and 2 pass from the states x_{k-1}, y_{k-1} to the states $y_k = v(x_{k-1}, y_{k-1}, k)$,

$x_k = u(x_{k-1}, y_k, k) = u(x_{k-1}, v(x_{k-1}, y_{k-1}, k), k)$. Thus to each situation $(u(\cdot), v(\cdot))$ uniquely correspond the paths of Players 1 and 2, $\overline{x} = (x_0, x_1, \ldots, x_k, \ldots, x_N)$ and $\overline{y} = (y_0, y_1, \ldots, y_k, \ldots, y_N)$; hence the payoff $K(u(\cdot), v(\cdot)) = F(x, y)$ is determined by (7.2.1).

This game depends on two parameters; the initial positions x_0, y_0 and the duration N. For this reason, we denote the game by $\Gamma(x_0, y_0, N)$. For the purposes of further discussion it is convenient to assign each game: $\Gamma(x_0, y_0, N)$ to the family of games $\Gamma(x, y, T)$ depending on parameters x, y, T.

7.2.2. The following result is a generalization of Theorem 4.2.1.

Theorem. *The game $\Gamma(x_0, y_0, N)$ has an equilibrium in pure strategies and the value of the game $V(x_0, y_0, N)$ satisfies the following functional equation*

$$V(x_0, y_0, k) = \max\left\{f(x_0, y_0),\right.$$
$$\left.\times \max_{y \in V_{y_0}} \min_{x \in U_{x_0}} V(x, y, k-1)\right\}, \quad k = 1, \ldots, N;$$
$$V(x, y, 0) = f(x, y). \tag{7.2.2}$$

Proof is carried out by induction for the number of steps. Let $N = 1$. Define the strategies $u^*(\cdot), v^*(\cdot)$, for the players in the game $\Gamma(x_0, y_0, 1)$ in the following way:

$$\min_{x \in U_{x_0}} f(x, y) = f(u^*(x_0, y, 1), y), \quad y \in V_{y_0}.$$

If $\max_{y \in V_{y_0}} \min_{x \in U_{x_0}} f(x, y) = f(u^*(x_0, y^*, 1), y^*)$ then $v^*(x_0, y_0, 1) = y^*$. Then

$$K(u^*(\cdot), v^*(\cdot)) = \max\left\{f(x_0, y_0), \max_{y \in V_{y_0}} \min_{x \in U_{x_0}} f(x, y)\right\}$$

and for any strategies $u(\cdot), v(\cdot)$ of the players in the game $\Gamma(x_0, y_0, 1)$

$$K(u^*(\cdot), v(\cdot)) \leq K(u^*(\cdot), v^*(\cdot)) \leq K(u(\cdot), v^*(\cdot)).$$

In view of this, the assertion of Theorem holds for $N \leq 1$.

We now assume that the assertion of Theorem holds for $N < n$. We shall prove it for $N = n+1$, i.e. for the game $\Gamma(x_0, y_0, n+1)$. Let us consider the family of games $\Gamma(x, y, n)$ $x \in U_{x_0}, y \in V_{y_0}$. Denote by $\overline{u}^n_{xy}(\cdot), \overline{v}^n_{xy}(\cdot)$ an equilibrium in the game $\Gamma(x, y, n)$. Then $K(\overline{u}^n_{xy}(\cdot), \overline{v}^n_{xy}(\cdot)) = V(x, y, n)$, where $V(x, y, n)$ is determined by relationships (7.2.2). Using the continuity of the function $f(x, y)$ and Lemma 5.2.1, we may prove the continuity of the function $V(x, y, n)$ in x, y. We define strategies $\overline{u}^{n+1}(\cdot), \overline{v}^{n+1}(\cdot)$ for players in the game $\Gamma(x_0, y_0, n+1)$ as follows:

$$\min_{x \in U_{x_0}} V(x, y, n) = V(\overline{u}^{n+1}(x_0, y, 1), y, n), \ \ y \in V_{y_0}.$$

If $\max_{y \in V_{y_0}} \min_{x \in U_{x_0}} V(x, y, n) = V(\overline{u}^{n+1}(x_0, \tilde{y}, 1), \tilde{y}, n)$, then $\overline{v}^{n+1}(x_0, y_0, 1) = \tilde{y}$ (for $x \neq x_0$, $y \neq y_0$ the functions $\overline{v}^{n+1}(x, y, 1)$ and $\overline{u}^{n+1}(x, y, 1)$ can be defined in an arbitrary way)

$$\overline{u}^{n+1}(\cdot, k) = \overline{u}^n_{x_1 y_1}(\cdot, k-1), \ \ k = 2, \ldots, n+1,$$

$$\overline{v}^{n+1}(\cdot, k) = \overline{v}^n_{x_1 y_1}(\cdot, k-1), \ \ k = 2, \ldots, n+1.$$

Here $x_1 \in U_{x_0}$, $y_1 \in V_{y_0}$ are the positions realized after the 1st step in the game $\Gamma(x_0, y_0, n+1)$. By construction,

$$K(\overline{u}^{n+1}(\cdot), \overline{v}^{n+1}(\cdot)) = \max\left\{ f(x_0, y_0), \max_{y \in V_{y_0}} \min_{x \in U_{x_0}} V(x, y, n) \right\}. \tag{7.2.3}$$

Let us fix an arbitrary strategy $u(\cdot)$ for Player 1 in the game $\Gamma(x_0, y_0, n+1)$. Let $u(x_0, \tilde{y}, 1) = x_1$ where $\tilde{y} = \overline{v}^{n+1}(x_0, y_0, 1)$ and $u^n_{xy}(\cdot)$ is the truncation of strategy $u(\cdot)$ to the game $\Gamma(x, y, n)$, $x \in U_{x_0}$, $y \in V_{y_0}$.

The following relationships are valid:

$$\begin{aligned} K(\overline{u}^{n+1}(\cdot), \overline{v}^{n+1}(\cdot)) &\leq \max\{f(x_0, y_0), V(x_1, \tilde{y}, n)\} \\ &= \max\{f(x_0, y_0), K(\overline{u}^n_{x_1 \tilde{y}}(\cdot), \overline{v}^n_{x_1 \tilde{y}}(\cdot)\} \\ &\leq \max\{f(x_0, y_0), K(u^n_{x_1 \tilde{y}}(\cdot), \overline{v}^n_{x_1 \tilde{y}}(\cdot))\} \\ &= K(u(\cdot), \overline{v}^{n+1}(\cdot)). \end{aligned} \tag{7.2.4}$$

In the same manner the inequality

$$K(\overline{u}^{n+1}(\cdot), \overline{v}^{n+1}(\cdot)) \geq K(\overline{u}^{n+1}(\cdot), v(\cdot)) \tag{7.2.5}$$

can be proved for any strategy $v(\cdot)$ of Player 2 in the game: $\Gamma(x_0, y_0, n+1)$. From relationships (7.2.3)–(7.2.5) it follows that the assertion of Theorem holds for $N = n+1$. This completes the proof of the theorem by induction.

We shall now consider the game $\overline{\Gamma}(x_0, y_0, N)$ which differs from the game $\Gamma(x_0, y_0, N)$ in that the information is provided by Player 1 about his choice. Now, at the step k in the game: $\overline{\Gamma}(x_0, y_0, N)$ Player 2 knows not only the states x_{k-1}, y_{k-1} and the step k, but also the state $x_k \in U_{x_{k-1}}$ chosen by Player 1. Similarly, to the Theorem 5.2.5. we may show that the *game* $\overline{\Gamma}(x_0, y_0, N)$ *has an equilibrium in pure strategies and the game value* $\overline{V}(x_0, y_0, N)$ *satisfies the equation*

$$\overline{V}(x_0, y_0, k) = \max\left\{f(x_0, y_0), \min_{x \in U_{x_0}} \max_{y \in V_{y_0}} \overline{V}(x, y, k-1)\right\},$$

$$k = 1, \ldots, N, \quad \overline{V}(x, y, 0) = f(x, y). \tag{7.2.6}$$

7.2.3. Let us consider the games $\Gamma'(x_0, y_0, N)$ and $\overline{\Gamma}'(x_0, y_0, N)$ that are distinguished from the games $\Gamma(x_0, y_0, N)$ and $\overline{\Gamma}(x_0, y_0, N)$ correspondingly in that the payoff function is equal to a distance between Players 1 and 2 at the final step of the game, i.e. $\rho(x_N, y_N)$. Then the assertion of Theorem 5.2.2 and its corollary are valid and, instead of relations (7.2.2),(7.2.6), the following equations hold:

$$V'(x, y, k) = \max_{y' \in V_y} \min_{x' \in U_x} V'(x', y', k-1), \;\; k = 1, \ldots, N,$$

$$V'(x, y, 0) = \rho(x, y); \tag{7.2.7}$$

$$\overline{V}'(x, y, k) = \min_{x' \in U_x} \max_{y' \in V_y} \overline{V}'(x', y', k-1), \;\; k = 1, \ldots, N,$$

$$\overline{V}'(x, y, 0) = \rho(x, y). \tag{7.2.8}$$

Example 3. Let us consider a discrete game of pursuit, in which the sets U_x are the circles of radius α centered at the point x, while the sets V_y are the circles of radius β centered at the point y $(\alpha > \beta)$. This corresponds to the game in which Player 2 (Evader) moves in a plane with the speed not exceeding β, while Player 1 (Pursuer) moves with a speed not exceeding α. The pursuer has a speed advantage and the second move is made by Player 1. The game of this type is called a discrete game of "simple pursuit" with discrimination against evader. The duration of the game is N steps and the payoff to Player 2 is equal to a distance between the players at the final step.

We shall find the value of the game and optimal strategies for the players by using the functional equation (7.2.7).

We have

$$V(x, y, 1) = \max_{y' \in V_y} \min_{x' \in U_x} \rho(x', y'). \tag{7.2.9}$$

Since U_x and V_y are the circles of radii α and β with centers at x and y, we have that if $U_x \supset V_y$, then $V(x, y, 1) = 0$, if, however, $U_x \not\supset V_y$, then $V(x, y, 1) = \rho(x, y) + \beta - \alpha = \rho(x, y) - (\alpha - \beta)$ (see Example 8 in 2.2.6). Thus,

$$V(x, y, 1) = \begin{cases} 0, & \text{if } U_x \supset V_y, \text{ i.e.} \\ & \quad \rho(x, y) - (\alpha - \beta) \leq 0, \\ \rho(x, y) - (\alpha - \beta), & \text{if } U_x \not\supset V_y, \end{cases}$$

in other words,

$$V(x, y, 1) = \max[0, \rho(x, y) - (\alpha - \beta)]. \tag{7.2.10}$$

Using induction on the number of steps k we shall prove that the following formula holds:

$$V(x, y, k) = \max[0, \rho(x, y) - k(\alpha - \beta)], \ k \geq 2. \tag{7.2.11}$$

Suppose (7.2.11) holds for $k = m - 1$. Show that this formula holds for $k = m$. Using equation (7.2.7) and relationships (7.2.9), (7.2.10) we obtain

$$\begin{aligned} V(x,y,m) &= \max_{y'\in V_y}\min_{x'\in U_x} V(x',y',m-1) \\ &= \max_{y'\in V_y}\min_{x'\in U_x}\{\max[0,\rho(x',y')-(m-1)(\alpha-\beta)]\} \\ &= \max\left[0, \max_{y'\in V_y}\min_{x'\in U_x}\{\rho(x',y')\}-(m-1)(\alpha-\beta)\right] \\ &= \max[0,\max\{0,\rho(x,y)-(\alpha-\beta)\}-(m-1)(\alpha-\beta)] \\ &= \max[0,\rho(x,y)-m(\alpha-\beta)], \end{aligned}$$

which is what we set out to prove.

If $V(x_0, y_0, m) = \rho(x_0, y_0) - m(\alpha - \beta)$, i.e. $\rho(x_0, y_0) - m(\alpha - \beta) > 0$, then the optimal strategy dictates Player 2 to choose at the kth step of the game the point y_k of intersection the line of centers x_{k-1}, y_{k-1} with the boundary $V_{y_{k-1}}$ that is the farthest from x_{k-1}. Here x_{k-1}, y_{k-1} are the players positions after the $(k-1)$ step, $k = 1, \ldots, N$. The optimal strategy for Player 1 dictates him to choose at the kth step of the game the point from the set $U_{x_{k-1}}$ that is the nearest to the point y_k. If both players are acting optimally, then the sequence of the chosen points $x_0, x_1, \ldots, x_N$, $y_0, y_1, \ldots, y_N$ lies along the straight line passing through x_0, y_0. If $V(x_0, y_0, m) = 0$, then an optimal strategy for Player 2 is arbitrary, while an optimal strategy for Player 1 remains unaffected. In this case, after some step k the equality $\max_{y\in V_{y_k}} \min_{x\in U_{x_k}} \rho(x,y) = 0$ is satisfied; therefore starting with the $(k+1)$ step the choices by Player 1 will repeat the choices by Player 2.

7.3 Existence of ϵ-Equilibria in Differential Games with Prescribed Duration

7.3.1. In this section, we shall prove existence of piecewise open-loop strategy ϵ-equilibria in differential games with prescribed duration (see Berkovitz (1964), Fleming (1961), Friedman (1971), Malafeyev (1980)). Let us discuss the case where the payoff to Player E is the distance $\rho(x(T), y(T))$ at the last instant of the game T.

Let the dynamics of the game be given by the following differential equations:

$$\text{for } P: \ \dot{x} = f(x, u); \tag{7.3.1}$$

$$\text{for } E: \ \dot{y} = g(y, v). \tag{7.3.2}$$

Here $x, y \in R^n$, $u \in U$, $v \in V$, where U, V are compact sets in the Euclidean spaces R^k and R^l, respectively, $t \in [0, \infty)$. Suppose the requirements in 5.1.6 are all satisfied.

Definition. Denote by $C_P^t(x_0)$ the set of points $x \in R^n$ for which there is a measurable open-loop control $u(t) \in U$ sending the point x_0 to x in time t, i.e. $x(t_0) = x_0$, $x(t_0 + t) = x$. The set $C_P^t(x_0)$ is called a reachability set for Player P from initial state x_0 in time t.

In this manner, we may also define the reachability set $C_E^t(y_0)$ for Player E from the initial state y_0 in time t.

We assume that the functions f, g are such that the reachability sets $C_P^t(x_0)$, $C_E^t(y_0)$ for Players P and E, respectively, satisfy the following conditions:

1. $C_P^t(x_0)$, $C_E^t(y_0)$ are defined for any $x_0, y_0 \in R^n$, $t_0, t \in [0, \infty)$ $(t_0 \leq t)$ and are compact sets of the space R^n;
2. the point to set map $C_P^t(x_0)$ is continuous in all its variables in Hausdorff metric, i.e. for every $\epsilon > 0$, $x_0' \in R^n$, $t \in [0, \infty)$ there is $\delta > 0$ such that if $|t - t'| < \delta$, $\rho(x_0, x_0') < \delta$, then $\rho^*(C_P^t(x_0), C_P^{t'}(x_0')) < \epsilon$. This also applies to $C_E^t(y_0)$.

Recall that the Hausdorff metric ρ^* in the space of compact sets R^n is given as follows:

$$\rho^*(A, B) \triangleq \max(\rho'(A, B), \rho'(B, A)), \quad \rho'(A, B) = \max_{a \in A} \rho(a, B)$$

and $\rho(a, B) = \min_{b \in B} \rho(a, b)$, where ρ is a standard metric in R^n.

We shall prove the existence theorem for the game of pursuit $\Gamma(x_0, y_0, T)$ with prescribed duration, where $x_0, y_0 \in R^n$ are initial positions for Players P and E respectively. T is the duration of the game. The game $\Gamma(x_0, y_0, T)$ proceeds as follows. Players

P and E at the time $t_0 = 0$ start their motions from positions x_0, y_0 in accordance with the chosen piecewise open-loop strategies. The game ends at the time $t = T$ and Player E receives from Player P an amount $\rho(x(T), y(T))$ (see 5.1.8). At each time instant $t \in [0, T]$ of the game $\Gamma(x_0, y_0, T)$ each player knows the instant of time t, his own position and the position of his opponent. Denote by $P(x_0, t_0, t)$ ($E(y_0, t_0, t)$) the set of trajectories of system (7.3.1), (7.3.2) emanating from the point $x_0(y_0)$ and defined on the interval $[t_0, t]$.

7.3.2. Let us fix some natural number $n \geq 1$. We set $\delta = T/2^n$ and introduce the games $\Gamma_i^\delta(x_0, y_0, T)$, $i = 1, 2, 3$, that are auxiliary with respect to the game $\Gamma(x_0, y_0, T)$.

The game $\Gamma_1^\delta(x_0, y_0, T)$ proceeds as follows. At the 1st step Player E in position y_0 chooses y_1 from the set $C_E^\delta(y_0)$. At this step Player P knows the choice of y_1 by Player E and, in position x_0, chooses the point $x_1 \in C_P^\delta(x_0)$. At the kth step $k = 2, 3, \ldots, 2^n$, Player E knows Player P's position $x_{k-1} \in C_P^\delta(x_{k-2})$ and his own position $y_{k-1} \in C_E^\delta(y_{k-2})$ and chooses the point $y_k \in C_E^\delta(y_{k-1})$. Player P knows x_{k-1}, y_{k-1}, y_k and chooses $x_k \in C_P^\delta(x_{k-1})$. At the 2^nth step the game ends and Player E receives an amount $\rho(x(T), y(T))$, where $x(T) = x^{2^n}$, $y(T) = y^{2^n}$.

Note that the players' choices at the k-th step of the points x_k, y_k from the rechability sets $C_P^\delta(x_{k-1})$, $C_E^\delta(y_{k-1})$ can be interpreted as their choices of the corresponding trajectories from the sets $P((x_{k-1}, k-1)\delta, k\delta)$, $E((y_{k-1}, k-1)\delta, k\delta)$, terminating in the points x_k, y_k at the time instant $t = k\delta$ (or the choice of controls $u(\cdot), v(\cdot)$ on $[(k-1)\delta, k\delta]$ to which these trajectories correspond according to (7.3.1), (7.3.2)).

The game $\Gamma_2^\delta(x_0, y_0, T)$ differs from the game $\Gamma_1^\delta(x_0, y_0, T)$ in that at the kth step Player P chooses $x_k \in C_P^\delta(x_{k-1})$ with a knowledge of x_{k-1}, y_{k-1}, while Player E, with an additional knowledge of x_k, chooses $y_k \in C_E^\delta(y_{k-1})$.

The game $\Gamma_3^\delta(x_0, y_0, T)$ differs from the game $\Gamma_2^\delta(x_0, y_0, T)$ in that at the 2^n step Player P chooses $x_{2^n} \in C_P^\delta(x_{2^n-1})$, then the game ends and Player E receives an amount $\rho(x(T), y(T-\delta))$, where $x(T) = x_{2^n}$, $y(T-\delta) = y_{2^n-1}$.

7.3.3. Lemma. *The games $\Gamma_i^\delta(x_0, y_0, T)$, $i = 1, 2, 3$, have an equilibrium for all $x_0, y_0, T < \infty$ and the value of the game $Val\Gamma_i^\delta(x_0, y_0, T)$ is a continuous function $x_0, y_0 \in R^n$. For any $n \geq 0$ there is*

$$Val\Gamma_1^\delta(x_0, y_0, T) \leq Val\Gamma_2^\delta(x_0, y_0, T),\ T = 2^n\delta. \tag{7.3.3}$$

Proof. The games $\Gamma_i^\delta(x_0, y_0, T)$, $i = 1, 2, 3$, belong to the class of multistage games defined in Sec. 5.2. The existence of an equilibrium in the games $\Gamma_i^\delta(x_0, y_0, T)$ and the continuity of functions $Val\Gamma_i^\delta(x_0, y_0, T)$ in x_0, y_0 immediately follows from Theorem in 5.2.2 and its corollary. The following recursion equations hold for the values of the games $\Gamma_i^\delta(x_0, y_0, T)$, $i = 1, 2$,

$$Val\Gamma_1^\delta(x_0, y_0, T) = \max_{y \in C_E^\delta(y_0)} \min_{x \in C_P^\delta(x_0)} Val\Gamma_1^\delta(x, y, T-\delta),$$

$$Val\Gamma_2^\delta(x_0, y_0, T) = \min_{x \in C_P^\delta(x_0)} \max_{y \in C_E^\delta(y_0)} Val\Gamma_2^\delta(x, y, T-\delta),$$

with the initial condition $Val\Gamma_1^\delta(x, y, 0) = Val\Gamma_2^\delta(x, y, 0) = \rho(x, y)$. Sequential application of Lemma 1.2.2 shows the validity of inequality (7.3.3).

7.3.4. Lemma. *For any integer $n \geq 0$ the following inequalities hold:*

$$Val\Gamma_1^{\delta_n}(x_0, y_0, T) \leq Val\Gamma_1^{\delta_{n+1}}(x_0, y_0, T),$$

$$Val\Gamma_2^{\delta_n}(x_0, y_0, T) \geq Val\Gamma_2^{\delta_{n+1}}(x_0, y_0, T),$$

where $\delta_k = T/2^k$

Proof. Show the validity of the first of these inequalities. The second inequality can be proved in a similar manner. To avoid cumbersome notation, let $C^k(y_i) = C_E^{\delta_k}(y_i)$, $C^k(x_i) = C_P^{\delta_k}(x_i)$,

$i = 0, 1, \ldots, 2^n - 1$. We have

$$Val\Gamma_1^{\delta_{n+1}}(x_0, y_0, T)$$
$$= \max_{y_1 \in C^{n+1}(y_0)} \min_{x_1 \in C^{n+1}(x_0)} \max_{y_2 \in C^{n+1}(y_1)} \min_{x_2 \in C^{n+1}(x_1)}$$
$$Val\Gamma_1^{\delta_{n+1}}(x_2, y_2, T - 2\delta_{n+1})$$
$$\geq \max_{y_1 \in C^{n+1}(y_0)} \max_{y_2 \in C^{n+1}(y_1)} \min_{x_1 \in C^{n+1}(x_0)} \min_{x_2 \in C^{n+1}(x_1)}$$
$$Val\Gamma_1^{\delta_{n+1}}(x_2, y_2, T - 2\delta_{n+1})$$
$$= \max_{y_1 \in C^n(y_0)} \min_{x_1 \in C^n(x_0)} Val\Gamma_1^{\delta_{n+1}}(x_1, y_1, T - \delta_n).$$

Continuation of this process yields

$$Val\Gamma_1^{\delta_{n+1}}(x_0, y_0, T)$$
$$\geq \max_{y_1 \in C^n(y_0)} \min_{x_1 \in C^n(x_0)} \cdots \max_{y_{2^n} \in C^n(y_{2^n-1})} \min_{x_{2^n} \in C^n(x_{2^n-1})} \rho(x_{2^n}, y_{2^n})$$
$$= Val\Gamma_1^{\delta_n}(x_0, y_0, T).$$

7.3.5. Theorem. *For all $x_0, y_0 \in R^n$, $T < \infty$ there is the limit equality*

$$\lim_{n\to\infty} Val\Gamma_1^{\delta_n}(x_0, y_0, T) = \lim_{n\to\infty} Val\Gamma_2^{\delta_n}(x_0, y_0, T),$$

where $\delta_n = T/2^n$.

Proof. Let us fix some $n \geq 0$. Let $u(\cdot), v(\cdot)$ be a pair of strategies in the game $\Gamma_2^{\delta_n}(x_0, y_0, T)$. This pair remains the same in the game $\Gamma_3^{\delta_n}(x_0, y_0, T)$. Suppose that the sequence $x_0, x_1, \ldots, x_{2^n}, y_0, y_1, \ldots, y_{2^n}$ is realized in situation $u(\cdot), v(\cdot)$. Denote the payoff functions in the games $\Gamma_2^{\delta_n}(x_0, y_0, T)$, $\Gamma_3^{\delta_n}(x_0, y_0, T)$ by $K_2(u(\cdot), v(\cdot)) = \rho(x_{2^n}, y_{2^n})$, $K_3(u(\cdot), v(\cdot)) = \rho(x_{2^n}, y_{2^n-1})$, respectively. Then

$$K_2(u(\cdot), v(\cdot)) \leq K_3(u(\cdot), v(\cdot)) + \rho(y_{2^n-1}, y_{2^n}).$$

Hence, by the arbitrariness of $u(\cdot), v(\cdot)$, we have:

$$Val\Gamma_2^{\delta_n}(x_0, y_0, T) \leq Val\Gamma_3^{\delta_n}(x_0, y_0, T) + \max_{y \in C_E^{T-\delta_n}(y_0)} \max_{y' \in C_E^{\delta_n}(y)} \rho(y, y'). \tag{7.3.4}$$

Let $y_1^{\delta_n} \in C_E^{\delta_n}(y_0)$, then $C_E^{T-\delta_n}(y_1^{\delta_n}) \subset C_E^T(y_0)$. We now write inequality (7.3.4) for the games with the initial state $x_0, y_1^{\delta_n}$. In view of the previous inclusion, we have

$$Val\Gamma_2^{\delta_n}(x_0, y_1^{\delta_n}, T) \leq Val\Gamma_3^{\delta_n}(x_0, y_1^{\delta_n}, T) + \max_{y \in C_E^T(y_0)} \max_{y' \in C_E^{\delta}(y)} \rho(y, y'). \tag{7.3.5}$$

From the definition of the games $\Gamma_1^{\delta_n}(x_0, y_0, T)$ and $\Gamma_3^{\delta_n}(x_0, y_0, T)$ follows the equality

$$Val\Gamma_1^{\delta_n}(x_0, y_0, T) = \max_{y_1^{\delta_n} \in C_E^{\delta_n}(y_0)} Val\Gamma_3^{\delta_n}(x_0, y_1^{\delta_n}, T). \tag{7.3.6}$$

Since the function $C_E^t(y)$ is continuous in t and the condition $C_E^0(y) = y$ is satisfied, the second term in (7.3.5) tends to zero as $n \to \infty$. Denote it by $\epsilon_1(n)$. From (7.3.5), (7.3.6) we obtain

$$Val\Gamma_1^{\delta_n}(x_0, y_0, T) \geq Val\Gamma_2^{\delta_n}(x_0, y_1^{\delta_n}, T) - \epsilon_1(n). \tag{7.3.7}$$

By the continuity of the function $Val\Gamma_2^{\delta_n}(x_0, y_0, T)$, from (7.3.7) we obtain

$$Val\Gamma_1^{\delta_n}(x_0, y_0, T) \geq Val\Gamma_2^{\delta_n}(x_0, y_0, T) - \epsilon_1(n) - \epsilon_2(n), \tag{7.3.8}$$

where $\epsilon_2(n) \to 0$ as $n \to \infty$. Passing in (7.3.8) to the limit as $n \to \infty$ (which is possible in terms of the Lemmas in 5.3.3, 5.3.4 and the limit existence theorem for a monotone bounded sequence) we obtain

$$\lim_{n \to \infty} Val\Gamma_1^{\delta_n}(x_0, y_0, T) \geq \lim_{n \to \infty} Val\Gamma_2^{\delta_n}(x_0, y_0, T). \tag{7.3.9}$$

From Lemma 5.3.3 the inverse inequality follows. Hence both limits in (7.3.9) coincide.

7.3.6. The statement of Theorem 5.3.5 is proved on the assumption that the partition sequence of the interval $[0, T]$

$$\sigma_n = \{t_0 = 0 < t_1 < \ldots < t_N = T\},\ n = 1, \ldots,$$

satisfies the condition $t_{j+1} - t_j = T/2^n$, $j = 0, 1, \ldots, 2^n - 1$. The statements of Theorem 5.3.5 and Lemmas 5.3.3, 5.3.4 hold for any sequence σ_n of refined partitions of the interval $[0, T]$, i.e. such that $\sigma_{n+1} \supset \sigma_n$ (this means that the partition σ_{n+1} is obtained from σ_n by adding new points)

$$\gamma(\sigma_n) = \max_i (t_{i+1} - t_i) \to_{n \to +\infty} 0.$$

We shall now consider any such partition sequences of the interval $[0, T]$ $\{\sigma_n\}$ and $\{\sigma'_n\}$.

Lemma. *The following equality holds*:

$$\lim_{n \to \infty} Val\Gamma_1^{\sigma_n}(x_0, y_0, T) = \lim_{n \to \infty} Val\Gamma_1^{\sigma'_n}(x_0, y_0, T),$$

where $x_0, y_0 \in R^n$. $T < \infty$.

Proof is carried out by *reductio ad absurdum*. Suppose the statement of this lemma is not true. For definiteness assume that the following inequality is satisfied:

$$\lim_{n \to \infty} Val\Gamma_1^{\sigma_n}(x_0, y_0, T) > \lim_{n \to \infty} Val\Gamma_1^{\sigma'_n}(x_0, y_0, T).$$

Then, by Theorem 5.3.5, we have

$$\lim_{n \to \infty} Val\Gamma_1^{\sigma_n}(x_0, y_0, T) > \lim_{n \to \infty} Val\Gamma_2^{\sigma'_n}(x_0, y_0, T).$$

Hence, we may find natural numbers m_1, n_1 such that the following inequality is satisfied:

$$Val\Gamma_1^{\sigma_{m_1}}(x_0, y_0, T) > Val\Gamma_2^{\sigma'_{n_1}}(x_0, y_0, T).$$

Denote by $\overline{\sigma}$ the partition of the interval $[0, T]$ by the points belonging to both the partitions σ_{m_1} and σ'_{n_1}. For this partition

$$\begin{aligned} Val\Gamma_2^{\overline{\sigma}}(x_0, y_0, T) &\leq Val\Gamma_2^{\sigma'_{n_1}}(x_0, y_0, T) < Val\Gamma_1^{\sigma_{m_1}}(x_0, y_0, T) \\ &\leq Val\Gamma_1^{\overline{\sigma}}(x_0, y_0, T), \end{aligned}$$

whence

$$Val\Gamma_2^{\overline{\sigma}}(x_0, y_0, T) < Val\Gamma_1^{\overline{\sigma}}(x_0, y_0, T).$$

This contradicts (7.3.3), hence the above assumption is not true and the statement of the lemma is true.

7.3.7 Theorem. *For all $x_0, y_0, T < \infty$ in the game $\Gamma(x_0, y_0, T)$ there exists an ϵ-equilibium for any $\epsilon > 0$. In this case*

$$Val\Gamma(x_0, y_0, T) = \lim_{n\to\infty} Val\Gamma_1^{\sigma_n}(x_0, y_0, T), \tag{7.3.10}$$

where $\{\sigma_n\}$ is any sequence of refined partitions of the interval $[0, T]$.

Proof. Let us specify an arbitrarily chosen number $\epsilon > 0$ and show that for Players P and E there are respective strategies $u_\epsilon(\cdot)$ and $v_\epsilon(\cdot)$ such that for all strategies $u(\cdot) \in P$ and $v(\cdot) \in E$ the following inequalities hold:

$$\begin{aligned} K(x_0, y_0, u_\epsilon(\cdot), v(\cdot)) - \epsilon &\leq K(x_0, y_0, u_\epsilon(\cdot), v_\epsilon(\cdot)) \\ &\leq K(x_0, y_0, u(\cdot), v_\epsilon(\cdot)) + \epsilon. \end{aligned} \tag{7.3.11}$$

By Theorem 5.3.5, there is a partition σ of the interval $[0, T]$ such that

$$Val\Gamma_2^{\sigma}(x_0, y_0, T) - \lim_{n\to\infty} Val\Gamma_2^{\sigma_n}(x_0, y_0, T) < \epsilon/2,$$

$$\lim_{n\to+\infty} Val\Gamma_1^{\sigma_n}(x_0, y_0, T) - Val\Gamma_1^{\sigma}(x_0, y_0, T) < \epsilon/2.$$

Let $u^\epsilon(\cdot) = (\sigma, a_{u^\epsilon})$, $v^\epsilon(\cdot) = (\sigma, b_{v^\epsilon})$, where a_{u^ϵ}, b_{v^ϵ} are the optimal strategies for the players P and E, respectively, in the games $\Gamma_2^\sigma(x_0, y_0, T)$ and $\Gamma_1^\sigma(x_0, y_0, T)$.

Then the following relationships are valid:

$$K(x_0, y_0, u^\epsilon(\cdot), v(\cdot)) \leq Val\Gamma_2^\sigma(x_0, y_0, T) < \lim_{n\to\infty} Val\Gamma_2^{\sigma_n}(x_0, y_0, T) + \frac{\epsilon}{2}, \quad v(\cdot) \in E, \tag{7.3.12}$$

$$K(x_0, y_0, u(\cdot), v^\epsilon(\cdot)) \geq Val\Gamma_1^\sigma(x_0, y_0, T) > \lim_{n\to\infty} Val\Gamma_1^{\sigma_n}(x_0, y_0, T) - \frac{\epsilon}{2}, \quad u(\cdot) \in P. \tag{7.3.13}$$

From (7.3.12), (7.3.13) and Theorem 5.3.5 we have

$$-\frac{\epsilon}{2} < K(x_0, y_0, u^\epsilon(\cdot), v^\epsilon(\cdot)) - \lim_{n\to\infty} Val\Gamma_1^{\sigma_n}(x_0, y_0, T) < \frac{\epsilon}{2}. \tag{7.3.14}$$

From relationships (7.3.12)–(7.3.14) follows (7.3.11). By the arbitrariness of ϵ from (7.3.14) follows (7.3.10). This completes the proof of theorem.

7.3.8. *Remark.* The specific type of the payoff was not used in the proof of the existence theorem. Only continuous dependence of the payoff on the realized trajectories is essential. Therefore, Theorem 5.3.7 holds if any continuous functional of the trajectories $x(t), y(t)$ is considered in place of $\rho(x(T), y(T))$. In particular, such a functional can be $\min_{0\leq t<T} \rho(x(t), y(t))$, i.e. a minimum distance between players during the game. Thus the result of this section also holds for the differential minimum result game of pursuit with prescribed duration.

7.4 Differential Time-Optimal Games of Pursuit

7.4.1. Differential *time-optimal games of pursuit* are special case of differential games with integral payoff as defined in 5.1.8. The classes of strategies P and E are the same as in the game with prescribed duration. We assume that the set $F = \{(x, y) : \rho(x, y) \leq l,\ l > 0\}$ is given in $R^n \times R^n$ and $x(t), y(t)$ are trajectories for Players P and E in situation $(u(\cdot), v(\cdot))$ from initial conditions x_0, y_0. Denote

$$t_n(x_0, y_0; u(\cdot), v(\cdot)) = \min\{t : (x(t), y(t)) \in F\}. \tag{7.4.1}$$

If there is no t such that $(x(t), y(t)) \in F$, then $t_n(x_0, y_0; u(\cdot), v(\cdot))$ is $+\infty$. In the differential time-optimal game of pursuit the payoff to Player E is

$$K(x_0, y_0; u(\cdot), v(\cdot)) = t_n(x_0, y_0; u(\cdot), v(\cdot)). \tag{7.4.2}$$

The game depends on the initial conditions x_0, y_0, therefore it is denoted by $\Gamma(x_0, y_0)$.

From the definition of the payoff function (7.4.2) it follows that the objective of Player E in the game $\Gamma(x_0, y_0)$ is to maximize the time of approaching Player P within a given distance $l \geq 0$. Conversely, Player P wishes to minimize this time.

7.4.2. There is a close relation between the time-optimal game of pursuit $\Gamma(x_0, y_0, T)$ and the minimum result game of pursuit with prescibed duration. Let $\Gamma(x_0, y_0, T)$ be the game of pursuit with prescribed duration T for achievement of a minimum result (the payoff to Player E is $\min_{0 \leq t < T} \rho(x(t), y(t))$). It was shown that for the games of this type there is an ϵ-equilibrium in the class of piecewise open-loop strategies for any $\epsilon > 0$ (see 5.3.8). Let $V(x_0, y_0, T)$ be the value of the game $\Gamma(x_0, y_0, T)$ and $V(x_0, y_0)$ be the value of the game $\Gamma(x_0, y_0)$ if it exists.

Lemma. *With x_0, y_0 fixed, the function $V(x_0, y_0, T)$ is continuous and does not increase in T on the interval $[0, \infty]$.*

Proof. Let $T_1 > T_2 > 0$. Denote by $v_\epsilon^{T_1}$ a strategy for Player E in the game $\Gamma(x_0, y_0, T)$ which guarantees that a distance between Players E and P on the interval $[0, T_1]$ will be at least $\max[0, V(x_0, y_0, T_1) - \epsilon]$. Hence it does ensure a distance $\max[0, V(x_0, y_0, T_1) - \epsilon]$ between the players on the interval $[0, T_2]$, where $T_2 < T_1$. Therefore

$$V(x_0, y_0, T_2) \geq \max[0, V(x_0, y_0, T_1) - \epsilon] \tag{7.4.3}$$

(the strategy ϵ-optimal in the game $\Gamma(x_0, y_0, T_1)$ is not necessarily ϵ-optimal in the game $\Gamma(x_0, y_0, T_2)$). Since ϵ can be chosen to be arbitrary, the statement of this Lemma follows from (7.4.3). The continuity of $V(x_0, y_0, T)$ in T will be left without proof. To be noted only is that this property can be obtained by using the continuity of $V(x_0, y_0, T)$ in x_0, y_0.

7.4.3. Let us consider the equation

$$V(x_0, y_0, T) = l \tag{7.4.4}$$

with respect to T. Three cases are possible here:

1) equation (7.4.4) has no roots;
2) it has a single root;
3) it has more than one root.

In case 3), the monotonicity and the continuity of the function $V(x_0, y_0, T)$ in T imply that the equation (7.4.4) has the whole segment of roots, the function $V(x_0, y_0, T)$, as a function of T, has a constancy interval. Let us consider each case individually.

Case 1. In this case the following is possible: a) $V(x_0, y_0, T) < l$ for all $T \geq 0$; b) $\inf_{T \geq 0} V(x_0, y_0, T) > l$; c) $\inf_{T \geq 0} V(x_0, y, T) = l$.

In case a) we have

$$V(x_0, y_0, 0) = \rho(x_0, y_0) < l,$$

i.e. $t_n(x_0, y_0; u(\cdot), v(\cdot)) = 0$ for all $u(\cdot), v(\cdot)$. The value of the game $\Gamma(x_0, y_0)$ is then $V(x_0, y_0) = 0$.

In case b) the following equality holds:

$$\inf_{T \geq 0} V(x_0, y_0, T) = \lim_{T \to \infty} V(x_0, y_0, T) > l.$$

Hence for any $T > 0$ (arbitrary large) Player E has a suitable strategy $v^T(\cdot) \in E$ which guarantees him l capture avoidance on the interval $[0, T]$. But Player P then has no strategy which could guarantee him

l-capture of Player E in finite time. However, we cannot claim that Player E has a strategy which ensures l-capture avoidance in finite time. The problem of finding initial states in which such a strategy exists reduces to solving the game of kind for player E. Thus, for $l < \lim_{T\to\infty} V(x_0, y_0, T)$ it can be merely stated that the value of the game $\Gamma(x_0, y_0)$, if any, is larger than any previously given T, i.e. it is $+\infty$.

c) is considered together with case 3).

Case 2. Let T_0 be a single root of equation (7.4.4). Then it follows from the monotonicity and the continuity of the function $V(x_0, y_0, T)$ in T that

$$V(x_0, y_0, T) > V(x_0, y_0, T_0) \text{ for all } T < T_0,$$

$$V(x_0, y_0, T) < V(x_0, y_0, T_0) \text{ for all } T > T_0, \tag{7.4.5}$$

$$\lim_{T\to T_0} V(x_0, y_0, T) = V(x_0, y_0, T_0). \tag{7.4.6}$$

Let us fix an arbitrary $T > T_0$ and consider the game of pursuit $\Gamma(x_0, y_0, T)$. The game has an ϵ-equilibrium in the class of piecewise open-loop strategies for any $\epsilon > 0$. This, in particular, means that for any $\epsilon > 0$ there is Player P's strategy $u_\epsilon(\cdot) \in P$ which ensures the capture of Player E within a distance $V(x_0, y_0, T) + \epsilon$, i.e.

$$K(u_\epsilon(\cdot), v(\cdot)) \leq V(x_0, y_0, T) + \epsilon, \quad v(\cdot) \in E, \tag{7.4.7}$$

where $K(u(\cdot), v(\cdot))$ is the payoff function in the game $\Gamma(x_0, y_0, T)$. Then (7.4.5), (7.4.6) imply the existence of $\overline{\epsilon} > 0$ such that for any $\epsilon < \overline{\epsilon}$ there is a number $\tilde{T}(\epsilon), T_0 < \tilde{T}(\epsilon) \leq T$ for which

$$\epsilon = V(x_0, y_0, T_0) - V(x_0, y_0, \tilde{T}(\epsilon)). \tag{7.4.8}$$

From (7.4.7), (7.4.8) it follows that for any $\epsilon < \overline{\epsilon}$

$$\begin{aligned} K(u_\epsilon(\cdot), v(\cdot)) &\leq V(x_0, y_0, T) + \epsilon \leq V(x_0, y_0, \tilde{T}(\epsilon)) + \epsilon \\ &= V(x_0, y_0, T_0) = l, \; v(\cdot) \in E, \end{aligned}$$

i.e. the strategy $u_\epsilon(\cdot)$ ensures l-capture in time T. Hence, by the arbitrariness of $T > T_0$, it follows that for any $T > T_0$ there is a corresponding strategy $u^T(\cdot) \in P$ which ensures l-capture in time T. In other words, for $\delta > 0$ there is $u_\delta(\cdot) \in P$ such that

$$t_n(x_0, y_0; u_\delta(\cdot), v(\cdot)) \leq T_0 + \delta \text{ for all } v(\cdot) \in E. \tag{7.4.9}$$

In a similar manner we may prove the existence of $v_\delta(\cdot) \in E$ such that

$$t_n(x_0, y_0; u(\cdot), v_\delta(\cdot)) \geq T_0 - \delta \text{ for all } u(\cdot) \in P. \tag{7.4.10}$$

It follows from (7.4.9), (7.4.10) that in the time-optimal game of pursuit $\Gamma(x_0, y_0)$ for any $\epsilon > 0$ there is an ϵ-equilibrium in piecewise open-loop strategies and the value of the game is equal to T_0, with T_0 as a single root of equation (7.4.4).

Case 3. Denote by T_0 the minimal root of equation (7.4.4). Generally speaking, we cannot now state that the value of the game $Val\Gamma(x_0, y_0) = T_0$. Indeed, $V(x_0, y_0, T_0) = l$ merely implies that in the game $\Gamma(x_0, y_0, T_0)$ for any $\epsilon > 0$ Player P has a strategy $u_\epsilon(\cdot)$ which ensures for him, in time T_0, the capture of Player E within a distance of at most $l + \epsilon$. From the existence of more than one root of equation (7.4.4), and from the monotonicity of $V(x_0, y_0, T)$ in T we obtain the existence of the interval of constancy of the function $V(x_0, y_0, T)$ in $T \in [T_0, T_1]$. Therefore, an increase in the duration of the game $\Gamma(x_0, y_0, T_0)$ by δ, where $\delta < T_1 - T_0$, does not involve a decrease in the guaranteed approach to Player E, i.e. for all $T \in [T_0, T_1]$ Player P can merely ensure approaching Player E within a distance $l + \epsilon$ (for any $\epsilon > 0$), and it is beyond reason to hope for this quantity to become zero for some $T \in [T_0, T_1]$. If the game $\Gamma(x_0, y_0, T_0)$ had an equilibrium (but not an ϵ-equilibrium), then the value of the game $\Gamma(x_0, y_0)$ would also be equal to T_0 in Case 3.

7.4.4. Let us modify the notion of an equilibrium in the game $\Gamma(x_0, y_0)$. Further, in this section it may be convenient to use the notation $\Gamma(x_0, y_0, l)$ instead of $\Gamma(x_0, y_0)$ emphasizing the fact that the game $\Gamma(x_0, y_0, l)$ terminates when the players come within a distance l of each other.

Let $t_n^l(x_0, y_0; u(\cdot), v(\cdot))$ be the time until coming within a distance l in situation $(u(\cdot), v(\cdot))$ and let there be $\epsilon \geq 0$, $\delta \geq 0$.

Definition. We say that the pair of strategies $\overline{u}_\epsilon^\delta(\cdot), \overline{v}_\epsilon^\delta(\cdot)$ constitutes an ϵ, δ-equilibrium in the game $\Gamma(x_0, y_0, l)$ if

$$\begin{aligned} t_n^{l+\delta}(x_0, y_0; u(\cdot), \overline{v}_\epsilon^\delta(\cdot)) + \epsilon \geq t_n^{l+\delta}(x_0, y_0; \overline{u}_\epsilon^\delta(\cdot), \overline{v}_\epsilon^\delta(\cdot)) \\ \geq t_n^{l+\delta}(x_0, y_0; \overline{u}_\epsilon^\delta(\cdot), v(\cdot)) - \epsilon \end{aligned}$$

for all strategies $u(\cdot) \in P$, $v(\cdot) \in E$.

Definition. Let there be a sequence $\{\delta_k\}$, $\delta_k \geq 0$, $\delta_k \to 0$ such that in all of the games $\Gamma(x_0, y_0, l + \delta_k)$ for every $\epsilon > 0$ there is an ϵ-equilibrium. Then the limit

$$\lim_{k \to \infty} V(x_0, y_0, l + \delta_k) = V'(x_0, y_0, l)$$

is called the value of the game $\Gamma(x_0, y_0, l)$ in the generalized sense.

Note that the quantity $V'(x_0, y_0, l)$ does not depend on the choice of a sequence $\{\delta_k\}$ because of the monotone decrease of the function $V(x_0, y_0, l)$ in l.

Definition. We say that the game $\Gamma(x_0, y_0, l)$ has the value in the generalized sense if there exists a sequence $\{\delta_k\}$, $\delta_k \to 0$ such that for every $\epsilon > 0$ and $\delta_k \in \{\delta_k\}$ in the game $\Gamma(x_0, y_0, l)$ there exists an ϵ, δ_k-equilibrium.

It can be shown that if the game $\Gamma(x_0, y_0, l)$ has the value in the ordinary sense, then its value $V'(x_0, y_0, l)$ (in the generalized sense) exists and is

$$\lim_{\epsilon \to 0, \delta_k \to 0} t_n^{l+\delta_k}(x_0, y_0; \overline{u}_\epsilon^\delta(\cdot), \overline{v}_\epsilon^\delta(\cdot)) = V'(x_0, y_0, l).$$

From the definition of the value and solution of the game $\Gamma(x_0, y_0, l)$ (in the generalized sense) it follows that if in the game $\Gamma(x_0, y_0, l)$ for every $\epsilon > 0$ there is an ϵ-equilibrium in the ordinary sense (i.e. the solution in the ordinary sense), then $V(x_0, y_0, l) = V'(x_0, y_0, l)$ (it suffices to take a sequence $\delta_k \equiv 0$ for all k).

Theorem. *Let equation* (7.4.4) *have more than one root and let T_0 be the least root, $T_0 < \infty$. Then there exists the value $V'(x_0, y_0, l)$ (in the generalized sense) of the time-optimal game of pursuit $\Gamma(x_0, y_0, l)$ and $V'(x_0, y_0, l) = T_0$.*

Proof. The monotonicity and continuity of the function $V(x_0, y_0, T)$ in T imply the existence of a sequence $T_k \to T_0$ on the left such that $V(x_0, y_0, T_k) \to V(x_0, y_0, T_0) = l$ and the function $V(x_0, y_0, T_k)$ is strictly monotone in the points T_k. Let

$$\delta_k = V(x_0, y_0, T_k) - l \geq 0.$$

The strict monotonicity of the function $V(x_0, y_0, T)$ in the points T_k implies that the equation $V(x_0, y_0, T) = l + \delta_k$ has a single root T_k. This means that for every $\delta_k \in \{\delta_k\}$ in the games $\Gamma(x_0, y_0, l+\delta_k)$ there is an ϵ-equilibrium for every $\epsilon > 0$ (see Case 2 in 5.4.3). The game $\Gamma(x_0, y_0, l)$ then has a solution in the generalized sense:

$$\lim_{k\to\infty} V(x_0, y_0, l + \delta_k) = \lim_{k\to\infty} T_k = T_0 = V'(x_0, y_0, l).$$

This completes the proof of the theorem.

We shall now consider Case 1c in 5.4.3. We have $\inf_T V(x_0, y_0, T) = l$. Let $T_k \to \infty$. Then $\lim_{k\to\infty} V(x_0, y_0, T_K) = l$. From the monotonicity and continuity of $V(x_0, y_0, T)$ in T it follows that the sequence $\{T_k\}$ can be chosen to be such that the function $V(x_0, y_0, T)$ is strictly monotone in the points T_k. Then, as in the proof of Theorem in 5.4.4, it can be shown that there exists a sequence $\{\delta_k\}$ such that

$$\lim_{k\to\infty} V(x_0, y_0, l + \delta_k) = \lim_{k\to\infty} T_k = T_0 = \infty.$$

Thus, in this case there also exists a generalized solution, while the generalized value of the game $\Gamma(x_0, y_0, l)$ is infinity.

7.4.5. It is often important to find out whether Player P can guarantee l-capture from the given initial positions x, y in finite time T. If it is impossible, then we have to find out whether Player E can guarantee l-capture avoidance within a specified period of time.

Let $V(x, y, T)$ be the value of the game with prescribed duration T from initial states $x, y \in R^n$ with the payoff $\min_{0 \leq t \leq T} \rho(x(t), y(t))$. Then the following alternatives are possible: 1) $V(x, y, T) > l$; 2) $V(x, y, T) \leq l$.

Case 1. From the definition of the function $V(x, y, T)$ it follows that for every $\epsilon > 0$ there is a strategy for Player E such that for all strategies $u(\cdot)$

$$K(x, y; u(\cdot), v_\epsilon^*(\cdot)) \geq V(x, y, T) - \epsilon.$$

Having chosen the ϵ to be sufficiently small we may ensure that

$$K(x, y; u(\cdot), v_\epsilon^*(\cdot)) \geq V(x, y, T) - \epsilon > l$$

holds for all strategies $u(\cdot) \in P$ of Player P. From the form of the payoff function K it follows that, by employing a strategy $v_\epsilon^*(\cdot)$, Player E can ensure that the inequality $\min_{0 \leq t \leq T_0} \rho(x(t), y(t)) > l$ would be satisfied no matter what Player P does. That is, in this case Player E ensures l-capture avoidance on the interval $[0, T]$ no matter what Player P does.

Case 2. Let T_0 be a minimal root of the equation $V(x, y, T) = l$ with x, y fixed (if $\rho(x, y) < l$, then T_0 is taken to be 0). From the definition of $V(x, y, T_0)$ it then follows that in the game $\Gamma(x, y, T_0)$ for every $\epsilon > 0$ Player P has a strategy u_ϵ^* which ensures that

$$K(x, y; u_\epsilon^*(\cdot), v(\cdot)) \leq V(x, y; T_0) + \epsilon = l + \epsilon$$

for all strategies $v(\cdot) \in E$ of Player E. From the form of the payoff function K it follows that, by employing a strategy $u_\epsilon^*(\cdot)$, Player P can ensure that the inequality $\min_{0 \leq t \leq T} \rho(x(t), y(t)) \leq l + \epsilon$ would be satisfied no matter what Player E does. Extending arbitrarily the strategy $u_\epsilon^*(\cdot)$ to the interval $[T_0, T]$ we have that, in Case 2, for every $\epsilon > 0$ Player P can ensure $(l + \epsilon)$-capture of Player E in time T no matter what the latter does.

This in fact proves the following theorem (of alternative).

Theorem. *For every $x, y \in R^n$, $T > 0$ one of the following assertions holds:*

1. *from initial conditions x, y Player E can ensure l-capture avoidance during the time T no matter what Player P does;*
2. *for any $\epsilon > 0$ Player P can ensure $(l+\epsilon)$-capture of Player E from initial states x, y during the time T no matter what Player E does.*

7.4.6. For each fixed $T > 0$ the entire space $R^n \times R^n$ is divided into three nonoverlapping regions: region $A = \{x, y : V(x, y, T) < l\}$ which is called the capture zone; region $B = \{x, y : V(x, y, T) > l\}$ which is naturally called the escape zone; region $C = \{x, y : V(x, y, T) = l\}$ is called the indifference zone.

Let $x, y \in A$. By the definition of A, for any $\epsilon > 0$ Player P has a strategy $u_\epsilon^*(\cdot)$ such that

$$K(x, y; u_\epsilon^*(\cdot), v(\cdot)) \leq V(x, y, T) + \epsilon$$

for all strategies $v(\cdot)$ of Player E. By a proper choice of $\epsilon > 0$ it is possible to ensure that the following inequality be satisfied:

$$K(x, y; u_\epsilon^*(\cdot), v(\cdot)) \leq V(x, y, T) + \epsilon < l.$$

This means that the strategy u_ϵ^* of Player P guarantees him l-capture of Player E from initial states during the time T. We thus obtain the following refinement of Theorem 5.4.5.

Theorem. *For every fixed $T > 0$ the entire space is divided into three nonoverlapping regions A, B, C possessing the following properties:*

1. *for any $x, y \in A$ Player P has a strategy $u_\epsilon^*(\cdot)$ which ensures l-capture of Player E on the interval $[0, T]$ no matter what the latter does;*
2. *for $x, y \in B$ Player E has a strategy $v_\epsilon^*(\cdot)$ which ensures l-capture avoidance of Player P on the interval $[0, T]$ no matter what the latter does;*

3. *if* $x, y \in C$ *and* $\epsilon > 0$, *then Player* P *has a strategy* $u^*_\epsilon(\cdot)$ *which ensures* $(l + \epsilon)$ *capture of Player* E *during the time* T *no matter what the latter does.*

7.5 Necessary and Sufficient Condition for Existence of Optimal Open-Loop Strategy for Evader

7.5.1. An important subclass of games of pursuit is represented by the games in which an optimal strategy for evader is a function of time only (this is what is called a regular case).

We shall restrict consideration to the games of pursuit with prescribed duration, although all of the results below can be extended to the time-optimal games of pursuit. Let $C^T_P(x)(C^T_E(y))$ be a reachability set for Player $P(E)$ from initial state $x(y)$ by the time T, i.e. the set of those positions at which Player $P(E)$ can arrive from the initial state $x(y)$ at the time T by employing all possible measurable open-loop controls $u(t), (v(t))$, $t \in [0, T]$ provided the motion occurs in terms of the system $\dot{x} = f(x, u)$ $(\dot{y} = g(y, v))$. Let us introduce the quantity

$$\hat{\rho}_T(x_0, y_0) = \max_{y \in C^T_E(y_0)} \min_{x \in C^T_P(x_0)} \rho(x, y), \tag{7.5.1}$$

which may at times also be called (see Krasovskii (1985), Krasovskii and Subbotin (1974)) a *hypothetical mismatch* of the sets $C^T_E(y_0)$ and $C^T_P(x_0)$ (see Example 6 in 2.2.6).

The function $\hat{\rho}_T(x_0, y_0)$ has the following properties:

1^0. $\hat{\rho}_T(x_0, y_0) \geq 0$, $\hat{\rho}_T(x_0, y_0)|_{T=0} = \rho(x_0, y_0)$;
2^0. $\hat{\rho}_T(x_0, y_0) = 0$ if $C^T_P(x_0) \supset C^T_E(y_0)$;
3^0. If $V(x_0, y_0, T)$ is the value of the game $\Gamma(x_0, y_0, T)$ with prescribed duration and terminal payoff $\rho(x(T), y(T))$, then

$$V(x_0, y_0, T) \geq \hat{\rho}_T(x_0, y_0).$$

Indeed, property 1^0 follows from non-negativity of the function $\rho(x, y)$. Let $C_P^T(x_0) \supset C_E^T(y_0)$. Then for every $y' \in C_E^T(y_0)$ there is $x' \in C_P^T(x_0)$ such that $\rho(x', y') = 0$, $(x' = y')$, whence follows 2^0. Property 3^0 follows from the fact that Player E can always guarantee himself an amount $\hat{\rho}_T(x_0, y_0)$ by choosing the motion directed towards the point $M \in C_E^T(y_0)$ for which

$$\hat{\rho}_T(x_0, y_0) = \min_{x \in C_P^T(x_0)} \rho(x, M).$$

The point M is called the *center of pursuit.*

7.5.2. Let $\Gamma_\delta(x_0, y_0, T)$ be a discrete game of pursuit with step δ $(\delta = t_{k+1} - t_k)$, prescribed duration T, discrimination against Player E, and initial states x_0, y_0. Then the following theorem holds.

Theorem. *In order for the following equality to hold for any $x_0, y_0 \in R^n$ and $T = \delta \cdot k$, $k = 1, 2, \dots$:*

$$\hat{\rho}_T(x_0, y_0) = Val\Gamma_\delta(x_0, y_0, T), \tag{7.5.2}$$

it is necessary and sufficient that for all $x_0, y_0 \in R^n$, $\delta > 0$ and $T = \delta \cdot k, k = 1, 2, \dots$, there be

$$\hat{\rho}_T(x_0, y_0) = \max_{y \in C_E^\delta(y_0)} \min_{x \in C_P^\delta(x_0)} \hat{\rho}_{T-\delta}(x, y) \tag{7.5.3}$$

($Val\Gamma_\delta(x_0, y_0, T)$ is the value of the game $\Gamma_\delta(x_0, y_0, T)$).

The proof of this theorem is based on the following result.

Lemma. *The following inequality holds for any $x_0, y_0 \in R^n$, $T \geq \delta$:*

$$\hat{\rho}_T(x_0, y_0) \leq \max_{y \in C_E^\delta(y_0)} \min_{x \in C_P^\delta(x_0)} \hat{\rho}_{T-\delta}(x, y).$$

Proof. By the definition of the function $\hat{\rho}_T$, we have

$$\max_{y \in C_E^\delta(y_0)} \min_{x \in C_P^\delta(x_0)} \hat{\rho}_{T-\delta}(x, y)$$
$$= \max_{y \in C_E^\delta(y_0)} \min_{x \in C_P^\delta(x_0)} \max_{\overline{y} \in C_E^{T-\delta}(y)} \min_{\overline{x} \in C_P^{T-\delta}(x)} \rho(\overline{x}, \overline{y}).$$

For all $x \in C_P^\delta(x_0)$ there is an inclusion $C_P^{T-\delta}(x) \subset C_P^T(x_0)$. Hence for any $x \in C_T^\delta(x_0)$, $\overline{y} \in C_E^{T-\delta}(y)$.

$$\min_{\overline{x} \in C_P^{T-\delta}(x)} \rho(\overline{x}, \overline{y}) \geq \min_{\overline{x} \in C_P^T(x_0)} \rho(\overline{x}, \overline{y}).$$

Then for all $x \in C_P^\delta(x_0)$, $y \in C_E^\delta(y)$

$$\max_{\overline{y} \in C_E^{T-\delta}(y)} \min_{\overline{x} \in C_P^{T-\delta}(x)} \rho(\overline{x}, \overline{y}) \geq \max_{\overline{y} \in C_E^{T-\delta}(y)} \min_{\overline{x} \in C_P^T(x_0)} \rho(\overline{x}, \overline{y})$$

and

$$\min_{x \in C_P^\delta(x_0)} \max_{\overline{y} \in C_E^{T-\delta}(y)} \min_{\overline{x} \in C_P^{T-\delta}(x)} \rho(\overline{x}, \overline{y}) \geq \max_{\overline{y} \in C_E^{T-\delta}(y)} \min_{\overline{x} \in C_P^T(x_0)} \rho(\overline{x}, \overline{y}).$$

Thus

$$\begin{aligned} \max_{y \in C_E^\delta(y_0)} \min_{x \in C_P^\delta(x_0)} \hat{\rho}_{T-\delta}(x, y) &\geq \max_{y \in C_E^\delta(y_0)} \max_{\overline{y} \in C_E^{T-\delta}(y)} \min_{\overline{x} \in C_P^T(x_0)} \rho(\overline{x}, \overline{y}) \\ &= \max_{y \in C_E^T(y_0)} \min_{x \in C_P^T(x_0)} \rho(x, y) = \hat{\rho}_T(x_0, y_0). \end{aligned}$$

This completes the proof of lemma.

We shall now prove the Theorem.

Necessity. Suppose that condition (7.5.2) is satisfied and condition (7.5.3) is not. Then, by Lemma, there exist $\delta > 0$, $x_0, y_0 \in R^n$, $T_0 = \delta k_0$, $k_0 \geq 1$ such that

$$\hat{\rho}_{T_0}(x_0, y_0) < \max_{y \in C_E^\delta(y_0)} \min_{x \in C_P^\delta(x_0)} \hat{\rho}_{T_0-\delta}(x, y). \tag{7.5.4}$$

Let $u^0(\cdot)$ be an optimal strategy for Player P in the game $\Gamma_\delta(x_0, y_0, T_0)$ and suppose that at the 1st step Player E chooses the point $y^* \in C_E^\delta(y_0)$ for which

$$\min_{x \in C_P^\delta(x_0)} \hat{\rho}_{T_0-\delta}(x, y^*) = \max_{y \in C_E^\delta(y_0)} \min_{x \in C_P^\delta(x_0)} \hat{\rho}_{T_0-\delta}(x, y). \tag{7.5.5}$$

Let $x^0(\delta)$ be the state to which Player P passes at the 1st step when he uses strategy $u^0(\cdot)$, and let $\overline{v}^0(\cdot)$ be an optimal strategy for Player E in the game $\Gamma_\delta(x_0(\delta), y^*, T_0 - \delta)$. Let us consider the strategy $\tilde{v}(\cdot)$ for Player E in the game $\Gamma_\delta(x_0, y_0, T_0)$: at the time instant $t = 0$ he chooses the point y^* and from the instant $t = \delta$ uses strategy $\overline{v}^0(\cdot)$.

Denote by $\hat{u}^0(\cdot)$ the truncation of strategy $u^0(\cdot)$ on the interval $[\delta, T_0]$. From (7.5.2), (7.5.4), (7.5.5) (by (7.5.2) $\hat{\rho}_T(x_0, y_0)$ is the value of the game $\Gamma_\delta(x_0, y_0, T)$) we find

$$\begin{aligned}\hat{\rho}_{T_0}(x_0, y_0) &\geq K(u^0(\cdot), \tilde{v}(\cdot); x_0, y_0, T_0)\\ &= K(\hat{u}^0(\cdot), \overline{v}^0(\cdot); x^0(\delta), y^*, T_0 - \delta)\\ &= \hat{\rho}_{T_0-\delta}(x^0(\delta), y^*) \geq \min_{x \in C_P^\delta(x_0)} \hat{\rho}_{T_0-\delta}(x, y^*)\\ &= \max_{y \in C_E^\delta(y_0)} \min_{x \in C_P^\delta(x_0)} \hat{\rho}_{T_0-\delta}(x, y) > \hat{\rho}_{T_0}(x_0, y_0).\end{aligned}$$

This contradiction proves the necessity of condition (7.5.3).

Sufficiency. Note that the condition (7.5.3), in conjunction with the condition $\hat{\rho}_T(x_0, y_0)|_{T=0} = \rho(x_0, y_0)$, shows that the function $\hat{\rho}_T(x_0, y_0)$ satisfies the functional equation for the function of the value of the game $\Gamma_\delta(x_0, y_0, T)$. As follows from the proof of Theorem in 5.2.2, this condition is sufficient for $\hat{\rho}_T(x_0, y_0)$ to be the value of the game $\Gamma_\delta(x_0, y_0, T)$.

7.5.3. Lemma. *In order for Player E's optimal open-loop strategy (i.e. the strategy which is the function of time only) to exist in the game $\Gamma(x_0, y_0, T)$ it is necessary and sufficient that*

$$Val\Gamma(x_0, y_0, T) = \hat{\rho}_T(x_0, y_0). \tag{7.5.6}$$

Proof. Sufficiency. Let $v^*(t)$, $t \in [0, T]$ be an admissible control for Player E which sends the point y_0 to a point M such that

$$\hat{\rho}_T(x_0, y_0) = \min_{x \in C_P^T(x_0)} \rho(x, M).$$

Denote $v^*(\cdot) = \{\sigma, v^*(t)\}$, where the partition σ of the interval $[0, T]$ consists of two points $t_0 = 0, t_1 = T$. Evidently, $v^*(\cdot) \in E$. By Theorem in 1.3.4, $v^*(\cdot) \in E$ is an optimal strategy for Player E

in the game $\Gamma(x_0, y_0, T)$ if

$$Val\Gamma(x_0, y_0, T) = \inf_{u(\cdot)\in P} K(u(\cdot), v^*(\cdot); x_0, y_0, T).$$

But this equality follows from (7.5.6), since

$$\inf_{u(\cdot)\in P} K(u(\cdot), v^*(\cdot); x_0, y_0, T) = \hat{\rho}_T(x_0, y_0).$$

Necessity. Suppose that in the game $\Gamma(x_0, y_0, T)$ there exists an optimal open-loop strategy for Player E. Then

$$\begin{aligned} Val\Gamma(x_0, y_0, T) &= \sup_{v(\cdot)\in E} \inf_{u(\cdot)\in P} K(u(\cdot), v(\cdot); x_0, y_0, T) \\ &= \max_{y\in C_E^T(y_0)} \inf_{u(\cdot)\in P} \rho(x(T), y) = \hat{\rho}_T(x_0, y_0). \end{aligned}$$

This completes the proof of lemma.

Theorem. *In order for Player E to have an optimal open-loop strategy for any $x_0, y_0 \in R^n$, $T > 0$ in the game $\Gamma(x_0, y_0, T)$ it is necessary and sufficient that for any $\delta > 0$, $x_0, y_0 \in R^n$, $T \geq \delta$*

$$\hat{\rho}_T(x_0, y_0) = \max_{y\in C_E^\delta(y_0)} \min_{x\in C_P^\delta(x_0)} \hat{\rho}_{T-\delta}(x, y). \tag{7.5.7}$$

Proof. Sufficiency. By Theorem in 5.5.2, condition (7.5.7) implies relationship (7.5.2) from which, by passing to the limit (see Theorem in 5.3.7) we obtain

$$\hat{\rho}_T(x_0, y_0) = Val\Gamma(x_0, y_0, T).$$

By Lemma in 5.5.3, this implies existence of an optimal open-loop strategy for Player E.

Necessity of condition (7.5.7) follows from Theorem in 5.5.2, since the existence of an optimal open-loop strategy for Player E in the game $\Gamma(x_0, y_0, T)$ involves the existence of such a strategy in all games $\Gamma_\delta(x_0, y_0, T)$, $T = \delta k$, $k \geq 1$ and the validity of relationship (7.5.3).

7.6 Fundamental Equation

In this section we will show that, under some particular conditions, the value function of the differential game satisfies a partial differential equation which is called fundamental. Although in monographic literatures R. Isaacs (1965) was the first to consider this equation, it is often referred to as the *Isaacs–Bellman equation.*

7.6.1. By employing Theorem in 5.5.3, we shall derive a partial differential equation for the value function of the differential game. We assume that the conditions of Theorem in 5.5.3 hold for the game $\Gamma(x, y, T)$. Then the function $\hat{\rho}_T(x, y)$ is the value of the game $\Gamma(x, y, T)$ of duration T from initial states x, y.

Suppose that in some domain Ω of the space $R^n \times R^n \times [0, \infty)$ the function $\hat{\rho}_T(x, y)$ has continuous partial derivatives in all its variables. We shall show that in this case the function $\hat{\rho}_T(x, y)$ in domain Ω satisfies the extremal differential equation

$$\frac{\partial \hat{\rho}}{\partial T} - \max_{v \in V} \sum_{i=1}^{n} \frac{\partial \hat{\rho}}{\partial y_i} g_i(y, v) - \min_{u \in U} \sum_{i=1}^{n} \frac{\partial \hat{\rho}}{\partial x_i} f_i(x, u) = 0, \tag{7.6.1}$$

where the functions $f_i(x, u), g_i(y, v)$, $i = 1, \ldots, n$ determine the behavior of players in the game Γ (see (7.3.1), (7.3.2)).

Suppose that (7.6.1) fails to hold in some point $(x, y, T) \in \Omega$. For definiteness, let

$$\frac{\partial \hat{\rho}}{\partial T} - \max_{v \in V} \sum_{i=1}^{n} \frac{\partial \hat{\rho}}{\partial y_i} g_i(y, v) - \min_{u \in U} \sum_{i=1}^{n} \frac{\partial \hat{\rho}}{\partial x_i} f_i(x, u) < 0.$$

Let $\overline{v} \in V$ be such that in the point involved, $(x, y, T) \in \Omega$, the following relationship is satisfied:

$$\sum_{i=1}^{n} \frac{\partial \hat{\rho}}{\partial y_i} g_i(y, \overline{v}) = \max_{v \in V} \sum_{i=1}^{n} \frac{\partial \hat{\rho}}{\partial y_i} g_i(y, v).$$

Then the following inequality holds for any $u \in U$ in the point $(x, y, T) \in \Omega$:

$$\frac{\partial \hat{\rho}}{\partial T} - \sum_{i=1}^{n} \frac{\partial \hat{\rho}}{\partial y_i} g_i(y, \overline{v}) - \sum_{i=1}^{n} \frac{\partial \hat{\rho}}{\partial x_i} f_i(x, u) < 0. \tag{7.6.2}$$

From the continuous differentiability of the function $\hat{\rho}$ in all its variables it follows that the inequality (7.6.2) also holds in some neighborhood S of the point (x, y, T). Let us choose a number $\delta > 0$ so small that the point $(x(\tau), y(\tau), T - \tau) \in S$ for all $\tau \in [0, \delta]$. Here

$$x(\tau) = x + \int_0^\tau f(x(t), u(t))dt,$$

$$y(\tau) = y + \int_0^\tau g(y(t), \overline{v}(t))dt$$

are the trajectories of systems (7.3.1), (7.3.2) corresponding to some admissible control $u(t)$ and $\overline{v}(t) \equiv \overline{v}$ and initial conditions $x(0) = x$, $y(0) = y$, respectively. Let us define the function

$$G(\tau) = \frac{\partial \hat{\rho}}{\partial T}|_{(x(\tau), y(\tau), T-\tau)} - \sum_{i=1}^{n} \frac{\partial \hat{\rho}}{\partial y_i}|_{(x(\tau), y(\tau), T-\tau)} g_i(y(\tau), \overline{v})$$

$$- \sum_{i=1}^{n} \frac{\partial \hat{\rho}}{\partial x_i}|_{(x(\tau), y(\tau), T-\tau)} f_i(x(\tau), u(\tau)), \quad \tau \in [0, \delta].$$

The function $G(\tau)$ is continuous in τ, therefore there is a number $c < 0$ such that $G(\tau) \leq c$ for $\tau \in [0, \delta]$. Hence we have

$$\int_0^\delta G(\tau)d\tau \leq c\delta. \tag{7.6.3}$$

It can be readily seen that

$$G(\tau) = -\frac{d\hat{\rho}}{d\tau}|_{(x(\tau), y(\tau), T-\tau)}.$$

From (7.6.3) we obtain

$$\hat{\rho}_T(x, y) - \hat{\rho}_{T-\delta}(x(\delta), y(\delta)) \leq c\delta.$$

Hence, by the arbitrariness of $u(t)$, it follows that

$$\hat{\rho}_T(x,y) < \max_{y' \in C_E^\delta(y)} \min_{x' \in C_P^\delta(x)} \hat{\rho}_{T-\delta}(x',y').$$

But this contradicts (7.5.7).

We have thus shown that in the case when Player E in the game $\Gamma(x,y,T)$ has an optimal open-loop strategy for any $x, y \in R^n$, $T > 0$, the value of the game $V(x,y,T)$ (it coincides with $\hat{\rho}_T(x,y)$ by Lemma in 5.5.3) in the domain of the space $R^n \times R^n \times [0,\infty)$, where this function has continuous partial derivatives, satisfies the equation

$$\frac{\partial V}{\partial T} = \max_{v \in V} \sum_{i=1}^{n} \frac{\partial V}{\partial y_i} g_i(y,v) + \min_{u \in U} \sum_{i=1}^{n} \frac{\partial V}{\partial x_i} f_i(x,u) \tag{7.6.4}$$

with the initial condition $V(x,y,T)|_{T=0} = \rho(x,y)$. Suppose we have defined $\overline{u}, \overline{v}$ computing max and min to (7.6.4) as functions of x, y and $\frac{\partial V}{\partial x}$, $\frac{\partial V}{\partial y}$ that is

$$\overline{u} = \overline{u}\left(x, \frac{\partial V}{\partial x}\right), \quad \overline{v} = \overline{v}\left(y, \frac{\partial V}{\partial y}\right). \tag{7.6.5}$$

Substituting expressions (7.6.5) into (7.6.4) we obtain

$$\sum_{i=1}^{n} \frac{\partial V}{\partial y_i} g_i\left(y, \overline{v}\left(y, \frac{\partial V}{\partial y}\right)\right) + \sum_{i=1}^{n} \frac{\partial V}{\partial x_i} f_i\left(x, \overline{u}\left(x, \frac{\partial V}{\partial x}\right)\right) = \frac{\partial V}{\partial T} \tag{7.6.6}$$

subject to

$$V(x,y,T)|_{T=0} = \rho(x,y). \tag{7.6.7}$$

Thus, to define $V(x,y,T)$ we have the initial value problem for the first-order partial differential equation (7.6.6) with the initial condition (7.6.7).

Remark. In the derivation of the functional equations (7.6.4), (7.6.6), and in the proof of Theorem in 5.5.3, no use was made of a specific payoff function, therefore this theorem holds for any continuous terminal payoff $H(x(T),y(T))$. In this case, however, instead of the quantity $\hat{\rho}_T(x, y)$ we have to consider the quantity

$$\hat{H}_T(x, y) = \max_{y' \in C_E^T(y)} \min_{x' \in C_P^T(x)} H(x', y').$$

Equation (7.6.4) also holds for the value of the differential game with prescribed duration and any terminal payoff, i.e. if in the differential game $\Gamma(x, y, T)$ with prescribed duration and terminal payoff $H(x(T), y(T))$ there is an optimal open-loop strategy for Player E, then the value of the game $V(x, y, T)$ in the domain of the space $R^n \times R^n \times [0, \infty)$, where there exist continuous partial derivatives, satisfies equation (7.6.4) with the initial condition $V(x, y, T)|_{T=0} = H(x, y)$ or equation (7.6.6) with the same initial condition.

7.6.2. We shall now consider the games of pursuit in which the payoff function is equal to the time-to-capture. For definiteness, we assume that terminal manifold F is a sphere $\rho(x, y) = l$, $l > 0$. We also assume that the sets $C_P^t(x)$ and $C_E^t(y)$ are t-continuous in zero uniformly with respect to x and y.

Suppose the following quanitity makes sense:

$$\theta(x, y, l) = \max_{v(t)} \min_{u(t)} t_n^l(x, y; u(t), v(t)),$$

where $t_n^l(x, y; u(t), v(t))$ is the time of approach within l-distance for the players P and E moving from initial points x, y and using measurable open-loop controls $u(t)$ and $v(t)$, respectively. Also, suppose the function $\theta(x, y, l)$ is continuous in all its independent variables.

Let us denote the time-optimal game by $\Gamma(x_0, y_0)$. As in Secs. 5.4, 5.5, we may derive necessary and sufficient conditions for existence of an optimal open-loop strategy for Player E in the time-optimal game. The following theorem holds.

Theorem. *In order for Player E to have an optimal open-loop strategy for any $x_0, y_0 \in R^n$ in the game $\Gamma(x_0, y_0)$ it is necessary and sufficient that for any $\delta > 0$ and any $x_0, y_0 \in R^n$*

$$\theta(x_0, y_0, l) = \delta + \max_{y' \in C_E^\delta(y_0)} \min_{x' \in C_P^\delta(x_0)} \theta(x', y', l).$$

For the time-optimal game of pursuit the equation (7.6.4) becomes

$$\max_{v\in V}\sum_{i=1}^{n}\frac{\partial\theta}{\partial y_i}g_i(y,v)+\min_{u\in U}\sum_{i=1}^{n}\frac{\partial\theta}{\partial x_i}f_i(x,u)=-1 \tag{7.6.8}$$

with the initial condition

$$\theta(x,y,l)|_{\rho(x,y)=l}=0. \tag{7.6.9}$$

Here it is assumed that there exist the first order continuous partial derivatives of the function $\theta(x,y,l)$ with respect to x,y. Assuming that the $\overline{u},\overline{v}$ sending max and min to (7.6.8) can be defined as functions of $x,y,\partial\theta/\partial x,\partial\theta/\partial y$, i.e. $\overline{u}=\overline{u}(x,\frac{\partial\theta}{\partial x}),\overline{v}=\overline{v}(y,\frac{\partial\theta}{\partial y})$, we can rewrite equation (7.6.8) as

$$\sum_{i=1}^{n}\frac{\partial\theta}{\partial y_i}g_i\left(y,\overline{v}\left(y,\frac{\partial\theta}{\partial y}\right)\right)+\sum_{i=1}^{n}\frac{\partial\theta}{\partial x_i}f_i\left(x,\overline{u}\left(x,\frac{\partial\theta}{\partial x}\right)\right)=-1 \tag{7.6.10}$$

subject to

$$\theta(x,y,l)|_{\rho(x,y)=l}=0. \tag{7.6.11}$$

The derivation of equation (7.6.8) is analogous to the derivation of equation (7.6.4) for the game of pursuit with prescribed duration.

Both initial value problems (7.6.4), (7.6.7) and (7.6.8), (7.6.9) are nonlinear in partial derivatives, therefore their solution presents serious difficulties.

7.6.3. We shall now derive equations of characteristics for (7.6.4). We assume that the function $V(x,y;T)$ has continuous mixed second derivatives over the entire space, the functions $g_i(y,v)$, $f_i(x,u)$ and the functions $\overline{u}=\overline{u}(x,\frac{\partial V}{\partial x})$, $\overline{v}=\overline{v}(y,\frac{\partial V}{\partial y})$ have continuous first derivatives with respect to all their variables, and the sets U,V have the aspect of parallelepipeds $a_m\le u_m\le b_m$, $m=1,\dots,k$ and $c_q\le v_q\le d_q$, $q=1,\dots,l$. where $u=(u_1,\dots,u_k)\in U$,

$v = (v_1, \dots, v_l) \in V$. Denote

$$B(x, y, T) = \frac{\partial V}{\partial T} - \sum_{i=1}^{n} \frac{\partial V}{\partial x_i} f_i(x, \overline{x}) - \sum_{i=1}^{n} \frac{\partial V}{\partial y_i} g_i(y, \overline{v}).$$

The function $B(x, y, T) \equiv 0$, thus taking partial derivatives with respect to $x_1, \dots, x_n$ we obtain

$$\begin{aligned} \frac{\partial B}{\partial x_k} &= \frac{\partial^2 V}{\partial T \partial x_k} - \sum_{i=1}^{n} \frac{\partial^2 V}{\partial x_i \partial x_k} f_i \\ &- \sum_{i=1}^{n} \frac{\partial V}{\partial x_i} \frac{\partial f_i}{\partial x_k} \\ &- \sum_{i=1}^{n} \frac{\partial^2 V}{\partial y_i \partial x_k} g_i - \sum_{m=1}^{k} \frac{\partial}{\partial u_m} \left(\sum_{i=1}^{n} \frac{\partial V}{\partial x_i} f_i \right) \frac{\partial \overline{u}_m}{\partial x_k} \\ &- \sum_{i=1}^{l} \frac{\partial}{\partial v_q} \left(\sum_{i=1}^{n} \frac{\partial V}{\partial y_i} g_i \right) \frac{\partial \overline{v}_q}{\partial x_k} = 0, \quad k = 1, \dots, n. \end{aligned} \tag{7.6.12}$$

For every fixed point $(x, y, T) \in R^n \times R^n \times [0, \infty)$ the maximizing value $\overline{v}$ and the minimizing value $\overline{u}$ in (7.6.4) lie either inside or on the boundary of the interval of constraints. If this is an interior point, then

$$\frac{\partial}{\partial u_m} \left(\sum_{i=1}^{n} \frac{\partial V}{\partial x_i} f_i \right) |_{u=\overline{u}} = 0, \quad \frac{\partial}{\partial v_q} \left(\sum_{i=1}^{n} \frac{\partial V}{\partial y_i} g_i \right) |_{v=\overline{v}} = 0.$$

If, however, $\overline{u}(\overline{v})$ is at the boundary, then two cases are possible. Let us discuss these cases for one of the components $\overline{u}_m(x, \frac{\partial V}{\partial x})$ of the vector $\overline{u}$. The other components of vector $\overline{u}$ and vector $\overline{v}$ can be investigated in a similar manner. For simplicity assume that at some point (x', y', T')

$$\overline{u}_m = \overline{u}_m \left(x', \frac{\partial V(x', y', T')}{\partial x} \right) = a_m.$$

Case 1. In the space R^n there exists a ball with its center at the point x' and the following equality holds for all points x:

$$\overline{u}_m = \overline{u}_m\left(x, \frac{\partial V(x, y', T')}{\partial x}\right) = a_m.$$

The function $\overline{u}_m$ assumes on the ball a constant value; therefore in the point x' we have

$$\frac{\partial \overline{u}_m}{\partial x_i} = 0, \quad i = 1, \ldots, n.$$

Case 2. Such a ball does not exist. Then there is a sequence of interior points x_r, $\lim_{r\to\infty} x_r = x'$ such that

$$\overline{u}_m\left(x_r, \frac{\partial V(x_r, y', T')}{\partial x}\right) \neq a_m.$$

Hence

$$\frac{\partial}{\partial u_m}\left(\sum_{i=1}^{n} \frac{\partial V}{\partial x_i}|_{(x_r, y', T')} f_i(x_r, \overline{u})\right) = 0.$$

From the continuity of derivatives $\partial V/\partial x_i$, $\partial f_i/\partial u_m$ and function $\overline{u} = \overline{u}(x, \frac{\partial V(x,y,T)}{\partial x})$ it follows that the preceding equality also holds in the point (x', y', T').

Thus, the last two terms in (7.6.12) are zero and the following equality holds for all $(x, y, T) \in R^n \times [0, \infty)$:

$$\frac{\partial B}{\partial x_k} = \frac{\partial^2 V}{\partial T \partial x_k} - \sum_{i=1}^{n} \frac{\partial^2 V}{\partial x_i \partial x_k} f_i(x, \overline{u})$$

$$- \sum_{i=1}^{n} \frac{\partial V}{\partial x_i}\frac{\partial f_i}{\partial x_k} - \sum_{i=1}^{n} \frac{\partial^2 V}{\partial y_i \partial x_k} g_i(y, \overline{v}) = 0, \quad k = 1, 2, \ldots, n.$$

Let $\overline{x}(t), \overline{y}(t)$, $t \in [0, T]$ be a solution of the system

$$\dot{x} = f\left(x, \overline{u}\left(x, \frac{\partial V(x, y, T-t)}{\partial x}\right)\right),$$

$$\dot{y} = g\left(y, \overline{v}\left(y, \frac{\partial V(x, y, T-t)}{\partial y}\right)\right)$$

with the initial condition $x(0) = x_0$, $y(0) = y_0$. Along the solution $\overline{x}(t), \overline{y}(t)$ we have

$$\frac{\partial^2 V(\overline{x}(t), \overline{y}(t), T-t)}{\partial T \partial x_k} - \sum_{i=1}^{n} \frac{\partial^2 V(\overline{x}(t), \overline{y}(t), T-t)}{\partial x_i \partial x_k} f_i(\overline{x}(t), \tilde{u}(t))$$
$$- \sum_{i=1}^{n} \frac{\partial V(\overline{x}(t), \overline{y}(t), T-t)}{\partial x_i} \frac{\partial f_i(\overline{x}(t), \tilde{u}(t))}{\partial x_k}$$
$$- \sum_{i=1}^{n} \frac{\partial^2 V(\overline{x}(t), \overline{y}(t), T-t)}{\partial y_i \partial x_k} g_i(\overline{y}(t), \tilde{v}(t)) = 0, \quad k = 1, \ldots, n, \tag{7.6.13}$$

where

$$\tilde{u}(t) = \overline{u}\left(\overline{x}(t), \frac{\partial V(\overline{x}(t), \overline{y}(t), T-t)}{\partial x}\right),$$
$$\tilde{v}(t) = \overline{v}\left(\overline{y}(t), \frac{\partial V(\overline{x}(t), \overline{y}(t), T-t)}{\partial y}\right).$$

However,

$$\frac{d}{dt}\left(\frac{\partial V(\overline{x}(t), \overline{y}(t), T-t)}{\partial x_k}\right) = \sum_{i=1}^{n} \frac{\partial^2 V(\overline{x}(t), \overline{y}(t), T-t)}{\partial x_k \partial x_i} f_i(\overline{x}(t), \tilde{u}(t))$$
$$+ \sum_{i=1}^{n} \frac{\partial^2 V(\overline{x}(t), \overline{y}(t), T-t)}{\partial x_k \partial y_i} g_i(\overline{y}(t), \tilde{v}(t))$$
$$- \frac{\partial^2 V(\overline{x}(t), \overline{y}(t), T-t)}{\partial x_k \partial T}, \qquad k = 1, \ldots, n. \tag{7.6.14}$$

Note that for the twice continuously differentiable function we may reverse the order of differentiation. Now (7.6.13) can be rewritten in terms of (7.6.14) as

$$\frac{d}{dt}\left(\frac{\partial V(\overline{x}(t), \overline{y}(t), T-t)}{\partial x_k}\right)$$
$$= - \sum_{i=1}^{n} \frac{\partial V(\overline{x}(t), \overline{y}(t), T-t)}{\partial x_i} \frac{\partial f_i(\overline{x}(t), \tilde{u}(t))}{\partial x_k}, \quad k = 1, \ldots, n.$$

In a similar manner we obtain the equations

$$\frac{d}{dt}\left(\frac{\partial V(\overline{x}(t),\overline{y}(t),T-t)}{\partial y_i}\right)$$
$$= -\sum_{j=1}^{n}\frac{\partial V(\overline{x}(t),\overline{y}(t),T-t)}{\partial y_j}\frac{\partial g_j(\overline{y}(t),\tilde{v}(t))}{\partial y_i},\ i=1,\dots,n.$$

Since for $t \in [0,T]$

$$V(\overline{x}(t),\overline{y}(t),T-t) = H(\overline{x}(T),\overline{y}(T)),$$

we have

$$\frac{d}{dt}\left(\frac{\partial V(\overline{x}(t),\overline{y}(t),T-t)}{\partial T}\right) = 0.$$

Let us introduce the following notation:

$$V_{x_i}(t) \triangleq \frac{\partial V(\overline{x}(t),\overline{y}(t),T-t)}{\partial x_i},$$
$$V_{y_i}(t) \triangleq \frac{\partial V(\overline{x}(t),\overline{y}(t),T-t)}{\partial y_i},\quad i=1,\dots,n;$$
$$V_x(t) \triangleq \{V_{x_i}(t)\},\quad V_y(t) \triangleq \{V_{y_i}(t)\},$$
$$V_T(t) \triangleq \frac{\partial V(\overline{x}(t),\overline{y}(t),T-t)}{\partial T}.$$

As a result we obtain the following system of ordinary differential equations for the functions $x(t), y(t), V_x(t), V_y(t)$:

$$\begin{aligned}
\dot{x}_i &= f_i(x,\overline{u}(x,V_x)),\\
\dot{y}_i &= g_i(y,\overline{v}(y,V_y)),\\
\dot{V}_{x_k} &= -\sum_{i=1}^{n} V_{x_i}\frac{\partial f_i(x,\overline{u}(x,V_x))}{\partial x_k},\\
\dot{V}_{y_k} &= -\sum_{i=1}^{n} V_{y_i}\frac{\partial g_i(y,\overline{v}(y,V_y))}{\partial y_k},\\
V_T &= 0,\quad i,k=1,\dots,n,
\end{aligned} \tag{7.6.15}$$

and, by (7.6.6), we have

$$V_T = \sum_{i=1}^{n} V_{y_i} g_i(y, \overline{v}(y, V_y)) + \sum_{i=1}^{n} V_{x_i} f_i(x, \overline{u}(x, V_x)).$$

In order to solve the system of nonlinear equations (7.6.15) with respect to the functions $x(t), y(t), V_{x_k}(t), V_{y_k}(t), V_T(t)$, we need to define initial conditions. For the function $V(\overline{x}(t), \overline{y}(t), T-t)$ such conditions are given at the time instant $t = T$, therefore we introduce the variable $\tau = T - t$ and write the equation of characteristics as a regression. Let us introduce the notation $\overset{\circ}{x} = -\dot{x}, \overset{\circ}{y} = -\dot{y}$. The equation of characteristics become

$$\begin{aligned}
\overset{\circ}{x}_i &= -f_i(x, \overline{u}), \\
\overset{\circ}{y}_i &= -g_i(y, \overline{v}), \\
\overset{\circ}{V}_{x_k} &= \sum_{i=1}^{n} V_{x_i} \frac{\partial f_i(x, \overline{u})}{\partial x_k}, \\
\overset{\circ}{V}_{y_k} &= \sum_{i=1}^{n} V_{y_i} \frac{\partial g_i(y, \overline{v})}{\partial y_k}, \\
\overset{\circ}{V}_T &= 0.
\end{aligned} \tag{7.6.16}$$

In the specification of initial conditions for system (7.6.16), use is made of the relationship $V(x, y, T)|_{T=0} = H(x, y)$. Let $x|_{\tau=0} = s$, $y_{\tau=0} = s'$. Then

$$\begin{aligned}
V_{x_i}|_{\tau=0} &= \frac{\partial H}{\partial x_i}|_{x=s, y=s'}, \\
V_{y_i}|_{\tau=0} &= \frac{\partial H}{\partial y_i}|_{x=s, y=s'}, \\
V_T|_{\tau=0} &= \sum_{i=1}^{n} V_{y_i}|_{\tau=0} g_i(s', \overline{v}(s', V_y|_{\tau=0})) \\
&\quad + \sum_{i=1}^{n} V_{x_i}|_{\tau=0} f_i(s, \overline{u}(s, V_x|_{\tau=0})).
\end{aligned} \tag{7.6.17}$$

Possible ways of solving system (7.6.16)–(7.6.17) are discussed in detail in Isaacs (1965).

In a similar manner, using equation (7.6.8) we may write the equation of characteristics for the problem of time-optimal pursuit.

7.7 Methods of Successive Approximations for Solving Differential Games of Pursuit

7.7.1. Let $\Gamma_\delta(x, y, T)$ be a discrete form of the differential game $\Gamma(x, y, T)$ of duration $T > 0$ with a fixed step of partition δ and discrimination against Player E for the time $\delta > 0$ in advance. Denote by $V_\delta(x, y, T)$ the value of the game $\Gamma_\delta(x, y, T)$.[1] Then

$$\lim_{\delta \to 0} V_\delta(x, y, T) = V(x, y, T)$$

and optimal strategies in the game $\Gamma_\delta(x, y, T)$ for sufficiently small δ can be efficiently used to construct ϵ-equilibria in the game $\Gamma(x, y, T)$.

7.7.2. The essence of the numerical method is to construct an algorithm of finding a solution of the game $\Gamma_\delta(x, y, T)$. We shall now expound this method.

Zero-order approximation. A zero-order approximation for the function of the value of the game $V_\delta(x, y, T)$ is taken to be the function

$$V_\delta^0(x, y, T) = \max_{\eta \in C_E^T(y)} \min_{\xi \in C_P^T(x)} \rho(\xi, \eta), \tag{7.7.1}$$

where $C_P^T(x), C_E^T(y)$ are reachability sets for the players P and E from initial states $x, y \in R^n$ by the time T.

The choice of the function $V_\delta^0(x, y, T)$ as an initial approximation is justified by the fact that in a sufficiently large class of games (what

[1] The terminal payoff is equal to $\rho(x(T), y(T))$, where $\rho(x, y)$ is a distance in R^n.

is called a regular case) it turns out to be the value of the game $\Gamma(x, y, T)$. The following approximations are constructed by the rule:

$$V_\delta^1(x,y,T) = \max_{\eta \in C_E^\delta(y)} \min_{\xi \in C_P^\delta(x)} V_\delta^0(\xi, \eta, T-\delta),$$

$$V_\delta^2(x,y,T) = \max_{\eta \in C_E^\delta(y)} \min_{\xi \in C_P^\delta(x)} V_\delta^1(\xi, \eta, T-\delta),$$

$$\dots$$

$$V_\delta^k(x,y,T) = \max_{\eta \in C_E^\delta(y)} \min_{\xi \in C_P^\delta(x)} V_\delta^{k-1}(\xi, \eta, T-\delta) \tag{7.7.2}$$

for $T > \delta$ and $V_\delta^k(x,y,T) = V_\delta^0(x,y,T)$ for $T \le \delta$, $k \ge 1$.

As may be seen from formulas (7.7.2), the max min operation is taken over the reachability sets $C_E^\delta(y), C_P^\delta(x)$ for the time δ, i.e. for one step of the discrete game $\Gamma_\delta(x, y, T)$.

7.7.3. Theorem. *For the fixed x, y, T, δ the numerical sequence $\{V_\delta^k(x,y,T)\}$ does not decrease with the growth of k.*

Proof. First we prove the inequality

$$V_\delta^1(x,y,T) \ge V_\delta^0(x,y,T).$$

For all $\xi \in C_P^\delta(x)$ there is $C_P^{T-\delta}(\xi) \subset C_P^T(x)$. For any $\overline{\eta} \in C_E^{T-\delta}(\eta)$, $\xi \in C_P^\delta(x)$ we have

$$\min_{\overline{\xi} \in C_P^{T-\delta}(\xi)} \rho(\overline{\xi}, \overline{\eta}) \ge \min_{\overline{\xi} \in C_P^T(x)} \rho(\overline{\xi}, \overline{\eta}).$$

Hence

$$\begin{aligned} V_\delta^1(x,y,T) &= \max_{\eta \in C_E^\delta(y)} \min_{\xi \in C_P^\delta(x)} \max_{\overline{\eta} \in C_E^{T-\delta}(\eta)} \min_{\overline{\xi} \in C_P^{T-\delta}(\xi)} \rho(\overline{\xi}, \overline{\eta}) \\ &\ge \max_{\eta \in C_E^\delta(y)} \max_{\overline{\eta} \in C_E^{T-\delta}(\eta)} \min_{\overline{\xi} \in C_P^T(x)} \rho(\overline{\xi}, \overline{\eta}) \\ &= \max_{\eta \in C_E^T(y)} \min_{\xi \in C_P^T(x)} \rho(\xi, \eta) = V_\delta^0(x,y,T). \end{aligned}$$

We now assume that for $l \le k$ there is

$$V_\delta^l(x,y,T) \ge V_\delta^{l-1}(x,y,T). \tag{7.7.3}$$

We prove this inequality for $l = k+1$. From relationships (7.7.2) and (7.7.3) it follows that

$$V_\delta^{k+1}(x,y,T) = \max_{\eta \in C_E^\delta(y)} \min_{\xi \in C_P^\delta(x)} V_\delta^k(\xi,\eta,T-\delta)$$

$$\geq \max_{\eta \in C_E^\delta(y)} \min_{\xi \in C_P^\delta(x)} V_\delta^{k-1}(\xi,\eta,T-\delta) = V_\delta^k(x,y,T).$$

Thus, in case $T > \delta$, by induction, the statement of the theorem is proved (in case $T \leq \delta$ the statement of the theorem is obvious).

7.7.4. Theorem. *The sequence $\{V_\delta^k(x,y,T)\}$ converges in a finite number of steps N, with the estimate $N \leq [\frac{T}{\delta}] + 1$, where the brackets stand for the integer part.*

Proof. Let $N = [T/\delta] + 1$. We show that

$$V_\delta^N(x,y,T) = V_\delta^{N+1}(x,y,T). \tag{7.7.4}$$

Equation (7.7.4) can be readily obtained from construction of the sequence $\{V_\delta^k(x,y,T)\}$. Indeed,

$$V_\delta^N(x,y,T) = \max_{\eta^1 \in C_E^\delta(y)} \min_{\xi^1 \in C_P^\delta(x)} V_\delta^{N+1}(\xi^1,\eta^1,T-\delta)$$

$$= \max_{\eta^1 \in C_E^\delta(y)} \min_{\xi^1 \in C_P^\delta(x)} \max_{\eta^2 \in C_E^\delta(\eta^1)} \cdots \max_{\eta^{N-1} \in C_E^\delta(\eta^{N-2})}$$

$$\times \min_{\xi^{N-1} \in C_P^\delta(\xi^{N-2})} V_\delta^1(\xi^{N-1},\eta^{N-1},T-(N-1)\delta).$$

Similarly we get

$$V_\delta^{N+1}(x,y,T) = \max_{\eta_1 \in C_E^\delta(y)} \min_{\xi^1 \in C_P^\delta(x)} \max_{\eta^2 \in C_E^\delta(\eta^1)} \cdots \max_{\eta^{N-1} \in C_E^\delta(\eta^{N-2})}$$

$$\times \min_{\xi^{N-1} \in C_P^\delta(\xi^{N-2})} V_\delta^2(\xi^{N-1},\eta^{N-1},T-(N-1)\delta).$$

But $T-(N-1)\delta = \alpha < \delta$, therefore

$$V_\delta^1(\xi^{N-1},\eta^{N-1},\alpha) = V_\delta^2(\xi^{N-1},\eta^{N-1},\alpha) = V_\delta^0(\xi^{N-1},\eta^{N-1},\alpha),$$

whence equality (7.7.4) follows.

The coincidence of members of the sequence V_δ^k for $k \geq N$ is derived from (7.7.4) by induction. This completes the proof of the theorem.

7.7.5. Theorem. *The limit of the sequence* $\{V_\delta^k(x,y,T)\}$ *coincides with the value of the game* $\Gamma_\delta(x,y,T)$.

Proof. This theorem is essentially a corollary to Theoren in 5.7.4. Indeed, let

$$V_\delta(x,y,T) = \lim_{k\to\infty} V_\delta^k(x,y,T).$$

Convergence takes place in a finite number of steps not exceeding $N = [T/\delta] + 1$; therefore in the recursion equation (7.7.2) we may pass to the limit as $k \to \infty$. The limiting function $V_\delta(x,y,T)$ satisfies the equation

$$V_\delta(x,y,T) = \max_{\eta \in C_E^\delta(y)} \min_{\xi \in C_P^\delta(x)} V_\delta(\xi,\eta,T-\delta) \tag{7.7.5}$$

with initial condition

$$V_\delta(x,y,T)|_{0\leq T\leq\delta} = \max_{\eta \in C_E^T(y)} \min_{\xi \in C_P^T(x)} \rho(\xi,\eta), \tag{7.7.6}$$

which is a sufficient condition for the function $V_\delta(x,y,T)$ to be the value of the game $\Gamma_\delta(x,y,T)$, (this is also a "regularity" criterion).

7.7.7. We shall now provide a modification of the method of successive approximation discussed above.

The initial approximation is taken to be the function $\tilde{V}_\delta^0(x,y,T) = V_\delta^0(x,y,T)$, where $V_\delta^0(x,y,T)$ is defined by (7.7.1). The following approximations are constructed by the rule:

$$\tilde{V}_\delta^{k+1}(x,y,T) = \max_{i\in[1:N]} \max_{\eta \in C_E^{i\delta}(y)} \min_{\xi \in C_P^{i\delta}(x)} \tilde{V}_\delta^k(\xi,\eta,T-i\delta)$$

for $T > \delta$, where $N = [T/\delta]$, and $\tilde{V}_\delta^{k+1}(x,y,T) = \tilde{V}_\delta^0(x,y,T)$ for $T \leq \delta$.

The statements of the theorems in 5.7.3–5.7.5 hold for the sequence of functions $\{\tilde{V}_\delta^k(x,y,T)\}$ and the sequence of functions $\{V_\delta^k(x,y,T)\}$.

The proof of these statements for the sequence of functions $\{\tilde{V}_\delta^k(x,y,T)\}$ is almost an exact replica of a similar argument for the sequence of functions $\{V_\delta^k(x,y,T)\}$. In the region $\{(x,y,T)|T>\delta\}$ the functional equation for the function of the value of the game $\Gamma_\delta(x,y,T)$ becomes

$$V_\delta(x,y,T) = \max_{i\in[1:N]} \max_{\eta\in C_E^{i\delta}(y)} \min_{\xi\in C_P^{i\delta}(x)} V_\delta(\xi,\eta,T-i\delta), \qquad (7.7.7)$$

where $N=[T/\delta]$, while the initial condition remains unaffected, i.e. it is of the form (7.7.6).

7.7.7. We shall now prove the equivalence of equations (7.7.5) and (7.7.7).

Theorem. *Equations* (7.7.5) *and* (7.7.7) *with initial condition* (7.7.6) *are equivalent.*

Proof. Suppose the function $V_\delta(x,y,T)$ satisfies equation (7.7.5) and initial condition (7.7.6). Show that this function satisfies equation (7.7.7) in the region $\{(x,y,T)|T>\delta\}$.

Indeed, the following relationships hold:

$$\begin{aligned}
V_\delta(x,y,T) &= \max_{\eta\in C_E^{\delta}(y)} \min_{\xi\in C_P^{\delta}(x)} V_\delta(\xi,\eta,T-\delta)\\
&= \max_{\eta\in C_E^{\delta}(y)} \min_{\xi\in C_P^{\delta}(x)} \max_{\overline{\eta} C_E^{\delta}(\eta)} \min_{\overline{\xi}\in C_P^{\delta}(\xi)} V_\delta(\overline{\xi},\overline{\eta},T-2\delta)\\
&\geq \max_{\eta\in C_E^{\delta}(y)} \max_{\overline{\eta}\in C_E^{\delta}(\eta)} \min_{\xi\in C_P^{\delta}(x)} \min_{\overline{\xi}\in C_P^{\delta}(\xi)} V_\delta(\overline{\xi},\overline{\eta},T-2\delta)\\
&= \max_{\eta\in C_E^{2\delta}(y)} \min_{\xi\in C_P^{2\delta}(x)} V_\delta(\xi,\eta,T-2\delta) \geq \dots\\
&\geq \max_{\eta\in C_E^{i\delta}(y)} \min_{\xi\in C_P^{i\delta}(x)} V_\delta(\xi,\eta,T-i\delta) \geq \dots .
\end{aligned}$$

When $i=1$ we have

$$V_\delta(x,y,T) = \max_{\eta\in C_E^{\delta}(y)} \min_{\xi\in C_P^{\delta}(x)} V_\delta(\xi,\eta,T-\delta),$$

hence

$$V_\delta(x,y,T) = \max_{i\in[1:N]} \max_{\eta\in C_E^{i\delta}(y)} \min_{\xi\in C_P^{i\delta}(x)} V_\delta(\xi,\eta,T-i\delta),$$

where $N=[T/\delta]$, which proves the statement.

Now suppose the function $V_\delta(x, y, T)$ in the region $\{(x, y, T)| T > \delta\}$ satisfies equation (7.7.7) and initial condition (7.7.6). Show that this function also satisfies equation (7.7.5). Suppose the opposite is true. Then the following inequality must hold in the region $\{(x, y, T)|T > \delta\}$:

$$V_\delta(x, y, T) > \max_{\eta \in C_E^\delta(y)} \min_{\xi \in C_P^\delta(x)} V_\delta(\xi, \eta, T - \delta).$$

However,

$$\begin{aligned}
&\max_{\eta \in C_E^\delta(y)} \min_{\xi \in C_P^\delta(x)} V_\delta(\xi, \eta, T - \delta) \\
&\quad = \max_{\eta \in C_E^\delta(y)} \min_{\xi \in C_P^\delta(x)} \max_{i \in [1:N-1]} \max_{\overline{\eta} \in C_E^{i\delta}(\eta)} \min_{\overline{\xi} \in C_P^{i\delta}(\xi)} V_\delta(\overline{\xi}, \overline{\eta}, T - (i+1)\delta) \\
&\quad \geq \max_{\eta \in C_E^\delta(y)} \max_{i \in [1:N-1]} \max_{\overline{\eta} \in C_E^{i\delta}(\eta)} \min_{\xi \in C_P^\delta(x)} \min_{\overline{\xi} \in C_P^{i\delta}(\xi)} V_\delta(\overline{\xi}, \overline{\eta}, T - (i+1)\delta) \\
&\quad = \max_{i \in [1:N-1]} \max_{\eta \in C_E^\delta(y)} \max_{\overline{\eta} \in C_E^{i\delta}(\eta)} \min_{\xi \in C_P^\delta(x)} \min_{\overline{\xi} \in C_P^{i\delta}(\xi)} V_\delta(\overline{\xi}, \overline{\eta}, T - (i+1)\delta) \\
&\quad = \max_{i \in [2:N]} \max_{\eta \in C_E^{i\delta}(y)} \min_{\xi \in C_P^{i\delta}(x)} V_\delta(\xi, \eta, T - i\delta) = V_\delta(x, y, T).
\end{aligned}$$

Since for $i = 1$ the strong inequality holds, this contradiction proves the theorem.

7.8 Examples of Solutions to Differential Games of Pursuit

7.8.1. *Example 4. Simple motion.* Let us consider the differential game $\Gamma(x_0, y_0, T)$ in which the motion by the Players P and E in the Euclidean space R^n is governed by the following equations:

$$\begin{aligned}
&\text{for } P: \ \dot{x} = \alpha u, \ \|u\| < 1, \ x(0) = x_0, \\
&\text{for } E: \ \dot{y} = \beta v, \ \|v\| < 1, \ y(0) = y_0,
\end{aligned} \tag{7.8.1}$$

where α, β are constants $\alpha > \beta > 0$, $x, y, u, v \in R^n$.

The payoff to Player E is

$$H(x(T), y(T)) = \|x(T) - y(T)\|.$$

Let $\Gamma_\delta(x, y, T)$ be a discrete form of the differential game $\Gamma(x, y, T)$ with the partition step $\delta > 0$ and discrimination against Player E. The game $\Gamma_\delta(x, y, T)$ has N steps, where $N = T/\delta$. By Sec. 5.2 (see Example in 5.2.3) the game $\Gamma_\delta(x, y, T)$ has the value

$$\begin{aligned} V_\delta(x, y, T) &= \max\{0, \|x - y\| - N \cdot \delta \cdot (\alpha - \beta)\} \\ &= \max\{0, \|x - y\| - T(\alpha - \beta)\}, \end{aligned}$$

and the optimal motion by players is along the straight line connecting the initial states x, y.

By the results of 5.3, the value of the original differential game

$$V(x, y, T) = \lim_{\delta \to 0} V_\delta(x, y, T) = \max\{0, \|x - y\| - T(\alpha - \beta)\}. \quad (7.8.2)$$

It can be seen that

$$V(x, y, T) = \max_{y' \in C_E^T(y)} \min_{x' \in C_P^T(x)} \|x' - y'\| = \hat{\rho}_T(x, y),$$

where $C_E^T(y) = S(y, \beta T)$ is the ball in R^n of radius βT with its center at the point y, similarly $C_P^T(x) = S(x, \alpha T)$. Thus, by Lemma in 5.5.3, Player E in the game $\Gamma(x_0, y_0, T)$ has the optimal open-loop strategy $v^*(t)$, $t \in [0, T]$, which leads Player E's trajectory to the point $y^* \in C_E^T(y_0)$ for which

$$\hat{\rho}_T(x_0, y_0) = \min_{x' \in C_P^T(x_0)} \|x' - y^*\|.$$

Evidently,

$$v^*(t) \equiv v^* = \begin{cases} \dfrac{y_0 - x_0}{\|y_0 - x_0\|}, & \text{with } y_0 \neq x_0, \\ v, & \text{with } y_0 = x_0, \end{cases}$$

where $v \in R^n$ is an arbitrary vector such that $\|v\| = 1$. From the results of 5.6 it follows that in the region Δ

$$\Delta = \{(x, y, T) : \|x - y\| - T(\alpha - \beta) > 0\},$$

where there exist continuous partial derivatives

$$\frac{\partial V}{\partial T} = -(\alpha - \beta), \quad \frac{\partial V}{\partial x} = -\frac{\partial V}{\partial y} = \frac{x - y}{\|x - y\|},$$

the function $V(x, y, T)$ satisfies equation (7.7.4):

$$\frac{\partial V}{\partial T} - \alpha \min_{\|u\|<1} \left(\frac{\partial V}{\partial x}, u\right) - \beta \max_{\|v\|\leq 1} \left(\frac{\partial V}{\partial y}, v\right) = 0. \tag{7.8.3}$$

In equation (7.8.3), minimum and maximum are achieved under controls

$$\overline{u}\left(x, \frac{\partial V}{\partial x}\right) = -\frac{\frac{\partial V}{\partial x}}{\|\frac{\partial V}{\partial x}\|} = \frac{y - x}{\|y - x\|}; \tag{7.8.4}$$

$$\overline{v}\left(y, \frac{\partial V}{\partial y}\right) = \frac{\frac{\partial V}{\partial y}}{\|\frac{\partial V}{\partial y}\|} = \frac{y - x}{\|y - x\|}. \tag{7.8.5}$$

Strategies (7.8.4), (7.8.5) are optimal in the differential game (7.8.1). The strategy $\overline{u}(x, y)$ determined by relationship (7.8.4) is called a "pursuit strategy", since at each instant of time for Player P using this strategy the vector of his velocity is pointing towards Evader E.

7.8.2. *Example 5. Game of pursuit with frictional forces.* The pursuit takes place over the plane. Equations of motion are of the form:

for P:

$$\begin{aligned} &\dot{q}_i = p_i, \\ &\dot{p}_i = \alpha u_i - k_P p_i, \quad i = 1, 2, \quad \|u\| \leq 1; \end{aligned} \tag{7.8.6}$$

for E:

$$\dot{r}_i = s_i,$$

$$\dot{s}_i = \beta v_i - k_E s_i, \quad i = 1, 2, \quad \|v\| \le 1; \tag{7.8.7}$$

$$q_i(0) = q_i^0,\ p_i(0) = p_i^0,\ r_i(0) = r_i^0,$$

$$s_i(0) = s_i^0, \quad i = 1, 2, \quad \alpha, \beta, k_E, k_P > 0. \tag{7.8.8}$$

Here $q = (q_1, q_2)$ and $r = (r_1, r_2)$ are positions on the plane for Players P and E, respectively; $p = (p_1, p_2)$ and $s = (s_1, s_2)$ are the players' momenta; k_P, k_E are some constants interpreted to mean friction coefficients.

The payoff to Player E is taken to be

$$H(q(T), r(T)) = \|q(T) - r(T)\|$$
$$= \sqrt{[q_1(T) - r_1(T)]^2 + [q_2(T) - r_2(T)]^2}.$$

In the plane $q = (q_1, q_2)$, the reachability set $C_P^T(q^0, p^0)$ for Player P from the initial states $p(0) = p^0$, $q(0) = q^0$ in the time T is the circle (Exercise 18) of radius

$$R_P(T) = \frac{\alpha}{k_P^2}(e^{-k_P T} + k_P T - 1)$$

with its center at the point

$$a(q^0, p^0, T) = q^0 + p^0 \frac{1 - e^{-k_P T}}{k_P}.$$

Similarly, the set $C_E^T(r^0, s^0)$ is the circle of radius

$$R_E(T) = \frac{\beta}{k_E^2}(e^{-k_E T} + k_E T - 1)$$

with its center at the point

$$b(r^0, s^0, T) = r^0 + s^O \frac{1 - e^{-k_E T}}{k_E}.$$

For the quantity $\hat{\rho}_T(q^0, p^0, r^0, s^0)$ determined by relationship (7.5.1), in this differential game there is

$$\hat{\rho}_T(q^0, p^0, r^0, s^0) = \max_{r \in C_E^T(r^0, s^0)} \min_{q \in C_P^T(q^0, p^0)} \|q - r\|.$$

Hence (see formula (7.2.10)) we have

$$\hat{\rho}_T(q, p, r, s) = \max\{0, \|a(q, p, T) - b(r, s, T)\| - (R_P(T) - R_E(T))\}$$

$$= \max\left\{0, \sqrt{\sum_{i=1}^{2}\left(q_i - r_i + p_i\frac{1 - e^{-k_P T}}{k_P} - s_i\frac{1 - e^{-k_E T}}{k_E}\right)^2}\right.$$

$$\left. - \left(\alpha\frac{e^{-k_P T} + k_P T - 1}{k_P^2} - \beta\frac{e^{-k_E T} + k_E T - 1}{k_E^2}\right)\right\}. \quad (7.8.9)$$

In particular, the conditions $\alpha > \beta$, $\frac{\alpha}{k_P} > \frac{\beta}{k_E}$ suffice to ensure that for any initial states q, p, r, s there is a suitable T for which $\hat{\rho}_T(q, p, r, s) = 0$.

The function $\hat{\rho}_T(q, p, r, s)$ satisfies the extremal differential equation (7.7.1) in the domain $\Omega = \{(q, p, r, s, T) : \hat{\rho}_T(q, p, r, s) > 0\}$. In fact, in the domain Ω there are continuous partial derivatives

$$\frac{\partial \hat{\rho}}{\partial T}, \frac{\partial \hat{\rho}}{\partial q_i}, \frac{\partial \hat{\rho}}{\partial p_i}, \frac{\partial \hat{\rho}}{\partial r_i}, \frac{\partial \hat{\rho}}{\partial s_i}, \quad i = 1, 2. \quad (7.8.10)$$

Equation (7.7.1) becomes

$$\frac{\partial \hat{\rho}}{\partial T} - \sum_{i=1}^{2}\left(\frac{\partial \hat{\rho}}{\partial q_i}p_i + \frac{\partial \hat{\rho}}{\partial r_i}s_i - \frac{\partial \hat{\rho}}{\partial p_i}k_P p_i - \frac{\partial \hat{\rho}}{\partial s_i}k_E s_i\right)$$

$$-\beta \max_{\|v\| \le 1} \sum_{i=1}^{2} \frac{\partial \hat{\rho}}{\partial s_i} v_i - \alpha \min_{\|u\| \le 1} \sum_{i=1}^{2} \frac{\partial \hat{\rho}}{\partial p_i} u_i = 0. \quad (7.8.11)$$

Here extrema are achieved on the controls $\overline{u}, \overline{v}$ determined by the following formulas:

$$\overline{u}_i = -\frac{\frac{\partial \hat{\rho}}{\partial p_i}}{\sqrt{(\frac{\partial \hat{\rho}}{\partial p_1})^2 + (\frac{\partial \hat{\rho}}{\partial p_2})^2}}, \tag{7.8.12}$$

$$\overline{v}_i = \frac{\frac{\partial \hat{\rho}}{\partial s_i}}{\sqrt{(\frac{\partial \hat{\rho}}{\partial s_1})^2 + (\frac{\partial \hat{\rho}}{\partial s_2})^2}}, \quad i = 1, 2. \tag{7.8.13}$$

Substituting these controls into (7.8.11) we obtain the nonlinear first-order partial differential equation

$$\frac{\partial \hat{\rho}}{\partial T} - \sum_{i=1}^{2} \left(\frac{\partial \hat{\rho}}{\partial q_i} p_i + \frac{\partial \hat{\rho}}{\partial r_i} s_i - \frac{\partial \hat{\rho}}{\partial p_i} k_P p_i - \frac{\partial \hat{\rho}}{\partial s_i} k_E s_i \right)$$
$$-\beta \sqrt{\left(\frac{\partial \hat{\rho}}{\partial s_1}\right)^2 + \left(\frac{\partial \hat{\rho}}{\partial s_2}\right)^2} + \alpha \sqrt{\left(\frac{\partial \hat{\rho}}{\partial p_1}\right)^2 + \left(\frac{\partial \hat{\rho}}{\partial p_2}\right)^2} = 0. \tag{7.8.14}$$

Computing the partial derivatives (7.8.10) we see that the function $\hat{\rho}_T(q, p, r, s)$ in the domain Ω satisfies equation (7.8.14). Note that the quantity $\hat{\rho}_T(q^0, p^0, r^0, s^0)$ is the value of the differential game (7.8.6)–(7.8.8) and the controls determined by relationships (7.8.12), (7.8.13) are optimal in the domain Ω.

From formulas (7.8.12), (7.8.13), (7.8.9) we find

$$\overline{u}_i = \frac{r_i - q_i + s_i \frac{1-e^{-k_E T}}{k_E} - p_i \frac{1-e^{-k_P T}}{k_P}}{\sqrt{\sum_{i=1}^{2} \left(r_i - q_i + s_i \frac{1-e^{-k_E T}}{k_E} - p_i \frac{1-e^{-k_P T}}{k_P} \right)^2}},$$
$$\overline{v}_i = \overline{u}_i, \quad i = 1, 2. \tag{7.8.15}$$

In the situation $\overline{u}, \overline{v}$ the force direction for each of the players is parallel to the line connecting the centers of reachability circles (as follows from formula (7.8.15)) and remains unaffected, since in this situation the centers of reachability circles move along the straight line.

7.9 Games of Pursuit with Delayed Information for Pursuer

7.9.1. In this chapter we investigated games examined conflict-controlled processes where each participant (player) has perfect information, i.e. at each current instant of time Player $P(E)$ is aware of his state $x(t)[y(t)]$ and the opponent's state $y(t)[x(t)]$. Existence theorems were obtained for pure strategy ϵ-equilibria in such games and various methods for constructing solutions were illustrated. This was made possible by the fact that the differential games with perfect information are the limiting case of multistage games with perfect information where the time interval between two sequential moves tends to zero. In differential games with incomplete information, where mixed strategies play an important role, we have a completely different situation. Without analyzing the entire problem we will deal with game of pursuit with prescribed duration, terminal payoff and delayed information for Player P on the phase state (state variable) of Player E, the time of delay being $l > 0$.

7.9.2. Let there be given some number $l > 0$ referred to as the information delay. For $0 \leq t \leq l$, Pursuer P at each instant of time t knows his own state $x(t)$, the time t and the initial position y_0 of Evader E. For $l \leq t \leq T$, Player P at each instant of time t knows his own state $x(t)$, the time t and the state $y(t-l)$ of Player E at the time instant $t-l$. Player E at each instant of time t knows his own state $y(t)$, the opponent's state $x(t)$ and the time t. His payoff is equal to a distance between the players at the time instant T, the payoff to Player P is equal to the payoff to Player E but opposite in sign (the game is zero-sum). Denote this game by $\Gamma(x_0, y_0, T)$.

Definition. The pure piecewise open-loop strategy $v(\cdot)$ for Player E means the pair $\{\tau, b\}$, where τ is a partitioning of the time interval $[0, T]$ by a finite number of points $0 = t_1 < \ldots < t_k = T$, and b is the map which places each state $x(t_i), y(t_i), t_i$ in correspondence with the measurable open-loop control $v(t)$ of Player E for $t \in [t_i, t_{i+1})$.

Definition. The pure piecewise open-loop strategy $u(\cdot)$ for Player P means the pair $\{\sigma, a\}$, where σ is an arbitrary partitioning of the time interval $[0, T)$ by a finite number of points $0 = t'_1 < t'_2 < \ldots < t'_s = T$, and a is the map which places each state $x(t'_i), y(t'_i - l), t'_i$ for $l \leq t'_i$ in correspondence with the segment of Player P's measurable open-loop control $u(t)$ for $t \in [t'_i, t'_{i+1})$. For $t'_i \leq l$, the map a places each state $x(t'_i), y_0, t'_i$ in correspondence with the segment of Player P's measurable control $u(t)$ for $t \in [t'_i, t'_{i+1})$.

The sets of all pure piecewise open-loop strategies for the players P and E are denoted by P and E, respectively.

Equations of motion are of the form

$$\begin{aligned} \dot{x} &= f(x, u), \ u \in U \subset R^p, \ x \in R^n, \\ \dot{y} &= g(y, v), \ v \in V \subset R^q, \ y \in R^n. \end{aligned} \tag{7.9.1}$$

We assume that the conditions which ensure the existence and uniqueness of a solution to system (7.9.1) for any pair of measurable open-loop controls $u(t), v(t)$ with the given initial conditions x_0, y_0 are satisfied. This ensures the existence of a unique solution to system (7.9.1) where the players P and E use piecewise open-loop strategies $u(\cdot) \in P$, $v(\cdot) \in E$ with the given initial conditions x_0, y_0. Thus, in any situation $(u(\cdot), v(\cdot))$ with the given initial conditions x_0, y_0 the payoff function for Player E is determined in a unique way

$$K(x_0, y_0; u(\cdot), v(\cdot)) = \rho(x(T), y(T)), \tag{7.9.2}$$

where $x(t), y(t)$ is a solution to system (7.9.1) with initial conditions x_0, y_0 in situation $(u(\cdot), v(\cdot))$, and ρ is the Euclidean distance.

7.9.3. We can demonstrate with simple examples that in the game under study $\Gamma(x_0, y_0, T)$ the ϵ-equilibria do not exist for all $\epsilon > 0$. For this reason, to construct equilibria, we shall follow the way proposed by John von Neumann and Oskar Morgenstern (1944) for finite positional games with incomplete information. The strategy spaces of the players P and E will be extended to what are called *mixed piecewise open-loop behavior strategies* (MPOLBS) which allows for a random choice of control at each step.

Example 6. Equations of motion are of the form

$$\begin{aligned} &\text{for } P: \ \dot{x} = u, \ \|u\| \leq \alpha, \\ &\text{for } E: \ \dot{y} = v, \ \|v\| \leq \beta, \\ &\alpha > \beta > 0, \ x, y, \in R^2, \ u, v \in R^2. \end{aligned} \tag{7.9.3}$$

The payoff to Player E is $\rho(x(T), y(T))$, where $x(t), y(t)$ is a solution to system (7.9.3) with the initial conditions $x(t_0) = x_0$, $y(t_0) = y_0$. Player P is informed only about the initial state y_0 of his opponent, while Player E is completely informed about Player P's state ($l = T$).

Let $\overline{v}(x, y, t)$ be some piecewise open-loop strategy for Player E. For each strategy $\overline{v}$ there is a strategy $\overline{u}(x, t)$ of Player P using only information about the initial position of Player E, his current position and the time from the start of the game, for which a payoff of $\rho(x(T), y(T)) \leq \epsilon$ for $T \geq \rho(x_0, y_0)/(\alpha - \beta)$. Indeed, let $u^*(x, y, t)$ be a strategy for Player P in the game with perfect information. The strategy is as follows: Player E is pursued until the capture time t_n (while the capture of E takes place) while for $t_n \leq t \leq T$ the point $x(t)$ is kept in some ϵ-neighborhood of the evading point. It is an easy matter to describe analytically such a strategy in the game with perfect information (see Example 4, 5.8.1). Let us construct the players' trajectories $\overline{x}(t), \overline{y}(t)$ in situation $(u^*(x, y, t), \overline{v}(x, y, t))$ from the initial states x_0, y_0. To do this, it suffices to integrate the system

$$\begin{aligned} &\dot{x} = u^*(x, y, t), \ \overline{x}(t_0) = x_0, \\ &\dot{y} = \overline{v}(x, y, t), \ \overline{y}(t_0) = y_0. \end{aligned} \tag{7.9.4}$$

By construction $\rho(\overline{x}(T), \overline{y}(T)) \leq \epsilon$. Now let $\tilde{u}(t) = u^*(\overline{x}(t), \overline{y}(t), t)$. Although the strategy $u^*(x, y, t)$ using the information about E's position is inadmissible, the strategy $\tilde{u}(t)$ is admissible since it uses only information about the time from the start of the game and information about the initial state of Player E. It is apparent that in situations $(\tilde{u}(t), \overline{v}(x, y, t))$ and $(u^*(x, y, t), \overline{v}(x, y, t))$ the players' paths coincide since the strategy $\overline{v}(x, y, t)$ responds to the strategy $u^*(x, y, t)$ and the strategy $\tilde{u}(t)$ by choosing the same control $\overline{v}(\overline{x}(t), \overline{y}(t), t)$.

We have thus shown that for each strategy $\overline{v}(x, y, t)$ there is an open-loop control $\tilde{u}(t)$ which is an admissible strategy in the game with incomplete information and is such that $\rho(\overline{x}(T), \overline{y}(T)) \leq \epsilon$, where $\overline{x}(t), \overline{y}(t)$ are the corresponding trajectories. The choice of $\overline{v}(x, y, t)$ is made in an arbitrary way, hence it follows that

$$\sup \inf \rho(x(T), y(T)) = 0, \tag{7.9.5}$$

where sup inf is taken over the players' strategy sets in the game with incomplete information.

For any strategy $u(x, t)$ of Player P, however, we may construct a strategy $v(x, y, t)$ for Player E such that in situation $(u(x, t), v(x, y, t))$ the payoff ρ to Player E will exceed βT. Indeed, let $\overline{u}(x, t)$ be a strategy for Player P. Since his motion is independent of $y(t)$, the path of Player P can be obtained by integrating the system

$$\dot{x} = \overline{u}(x, t), \quad x(t_0) = x_0 \tag{7.9.6}$$

irrespective of what motion is made by Player E. Let $\overline{x}(t)$ be a trajectory resulting from integration of system (7.9.6). The points $\overline{x}(T)$ and y_0 are connected and the motion by Player E is oriented along the straight line $[\overline{x}(T), y_0]$ away from the point $\overline{x}(T)$. His speed is taken to be maximum. Evidently, the motion by Player E ensures a distance between him and the point $\overline{x}(T)$ which is greater than or equal to βT. Denote the thus constructed strategy for Player E by $\overline{v}(t)$. In the situation $(\overline{u}(x, t), \overline{v}(t))$, the payoff to Player E is then greater than or equal to βT. From this it follows that

$$\inf \sup \rho(x(T), y(T)) \geq \beta T, \tag{7.9.7}$$

where inf sup is taken over the players' strategy sets in the game with incomplete information.

It follows from (7.9.5) and (7.9.7) that the value of the game in the class of pure strategies does not exist in the game under study.

7.9.4. Definition. The MPOLBS for Player P means the pair $\mu(\cdot) = \{\tau, d\}$, where τ is an arbitrary partitioning of the time interval $[0, T]$ by a finite number of points $0 = t_1 < t_2 < \ldots < t_k = T$, and d is the map which places each state $x(t_i), y(t_i - l), t_i$ for $t_i > l$ and the state $x(t_i), y_0, t_i$ for $t_i \leq l$ in correspondence with the probability distribution $\mu_i'(\cdot)$ concentrated on a finite number of measurble open-loop controls $u(t)$ for $t \in [t_i, t_{i+1})$.

Similarly, MPOLBS for Player E means the pair $\nu(\cdot) = \{\sigma, c\}$, where σ is an arbitrary partitioning of the time interval $[0, T]$ by a finite number of points $0 = t_1' < t_2' < \ldots < t_s' = T$, and c is the map which places the state $x(t_i'), y(t_i'), t_i'$ in correspondence with the probability distribution $\nu_i'(\cdot)$ concentrated on a finite number of measurable open-loop controls $v(t)$ for $t \in [t_i, t_{i+1})$.

MPOLBS for the players P and E are denoted respectively by $\overline{P}$ and $\overline{E}$ (compare these strategies with "behavior strategies" in 4.8.3).

Each pair of MPOLBS $\mu(\cdot), \nu(\cdot)$ induces the probability distribution over the space of trajectories $x(t), x(0) = x_0$; $y(t), y(0) = y_0$. For this reason, the payoff $\overline{K}(x_0, y_0; \mu(\cdot), \nu(\cdot))$ in MPOLBS is interpreted to mean the mathematical expectation of the payoff averaged over the distributions over the trajectory spaces that are induced by MPOLBS $\mu(\cdot), \nu(\cdot)$. Having determined the strategy spaces $\overline{P}, \overline{E}$ and the payoff $\overline{K}$ we have determined the mixed extension $\overline{\Gamma}(x_0, y_0, T)$ of the game $\Gamma(x_0, y_0, T)$.

7.9.5. Denote by $C_P^T(x)$ and $C_E^T(y)$ the respective reachability sets of the players P and E from initial states x and y at the instant of time T, and by $\overline{C}_E^T(y)$ the convex hull of the set $C_E^T(y)$. We assume that the reachability sets are compact, and introduce the quantity

$$\gamma(y, T) = \min_{\xi \in \overline{C}_E^T(y)} \max_{\eta \in C_E^T(y)} \rho(\xi, \eta).$$

Let $\gamma(y, T) = \rho(\tilde{y}, \overline{y})$, where $y \in \overline{C}_E^T(y)$, $\overline{y} \in C_E^T(y)$. From the definition of the point $\tilde{y}$ it follows that it is a center of the minimal sphere containing the set $C_E^T(y)$. Hence it follows that this point is unique. At the same time, there exist at least two points of tangency of the set $C_E^T(y)$ to the minimal sphere containing it, these points coinciding with the points $\overline{y}$.

Let $y(t)$ be a trajectory ($y(0) = y_0$) of Player E for $0 \leq t \leq T$. When Player E moves along this trajectory the value of the quanitity $\gamma(y(t), T - t)$ changes, the point $\tilde{y}$ also changes. Let $\tilde{y}(t)$ be a trajectory of the point $\tilde{y}$ corresponding to the trajectory $y(t)$. The point $M \in C_E^{T-l}(y_0)$ will be referred to as the *center of pursuit* if

$$\gamma(M, l) = \max_{y' \in C_E^{T-l}(y_0)} \gamma(y', l).$$

7.9.6. We shall now consider an auxiliary simultaneous zero-sum game of pursuit over a convex hull of the set $C_E^T(y)$. Pursuer chooses a point $\xi \in \overline{C}_E^T(y)$ and Evader chooses a point $\eta \in C_E^T(y)$. The choices are made simultaneously. When choosing the point ξ, Player P has no information on the choice of η by Player E, and conversely. Player E receives a payoff $\rho(\xi, \eta)$. We denote the value of this game by $V(y, T)$ in order to emphasize the dependence of the game value on the parameters y and T which determine the strategy sets $\overline{C}_E^T(y)$ and $C_E^T(y)$ for players P and E, respectively. The game in normal form can be written as follows:

$$\Gamma(y, T) = \langle \overline{C}_E^T(y), C_E^T(y), \rho(y', y'') \rangle.$$

The strategy set of the minimizing player P is convex, the function $\rho(y', y'')$ is also convex in its independent variables and is continuous. Theorem in 2.5.5 can be applied to such games. Therefore the game $\Gamma(y, T)$ has an equilibrium in mixed strategies. An optimal strategy for Player P is pure, and an optimal strategy for Player E assigns the positive probability to at most $(n + 1)$ points from the set $C_E^T(y)$, with $V(y, T) = \gamma(y, T)$. An optimal strategy for Player P in the game $\Gamma(y, T)$ is the choice of a center of the minimal sphere $\tilde{y}$ containing the set $C_E^T(y)$. An optimal strategy for Player E assigns the positive probabilities to at most $(n+1)$ points among the points of tangency of the sphere to the set $C_E^T(y)$ (here n is the dimension of the space of y). The value of the game is equal to the radius of this sphere (see Example 11 in 2.5.5).

7.9.7. We shall now consider a simultaneous game $\Gamma(M, l)$, where M is the center of pursuit. Denote by $\overline{y}_1(M), \dots, \overline{y}_{n+1}(M)$ the points from the set $C_E^l(M)$ appearing in the spectrum of an optimal mixed strategy for Player E in the game $\Gamma(M, l)$ and by $y(M)$ an optimal strategy for Player P in this game.

Definition. The trajectory $y^*(t)$ is called conditionally optimal if $y^*(0) = y_0$, $y^*(T - l) = M$, $y^*(T) = \overline{y}_i(M)$ for some i from the numbers $1, \dots, n + 1$.

For each i there can be several conditionally optimal strategies of Player E.

Theorem. *Let $T \geq l$ and suppose that for any number $\epsilon > 0$ Player P can ensure by the time T the ϵ-capture of the center $\tilde{y}(T)$ of the minimal sphere containing the set $C_E^l(y(T - l))$. Then the game $\overline{\Gamma}(x_0, y_0, T)$ has the value $\gamma(M, l)$, and the ϵ-optimal strategy of Player P is pure and coincides with any one of his strategies which may ensure the $\epsilon/2$-capture of the point $\tilde{y}(T)$. An optimal strategy for Player E is mixed: during the time $0 \leq t \leq T - l$ he must move to the point M along any conditionally optimal trajectory $y^*(t)$ and then, with probabilities $p_1, \dots, p_{n+1}$ (the optimal strategy for Player E in the game $\Gamma(M, l)$), he must choose one of the conditionally optimal trajectories sending the point $y^*(T - l)$ to the points $\overline{y}_i(M)$, $i = 1, \dots, n+1$ which appear in the spectrum of an optimal mixed strategy for Player E in the game $\Gamma(M, l)$.*

Proof. Denote by $u_\epsilon(\cdot), \nu_*(\cdot)$ the strategies mentioned in Theorem whose optimality is to be proved. In order to prove Theorem, it suffices to verify the validity of the following relationships:

$$\overline{K}(x_0, y_0; \mu(\cdot), \nu_*(\cdot)) + \epsilon \geq \overline{K}(x_0, y_0; u_\epsilon(\cdot), \nu_*(\cdot))$$
$$\geq \overline{K}(x_0, y_0; u_\epsilon(\cdot), \nu(\cdot)) - \epsilon, \quad \mu(\cdot) \in \overline{P},\ \nu(\cdot) \in \overline{E}, \tag{7.9.8}$$

$$\lim_{\epsilon \to 0} \overline{K}(x_0, y_0; u_\epsilon(\cdot), \nu_*(\cdot)) = \gamma(M, l). \tag{7.9.9}$$

The left-hand side of inequality (7.9.8) follows from the definition of strategy $u_\epsilon(\cdot)$ by which for any piecewise open-loop strategy $u(\cdot) \in P$

$$\overline{K}(x_0, y_0; u(\cdot), \nu_*(\cdot)) + \epsilon \geq \overline{K}(x_0, y_0; u_\epsilon(\cdot), \nu_*(\cdot)).$$

Denote by $x^*(t)$ Pursuer's trajectory in situation $(u_\epsilon(\cdot), \nu_*(\cdot))$. Then

$$\overline{K}(x_0, y_0; u_\epsilon(\cdot), \nu_*(\cdot)) = \sum_{i=1}^{n+1} p_i \rho(x^*(T), \overline{y}_i(M)). \qquad (7.9.10)$$

Let R be a radius of the minimal sphere containing the set $C_E^l(M)$, i.e. $R = \gamma(M, l)$. Then $R - \epsilon/2 \leq \rho(x^*(T), \overline{y}_i(M)) \leq R + \epsilon/2$ for all $i = 1, \dots, n+1$, since the point $x^*(T)$ belongs to the $\epsilon/2$-neighborhood of the point $y(M)$. Since $\sum_{i=1}^{n+1} p_i = 1$, $p_i \geq 0$, from (7.9.10) we get

$$R - \epsilon/2 \leq \overline{K}(x_0, y_0; u_\epsilon(\cdot), \nu_*(\cdot)) \leq R + \epsilon/2, \qquad (7.9.11)$$

and this proves (7.9.9).

Suppose the state $x(T), y(T-l)$ have been realized in situation $(u_\epsilon(\cdot), \nu(\cdot))$ and $Q(\cdot)$ is the probability measure induced on the set $C_E^l(y(T-l))$. From the optimality of the mixed strategy $p = (p_1, \dots, p_{n+1})$ in the game $\Gamma(M, l)$ we have

$$\begin{aligned} R &= \sum_{i=1}^{n+1} p_i \rho(y(M), \overline{y}_i(M)) \geq \gamma(y(T-l), l) = Val\Gamma(y(T-l), l) \\ &\geq \int_{C_E^l(y(T-l))} \rho(\tilde{y}[y(T-l)], y) dQ, \qquad (7.9.12) \end{aligned}$$

where $\tilde{y}[y(T-l)]$ is the center of the minimal sphere containing the set $C_E^l(y(T-l))$.

However, $\rho(x(T), y[y(T-l)]) \leq \epsilon/2$, therefore for $y \in C_E^l(y(T-l))$ we have

$$\rho(x(T), y) \leq \frac{\epsilon}{2} + \rho(\tilde{y}[y(T-l)], y) \leq R + \epsilon/2. \qquad (7.9.13)$$

From inequalities (7.9.11)–(7.9.13) it follows that

$$\overline{K}(x_0, y_0; u_\epsilon(\cdot), \nu_*(\cdot)) \geq \int_{C_E^l(y(T-l))} \rho(x(T), y)dQ - \epsilon, \qquad (7.9.14)$$

but

$$\int_{C_E^l(y(T-l))} \rho(x(T), y)dQ = \overline{K}(x_0, y_0; u_\epsilon(\cdot), \nu(\cdot)). \qquad (7.9.15)$$

From formulas (7.9.14) and (7.9.15) we obtain the right-hand side of inequality (7.9.8). This completes the proof of the theorem.

For $T < l$ the solution of the game does not differ essentially from the case $T \geq l$ and Theorem holds if we consider $C_E^T(y_0)$, $\overline{C}_E^T(y_0)$, $\gamma(M, T)$, y_0 instead of $C_E^l(y_0)$, $\overline{C}_E^l(y_0)$, $\gamma(M, l)$, $y(T - l)$, respectively.

The diameter of the set $C_E^l(M)$ tends to zero as $l \to 0$, which is why the value of the auxiliary game $\Gamma(M, l)$ also tends to zero. But the value of this auxiliary game is equal to the value $V_l(x_0, y_0, T)$ of the game of pursuit with delayed information $\overline{\Gamma}(x_0, y_0, T)$ (here index l indicates the information delay). The optimal mixed strategy for Player E in $\Gamma(M, l)$ concentrating its mass on at most $n + 1$ points from $C_E^l(M)$ concentrates in the limit its entire mass in one point M, i.e. it becomes a pure strategy. This agrees with the fact that the game $\Gamma(x_0, y_0, T)$ becomes the game with perfect information as $l \to 0$.

Example 7. Equations of motion are of the form

$$\dot{x} = u, \ \|u\| \leq \alpha; \quad \dot{y} = v, \ \|v\| \leq \beta, \quad \alpha > \beta, \ x, y \in R^2.$$

Suppose the time T satisfies the condition $T > \rho(x_0, y_0)/(\alpha - \beta) + l$. The reachability set $C_E^l(y_0) = \overline{C}_E^l(y_0)$ and coincides with the circle of radius βl with its center at y_0. The value of the game $\Gamma(y, l)$ is equal to the radius of the circle $C_E^l(y)$, i.e. $V(y, l) = \beta l$.

Since $V(y, l)$ is now independent of y, any point of the set $C_E^{T-l}(y_0)$ can be the center of pursuit M. An optimal strategy for Player P in the game $\Gamma(y, l)$ is the choice of point y, and an optimal strategy for Player E is mixed and is the choice of any two diametrically opposite points of the circle $C_E^l(y)$ with probabilities $(1/2, 1/2)$. Accordingly, an optimal strategy for Pursuer in the game $\overline{\Gamma}(x_0, y_0, T)$ is the linear pursuit of the point $y(t-l)$ for $l \le t \le T$ (the point y_0 for $0 \le t \le l$) until the capture of this point; moreover, it must remain in $\epsilon/2$-neighborhood of this point. An optimal strategy for Player E (MPOLBS) is the transition from the point y_0 to an arbitrary point $M \in C_E^{T-l}(y_0)$ during the time $T-l$ and then the equiprobable choice of a direction towards one of the two diametrically opposite points of the circle $C_E^l(M)$. In this case $Val\overline{\Gamma}(x_0, y_0, T) = \beta l$.

7.10 Exercises and Problems

1. Construct a reachability set for Player P and E in a "simple motion" game.
2. Suppose Player E moves from the point $y_0 = (y_1^0, y_2^0)$ with velocity β which is constant in value and direction. Show that for each motion there is a unique motion by Player P from the point $x_0 = (x_1^0, x_2^0)$ with constant velocity α ($\alpha > \beta$) which realizes the capture (l-capture) of Player E within a minimal period of time. Such a motion by Player P is called a time-optimal response with respect to a capture point.
3. Suppose Player E moves from the point $y_0 = (y_1^0, y_2^0)$ with velocity β that is constant in value and direction. At the same time, Player P responds immediately by moving towards a capture point. Construct a capture point for each pair of motions by the players P and E. Show that the obtained locus of capture points for players E and P is the *Apollonius circle*, and write its equation.
4. Under conditions of the preceding exercise, construct the set of l-capture points for the players E and P.
5. Denote by $A(x_0, y_0)$ the set of capture points with respect to initial states x_0, y_0 of the players P and E (the Apollonius circle).

Suppose that for some instant of time τ (τ is less than the time-to-capture) the players E and P move along straight lines with maximum velocities towards the capture point M. Construct a new set of capture points $A(x(\tau), y(\tau))$ with respect to the states $x(\tau), y(\tau)$ that are initial at the instant of time τ. This is a new Apollonius circle.

Show that the circles $A(x_0, y_0)$ and $A(x(\tau), y(\tau))$ are tangent to one another at the point M, hence $A(x(\tau), y(\tau))$ are contained in the circle $\overline{A}(x_0, y_0)$ bounded by the circumference $A(x_0, y_0)$.

6. Suppose that Player E moves from the point y_0 along some smooth curve $y(t)$ with maximum velocity β. Player P moves with maximum velocity α; at each instant of time τ he knows Player E's position $y(t)$ and the direction of the velocity vector $v(\tau) = \{v_1(\tau), v_2(\tau)\}$, $(v_1^2(\tau) + v_2^2(\tau) = \beta^2)$. Construct Π-strategy for Player P. In accordance with this strategy he chooses the direction of the velocity vector towards the capture point M assuming that on the time interval $[\tau, \infty)$ Player E follows a constant direction $\{v_1(\tau), v_2(\tau)\}$ (he moves along the ray with constant velocity β).

Show that if Player P uses Π-strategy, then the line segment $[x(\tau), y(\tau)]$ connecting the current positions of the players is kept parallel to the segment $[x_0, y_0]$ until the time of capture.

7. Suppose that Player E moves from y_0 along some smooth curve $y(\tau)$ with maximum velocity β. Write an analytical expression for Π-strategy of Player P.

8. Show that when Player P uses Π-strategy, the capture point is always contained in the set $\overline{A}(x_0, y_0)$ bounded by the Apollonius circle $A(x_0, y_0)$.

Hint. The proof is carried out for the motions by Player E along k-vertex broken lines in terms of the statement of Exercise 5, then a passage to the limit is made.

9. *"Driver the Killer" game.* In order to write equations of motion for the players in this game, it suffices to specify five phase coordinates: two coordinates to identify the position of Player P

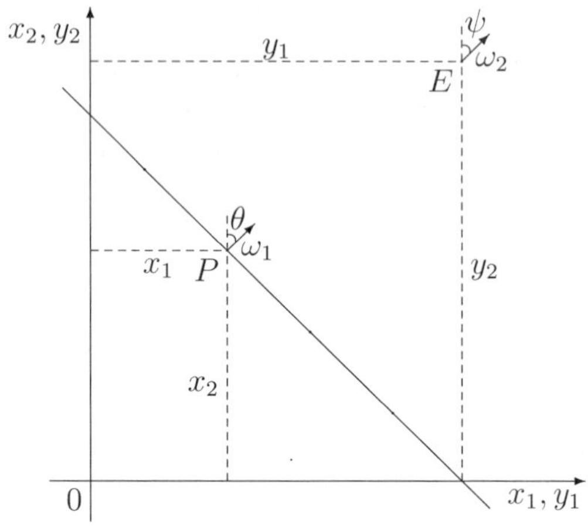

Figure 7.1 "Driver the Killer" game

(a motor vehicle), two coordinates to identify the position of Player E (a pedestrian), and one coordinate to indicate the direction of pursuit. Denote these coordinates by $x_1, x_2, y_1, y_2, \theta$ (Fig. 7.1). The state of the game at each instant of time is determined completely and uniquely by specifying these phase coordinates.

The control for Player E is simple. In order to describe the direction of his motion, it suffices to specify the angle ψ (see Fig. 7.1).

Let us choose the control for Player P. We draw through the point P the straight line CC' ($|C'P| = |PC| = R$) that is perpendicular to the velocity vector of pursuit. Player P may choose the instantaneous center of curvature of his trajectory at any point, say, at the point C_1 lying on this straight line outside the interval $C'C$.

The control u is taken to be equal to $R/|PC_1|$ in absolute value, positive for the points C_1 to the left of P and negative to the right of P; thus, $-1 \leq u \leq 1$. Prove that the equation of motion is:

$$\dot{x}_1 = \omega_1 \sin\theta, \;\; \dot{x}_2 = \omega_1 \cos\theta,$$

$$\dot{y}_1 = \omega_2 \sin\varphi, \;\; \dot{y}_2 = \omega_2 \cos\varphi, \;\; \dot{\theta} = \omega_1/R \cdot u.$$

10. *"Driver The Killer" game. Reduction of dimension.* We assume that a moving coordinate system related to motor vehicle P is chosen on the plane. In this system the coordinates y_1, y_2 of pedestrian can be regarded as components of a single variable vector x; the x_2 axis is taken to be always directed along the velocity vector of motor vehicle.

Suppose that Player P chooses at the instant of time t the curvature of his path to be centered at the point $C = (R/u, 0)$, and the distance CE is equal to d (Fig. 7.2). Then the rotation of Player P around the point C is equivalent to the rotation of x around C in the opposite sense, but with the same angular velocity. Thus the vector x moves with velocity that is equal to $\omega_1(du/R)$ in absolute value and perpendicular to CE. The components of his velocity are obtained by multiplying the modulus by $-x_2/d$ and $(x_1 - R/\varphi)/d$, respectively.

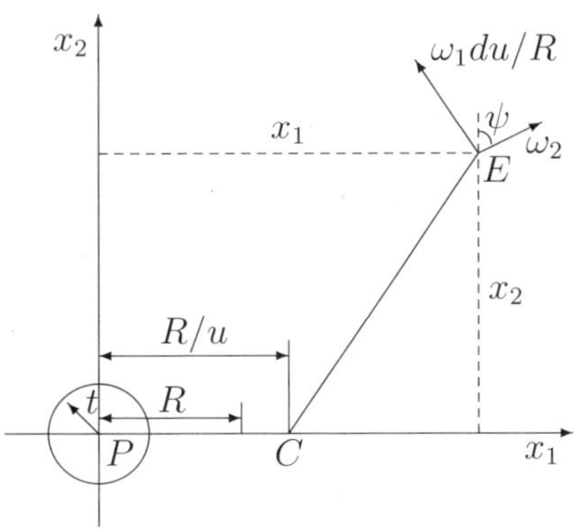

Figure 7.2 "Driver the Killer" game in the reduced space

Show that the equations of motion are:

$$\dot{x}_1 = -\frac{\omega_1}{R}x_2u + \omega_2 \sin\psi,$$
$$\dot{x}_2 = \frac{\omega_1}{R}x_1u - \omega_1 + \omega_2 \cos\psi,$$
$$-1 \le u \le 1, \quad 0 \le \psi \le 2\pi.$$

11. Let a and b be numbers such that $\rho = \sqrt{a^2+b^2} > 0$. Show that $\max_\psi(a\cos\psi + b\sin\psi)$ is attained in the point $\overline{\psi}$ such that $\cos\overline{\psi} = a/\rho$, $\sin\overline{\psi} = b/\rho$, and this maximum is equal to ρ.

12. Let the payoff be terminal and the equations of motion be

$$\dot{x}_1 = av + \omega \sin u,$$
$$\dot{x}_2 = -1 + \omega \cos u,$$
$$0 \le u \le 2\pi, \quad -1 \le v \le 1,$$

where a and ω are positive smooth functions of x_1 and x_2. Write the equation for the value V of the game in form (7.5.64) and (7.5.66) and show that the equation in form (7.5.69) is

$$aV_{x_1}\overline{v} - \omega\rho - V_{x_2} = 0,$$

where

$$\rho = \sqrt{V_{x_1}^2 + V_{x_2}^2}, \ \overline{v} = \mathrm{sgn}V_{x_1}, \ \sin\overline{u} = -V_x/\rho, \ \cos\overline{u} = -V_y/\rho.$$

Hint. Make use of Exercise 11.

13. *"Driver The Killer" game.* Write the main equation in form (7.7.8) and (7.7.10) for equations of motion in the natural space (Exercise 9) and in the reduced space (Exercise 10). In the first case, for v_x, v_y, v we introduce the notation v_1, v_2, v_3, v_4, v_5, where the indices refer to the relevant phase coordinates following the order in which they appear in equations of motion.

14. Find the equation of characteristics as a regression in the natural space for the "Driver The Killer" game. Here the main equation (7.7.10) becomes

$$\omega_1(v_1 \sin\theta + v_2 \cos\theta) + \omega_2\rho + \frac{\omega_1}{R} v_5 \overline{u} + 1 = 0,$$

where

$$\rho = \sqrt{v_3^2 + v_4^2},\ \overline{u} = -\mathrm{sgn} v_5,\ \sin\overline{\varphi} = v_3/\rho,\ \cos\overline{\varphi} = v_4/\rho.$$

15. Make use of a solution to Exercise 14 and show that the solution in the small of the "Driver The Killer" game for Player P is to make right-left turns as sharp as possible and the solution for Player E is to move along the straight line.

16. Write and illustrate equation (7.7.6) for the "pulling" game

$$\dot{x}_1 = u + v,\ |u| \leq \alpha, \quad \dot{x}_2 = u + v,\ |v| \leq \beta, \quad x(0) = x_0$$

with the terminal payoff $\rho(x(T), A)$, where A is some point, $A \in R^2$, lying outside the system reachability set by the time-instant T from the initial state x_0.

17. Write explicit expressions for optimal strategies in the game as in Exercise 16 and for its modification, where the duration of the game is not prefixed and the payoff to Player E is taken to be equal to the time of arrival at the origin of coordinates.

18. Prove that the reachability set of the controlled system

$$\dot{q}_i = p_i, \quad \dot{p}_i = \alpha u_i - k p_i,$$
$$q_i(0) = q_i^0,\ p_i(0) = p_i^0, \quad u_1^2 + u_2^2 \leq 1,\ i = 1, 2$$

in the space of geometric coordinates (q_1, q_2) is the circle of radius $R = \alpha(e^{-kT} + kT - 1)/k^2$ with its center at the point $\overline{q} = q^0 + p^0(1 - e^{-kT})/k$.

19. Prove that the function $\hat{\rho}_T(q, p, r, s)$ satisfies equation (7.7.6) written for this case.

20. The pursuit is carried out on the plane and equations of motion are of the form:
for P:

$$\dot{q}_i = p_i, \quad \dot{p}_i = \alpha u_i - k_P p_i, \quad |u| \leq 1, \; i = 1, 2,$$

for E:

$$\dot{y}_i = \beta v_i, \quad |v| \leq 1, \; i = 1, 2.$$

Here q, y are positions of the players P and E respectively and p is the momentum of Player P. Now in this case Player E moves in accordance with a "simple motion", while Player P represented by a material unit mass point moves under the frictional force α.

The payoff to a player is defined to be the distance between geometric positions of the players by the time T when the game ends:

$$H(q(T), y(T)) = \rho(q(T), y(T)) = \sqrt{\sum_{i=1}^{2}(q_i(T) - y_i(T))^2}.$$

Find the quantity $\hat{\rho}_T(q, y)$.

21. Derive equation (7.7.6) for the problem from Exercise 20.

22. Consider the game of "simple-pursuit" with prescribed duration T in the half-plane F, i.e. under the additional assumption that in the process of pursuit the players cannot leave the set F. Construct reachability sets for the players.

23. Find the quantity $\hat{\rho}_T(x, y)$ for the game of "simple pursuit" in the half-plane with prescribed duration.

24. Consider a zero-sum game of "simple pursuit" with prescribed duration between the two pursuers $P = \{P_1, P_2\}$ acting as one player and the evading player E. Equations of motion are of the form:

$$\begin{aligned}
\dot{x}^1 &= u^1, \quad |u^1| \le \alpha_1, \quad \beta < \min\{\alpha_1, \alpha_2\}, \\
\dot{x}^2 &= u^2, \quad |u^2| \le \alpha_2, \quad x^1, x^2, y \in R^2, \\
\dot{y} &= v, \quad |v| < \beta, \quad u^1, u^2, v \in R^2, \\
x^1(0) &= x_0^1, \quad x^2(0) = x_0^2, \quad y(0) = y_0.
\end{aligned}$$

The payoff to Player E is

$$\min_{i=1,2} \rho(x^i(T), y(T)),$$

i.e. Player E is interested in maximizing the distance to the nearest of the pursuers by the time the game ends.

Construct the rechability sets of the players and determine geometrically the maximin distance $\hat{\rho}_T(x_0^1, x_0^2, y)$ between these sets.

25. Extend Theorem in 5.9.7 to the case where the participants are several pursuers $P_1, \ldots, P_n$ acting as one player, and one evading player E.

Bibliography

Ashmanov, S. A. *Linear Programming*, p. 198. Nauka, Moscow, 1981.

Aumann, R. In Luce R. D. and Tucker A. W. editors, *Contributions to the Theory of Games IV*, Princeton: Princeton University Press, 1959.

Basar, T. *SIAM Journal of Control and Optimization*, 15, 1977.

Basar, T. and I. Olsder. *Dynamic Noncooperative Game Theory*, p. 233. Academic Press, London, 1984.

Bellman, R. *Rendiconty del Circolo Mathematico di Palermo*. ser. 2.1, N2. 1952.

Bellman, R. *Dynamic Programming*. Princeton University Press, NJ, 1957.

Berge, C. *Theorie generale des jeux a N personnes*, p. 114. Gauthier–Villars, Paris, 1957.

Berkovitz, L. D. (1964). *A Variational Approach to Differential Games*, p. 295. "*Ann. Math. Study 52*" Princeton, 1964.

Bertrand, J. *Journal des Savants*, 1983.

Blackwell, D. and M. Girshick. *Theory of Games and Statistical Decisions*, p. 330. Chapman & Hall, N.Y., Wiley, London, 1954.

Braess, D. *Unternehmensforschung*, 12, 1968.

Burger, E. *Introduction to the Theory of Games*, p. 211. Prentice-Hall, Englewood Cliffs, N.Y., 1963.

Danskin, J. M. *The Theory of Maximin and its Applications to Weapons Allocation Problems*, p. 126. Springer-Verlag, Berlin, 1967.

Davidov, E. G. *Methods and Models in Theory of Antagonistic Games*, p. 135. Moscow State Univ. Publ., Moscow, 1978.

Demyanov, V. F. and V. N. Malozemov *Introduction to minimax*, p. 307. Dover, NY, 1990

Diubin, G. N. and V. G. Suzdal. *Introduction to Applied Theory of Games*, p. 311. Nauka, Moscow, 1981.

Dixit, A. K. and S. Skeath. *Games of Strategy*. Norton, London, 1999.

Dresher, M. *Games of Strategy. Theory and Applications*, p. 186. Prentice-Hall, Englewood Cliffs, N.Y., 1961.

Eichberger, J. *Game Theory for Economists*, p. 315. Academic Press, London, 1993.

Feller, V. *Introduction to Probability Theory and Its Applications*, p. 1230. John Wiley & Sons Inc., N.Y.–London–Sydney–Toronto, 1971.

Fleming, W. H. *J. Math. Annal. and Appl.*, 3, 1961.

Friedman, A. *Differential Games*, p. 350. Wiley, N.Y., 1971.

Fudenberg, D. and J. Tirole. *Game Theory*, p. 580. The MIT Press, Cambridge, 1992.

Gale, D. *The Theory of Linear Economic Models*, p. 330. McGraw-Hill Book Comp., Inc., N.Y., London, 1960.

Gao, H., L. A. Petrosyan and A. A. Sedakov. *Procedia Computer Science*, 31, 255–264, 2014.

Germain, M., P. Toint, H. Tulkens, F. J. de Zeeuv, *CLIMNEG, Working paper*, 6, 1998.

Gladkova, M. A., A. A. Sorokina and N. A. Zenkevich. *Game Theory and Applications*, 16(7), 2013.

Haigh, J. *Advances in Applied Probability*, 7, 8–11, 1975.

Harsanyi, J. C. *International Economic Review*, 4, 235–246, 1963.

Harsanyi, J. C. *Papers in Game Theory*, p. 367. Reidel, Dordrecht, 1982.

Harsanyi, J. C. and R. Selten. *Management Science*, 18, 80–106, 1972.

Hart, S. and A. Mas-Colell. In A. E. Roth, editor, *The Shapley Value*. Cambridge University Press, Cambridge, 1988.

Haurie, A. *Journal of Optimization Theory and Application*, 18, 445–454, 1976.

Hu, T. *Integer Programming and Network Flows*, p. 411. Addison-Wesley Publ. Comp., Mento Park, Calif.-London-Dou Hills, 1970.

Isaacs, R. *Differential Games*, p. 384. Wiley, N.Y., 1965.

Jackson, M. and A. Watts. *Games and Economic Behavior*, 41(2), 265–291, 2002.

Jorgensen, S. and G. Sorger. *Journal of Optimization Theory and Applications*, 64, 1990.

Kalai, E. and M. Smorodinsky. *Econometrica*, 43, 513–518, 1975.

Karlin, S. *Mathematical Methods and Theory in Games, Programming and Economics*, p. 840. Pergamon Press, London, 1959.

Karlin, S. and R. Restrepo. In H. Kuhn and A. Tucker, editors, *Contributions to the Theory of Games*. Princeton University Press, Princeton, N.Y., 1957.

Kleimenov, A. F. Non Zero-Sum Positional Differential Games, p. 252, Ekaterinburg, Nauka, 1993.

Kolmogorov, A. N. and S. V. Fomin. *Elements of Theory of Functions and Functional Analysis*, p. 389. Nauka, Moscow, 1981.

Kozlovskaya, N. V. and N. A. Zenkevich. *International Game Theory Review*, 12(4), 2010.

Krasovskii, N. N. *Control of Dynamical System. Problem of Guaranteed Minimum Result*, p. 469. Nauka, Moscow, 1985.

Krasovskii, N. N. and A. I. Subbotin. *Positional Differential Games*, p. 456. Nauka, Moscow, 1974.

Kuhn, H. In *Annals of Mathematics Studies.* Princeton University Press, Princeton, 1953.

Leitmann, G. *Cooperative and Non-Cooperative Many Players Differential Games.* Springer-Verlag, New York, 1974.

Leitmann, G. and W. E. Schmitendorf. *Transactions on Automatic Control*, 23, 648–650, 1978.

Lemaire, J. *Cooperative Game Theory and Its Insurance Applications*, pp. 17–40. ASTIN Bulletin, 1991.

Luce, R. D. and H. Raiffa. *Games and Decisions. Introduction and Critical Survey*, p. 509. Wiley, N.Y., 1957.

Malafeyev, O. A. *Vestnik of the Leningrad State University*, 7, 1980.

Maynard, S. J. and G. R. Price. *The Logic of Animal Conflict.* Nature, London, 1973.

Mazalov, V. V. *Mathematical Game Theory and Applications.* John Wiley, Sons, 2014.

McKinsey, J. C. *Introduction to the Theory of Games*, p. 371. McGraw-Hill, N.Y., 1952.

Mehlmann, A. and R. Willing, *Optimization*, 15, 1984.

Moulin, H. *Theorie des jeux pour l'economie et la politique*, p. 200. Hermann, Paris, 1981.

Moulin, H. *Game Theory for the Social Sciences*, p. 465. N.Y. University Press, N.Y., 2nd edition, 1986.

Myerson, R. B. *International Journal of Game Theory*, 7, 1978.

Myerson, R. B. *Econometrica*, 45, 1981.

Nash J. F. Jr. *Proceedings National Academy Science USA*, 36, 1950.

Nash, J. F. *Econometrica*, 18, 1950b.

Nash, J. F. *Econometrica*, 21, 1953.

Owen, G. *Game Theory*, p. 230. Saunders, Philadelphia, 1968.

Owen, G. *Game Theory*, p. 230. Academic Press, N.Y., 2nd edition, 1982.

Parthasarathy, T. and T.E.S. Raghavan. *Some Topics in Two-Person Games*, p. 259. Amer. Elsevier, N.Y., 1971.

Peck, J. E. L. and A. L. Dulmage. *Canadian Journal of Mathematics*, 9(3), 1957.

Peters, H. *Game Theory*, p. 364. Springer, New York, 2008.

Petrosyan, L. A. *Doklady Akademii Nauk* (*Matematika*) (in Russian), 161(1), 1965.

Petrosyan, L. A. *Wissenshaftliche Zeitshrift der TU Dresden*, 4, 1968.

Petrosyan, L. A. *Doklady Akademii Nauk* (*Matematika*) (in Russian), 195(3), 1970.

Petrosyan, L. A. *Vestnik of the Leningrad State University*, 19, 1972.

Petrosyan, L. A. *Vestnik of the Leningrad State University*, 13, 1977.

Petrosyan, L. A. *Vestnik of the Leningrad State University*, 2, 1992.

Petrosyan, L. A. *Differential Games of Pursuit*, p. 325. World Scientific, Singapore, 1993.

Petrosyan, L. A. *Game Theory and Algebra*, 7, 1997.

Petrosyan, L. A. In Thuijsman, F., Wagener, F. (eds) *Advances in Dynamic and Evolutionary Games: Theory, Applications, and Numerical Methods*. Birkhauser, Boston, 2015.

Petrosyan, L. A., A. Azamov and H. Satimov. *Controlled Systems*, 13, 1974.

Petrosyan, L. A. and N. N. Danilov. *Vestnik of the Leningrad State University*, 1, 1979.

Petrosyan, L. A. and N. N. Danilov. *Cooperativbe Differential Games and Their Applications*. Izd. Tomskogo University, Tomsk, 1982.

Petrosyan, L. A. and Yu. Garnaev. *Search Games*, p. 217. Len. State Univ. Publ., Leningrad, 1992.

Petrosyan L. A. and A. A. Sedakov. *Vestnik of the St. Petersburg State University, Ser.10*, 4, 2014.

Petrosyan L. A., A. A. Sedakov and A. O. Bochkarev. *Mathematical Game Theory and Applications*, 5(4), 2013 (in Russian).

Petrosyan, L. A. and G. Zaccour. *Journal of Economic Dynamics and Control*, 27(3), 2003.

Petrosyan, L. A. and V. V. Zakharov. *Introduction to Mathematical Ecology*, p. 295. Len. State Univ. Publ., Leningrad, 1986.

Petrosyan, L. A. and N. A. Zenkevich. *The Mathematical Games Theory and its Applications*, 1(1), 2009.

Petrosyan, L. A. and N. A. Zenkevich. *Optimal Search in Conflict Conditions*, p. 96. Len. State Univ. Publ., Leningrad, 1986.

Petrosyan, L. A. and N. A. Zenkevich. *Game Theory*. World Scientific Ltd, Singapore, 1996.

Petrosyan, L. A. and N. A. Zenkevich. *Automation and Remote Control*, 72(8), 2015.

Petrosyan, L. A., N. A. Zenkevich and E. V. Shevkoplyas. *Game Theory*, p. 424. BCV-Press, Saint-Petersburg, 2012.

Pontryagin, L. S., V. G. Boltyanskii, R. V. Gamkrelidze and E. F. Mishchenko. *The Mathematical Theory of Optimal Processes*, Interscience Publishers, New York, NY, 1962.

Prokhorov, Y. V. and Y. A. Riazanov. *Probability Theory. Basic Notings. Central Limit Theorems. Random Processes*, p. 358. Nauka, Moscow, 1967.

Robinson, G. B. *An iteration method of solving a game*, volume P-154, p. 9. RAND Corp., 1950.

Rochet, J. C. *Selection of Unique Equilibrium Payoff for Extension Games with Perfect Information.* Mimeo, Universite de Paris, ix, 1980.

Rockafellar, R. T. *Convex Analysis*, p. 470. Princeton University Press, Princeton, 1970.

Roth, A. E. *Mathematics of Operations Research*, 2, 1977.

Rozenmuller, J. *Cooperative Games and Markets*, p. 115. Springer-Verlag, Berlin, 1971.

Sadovsky, A. L. *Doklady Akademii Nauk (Matematika)* (in Russian), 238(3), 1978.

Sakaguchi, M. *Operations Research Society of Japan*, 16, 1973.

Sansone, G. *Ordinary Differential Equaitons*, volume 2, p. 269. I. L., Moscow, 1954.

Selten, R. *International Journal of Game Theory*, 4, 1975.

Selten, S. *Math. Social Sci.*, vol. 5, 1983

Shapley, L. S. In H. Kuhn and A. Tucker, editors, *Contributions to the Theory of Games, II.* Princeton University Press, Princeton, 1953.

Shubik, M. *Game Theory in the Social Sciences: Concepts and Solutions*, p. 514. Cambridge, Mass., MIT Press, 1982.

Sion, M. and Ph. Wolfe. In M. Dresher, A. Tucker and P. Wolfe, editors, *Contributions to the Theory of Games, III.* Princeton University Press, Princeton, 1957.

Sorger, G. *Journal of Economic Dynamics and Control*, 13, 1989.

Thomson, W. and T. Lensberg. *Axiomatic Theory of Bargaining with a Variable Number of Agents.* Cambridge University Press, Cambridge, 1989.

Tirole, J. *The Theory of Industrial Organization*, p. 325. Cambridge, Mass., MIT Press, 2003.

Tsytsui, S. and K. J. Mino. *Journal of Economical Theory*, 52, 1990.

Tyniansky, N. T. and V. I. Zhukovsky. *Finding in Science and Engeneering. Mathematical Analysis*, volume 10. VINITI, Moscow, 1979.

Vaisbord, E. M. and V. I. Zhukovsky. *Introduction to Differential Multiperson Games and Their Appilcations*, p. 303. Sov. Radio, Moscow, 1980.

van Damme, E. *Stability and Perfection of Nash Equilibria*, p. 215. Springer-Verlag, Berlin, 1991.

von Neumann, J. *Mathematische Annalen*, 100, 1928.

von Neumann, J. and O. Morgenstern. *Theory of Games and Economic Behavior*, p. 625. Wiley, N.Y., 1944.

Vorobjev, N. N. *Game Theory Lectures for Economists and System Scientists*, p. 178. Springer, N.Y., 1977.

Vorobjev, N. N. *Fundamentals of Game Theory. Non-cooperative Games*, p. 496. Radio & Sviaz, N.Y., 1984.

Yanovskaya, E. B. *Bulletin of Soviet Academy!" of Science and Engineering Cybernetics*, 6, 1973.

Yeung, D. W. K. *Journal of Optimization Theory and Applications*, 62, 1989.

Yeung, D. W. K. and L. A. Petrosyan. In E.B. Yanovskaya, editor, *International Conference on Logic, Game Theory and Social Choice*. St Petersburg State Unviersity, 2001.

Yeung, D. W. K. and L. A. Petrosyan. *Cooperative Stochastic Differential Games*. Springer, New York, 2006.

Yeung, D. W. K. and L. A. Petrosyan. *Subgame Consistent Economic Optimization*, p. 396. Birkhauser, New York, 2012.

Zaccour, G., *INFOR* 46, 1 (2008).

Zenkevich, N. A. and I. V. Marchenko. In *Mathematical Methods of Optimization and Control in Systems*. Kal. State Univ. Publ., Kalinin, 1987.

Zenkevich, N. A. and I. V. Marchenko. *Threat Strategies in Two-person Games with Dependent Sets of Strategies*, p. 21. VINITI, Leningrad, 1990.

Zenkevich, N. A., L. A. Petrosyan and D. W. K Yeung. *Dynamic Games and Management Applications*, p. 415. St. Petersburg State Univ. Publ., St. Petersburg, 2009.

Zenkevich, N. A. and S. N. Voznyuk. In V. V. Mazalov and L. A. Petrosyan, editors, *Year-book on Game Theory and Applications*, volume 2. Nauka (Novosibirsk) and Nova Science Publ., N.Y., 1994a.

Zenkevich, N. A. and A. V. Zyatchin. *International Game Theory Review*, 16(4), 2014.

Index